# Building and Surveying Series

*(continued overleaf)*

List continued from previous page

**Building and Surveying Series**
Series Standing Order
ISBN 0–333–71692–2 hardcover
ISBN 0–333–69333–7 paperback
*(outside North America only)*

You can receive future titles in this series as they are published by placing a
standing order. Please contact your bookseller or, in the case of difficulty, write
to us at the address below with your name and address, the title of the series
and the ISBN quoted above.

Customer Services Department, Macmillan Distribution Ltd
Houndmills, Basingstoke, Hampshire RG21 6XS, England

# Quantity Surveying Practice

**Ivor H. Seeley**

*BSc (Est Man), MA, PhD, FRICS, CEng, FICE, FCIOB, FCIH*

*Emeritus Professor of The Nottingham Trent University*
*Chartered Quantity Surveyor*

**Second Edition**

palgrave
macmillan

First edition 1984
Second edition 1997

Published by
PALGRAVE MACMILLAN
Houndmills, Basingstoke, Hampshire RG21 6XS and
175 Fifth Avenue, New York, N.Y. 10010
Companies and representatives throughout the world

PALGRAVE MACMILLAN is the global academic imprint of the Palgrave Macmillan division of St. Martin's Press, LLC and of Palgrave Macmillan Ltd. Macmillan® is a registered trademark in the United States, United Kingdom and other countries. Palgrave is a registered trademark in the European Union and other countries.

ISBN-13: 978-0-3336-8907-3
ISBN-10: 0-3336-8907-0

A catalogue record for this book is available from the British Library

This book is printed on paper suitable for recycling and made from fully managed and sustained forest sources. Logging, pulping and manufacturing processes are expected to conform to the environmental regulations of the country of origin.

11
08

Printed in Great Britain by
Antony Rowe, Chippenham, Wiltshire

'There is a tide in the affairs of men
Which, taken at the flood leads on to fortune;
Omitted, all the voyage of their life
Is bound in the shallows and in miseries.
On such a full sea are we now afloat,
And we must take the current when it serves,
Or lose our ventures'.

William Shakespeare

*This book is dedicated to quantity surveyors in private practice, public offices and contracting organisations throughout the world.*

# Contents

marketing strategy; in-house marketing team v. external
consultants; corporate identity; promotions and public
relations; press, radio and television; sponsorship and
scholarships; client entertainment; exhibitions;
presentations; seminars and lectures; reports, newsletters
and authorships; personal contacts; advertising; brochures;
marketing audit, conclusions.

# Preface

This book is aimed primarily at quantity surveying degree students, although quantity surveying students on higher diploma courses and the higher level NVQs will also find it a valuable aid to their studies. Practising quantity surveyors and members and students of kindred professions could find much of its contents of interest and value.

The new edition has been updated throughout and extended substantially to take account of the many changes and developments that have occurred since the writing of the first edition, and to examine likely future developments and trends. Hence the many and increasingly diverse aspects which impinge upon quantity surveying practice are examined, described and illustrated with supporting examples, where it was felt that they would be of particular benefit to the student. The principal aim is to increase the usefulness and value of the book to all readers.

A study of the construction industry provides a good backcloth against which the activities of the quantity surveying profession can be considered. The nature and interrelationships of the parties to construction contracts, associated professional and other bodies, the operation of the design process, selection of procurement method, a comparison of contractual and tendering arrangements, and appropriate provisions of the main standard forms of contract are all considered in sufficient detail to meet the students' needs.

These explanatory chapters lead logically into an analysis of the principal activities of the quantity surveyor, encompassing the preparation of contract documentation, tender invitation and scrutiny, valuation of work in progress, pricing of variations, settlement of claims, cost control of projects and preparation of final accounts. The measurement for and preparation of bills of quantities is largely omitted from the book, as this subject is adequately covered in three other books written by the author, embracing both building and civil engineering measurement.

Where the term 'quantity surveyor' is used, this refers to the employer's quantity surveyor as distinct from the contractor's quantity surveyor. Frequent references to the JCT Standard Form relate to the *Standard Form of Building Contract 1980 Edition, Private with Quantities*. The term 'employer' is generally used in preference to 'client' in order to maintain the same terminology as employed in the JCT Standard Form.

The newly developing areas of project management and value management are covered in sufficient depth to give the student a good understanding of their main characteristics, benefits and applications in practice. The role of the quantity surveyor in connection with mechanical and electrical services, civil engineering, and heavy and process engineering is examined together with the practical implications. In addition, chapter 13 covers a diverse range of activities, many of which are from time to time undertaken by quantity surveyors, as their services are sought in a wider context than that normally associated with the quantity surveying function. On the wider front, the reader is directed towards the major areas of development, redevelopment and environmental management, all with a significant impact on the work of the quantity surveyor.

An extensive chapter covers quantity surveying organisation and practice, including office administration and the different approaches adopted in public and private offices and in contracting. This chapter encompasses such important matters as the appointment of quantity surveyors, opportunities for women in quantity surveying, salaries and benefits, partnerships, quality assurance, information technology and advertising. A new chapter has been introduced dealing with quantity surveying activities on a global front, in order that the reader may appreciate the significant growth of the profession in overseas countries and the opportunities available. The need for successful marketing of a quantity surveyor's services is now generally recognised and hence a new chapter has been included to examine the variety of marketing techniques that are available and to identify those with the greatest potential.

The growing need for a clear understanding of professional ethics, standards and conduct by all surveying students has been recognised by the Institution's Education and Membership Committee, and a chapter has accordingly been included to meet this need. In like manner, a study of quantity surveying education, training, professional development and research is incorporated to show the interrelationship of education, training and continuing professional development throughout a quantity surveyor's career in a changing world.

The final chapter of the book analyses the latest and likely future developments and trends in quantity surveying, encompassing selecting the most appropriate construction procurement method; improving value for money in construction; extension of multi-disciplinary practices and mergers of quantity surveying practices; greater use of partnering in construction; undertaking a possible management consultancy role; greater emphasis on risk analysis and management; increasing importance of government, EC and private funding; impact of post-Latham developments; a more detailed comparison of UK and continental practices and performance with its consequences for the UK; and culminating in the changing role of the quantity surveyor. These previously largely uncharted areas will compel the quantity surveyor to diversify and/or specialise to a much greater extent than hitherto, and will necessitate those practising to acquire a much extended knowledge base, to employ new skills and exercise increased initiative and ingenuity.

Useful appendices incorporate NJCC, RICS and RIBA contract administration forms, completed as appropriate, an initial cost plan and record of cost checks and a detailed elemental cost analysis.

IVOR H. SEELEY

Nottingham
Spring 1996

# Acknowledgements

The author is indebted to many quantity surveying offices and individual quantity surveyors who supplied valuable information in such a ready and helpful way. The following persons and organisations deserve special mention.

Christopher J. Vickers FRICS, E. C. Harris, London

Giles P. Harrison BSc FRICS, Currie and Brown, Redbourn, St Albans

Thomas N. L. Harrison BSc ARICS Dip Proj Man, Turner and Townsend, Darlington

Donald G. Jones FRICS, D. G. Jones and Partners, Richmond, Surrey

Nicolas A. Davis MA FRICS, Davis Langdon and Everest, London

Michael J. Clayton FRICS Dip Proj Man, Heery International Ltd, Walton-on-Thames

Geoffrey D. Avery FRICS, Higgs and Hill, New Malden

Samson H. Emiowele Dip QS (Nottm. Trent) FRICS FNIQS, Ekeoba and Emiowele Partnership, Lagos, Nigeria (President of the NIQS, 1994–96)

Brian Barton FRICS FCIArb, Barker and Barton, Nairobi, Kenya (RICS Kenya Group Chairman)

Anthony E. Netto LLB FRICS ACIArb MSISV MACostE, National University of Singapore

Anthony Hordern MIS (Aust) AFAIM, Executive Officer, Australian Institute of Quantity Surveyors

J. Z. Botha, Executive Director, Association of South African Quantity Surveyors

Barry W. Probert BSc (QS) RQS ARICS MAQS AAArb, Rousseau Probert Elliott, Port Elizabeth, South Africa (President of the ASAQS)

Professor Gaye le Roux RQS MAQS Cert QS (Natal), University of Port Elizabeth, South Africa

Zakaria B. Hashim FRICS FIS(M), Zakaria-Lee and Partners, Kuala Lumpur, Malaysia (QS) Sectional Chairman, Institution of Surveyors, Malaysia

Associate Professor Adrian J. Smith MEd FRICS, City University of Hong Kong

Brian R. Norton BSc ARICS, formerly of Currie and Brown Inc, Morriston, Jersey, USA (figures 9.1 and 9.3)

Joseph L. N. Martin DMS MBIM FRICS, Executive Director, BCIS (Appendix M and table 8.2)

Timothy G. Carter MSc FRICS FAPM, Davis Langdon and Everest, Chester, for providing much of the case study material for chapters 9 and 12, including figures 9.4, 9.5, 9.6, 12.1 and 12.2

Brian Cooke MSc MCIOB, Liverpool John Moores University (figure 5.1)

David W. Hoar FRICS, Quantity Surveying Practice Manager, Nottinghamshire County Council (cover design)

J. Alan Park FRICS MCIOB ACIArb, Stride Treglown Management, Bristol (figure 13.3)

DoE (figures 1.2 and 1.4)

Society of Chief Quantity Surveyors in Local Government (table 5.4)

Roger A. Waterhouse MSc FCIOB MIMgt MAPM MSIB, College of Estate Management (table 12.1)

E. C. Harris (tables 12.2 and 14.1)

Chapman Hendy Associates (table 15.1)

RICS (figures 17.1, 18.1 and 18.2 and tables 10.1 and 19.1)

Keith I. Hogg BSc ARICS, The Nottingham Trent University (tables 18.1 and 18.2)

Building Design Partnership (figure 19.2)

Geoffrey A. Quaife ARICS, Bucknall Austin, London (figure 19.1)

Professor Roger Flanagan MSc PhD FRICS FCIOB, University of Reading, and Dr Susan D. Stevens BSc PhD ARICS, English Heritage (figure 19.3)

BRE (figure 19.5 and tables 8.3 and 10.2)

Much valuable information has been obtained from RICS journals, *Building, Building Economist, New Builder* and *Building Technology and Management,* and from publications of the RICS, CIOB, DoE, the Treasury and BRE for which the author is most grateful.

Macmillan Press Ltd kindly gave permission to quote from *Building Economics, Building Maintenance* and *Civil Engineering Contract Administration and Control,* all of which were written by the same author.

The author also expresses his gratitude to the Royal Institution of Chartered Surveyors (Surveyors Publications) for kind permission to reproduce the contract administration forms in appendices E and F, for which the Institution owns the copyright, and to the National Joint Consultative Committee for Building to reproduce the tender forms in appendices A, B and C. Reproduction of extracts from the JCT80 Standard Form of Building Contract and RIBA contract administration forms in appendices G, H, J and K is made with the kind permission of the copyright holder RIBA Publications Ltd. The Business Round Table kindly agreed to the reproduction of the procurement selection document in figure 3.1.

My grateful thanks are due to Malcolm Stewart of Macmillan Press Ltd for his constant help so kindly given throughout and to my wife for her continual support and understanding.

# List of Figures

# List of Tables

# 1 The Construction Industry

It will be helpful to the reader to start by examining the structure, size and scope of the construction industry and its significance in the national economy. Other matters deserving attention include the effect of government action on the work of the industry, clients' needs, statutory provisions, changing techniques, productivity, constructability, quality assurance, safety aspects, relevant official bodies and sources and use of technical information.

## NATURE AND FORM OF THE INDUSTRY

The construction industry embraces a wide range of loosely integrated organisations that collectively construct, alter, refurbish and repair a wide range of different building and civil engineering structures. The industry has certain unique characteristics, stemming mainly from the physical nature of the construction product and its demand (Balchin *et al.*, 1994). No two projects are identical and site characteristics also vary extensively.

The construction industry is essentially an assembly industry, assembling on site the products of other industries. The designer's intentions are portrayed in drawings and other documents, and skilled operatives undertake the work of construction and assembly of components on the site. Construction work is mainly carried out on site and is thus subject to the vagaries of the weather and of ground conditions.

A wide range of economic factors influences the extent of activity in the industry, and these include the general economic climate, interest rates, rate of inflation, credit availability and extent of control of public sector spending. It is generally believed that housing tends to reflect the general position of the industry. In the late nineteen eighties and early nineteen nineties, new housing in the United Kingdom was on the decline, particularly in the public sector, although this was partially offset by the increase in refurbishment and alteration work.

In the early nineteen nineties it was estimated that there were around 200 000 building firms in Great Britain, but the majority of these were relatively small, as shown in figure 1.1, although collectively they employed less than half the total workforce.

The approximate distribution of construction work in the different sectors is shown in figure 1.2.

It is common for as much as 50 per cent of the construction industry labour force to be engaged in work of building alterations, maintenance and repairs. This creates a need for numerous small local contractors employing few operatives and little plant. At the other end of the scale

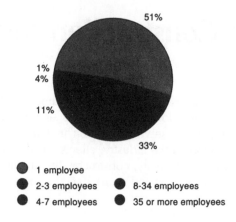

**Figure 1.1**  *Number of construction firms by size, Great Britain (1994)*
*Source*: Housing and Construction Statistics, DoE.

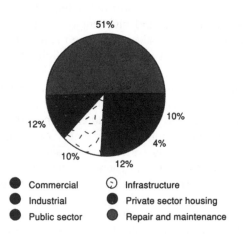

**Figure 1.2**  *Value of construction by sector, Great Britain (1994)*
*Source*: DoE.

are the very large national contractors undertaking multi-million pound projects at home and overseas.

*QS 2000* (1991) highlighted the rise in specialist subcontracting and the shift of general contractors from building to management and coordination and the extension of their role into design and management of the construction process, as opposed to the traditional contracting approach as described in chapter 2. Figure 1.3 shows construction employment in Great Britain from 1989 to 1997, with employment steadying in the mid 1990s, although the estimated total for 1997 showed a further decline. In the early 1990s there was a significant increase in output per employee.

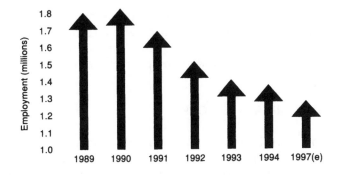

**Figure 1.3** *Construction employment, Great Britain (1989–97)*

*Source*: RICS, DoE.

## NATURE AND SCOPE OF CONSTRUCTION WORK

The construction industry can be subdivided into two major areas of activity, namely building and civil engineering; they tend to complement each other and many building contracts include some civil engineering work and vice versa. In general, building work satisfies man's need for shelter and includes such diverse buildings as houses, flats, schools, hospitals, shops, offices, factories and warehouses. Civil engineering work encompasses the essential services needed to make the buildings operative. Typical examples of civil engineering work include roads, bridges, reservoirs, waste water systems, railways, power stations, harbours and dams.

In 1993 the building workload of the construction industry in the United Kingdom was estimated by the DoE (Housing and Construction Statistics) at about £26bn on new work and around £20bn on repairs and maintenance. The proportional expenditure on housing rehabilitation and maintenance is likely to increase significantly as the DoE house condition survey for England in 1991 showed that 1.5 million houses were unfit for habitation, and several million more had major defects. Insufficient attention was still being paid to house improvements with a resultant deterioration of the housing stock and a construction industry working at far below capacity. In addition large sums need to be spent on a worn out infrastructure, urgently requiring the costly replacement of waste disposal and water supply networks and an inadequate transportation system.

In the Far East with its concentration of high rise developments, particularly in Hong Kong and Singapore, the provision of building services, lifts and escalators often represents a major part of the cost of a building and requires a high level of integration. These countries also have to import large quantities of materials and components. For example, Hong Kong imports cement from China, steel from Japan, ceramic tiles from Germany, plastic tiles from Britain, acoustic materials from the United States and services equipment from a number of industrial nations. Countries likely to have much increased construction programmes by the year 2000 include China, the Soviet Union, many eastern and western European countries and those in the Pacific Rim and parts of the Middle East.

## PUBLIC AND PRIVATE SECTORS

Construction work is undertaken in both the public and private sectors and the relative proportions vary over time. For instance in 1977, the public sector accounted for 55 per cent and the private sector for 45 per cent. The position changed significantly in the early nineteen-eighties when the Government was imposing substantial cuts in public expenditure, and by 1982, the public sector was generating only about 40 per cent of all new construction work, and by 1994 public sector new construction work had dropped to about 24 per cent, resulting from privatisation of publicly operated organisations and severe cost reductions in the public sector.

The main instigators of work in the two sectors are as follows:

| *Public Sector* | *Private Sector* |
| --- | --- |
| Central government | Developers |
| Local government | Financial institutions |
| Development corporations | Industry and commerce |
| Public corporations | Building societies |
| | Individual promoters |
| | Privatised companies |

Construction work may be carried out by private contractors or direct labour organisations (DLOs). In the past, local authorities have tended to use direct labour organisations for extensive areas of work. They certainly offer advantages for emergency work, particularly in the maintenance field. The Local Government Planning and Land Act 1980 requires these organisations to be cost effective and this generally entails the preparation of comparable schedules of rates to ensure competitiveness with private contractors. In the 1990s many local authority service departments were required to bid for work in competition with outside contractors.

The larger public construction projects mainly centred around education, health care, leisure and transportation schemes, as the momentum on housing provision reduced to minimal proportions with housing associations becoming the main providers of social housing. The private sector encompassed industrial and commercial projects, housing schemes, water supply and waste disposal systems and many refurbishment projects. In the mid 1990s, the Government was endeavouring to promote public and private partnerships through the private finance initiative (PFI), as described in chapter 19.

## TYPES OF CONTRACTOR

There is no general optimum size of construction firm, as the size is affected by the nature of the work, the conditions under which it needs to be carried out, and the nature of the organisation and the ability of the management. Much of the maintenance work and constructional work to small buildings requires so few operatives at a time that it cannot be attractive to large firms. As projects increase in size they require more labour and plant, and improved organisational ability. Much of the plant and management resources can only be obtained in individual units and they need to be kept continually employed on large projects to remain viable.

General builders, as distinct from specialists, may be classified broadly into five main groups, each of which is now described.

### (i) Large Contractors

Opinions vary as to how to classify a large firm and the parameters vary from in excess of 300 employees to over 1200 employees. These firms have the capacity to undertake large building and civil engineering contracts throughout the United Kingdom and often overseas. They usually have a head office and a number of branch offices in different parts of the country. They generally show the highest output per employee and are better equipped, financed and organised, but they tend to experience problems stemming from centralised supervision and management, with work being undertaken on widely dispersed and ever-changing sites. A rapidly expanding firm can outgrow the capacity of its management, although there is the advantage of economies of scale.

In the 1990 to 1995 recession, a significant proportion of the larger contractors was experiencing difficulties with regional subsidiaries, particularly in the depressed economic regions. Faced with a deteriorating business climate, not all of the larger firms encouraged their subsidiaries to adapt to the changing situation and enter new markets. Some appeared unwilling to abandon the centrally imposed policies and practices and, in consequence, the workloads of the subsidiaries dropped appreciably, some ceased to trade and others were technically bankrupt. The position has however improved as a new generation of managers emerged with a greater understanding of local conditions and needs. Some of the larger contractors suffered substantial losses in the period of recession and others were forced out of business but, in general, they can withstand better the adverse climate than the smaller firms.

### (ii) Medium-sized Contractors

These contractors generally employ between 50 and 300 operatives and are most likely to operate on a regional basis. They can undertake quite large contracts, usually within about 50 km radius of their head offices, and are often prepared to undertake civil engineering as well as building work.

The recession in the industry caused considerable problems for many firms in this category and there were many insolvencies in this group. There is evidence of some degree of polarisation towards large and small firms. Survival requires good management, and sound management is based on effective training, knowledge, skill and hard work.

### (iii) Small Contractors

Small contractors rarely employ more than 50 operatives and usually many fewer. They prefer to operate within a reasonable distance of their offices and travel further afield only under special circumstances. These are often well-established firms but problems may arise where the attributes of the founder are not possessed by the son or grandson in control of the organisation.

The work undertaken by this class of contractor consists mainly of extensions to existing buildings, refurbishment, repairs and maintenance and small new building projects, particularly houses.

### (iv) Jobbing Builders

This class of builder forms the largest part of the industry. He may employ up to 10 operatives, but more usually employs only very few and, in many cases, none at all. The principal often works at the tools himself. The work undertaken is mainly building repairs and maintenance, although in some cases a limited amount of small contract work will also be carried out.

The jobbing builder usually prepares his own estimates of cost and the accounts to be rendered to clients and keeps his own books of account. He does therefore need a very wide knowledge and experience and he performs a very important function. There is normally at least one in every sizeable village.

### (v) Speculative Builders

These generally purchase land for residential development, construct the necessary roads and sewers, and erect the dwellings. The dwellings are generally sold to the public through an estate agent. Housing associations also make use of their services, although in the mid 1990s the associations were suffering from reduced Housing Corporation grants and having to cut back on the scale of development.

The houses are sometimes architect designed, and many architects are prepared to provide plans only, for an agreed fee. Most of the builders are members of the National House Building Council, and the Council arranges for the houses to be inspected during construction and issues a certificate on completion where the house conforms to an acceptable standard of construction with a ten-year guarantee.

Hence the particular contractors selected to tender for a given project will be determined largely by its nature and scope. For example, a national contractor would not be engaged to build a small bungalow in a remote part of Suffolk. A more likely approach would be to select a number of small local builders with the necessary experience and resources and invite them to tender for the work.

### (vi) Specialist Subcontractors

A century ago, it was customary for a building to be designed entirely by one architect and for the building to be erected by a single contractor, employing all the necessary craftsmen and labourers. The trades were relatively few in number and with long-established practice and stable conditions, a high standard of craftsmanship was achieved.

In the post-war years the use of new materials and components coupled with new techniques and methods of construction and increased mechanisation have had a dramatic effect on the building process. Specialist nominated subcontractors have been engaged to an increasing extent to undertake the various specialist activities. On modern large, complex projects it is not unusual for subcontractors collectively to carry out up to two-thirds of the total work in the contract, and their selection and coordination have become an important and complicated activity.

It is virtually impossible for a single architect or builder to possess the specialised knowledge and experience required to deal effectively with all aspects of modern building schemes. The specialist subcontractor is able to concentrate on one particular aspect of building or related work and so perfect his expertise, train and engage operatives with high technical experience and skill, fully utilise specialist equipment, and achieve high standards of performance and efficiency with a good quality finished product.

The normal contractual procedure is for the employer (client) to enter into a contract with the main contractor, under which the main contractor undertakes to employ, as subcontractors, the specialists nominated by the architect. There is, in consequence, no contractual relationship between the employer and the nominated subcontractor. The nominated subcontractor enters into a subcontract with the main contractor, whereby the main contractor can be assured of the full cooperation of the subcontractor. The subcontractor is required to indemnify the main contractor against the same liabilities and obligations under the subcontract as those for which the main contractor indemnifies the employer under the main contract. The nominated subcontractor is responsible for the actions of his employees on the site and is entitled to make use of the general facilities provided by the main contractor. The term 'employer' is used instead of 'client' in the Standard Forms of Contract, relating to the fact that the client employs the contractor to construct a building. The author prefers the term 'client' as both parties are employers.

Specialist subcontractors can be broadly classified into three main categories:

(i) Craft firms who specialise in one of the building crafts, such as plumbing, plastering, tiling and painting.
(ii) Constructional firms who specialise in structural work, such as piling, structural steelwork, reinforced concrete and patent floor construction.
(iii) Firms who specialise in the mechanical and electrical equipment of buildings, such as electrical installations, heating, ventilating and air conditioning plant, lifts, escalators and associated services.

Certain contractors favour the use of labour-only subcontractors, sometimes referred to as 'lump labour', because these self employed operatives contract their labour to fix the contractor's materials for a lump sum payment. Labour-only subcontracting has been a cause of considerable controversy over the years. The trade unions allege that it leads to poor standards of workmanship, tax avoidance and substandard conditions of work for the operatives. Supporters of the system claim that it enables contractors to effectively accommodate the considerable fluctuations in demand and cope with the high cost of employing permanent labour within the industry. Government action in the introduction of the Employment Protection Act and the imposition of various employment taxes encouraged contractors to make use of labour only subcontractors, but the Government subsequently tightened up on this category of worker and made them less attractive.

### (vii) Civil Engineering Contractors

There are some firms who concentrate on civil engineering work at home and overseas, and they specialise in such work as highways, harbours and docks, power stations, pipelines and treatment plant. They are often supported by specialist subcontractors such as those involved with the supply and installation of pumps and associated switch gear.

### EFFECT OF THE CONSTRUCTION INDUSTRY ON THE NATIONAL ECONOMY

The fortunes of the British construction industry are closely related to the state of the British economy as a whole. The two are inextricably bound together, and thus the industry suffered its worst post-war problems in the severe financial recession of the early nineteen nineties. Equally there is probably no other industry that stands to benefit more directly from future genuine economic growth.

The output of the construction industry, at about £48bn in 1994 accounted for approximately 8 per cent of the UK Gross Domestic Product (GDP), as compared with 7.5 per cent in the 1988–89 boom period, and almost one-quarter of the nation's fixed capital investment. Of the nation's labour force about 5 per cent (1.35 million people) were either employed or seeking employment in the construction industry, excluding the associated professions and material suppliers. Its output in 1994 in volume terms is little more than 80 per cent of that recorded in 1988 and there were about 450 000 fewer people employed in the construction industry in 1994. Despite this downturn in the industry's fortunes it is evident that the construction industry has a very important role in the national economy. It seems certain that the demand for construction work would increase substantially in a period of economic growth, with increased demand for buildings and improved employment prospects in the industry.

There was a rise in output per worker in the early nineteen-nineties despite some under-utilisation of labour. Contracting firms tend to retain operatives even though there is insufficient work to keep them fully occupied, while hoping for a later upsurge. A further difficulty is that output per worker is expressed in monetary terms, but not in terms of actual work done, because contractors have submitted low tenders in order to obtain contracts.

Any changes in an industry as large as construction have a significant effect on the rest of the economy, and its influence is probably greater than the statistics indicate. The future prosperity of the country depends to a considerable extent on investment levels and on the performance of industries such as construction, assisted by work undertaken in overseas markets. Furthermore, changes in investment have a multiplier effect on the level of national income.

To obtain a more correct assessment of the construction industry's contribution as a source of employment, it is also necessary to examine the number of people employed in associated industries. The industry uses a wide range of components, such as bricks, steel sections, floor and wall tiles, roofing tiles and slates, joinery fittings, sanitary appliances and pipes. When these industries are taken into account the numbers employed double and represent approximately 10 per cent of the working population. In like manner the construction industry's contribution to the Gross Domestic Product (GDP) is also doubled to over 10 per cent, even in the 1990–94 severely depressed market for the industry.

Hence the construction industry in the United Kingdom is of key economic importance because it is a major client of the government, a large contributor to the national stock of investment goods, the provider of buildings to accommodate a wide range of essential activities, with a very large output and employing large numbers of people either directly or indirectly.

The products of the construction industry, with the exception of maintenance and repairs, are paid for out of capital. The capital may belong to the owner of the new building or, more frequently, may be borrowed from another source. Because of the long life of buildings, the stock is large in relation to the annual production.

Some buildings create a capital asset, some create a durable consumer good, and some provide both. The difference between capital and a consumer good is a subjective one; that is to say, it is a difference in the mind of the person or group of persons exerting the demand upon resources, whether it is done in order to generate income or to obtain a specific form of direct satisfaction. Hence the same house may be a capital good to the developer and a durable consumer good to the purchaser of the house for owner occupation.

## EFFECT OF GOVERNMENT ACTION ON THE CONSTRUCTION INDUSTRY

The Government has a crucial role in determining demand for the construction industry's output and its growth prospects, because public authorities buy about 30–35 per cent of its output and general economic measures have a powerful influence on the demand for private housing, and commercial and industrial building. A steady, rather than wildly fluctuating demand is particularly important for the industry if it is to plan its work ahead, make sure of supplies and deploy its resources effectively. Because the industry is under-capitalised it soon experiences difficulties when monetary policies are introduced to check the economy as a whole, as became very evident in the late 1970s, early 1980s and early 1990s.

Over the years successive governments have found it necessary sometimes to restrain and at other times to stimulate the economy, and the construction industry is invariably caught up in this process. When policies of restraint operate, there is usually a reduction in the volume of public building work, particularly house building, and projects such as motorways and town centre redevelopment schemes are likely to be cut back or postponed. The adverse effects on contractors will not be immediate as contracts in hand will normally be completed. In the longer term however the results can be serious, resulting in:

(1) unemployment of building operatives;
(2) smaller building firms being forced out of business;
(3) larger construction firms being reluctant to invest large sums in new plant and equipment or to experiment with new techniques;
(4) suppliers of materials and components being unlikely to extend their plant and tending to reduce production and stocks drastically;
(5) recruitment of persons into the industry at all levels being made more difficult;
(6) the lack of continuity of construction work making for increased building costs and reduced efficiency; and
(7) training, education and research and development (R&D) will all suffer.

Every industry is affected for good or ill by the state of the national economy, and this is particularly so in an industry that is both home based and labour intensive. The construction industry's unattractive position in the late 1970s, early 1980s and early 1990s was due largely to the poor economic condition of the country. The industry was particularly vulnerable, with almost one-half of the national capital formation in construction and 50 per cent of construction occurring in the public sector in the 1970s, reducing to 30 per cent in the 1990s.

The increasing number of insolvencies and bankruptcies in the construction industry is a matter of great concern and is illustrated in figure 1.4.

Capital investments take several years to reach fruition and often many more years to become profitable. Unfortunately politicians, who set the scene, work on a much shorter time scale than those who operate within

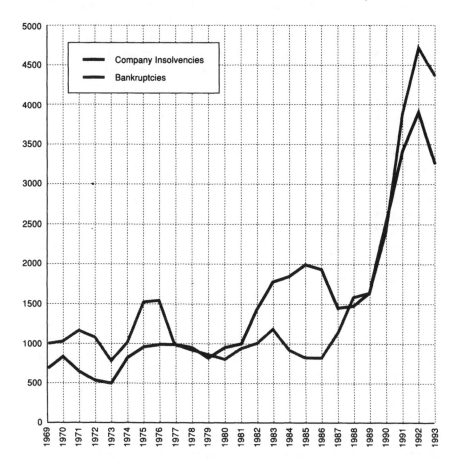

**Figure 1.4**   *Insolvencies and bankruptcies in the construction industry in England and Wales (1969–93)*

*Source*: DoE Digest of Data for the Construction Industry (1995).

it. There are a number of public sector capital schemes that can be post-poned until the country is in better shape without causing too many social problems and little, if any, damage to the national earning capacity. But there are other public sector capital projects that are potential wealth creators, and which, if cut too sharply, will permanently damage the national abil-ity to compete in world markets.

In 1977 the group of eight, representing the professionals, contractors' operatives and material producers was established to press the case for more construction work with Ministers and subsequently Members of Par-liament. This group was subsequently disbanded by the Government and the main concerted construction/Government voice in the 1990s rested with the Construction Industry Council (CIC). The greatest pressure was on the need for more public spending to improve the infrastructure and to raise the efficiency of the economy through increased building invest-ment, and to reduce high unemployment and the threat of developing social problems. Because of the construction industry's fragmented nature, both in terms of diversity within the construction team and the large number of sites and firms scattered throughout the country, the industry has tradi-tionally suffered from lack of cohesion and restricted political influence. The formation of the group of eight was hoped to help rectify these de-ficiencies but its success rate throughout its lifespan was very limited.

The Government's initial argument that the private sector would move in to fill the vacuum created by the cuts in public work was never a serious proposition, for much of the construction industry's workload is either provided to users whose needs can only be catered for by the State or, as is the case with many civil engineering projects, has such long lead times as to be of little interest to private financiers who require fast re-turns and low risks. However, to make any real headway with the Govern-ment, the industry has to convince its Treasury critics that the increase in construction employment, output and efficiency which will result from extra public spending on construction will cancel out the negative effects on public sector borrowing, interest rates, inflation and the balance of payments.

In the 1980s and 1990s there was an increasing urgent need to replace a collapsing infrastructure, particularly sewers, and to increase expendi-ture on energy conservation and motorway maintenance, as well as to construct some major new projects such as the Channel Tunnel and the Severn Barrage Scheme. Overcrowded highways and prisons were further causes for concern.

Three studies published in 1981 by the Trades Union Congress, The Economist Intelligence Unit and Cambridge Econometrics Ltd all attempted to show that greater public spending on construction would result in in-creased output, employment and efficiency, which would, in their turn, cancel out many of the negative effects on the public sector borrowing requirement, interest rates, inflation and the balance of payments. There would also be a reduction in unemployment benefits and greater tax payments. The Government, however, argued that increased public expenditure on public works, if financed by borrowing that the Government could not afford, could only lead to higher inflation and so reduce the industry's competitiveness, to higher taxation reducing the industry's cash flow and to 'crisis' public expenditure cuts which would subsequently plunge the industry into still deeper recession. This provides the two extreme views of the situation and highlights the dilemma facing the industry.

Unfortunately, even when public money is allocated to construction, it is not always spent and, for example, underspending became an increasing problem in the early 1980s. In 1981/82 local authorities in England had an estimated capital expenditure on housing of £1875m, which showed an underspend of £430m on the combined total of their housing capital allocation and the prescribed portion of their housing capital receipts. In fact the underspend equated to the capital receipts. Many authorities chose to spend their receipts on maintaining existing staffing levels and other forms of current spending, or to reduce borrowing.

Most types of new construction generate future current spending commitments; for example, new hospitals and schools entail considerable operating and maintenance costs. In the early 1980s and 1990s with pressure increasing on local authorities to reduce current spending, many were reluctant to increase future commitments and possibly become subject to severe central government penalties. Government policy also tends to vacillate.

Even when underspending seems likely, it is often difficult to prevent it because of the time lag between deciding to increase construction spending and obtaining tenders. The conversion of orders into output and increased spending also takes time. One way of overcoming this problem is to operate a rolling programme of projects, although the longer projects are delayed the greater the number of modifications that may be required. It is anticipated that urban regeneration and renewal programmes will form an important feature of construction work in the 1990s.

Hence it is evident that Government spending on capital investment, such as schools, hospitals, roads and public housing can be curtailed with relative ease, and these cuts have the greatest impact on the work of the construction industry. A change of priorities by the Government, such as a transfer of funds from housing to motorways, creates problems for the construction industry, as these two activities are largely undertaken by different types and sizes of firms using substantially different forms of plant. Thus, it is not only the level or volume of construction activity that needs considering but also its composition. For example, substantial increases in the allocation of improvement grants will result in a significant rise in the volume of work available to small builders (Seeley, 1996).

The Housing Corporation's approved development programme of social housing for 1995–96 was cut by 20 per cent from £1.48bn to £1.18bn, to £1.062bn in 1996–97 and £932m in 1997–98. The National Federation of Housing Associations (NFHA), changed to National Housing Federation (NHF) in 1996, warned that these cuts would lead to the number of housing association new rented and low cost homes starting on site falling by about 40 per cent from 45 000 to 27 400 in 1995, with a loss of 4000 construction jobs in 1995 and a further 7900 in 1996–97. The Government argument that housing associations could build more dwellings for less money was no longer valid because of higher land and building prices, and there was a real danger that quality would continue to suffer and the higher rents cease to be affordable.

Other forms of control of construction work by the Government include statutory regulations, which will be considered later in this chapter, monetary controls and taxation. The Government uses various controls, such as bank base rate, open market operations and hire purchase restrictions, to alter the level of interest rates, and to control the amount of credit available and the terms on which it can be obtained. Credit restrictions strike at contractors both directly, through banks and lending institutions, and indirectly, as through builders' suppliers. Long term high interest rates can result in the postponement or abandonment of projects. Rent controls on residential properties may render them uneconomic for landlords and hasten their deterioration.

Taxation is an important tool in a government's fiscal policy. An increase in capital tax on the value of property or on gains made from the sale of buildings or land may result in decreased demand, while an increase in tax levied on property income or use is likely to result in reduced demand for new buildings, unless it is offset by higher rents or profits. A practical example was the Development Land Tax Act 1976, which introduced a charge in the majority of cases whenever an owner of land realised its development value by disposing of his interest or carrying out material development. In 1984, the first £75 000 of development chargeable to any person in a financial year was exempt, while the remainder was chargeable at 60 per cent. These provisions were however repealed in 1985. Value added tax (VAT) was introduced in Great Britain in 1973, and operates at each stage of manufacture with the tax being levied at a flat rate per cent on the increase in value of the article accruing from the manufacturing process. New building work has a zero rating, but maintenance and repairs, alterations and extensions are subject to tax at the standard rate (17 ½ per cent in 1995). It is possible that in the future VAT at standard rate will be extended to new buildings as in other EU countries. The introduction of VAT on energy bills in 1994 increased building occupancy costs.

## CLIENTS' NEEDS

The range of building clients is extensive from central and local government and housing associations to private organisations concerned with a wide variety of buildings. Their policies and procedures vary considerably but they all want soundly constructed buildings, which will function effectively, look attractive and be economical in cost. Ideally, the architect should give his client not merely what he demanded in his brief, but what he never though of asking for, yet found he wanted all along.

A satisfied client is the object of all industries and is their best advertisement. The larger the financial outlay the greater is the difficulty to satisfy. Consequently the construction industry has possibly a higher proportion of critical clients than most industries. The client expects very high standards whether he be purchasing a small house or investing in a multi-million pound development. He is often very disturbed by the amount of time it takes to acquire a suitable site, prepare designs and supporting documentation, obtain the necessary consents and finally secure completion of the project.

A report by the Building Research Establishment (1978) indicated that although the preparation of the brief and the layout design are crucial, all aspects of the design, construction and use of the building have to be dealt with effectively to ensure quality and value. Quality is concerned with the totality of the attributes of a building which enable it to satisfy needs. It encompasses three main aspects:

(1) External attributes: the effects of the building on its surroundings and vice-versa, such as appearance, compatibility and safety.
(2) Performance attributes: aspects of the building that make it operationally efficient and provide reasonable conditions for users, such as size and layout of spaces, services, environment, safety, security, maintenance, adaptability and longevity.
(3) Aesthetics and amenity: they include the external appearance of the building and any landscaping externally, while internally standards of comfort, convenience and visual attraction are important.

In general the target life for the primary structure of buildings should be about 100 years with lower values for components and services. The building should also be sufficiently adaptable to accommodate the changes in use that will arise over the decades (BRE, 1978).

Some of the more important factors impinging on construction clients' requirements, developed from Harvey and Ashworth (1993), are as follows:

*Contractual and time provisions*
• selection of most suitable procurement method
• length of design period

- clearly defined tender free from conditions and ambiguities
- acceptable commencement date on site and contract period
- satisfactory completion of work by the date stated in the contract, with subsequent rectification of defects

*Cost and value*
- close relationship between budget, tender and final account
- value for money in terms of costs incurred and good functional and quality attributes
- effective and acceptable life cycle cost approach

*Performance*
- comprehensive, well considered and realistic brief
- satisfactory design encompassing good standard of constructability, appearance, quality, function and in-built flexibility
- sound, trouble-free construction with no latent defects and adequate guarantees and after sales service
- effective durability, reliability and maintainability of building(s)

*Management*
- harmonious and trusting business relationships with contractor free from conflict
- well defined allocation of responsibilities between parties to contract
- full and clearly detailed accountability, particularly in the public sector
- minimum exposure of the client to risks

A valuable comparison of construction performance and cost in the UK and mainland Europe (Business Round Table, 1994) confined consideration in the case studies to clients who adopted a 'hands-on' approach, as this produced great benefits when the clients and their design and construction teams are working in new territories and have to adapt to unfamiliar local practice. This enables emerging problems to be tackled and solved together thereby eliminating confrontation and disputes. The close collaborative working resulted in few surprises for the client and greater effort by the construction team to get the work done.

From 1989–93 contractors and subcontractors generally worked to negative margins, with the result that building prices in 1994 were still on average 8 per cent lower than in 1988 despite the intervening inflation. Although tender prices had been forced down by fierce competition, there were many cases where project out-turn costs had overrun due to claims from contractors trying to recover their losses, with insurers predicting more claims to come. Hence it could prove wiser and cheaper to grant contractors, subcontractors and suppliers realistic prices initially. This is further reflected in clients often seeking increased efficiency on site and a higher quality end product, and they will only get what they want if they are prepared to pay for it.

One Regional Health Authority found that the requirements for teaching hospitals were so complicated that the normal practice whereby the client prepares the brief and then appoints the designer was no longer feasible. It appeared that a planning team was necessary to prepare the brief. For certain complicated briefing problems, the Department of the Environment developed an activity data method in an attempt to find out what the client really requires. For the normal client this would be formidable since he probably does not have the staff for these researches.

With shopping, the requirements and opportunities are always changing. Major influences on shoppers' choice are the amenities offered such as the variety and size of shops and ease of car parking, but an equally strong, and often deciding factor, is the environment of the shopping centre itself. There is a growing concern that many of the enclosed shopping centres built in the nineteen-sixties and seventies are too large and unsightly. It is not easy to graft large shopping complexes into the heart of towns and cities, which were built on a small scale over many years (Jones, 1983).

It is now generally believed that a more flexible and imaginative approach is needed and that new shopping schemes should blend in with the existing townscape, rather than dominate it. Low maintenance and improved energy conservation have become increasingly important aspects of shopping design from both the developer's and lessee's points of view.

An instructive CIOB clients' guide (1980) dealing with building for industry and commerce recommended the setting up of a client project group (CPG) of experienced people to examine alternative strategies and identify needs, with the aid of an extensive list of relevant criteria. The group will subsequently feed into the project team who will develop the brief up to the feasibility stage.

A highly critical review which deserves careful consideration was given by Newens (1994), director of property at the National Westminster Bank, who spends £170m per annum on the bank's portfolio of 3400 buildings. He believed that clients were often confused by an increasing number of participants with each person in the construction team wanting authority over the project, but no one willing to take financial responsibility. He declared that 'the construction industry is too complex, costs me too much money and does not deliver what I expect it to deliver.'

A NEDO report (1989) identified major influences on the performance of projects as client participation, design quality and information, contractor's control over the site operation and integration of subcontractors with design and construction. Speed and punctuality were considered to be good measures of a successful project. However, speed of construction should not be the sole objective, important though it is.

With local authority housing an authoritative Government working group report wisely recommended a corporate approach to housing development, setting a cost target for each scheme and working within it, monitoring the overall costs of schemes from their inception, checking for value for money against established criteria, not having preconceived ideas about how to arrive at the end product, using good simple designs frequently rather than one-off plans, and considering the builder's advice on what can be produced most efficiently (DoE, 1978).

The former Greater London Council (1981) with its vast experience of public housing provision published a range of preferred dwelling plans, but it emphasised that the first aim of any design must be to make proposals appropriate to the specific needs of each individual brief and site, and that the preferred plans will be used only on sites that are considered suitable. Another valuable GLC study (1978) examined many aspects of housing layouts.

## STATUTORY AND OTHER REQUIREMENTS

Planning permission is required for most forms of development and this is obtained by making application to the local planning authority in the prescribed manner. The local planning authority can give unconditional permission, permission subject to conditions or refuse permission altogether and there is a right of appeal from this decision to the Minister. Applications can be in outline giving brief particulars of the development in order to secure permission in principle possibly before the land is purchased, or they can be detailed applications. The development has to commence within five years of the first approval.

*Development* is defined in the Town and Country Planning Act 1990 as 'the carrying out of building, engineering, mining or other operations in, on, over or under land, or the making of any material change in the use of any buildings or other land. The Town and Country Planning (Use Classes) Order 1987 prescribed a number of classes and where a change of use keeps within the same use class it does not constitute development and does not therefore require planning permission. A change from one use class to another, such as from an office to a shop, will require permission as the change is then material. The conversion of a single dwelling into two or more flats also constitutes development. Certain types of development are, however, expressly excluded from planning control and are deemed to be *permitted development*. Typical examples are the enlargement, improvement or other alteration of a dwelling house within certain limits; provision on land of temporary buildings and plant; replacement of underground services; access to certain highways; exterior painting; and erection of gates, fences and boundary walls up to certain specified heights. Permission for development is often restricted to a specified density which, in the case of residential development, is usually expressed in either persons or habitable rooms per hectare of site. Readers requiring more detailed information on planning legislation and its applications and implications are referred to Heap (1991) and Telling and Duxberry (1993).

The Department of the Environment has compiled a list of buildings of special architectural or historic interest, and these buildings cannot be demolished, extended or altered so as to affect their character without obtaining a listed building consent from the local planning authority. The only exceptions are where the works are urgently needed for the safety or health of persons, or the preservation of the buildings.

Other statutory requirements include obtaining consent under the current Building Regulations, to ensure that the building work carried out conforms to certain minimum standards of construction, which include structural stability, weatherproofing, hygiene and energy conservation. In addition, industrial buildings have to conform to the requirements of the Factories Act 1961 and the Clean Air Acts, 1956 and 1968, as amended by the Local Government Planning and Land Act 1980. New shops and offices have to comply with the Offices, Shops and Railway Premises Act, 1963, while there are a number of other statutes which lay down requirements for the construction of certain other types of building. The requirements of The Health and Safety at Work Act 1974 and the Environment Protection Act 1990 also require consideration. Hence, depending on the type of building, there can be a number of relevant statutory requirements that have to be satisfied.

In addition to statutory requirements, a developer may find that the development site is subject to natural rights of adjoining owners and occupiers, restrictive covenants or easements, such as rights of light, support, way and drainage. These rights may result in additional expense in the development of the site or even restrict the development on some parts of the site. They do accordingly justify careful scrutiny in any feasibility study.

## EC Directives and Eurocodes

The public works directive 71/305 has been in existence since 1971 and requires national public works authorities to advertise all tenders for construction over a specified value in the *European Journal*. It also permits consultants/contractors from other member states to submit alternative bids. The directive has been revised to raise the level of contract values and to require public works authorities to accept for consideration alternative tenders based upon European codes and standards. James (1994) details the three EC (EU) directives which affect liability for construction products with their UK statutory counterparts. European codes will replace BSI codes following approval and an appropriate transitional period. The objective of Eurocodes is to facilitate trade in professional services across member states.

In principle, UK consultants will find it easier to obtain local approval of their designs in other member states when they use Eurocodes. In practice, the problem of code approval is only one factor in the overall difficulty of operating in other European countries. Another factor is language, as British professionals are not noted for their fluency in other languages, and the competition from non-European countries, such as Japan and South Korea (Seeley, 1992). More information about the problems of operating in mainland Europe is given in chapter 14.

## CHANGING CONSTRUCTIONAL TECHNIQUES

There have been considerable changes in constructional techniques since the last war, accompanied by increasing mechanisation of work on sites and improved planning and organisation of the work. Structural frames of steel and reinforced concrete superseded load-bearing brickwork for many structures and so setting out became more complex and demanded greater accuracy, and higher standards of safety were required. From this emerged another development when the part of the frame that supported the curtain walling and part of the floors was hung from a reinforced concrete structure instead of rising from the ground and resting on foundations.

Next the treatment of the external walls changed. Curtain walling of glass and metal gave way to large precast concrete cladding panels, which created new problems in waterproofing of joints, controlling expansion and contraction, and hoisting and placing the panels themselves, with their implications for crane selection and location.

Another interesting development was the adaptation to building of the civil engineering technique of slip forming (vertical continuous climbing shuttering). Adaptations were necessary to accommodate window and door openings and other features and there were certain limitations on its use. It did, however, save a significant amount of construction time. Another innovation – the 'lift slab' technique – also saved time and produced other advantages such as eliminating soffit shuttering, installing services in slabs before pouring concrete, reducing the vertical handling of materials and components, and reducing interference by bad weather.

Probably the largest and most significant change was the ever increasing demand for more varied and sophisticated mechanical and electrical services in nearly every type of building, which increased from about 25 to 50 per cent of total contract value on larger projects. The main contractor frequently finds it necessary to have on his site staff a qualified mechanical and electrical engineer to liaise with services consultants and subcontractors.

One area that has not been quite such a success story has been that of industrialised building. The Association of Metropolitan Authorities, (1983) has made a critical appraisal of non-traditional dwellings constructed in the nineteen-forties and fifties and estimated that the likely total cost of repair, demolition and replacement could be around £5000m in the 1980s.

The Centre for Strategic Studies in Construction at Reading University (1988), in considering likely changes in building by year 2001, foresees that traditional craft skills will be largely replaced by skills in fixing techniques, integration between specialists responsible for structures and claddings to provide a complete shell and core, and increasing mechanisation will help to alleviate skill shortages, with most of the work moved off site to modern factories capable of responding economically to variable demands for building.

The Centre also envisaged an increasing proportion of building work being in mechanical and electrical services, with much greater modularisation of components as one-off designing is likely to be replaced by flexible servicing with short life plug-in components, and greater integration of services and dry finishes.

Modern office buildings will require greater investment in information technology facilities and associated sophisticated systems, while intelligent buildings will become more common to satisfy the more searching needs of the occupants. Greater use of robotics in construction should result in increased productivity and improved quality (Harvey and Ashworth, 1993).

Japanese housebuilders have introduced standardised components, modular building and a semi-automated construction process, which has speeded up production and halved labour costs. This has allowed more money to be spent on better quality materials and designs that will last longer and look better. Furthermore, factory production ensures tight quality control. The greater use of research and development by the Japanese should result in a better understanding of client needs and producing what the client wants rather than what is cheapest to build (Thompson, 1994).

Despite the many weaknesses of the immediate post-war prefabricated developments, it has now been shown both in the UK and in Japan that modular or volumetric construction is well suited to hotels, student accommodation, MoD buildings and the education sector. The common denominator is that each lends itself to repetition. Modern modular buildings generally have engineered timber, or more commonly, lightweight steel frames, sophisticated, fire-resistant panels and insulation, are fully serviced and, in some cases, fully furnished. Crucially, off-site fabrication entails construction in a controlled environment, protected from the vagaries of the weather (Pettipher, 1994).

## PRODUCTIVITY

At the launch of the UK construction productivity network (CPN) in 1995, it was announced that UK construction productivity improved by 50 per cent between 1980 and 1992 compared with 32 per cent in France, 31 per cent in Japan and 17 per cent in the former West Germany. However, per capita productivity is much lower in the UK than in the other countries and hence there is a vital need to increase performance and to assist in achieving the Latham (1994) target of 30 per cent reduction in real costs by the year 2000.

To effect control, productivity must be forecast under any given set of conditions and variances from that forecast monitored. In the future predictive models may be developed assisted by the increasing use of microprocessors, although both Pilcher (1992) and Warren (1993) have illustrated the difficulties encountered in measuring productivity.

One approach is to introduce incentive schemes which by 1979 were being operated for about one-half of the workers in the industry, although the greater proportion of these were employed by the very large firms. All these schemes aim to secure greater productivity for the employer with employees also receiving additional payments for their increased output. The basis for determining a target must be some ascertainable standard, normally obtained from the analysis of data from site. This must be considered very carefully because one factor that is difficult to determine is the rate at which the operatives were working when the work was carried out.

Oxley (1978) identified the main features of a good incentive scheme as:

(1) The amount of bonus paid to the operative is in direct proportion to the time saved and there is no limit to the amount that can be earned.
(2) Targets are issued wherever possible for all operations before work commences. The extent and nature of the operation should be explained clearly. Incentive packages are grouped to enable gangs to complete them in a relatively short time period.
(3) Targets are set for small gangs (except in repetitive work, where group bonuses may be more effective).
(4) Targets are not altered during an operation without agreement of both parties.
(5) Systems used for calculating bonus earnings are clear to all concerned.
(6) Arrangements are made for dealing with time lost due to reasons outside the operative's control.
(7) Bonus payments are made weekly.

Pilcher (1992) identified a realistic approach through the techniques of work study, incorporating work measurement and sampling, field and productivity ratings, time study, method study, process charts and multiple activity charts, and concluding with incentive schemes.

## CONSTRUCTABILITY AND CONTROL OF WASTE

The relative simplicity of constructing a building will influence the cost of a project. Constructability, sometimes described as buildability, has been defined as the extent to which the design of a building facilitates the ease of construction, subject to the overall requirements for the completed building (CIRIA, 1983). Hence the designer should have comparative ease of construction in mind at every stage of the design process, particularly in the early stages, taking a very practical approach. This necessitates a detailed knowledge of construction processes and techniques and the operational work on site, and is made much easier with the early appointment of the contractor. The principal aim is to make the construction as easy and simple as possible and to reduce waste, such as excessive cutting of components. Another aim is to make the maximum use of site plant and to increase productivity. The design details should encourage a good, logical sequence of constructional activities, and the contract documentation should show clearly that the design has been prepared to permit ease of building and the achievement of an economical project, often involving the consideration of alternative methods.

Readers requiring more detailed information on this important concept and its implementation are referred to *Constructability in Building and Engineering Projects* by Griffith and Sidwell (1995).

On occasions a conflict may arise between ease of building and the quality of construction and aesthetic requirements, and in these cases an acceptable compromise should be sought. It will be recognised that the economics of building work is only one of the criteria to be considered and the client's best interests are served by securing the optimum balance between the various design objectives.

Skoyles (1982) has described how not all materials delivered to construction sites are used for the purposes for which they are ordered, and how builders frequently use more materials than those for which they receive payment. These materials are either lost or they are used in the building process in ways that are not recognised by estimators. These differences have never been clearly defined and are both known as 'waste'. Materials and the labour spent on handling them to the fixing position account for nearly one-half the cost of traditionally constructed buildings, so any appreciable waste is a significant item of cost and a loss of resources.

The allowances for waste used by the estimator are generally known as 'norms'. Studies at the Building Research Establishment have shown that the waste that occurs on site bears little relationship to the norms used in estimating, of which examples are given in table 1.1.

Waste can be classified under two broad headings – direct waste (a total loss of materials which can occur every time materials are handled, moved, stacked or stored) and indirect waste (a monetary loss related to materials and sometimes to the way they are measured, and can arise

Table 1.1    Wastage of materials on sites

| Materials | Overall waste on sites investigated by BRE (%) | Estimators' normal allowance (%) |
|---|---|---|
| Common bricks | 8 | 4 |
| Facing bricks | 12 | 5 |
| Blocks | 9 to 10 | 5 |

*Source*: CIOB Technical Information Service No. 15 (1982).

from substitution, production and operational loss and negligence). These aspects highlight the need for designers to pay more attention to ensure that dimensions fit sizes of materials available on the market, for estimators to take a fresh look at estimating practice, for buyers to adopt a more rigorous approach and for site management to exercise more effective materials control, as described in CIOB Technical Information Service No. 87 (1987).

## QUALITY ASSURANCE AND MANAGEMENT

### General Background

In the late 1980s increasing concern was being expressed at the low standards of performance and quality often achieved in UK building work. This highlighted the need for structured and formal systems of construction management culminating in the establishment of the total performance concept for buildings. This concept gives clients the opportunity, from design brief through to commissioning and operating stages, to establish minimum standards of performance for all aspects of buildings, through quality assurance procedures.

The CIOB (1989) defined quality assurance as 'an objective demonstration of a builder's ability to produce building work in a cost effective way to meet the customer's requirements', while the RICS (1989) considered that quality assurance was a management process designed to give confidence to the client by consistently meeting stated objectives. BS 5750: 1987 defined quality management as the organisation structure, responsibilities, activities, resources and events appertaining to a firm that together provide organised procedures and methods of implementation to ensure the capability of the firm to meet quality requirements in accordance with Parts 1, 2 or 3 of BS 5750. This UK national standard for quality systems started the trend towards the registration of quality systems by certification and formed the basis of the European Standard for Quality Assurance (EN 29000), subsequently leading to BS EN ISO 9000: 1994.

Quality assurance requires appropriate systems, sound procedures, clear communication and documentation that is accurate and easily understood, and prescribed standards must be set and achieved. It impinges on every aspect of the total building process and all building professionals should promote quality assurance as a common objective. Within the construction industry there are five broad sectors where quality assurance is applicable.

(1) Client: in the project brief.
(2) Designer: in the design and specification.
(3) Manufacturers: in the supply of materials products and components.
(4) Contractors and subcontractors: in construction, supervision and management.
(5) User: in use of new building, its upkeep and repair (Griffith, 1990)

The respective duties, responsibilities and work methods encompassed by quality assurance procedures are contained within a *quality plan*, which prescribes the activities and events required to achieve the quality goal throughout the building process, and involves the coordination of each of the participants (CIOB, 1989). The Latham Report (1994) recommended that quality assurance certification should continue to be encouraged within the construction industry as a potentially useful tool for improving corporate management systems.

**Principal Applications**

At the outset, quality assurance procedures should ensure that the client is made aware of the range of methods and services entailed in a building project and that building is a sequential process, requiring a timely decision at each stage in order to avoid delays and additional costs caused by pursuing inappropriate options.

The formulation of the quality plan at the conceptual design stage should incorporate regulatory requirements, relevant technical standards, design responsibilities, lines of communication, and verification procedures.

Contractors short listed should have proven experience in the development of quality systems on similar projects. Each tendering contractor should submit proposals covering such matters as:

(1) the relationship between time, cost and quality in successfully achieving the requirements of the client;
(2) a project quality plan, describing the methods by which the contractor's responsibilities will be monitored throughout the project, the management resources that will be employed and records that will be available;
(3) a description of how the quality plan will interface with the design team quality plans and those of individual suppliers and subcontractors (CIOB, 1989).

In the purchase of materials, services and equipment, it is essential that only satisfactory sources are selected in cost effectiveness terms and in the ability to meet specified requirements. Materials and services should be procured from organisations which:

(1) give confidence that suppliers and subcontractors have effective quality systems;
(2) ensure that every supplier and subcontractor understands that supplies or services must be to the specified requirements;
(3) ensure that all information given in the order is up to date and accurate.

All building sites require formal, simple and effective documentation systems to improve their efficiency, and to provide evidence that the specified quality has been achieved at first attempt. Quality assurance introduces a series of checks, usually in the form of a checklist, identifying the stages at which it can be verified that the works comply with the specified requirements, and each stage must be agreed before proceeding to the next. A master programme should be prepared in which the activities of the client, contractor and subcontractors are coordinated (CIOB, 1989).

The quality assurance procedures for cost control ensure compliance with the client's requirements for comparison against the cost plan. The procedures define who has authority to issue instructions under the contract, the limits of financial authority for each party and the manner in which proposals are to be costed and assessed alongside other options before a financial commitment is made. They also define the method and frequency of cost reporting to the client.

Prior to advising the client's authorised representative that he considers the work to be complete, the contractor should carry out his own detailed inspection to ensure compliance with his own quality plan. Upon being so notified, the nominated representative carries out his own inspections. If he considers that the works do not comply in all respects with the contract documents, the client's nominated representative issues a *discrepancy report*, which is commonly termed a snagging list (CIOB, 1989).

## SAFETY ASPECTS

### General Background

Construction sites often create potentially dangerous situations and 60 to 100 persons are killed on them each year. Some activities are more vulnerable than others. For example, steel erection and dismantling account for many fatalities and serious accidents. Many site injuries result from operatives falling from structures or being hit by falling objects. Many others are caused by the misuse of mechanical plant and transport.

The site manager has a duty to protect the health and safety of employees and must take every reasonable precaution to prevent accidents, improve standards and comply with statutory requirements. He must encourage responsible attitudes to health and safety and increase operatives' awareness of the dangers (Fryer, 1990).

## CDM Regulations

The Construction (Design and Management) Regulations, commonly known as CDM or Condam Regulations, came into force on 31 March 1995. These regulations placed onerous responsibilities in respect of health and safety on construction projects on the client, designer and contractor.

The client has an obligation to ensure there is enough time as well as money to construct the building(s); to make all known facts about the site available to the design team; to appoint competent professionals and contractors; to appoint the *planning supervisor* and keep the *health and safety file.*

The designer must not carry out design work for construction purposes unless he has taken reasonable steps to ensure that the client is aware of the CDM Regulations. He must be able to demonstrate that he has taken safety into consideration throughout the design, and also taken action to prevent future accidents during occupation, maintenance and ultimate demolition. The Approved Code of Practice (ACOP), accompanying the Regulations, requires design documents to specify the essential maintenance procedures designers have agreed with clients at the design stage. The designer will also be expected to draw the contractor's attention to areas of possible danger and to show that he had considered alternatives.

The *planning supervisor* is responsible for producing the *design safety plan*, probably as an addition to the preliminaries, and drawing the contractor's attention to the requirements for method statements. Further duties include checking the contractor's *health and safety plan* and notifying the client of its adequacy, and producing and keeping the *health and safety file* and the final document for the client's retention.

The main contractor will probably be designated as the principal contractor for health and safety, who will be required to produce a detailed *health and safety plan* and method statements as part of the tender and to ensure that all subcontractors adhere to the safety requirements. He will also be expected to work with the *planning supervisor* to produce the *health and safety file*, to be handed to the client on completion, and to supply all necessary information about the materials supplied and their uses.

The construction industry voiced its fears about the high cost of implementing the CDM Regulations and architects objected to the substantial safety requirements during the design stage which it was argued did not apply on mainland Europe. In 1995 the Government (department of employment) estimated that the construction industry faced a £7.5m annual training bill for planning supervisors on top of a one off cost of £150m to comply with site safety laws. At the same time the Government forecast an annual saving to the industry of £580m through the effects of fewer accidents and improved health among the workforce. These estimates do however seem to be flawed and more reliable figures should be available after the regulations have been operating for a reasonable period.

. Consultants were still concerned as they feared that the cost of introducing the regulations would hit small companies disproportionately harder, with the role of the *planning supervisor* falling to the small engineer or architect as part of his normal fee. It was felt that architects may not be able to absorb the extra work of *planning supervisors* and would then attempt to pass the costs on to clients in higher fee bids.

## SEX DISCRIMINATION

Some readers may feel that this particular topic does not warrant mention in this book, but the author believes that in these more enlightened times, having regard to equal opportunities legislation and the important role that women can play, it deserves attention.

Discrimination and harassment still unfortunately exist in the construction industry despite attempts by the largest firms to implement equal opportunities policies and the operation of relevant legislation. Furthermore, DoE figures showed in 1994 there were fewer women operatives and managers in construction than in any other industrial sector. Of the 11 million working women, only 2000 were employed in construction.

The Latham Report (1994) expressed concern about the low numbers of women employed in the industry and called for the establishment of an industry-wide equal opportunities action plan to eliminate discrimination. It stated that there was no obvious reason why women should be seriously unrepresented at professional consultant level, while the traditional excuses offered in respect of site operatives are becoming less relevant as the building process becomes more mechanised.

Latham subsequently urged construction firms to produce recruitment policies specifically aimed at women and ethnic minorities and to consider producing a model equal opportunities advertisement. Appointments should be made on merit, but giving interviews will encourage women and increase their understanding of the skills employers are seeking.

An unpublished CIOB membership survey conducted in 1993 showed 48 per cent of women respondents complaining of discrimination hampering their promotion prospects. This deplorable situation must be rectified as a matter of urgency if the industry is to be able to compete effectively in the recruitment of more women and particularly the more talented ones.

## TRADE UNIONS

The introduction of the Employment Protection Act 1975 gave, for the first time in employment law, legal rights to employees in connection with their membership of and work for trade unions. Previously there were no such legal rights, although many good employers made agreements on a voluntary basis with trade unions whereby employees could take part in trade union activities. They did this in the belief that this arrangement would make it easier to negotiate with their employees, and also provide a means whereby representatives of the employees could be involved in matters of common interest, such as safety, health and welfare. The main functions of trade unions are to negotiate with employers over conditions of service and rates of remuneration.

Trade unions, provided they are independent trade unions, may claim recognition under the Employment Protection Act 1975, and so secure the right to represent the employees in a company. An independent trade union is one that is not under the control or domination of the employer, and which is not subject to interference in any form tending towards such control. An important exclusion so far as the construction industry is concerned is that of self-employed workers. The definition of an employee is not limited to manual workers, and encompasses those in clerical, administrative, supervisory and technical occupations – all have the basic right to belong to a trade union.

An employee is given the legal right to be a member or seek to be a member of an independent trade union and to take part in the union's activities, which may be paid time in working hours, if the employer has agreed to this. The practice of having shop stewards who spend a large part of their time in working hours on union activities is not uncommon with large employers. The employers agree to this practice believing that the ready availability of a shop steward to attend to matters in working hours will prevent disputes arising unnecessarily or leading to serious industrial action. Furthermore, a union or unions that are included in a closed shop agreement have the exclusive right to conduct trade union activities in working hours on the employer's premises.

The rights of employees on union matters are a subject on which complaints may be made to an industrial tribunal. Industrial tribunals are located throughout the country and consist of a legally qualified chairman with two other members, of whom one is selected from employers' organisations and the other from trade unions. When a tribunal finds a complaint to be well-founded it makes a declaration to that effect and then makes an award of compensation.

There are a number of trade unions associated with the construction industry of which the largest ones are the Union of Construction, Allied Trades and Technicians (UCATT); the Transport and General Workers Union (TGWU); General and Municipal and Boilermakers and Allied Trades Union

(GMBATU); Furniture, Timber and Allied Trades Union (FTAT); and Electrical, Electronic, Telecommunications and Plumbing Union (EETPU).

## ORGANISATIONS CONNECTED WITH THE CONSTRUCTION INDUSTRY

There are a large number of organisations, industry, trade and government based, which have an important role in the affairs of the industry, and each of the main categories of organisation will now be examined.

Firstly there are the employers' organisations of which the *Building Employers' Confederation* (BEC), formerly the National Federation of Building Trades Employers (NFBTE), is the largest organisation representing employers in the building industry. Its primary objectives are to represent its members, safeguard their interests, keep them adequately informed and provide advice where necessary. It operates at national, regional and local level. Membership is open to all employers of building labour, be they main contractors, subcontractors or specialists, provided they have a sound reputation and are prepared to conform to the Confederation's rules and code of conduct. The Confederation also protects the interests of affiliated organisations such as the House Builders' Federation and the National Federation of Painting and Decorating Contractors.

Another building employers' organisation is the *Federation of Master Builders* (FMB) whose primary aim is to represent the interests of small and medium-sized builders. On the civil engineering side most contractors belong to the *Federation of Civil Engineering Contractors* (FCEC) which was formed in 1919 with the worthy aims of protecting the interests of its members, establishing amicable arrangements and relations between members and their workpeople, regulating wages and working conditions in the industry through the Civil Engineering Construction Conciliation Board, maintaining a high standard of conduct, combating unfair practices and encouraging efficiency among its members, and formulating and securing the adoption in civil engineering contracts of a standard form of contract embodying equitable conditions.

In 1994 the BEC and FCEC, supported by major construction firms, commissioned a study by consultants about the possible formation of a new construction employers' organisation with a provisional title of 'Newco', with the main objectives of securing reduced running costs and the establishment of an effective, unified voice for the industry as compared with the current fragmented approach. The estimated savings of 20 per cent would be achieved by eliminating the wasteful duplication of effort and resources by eventually bringing the BEC, FCEC and the Export Group for the Constructional Industries (EGCI) together in one building and sharing central administrative and other services. A confederate structure was proposed in 1994 with six affiliates representing housebuilders (HBF), builders (BEC), civil engineers (FCEC), exporters (EGCI), major contractors, and specialist contractors (FBCS). It was claimed that the new body would still encourage and support the Construction Industry Employers Council (CIEC) to which both BEC and FCEC belonged.

Not all construction employers' organisations supported the merger concept because of a feared loss of influence for medium sized builders and the House Builders Federation. Many industry groups wanted the CIEC to continue as the main representative body for the industry. For example, the Federation of Master Builders (FMB) believed that the CIEC could be expanded to include the Construction Industry Council (CIC), the Constructors Liaison Group (CLG) and the Major Contractors Group to form a lobby which would be as effective as 'Newco'. Early in 1995 it was decided not to proceed with 'Newco', because of irreconcilable differences between BEC and FCEC, and to strengthen CIEC.

The membership of the Construction Industry Council (CIC) includes the chartered professional bodies, such as RICS, RIBA, ICE, IStructE, RTPI, CIOB and CIBSE, other non-chartered professional associations connected with the construction industry, and a number of business organisations such as the Association of Consultant Architects (ACA), the Association of Consulting Engineers (ACE) and the Consultant Quantity Surveyors Association (CQSA). There are also associate members including the Association of Heads of Surveying and BRE. In 1994 the Council had 37 members in total.

Major objectives of the CIC were to provide a forum to address areas of common interest related to construction and to present a unified voice to Government on matters of policy. The second objective was blunted when major bodies representing contracting and building materials interests decided that they did not wish to participate and subsequently formed their own grouping in the Construction Industry Employers Council (CIEC). However the two bodies acted in a very responsible way in 1992 when they issued a joint statement indicating their formal agreement to 'liaise and work together wherever possible for the benefit of the construction industry as a whole' – the first priority being to examine the nation's construction needs and to prepare proposals to put to Government.

The membership of CIEC comprises the BCE, FCEC, FMB, NCBMP (National Council of Building Materials Producers) and NSCC (National Specialist Contractors Council). Both bodies were consulted extensively during the discussions preceding the publication of the Latham Report (1994). CIC was identified in the report as the body best suited to coordinate the published recommendations on professional education and to issue a guide to briefing for clients. CIC also made a proposal to the Latham Review Implementation Forum (RIF) that a Joint Contracts Board (JCB) should be formed to influence the development of all standard forms of contract and that members of it should represent CIC, CIEC, CLG (Constructors Liaison Group) and CCF (Construction Clients' Forum). Other contracting employers organisations who were approached by Latham included CLG and NCG (National Contractors Group), the latter being one of the five sectors of BEC and comprising the largest companies in membership of the BEC.

In 1993 the CIOB introduced its Chartered Building Company Scheme in an attempt to improve the generally accepted poor image of the construction industry. To be eligible for membership a company must have at least 75 per cent of its directors professionally qualified and half of these must be members of the CIOB. An independent board within the CIOB administers the scheme and a code of conduct helps to ensure that member companies operate professionally and offer a quality service (Harvey & Ashworth, 1993).

The more specialist contractor groupings include the National Specialist Contractors Council (NSCC) made up of the Federation of Associations of Specialists and Subcontractors (FASS) and the Federation of Building Specialist Contractors (FBSC); and the Specialist Engineering Contractors' Group (SECG). The latter group comprised the Electrical Contractors Association, the Electrical Contractors Association of Scotland, the Heating and Ventilating Contractors Association, the National Association of Lift Makers, and the National Association of Plumbing, Heating and Mechanical Services Contractors. In 1993 the NSCC and SECG came together with the Building Structures Group to form the Constructors Liaison Group (CLG) as a reference and liaison centre for matters affecting the common interests of their member trade associations. All these associations and the groups continue to operate separately.

The main organisations representing construction clients comprise the Construction Clients' Forum (CCF) and the British Property Federation (BPF). The Federation issued a manual in 1983 which described in detail the way in which it intended projects under its members' control to be managed and this manual is examined further in chapter 3. Subsequently the Construction Industry Board was set up, as described in chapter 19.

Another important organisation is the *National Joint Council for the Building Industry* (NJCBI) which comprises equal representation from employer organisations and trade unions. Its main functions are to:

(1)  determine wages and working conditions of building trades operatives through the National Working Rules for the Building Industry;
(2)  settle grievances and disputes between employers and operatives; and
(3)  administer the National Joint Training Scheme for skilled building occupations.

The *National Working Rules* (NWR) are very comprehensive and prescribe very clear working arrangements in respect of wages, working hours, guaranteed weekly wage, conditions of service and termination of employment, extra payments in respect of working in unusual conditions and tool allowances, overtime payments, shift work and night work, travelling and lodging, trade union recognition and procedures, payment for absence due to sickness or injury, grievance disputes and differences, register of employers and death benefit cover.

In addition to the public holidays for which payment is made direct by the employer, an operative is entitled, under the *Annual Holidays Agreement for the Building and Civil Engineering Industries* (AHABCEI), to three weeks' annual holiday, one week of which is to be taken in the winter period at a time decided by the National Joint Council for the Building Industry. However, because of the mobility of the workforce in the industry, operatives may work for several employers during the course of a year. The problem has been resolved by the operation of the 'holidays with pay scheme' administered by the Holidays Management Company. Under the scheme employers purchase holiday stamps from the company and affix one stamp to the employee's 'holiday stamp card' for each week worked. Sections of the card are surrendered for each week of an operative's annual holiday.

Another noteworthy organisation is the *National Joint Consultative Committee for Building* (NJCC) which monitors procedures and promotes good practice within the building industry. It consists of representatives of the professions concerned with construction and the main contractors' organisations. For example it has produced some excellent guidelines for tendering procedures which are referred to in more detail in chapter 4.

On the training side, the *Construction Industry Training Board* (CITB) was established under the Industrial Training Act 1964, to ensure that there was an adequate supply of properly trained men and women at all levels and to improve the quality and efficiency of industrial training. A levy is imposed on all employers who, in their turn, can receive grants for sending employees on approved training courses.

The *Building Centre* has been established in Store Street, London, with branches in various provincial centres. It is supported by a number of building material manufacturers and, in addition to maintaining permanent exhibitions of an extensive range of building materials and components, it collates details of manufacturers and suppliers of trade literature, which are supplied to interested persons, and contains an excellent building bookshop stocking a wide range of relevant books and other publications.

*Trade Associations* have been established to promote materials and components used by the construction industry. Their main function is to supply technical information and literature, but some undertake extensive research and development activities. Trade associations in the United Kingdom include such well known organisations as the Aluminium Window Association, Brick Development Association, British Ceramic Tile Council, Cement and Concrete Association, Clay Pipe Development Association Ltd, Copper Development Association, Gypsum Products Development Association, Paintmakers' Association of Great Britain Ltd and the Timber Trade Federation.

Many building materials and components and their installation on site are the subject of British Standards and associated Codes of Practice, issued by the *British Standards Institution*. This is a private body representing all interests with financial support from the government. The work of the Institution is mainly performed through technical committees. The use of these Standards ensures good quality materials and work, and they embrace the results of latest research. Compliance with a British Standard is generally accepted as satisfying Building Regulations. Furthermore, their use reduces the work involved in specification writing and makes for greater uniformity in requirements, which helps the contractor. In the years ahead British Standards are likely to be replaced by Eurocodes (Seeley, 1995).

In post-war years the construction industry has become increasingly involved in the use and development of a perplexing array of new materials, products and techniques. The proliferation of unfamiliar and novel products makes evaluation and selection an increasingly difficult process. To overcome this problem, the *British Board of Agrément* (BBA) was established by the Government in 1966 to investigate new materials, products, components and processes, and new uses of established products, and to issue certificates of worthiness where appropriate. BBA is a member of the European Organisation for Technical Approvals (EOTA).

The *Building Research Establishment* (BRE) is a government research and development centre at Garston, near Watford, Hertfordshire, whose main objective is to improve construction techniques, and assist the industry through publications such as digests and research papers, tests and general advice. The BRE digests assist considerably in widening our knowledge and understanding of the use of building materials, the reasons for failures and ways of avoiding them. In 1996 the Government was considering proposals to privatise BRE, but the main representative bodies for consultants, contractors and subcontractors combined to submit an alternative strategy, whereby a reformed BRE would be governed by a council made up of representatives from Government, clients, industry and academia to make it more responsive to the needs of industry and Government.

## SOURCES OF TECHNICAL INFORMATION

The range of sources of technical information for those engaged in the construction industry is enormous and is growing rapidly. Computerised information retrieval and the extension of teletext and viewdata facilities will assist in the retrieval and assimilation of information in the future.

The main sources of information are libraries, professional and technical journals, trade literature, technical information and standards, government publications, legislation and statistics from a variety of sources. Most organisations build up their own libraries or information centres and some of the data is generated in-house as described in chapter 15. The British

Standards Institution, British Board of Agrément, Building Research Establishment and Department of the Environment supply much authoritative information which is continually in demand by all members of the construction team. There are also a variety of handbooks, directories and year-books which form useful sources of reference. On building costs there are the standard price books, which are becoming yearly more comprehensive, and the services offered by BCIS and BMI, as described in chapter 8.

For building materials, the trade associations and building centres have already been identified and there are other sources such as the yearly publication 'Specification', National Building Specification, RIBA Standard Catalogue and the Barbour Compendium of Building Products. Packaged information and library services are also provided by Barbour Index Ltd.

# 2 The Building Team and the Design Process

This chapter extends the general background to the quantity surveyor's work by describing the various members of the building team and their interrelationships, the part played by the relevant professional bodies, the design process and the quantity surveyor's role within it.

## THE BUILDING TEAM

### The Employer

The employer, who is also called the client or building owner, is the organisation or person who commissions the construction project and who pays the cost of the work, as certified by the architect's certificates. He does not usually come into contact with the various members of the building team, apart from the architect, to any great extent, and is not strictly part of it, although he is very much concerned with all that it does.

Much of the success or otherwise of a building contract will depend on the employer. To ensure a satisfactory outcome, he should select a competent architect and cooperate effectively with him. Unfortunately these conditions do not often operate in practice, as the employer often deliberates at great length as to whether he should proceed with a particular scheme, and, having eventually decided to do so, impatiently demands drawings and estimates of cost, and then requires tenders and finally a completed building in an incredibly short space of time. Inadequate drawings and other contract documents, prepared in haste, will result in many problems when the work is underway on the site. Sufficient time must be allowed for the proper planning of the project and for contractors to price and return the bills of quantities and form of tender, otherwise contractors have to price high to cover themselves against unidentified risks.

The employer should, as far as possible, refrain from requesting changes to the original design as the work proceeds. Such variations often result in delays and disorganisation of the work and can give rise to increased costs and claims from the contractor. It is most important that the employer shall honour the certificates issued by the architect within the period inserted in the appendix to the form of contract. Contractors have large sums of money tied up in building contracts and they require prompt payment to prevent cash flow problems. Finally, it should be borne in mind that most employers want what they need, when they want it, and at a price they can afford.

Latham (1994) described how clients are the core of the construction process, but they are dispersed and vary greatly in character. While developments in the public sector have fragmented the client base even further and made it a much less dominant client than hitherto. Some previously substantial programmes such as local authority housebuilding have been reduced significantly while other work is now partially funded by private investment or has been totally privatised. Privatisation has resulted in the transfer of many professional services from 'in house' operations in government departments and local authorities to the private sector. New procurers include Executive Agencies, National Health Trusts and Colleges of Further Education and the demise of the Property Services Agency (PSA) has given rise to a wide range of procurement techniques. In 1994 there were approximately 90 separate government procurement bodies. In like manner the FCEC reported significant changes to the client base in 1991/92 with 60 per cent of the workload being carried out by public sector clients as compared with 90 per cent in the late 1970s, and the proportion in the public sector is continuing to fall.

Latham (1994) stressed the responsibility of all construction clients, but especially those in the public sector, to commission projects of which present and future generations can be justly proud in terms of good design. A well designed building will increase the satisfaction, comfort and well-being of its occupants and, in the case of commercial buildings, improve productivity and performance.

## The Architect

The architect is often regarded as the leader of the building team, although the inroads of project managers and other professionals are tending to change the traditional approach. The architect often receives the commission to design and supervise the erection of the building. The amount of specialised knowledge required for the design of a modern, complex building is so great that the architect will almost certainly need assistance from specialists. For example, structural engineers may be required to design the structural frame, mechanical and electrical engineers to design the M&E services, and quantity surveyors to advise on contractual and cost aspects and prepare bills of quantities and other contract documentation. The architect may also need advice on ground investigations, landscaping and other aspects.

The architect is usually wholly and entirely responsible for the preparation of the contract, including the formulation of designs for the project. He is virtually in sole control of the project until the contract is signed, but should make it clear to all tenderers that he is acting on behalf of the employer. Once work has started on the site, he is responsible for ensuring that the contractor carries out the whole of the work in accordance with the contract and to the architect's reasonable satisfaction. He is without doubt, one of the principal parties to the contract and he comes into direct contact with the majority of persons connected with the work at one stage or another. Much of the success or otherwise of a project depends on the way in which the architect performs his functions.

The architect is normally the only member of the building team with an overall view of the project, and his functions usually include liaison with the employer, representatives of local authorities and statutory bodies, and with consultants and specialists. Liaison with the contractor will start at an earlier or later stage depending on the type of contract.

The architect's first task, after appointment, is to discuss with the employer his building requirements. This often emerges as a list of needs, commonly referred to as 'the brief'. It sometimes happens that the employer is uncertain of his requirements, and the architect helps in formulating them. When both site details and the principal building requirements are known, preliminary designs can be prepared. At this stage the architect, if he has not already done so, will begin to select his team. In preparing preliminary schemes it is generally necessary to consider the comparative costs of alternative proposals, and a quantity surveyor should be appointed. This should desirably be a direct appointment by the employer on the recommendation of the architect. In complex buildings it may also be advisable at this stage to bring in consulting engineers for the structure and services.

The architect's procedure then follows a logical sequence, often on the following lines:

(1) Preparation of preliminary schemes, including estimates, as part of a feasibility study.
(2) After approval by the employer, an outline planning application will be submitted, and an application sent to the freeholder where the site is leasehold.
(3) Preparation of sketch plans and approximate estimates.
(4) Preparation of structural and services schemes either by consultants or specialist subcontractors.
(5) Preparation of cost plan in consultation with the quantity surveyor.
(6) Preparation of working drawings and invitations for tenders from specialist suppliers and subcontractors, if these items are to be dealt with as prime cost sums.
(7) Preparation of bills of quantities, wherein the architect should maintain close contact with the quantity surveyor.
(8) During the period of bill preparation, suitable contractors will be invited to tender. To enable them to reach a firm decision, they will be informed of the general nature of the work, the date when bills will be available, and dates for submission of tenders and completion of work.

(9)  Formal invitation to tender with full documentation.
(10) Receipt of tenders, advice to employer on selection of contractor and preparation of contract documents.
(11) Supervision of construction work.
(12) Certifying payments to the contractor throughout the contract.
(13) Issuing variation orders and other architect's instructions as necessary.
(14) Directing how provisional sums are to be spent.
(15) Securing the remedying of defects at the end of the defects liability period.
(16) Certifying the final account.

The architect may give the contractor such instructions as he considers necessary in relation to the execution of the work and any variations required, but he cannot insist on an unreasonably high standard of workmanship, going far beyond that which could reasonably be contemplated from the contract documents.

The architect acts as expert adviser and agent for the employer. In his capacity as agent for the employer, the architect enters into many contractual obligations, ranging from the acceptance of a tender to variations of drawings giving details of new doors or changes to a drainage layout. Where the architect is named in the articles of agreement, the contractor is justified in treating every order received from the architect as a direct order from the employer. It is, however, unwise for the architect to issue a variation order covering a substantial variation to the contract, without receiving the employer's prior approval. In the preparation of designs, supervision of work and associated activities, the architect owes the employer a duty to exercise such reasonable and proper care, skill and judgement, as could reasonably be expected from a professional man in his position.

Hence although the role of the architect is primarily as a designer, he is nevertheless involved in the production of a building from inception to completion – from pure design, through production drawings and details, to supervising the contractor. He also has the important task of coordinating the activities of everyone else involved in the project.

It is a far cry from the relatively simple and straightforward relations between architect and builder, so admirably described by Creswell in 1930 in *The Honeywood File and Settlement.* The complexity of modern buildings, constructional techniques and employers' requirements, and the vastly increased number of people involved in the execution of the work, have necessitated a changed attitude and role from the architect. Some have identified the need for the architect to acquire different skills in business and management accompanied by a measure of specialisation.

Some contractors have expressed the view that architects tend to lack appreciation of the practical implications of their designs and see merit in young architects having a period of training or exposure with contractors. It has also become increasingly difficult to extract any form of specification from the majority of architects. Some constructional faults stem from poor detailing and problems can result from the use of new materials, increased vibration from traffic, and problems with sophisticated services arrangements. Trotter (1983) was even more scathing in his attack on architects and accused many of them of supplying inadequate details, working to unrealistic programmes and making excessive changes to design during construction. The architect faces a daunting task and needs plenty of skill, resourcefulness and tact, apart from proven design expertise.

In the late 1980s some of the problems facing architects in the design of buildings and their solutions to them were examined in public. HRH the Prince of Wales in 1989 continued his crusade to improve the quality of modern buildings and the urban environment in 'A *Vision of Britain,'* in a passionately described and attractively illustrated volume. His main themes were harmony (each building in tune with its neighbour); enclosure giving a sense of well-being, neighbourliness, cohesion and continuity; materials (using local materials as far as practicable to retain and enhance local character); community (providing the right sort of surroundings to create a community spirit, with community involvement – more particularly related to housing); and quality of character, fostered by attention to detail and human scale. Glancey (1989) gives an interesting and illustrated overview of the differing faces of modern British architecture in a very striking and informative way.

**The Quantity Surveyor**

Construction cost, construction management and construction communications are all key problem areas for an employer who has commissioned an important building or engineering project. A quantity surveyor is professionally trained, qualified and experienced in dealing with these problems on behalf of the employer. He is essentially a cost expert whose prime task is to ensure that the project is kept within the agreed budget and that the employer obtains value for money.

The detailed services that he offers are fully described later in this chapter, and cover such aspects as preliminary cost advice and cost planning, preparation of tender documents, advice on type of contract and method of obtaining tenders, negotiations with contractors, valuation of work in progress and settlement of the final account. He will at all times need to collaborate very closely with the architect.

**Consulting Engineers**

On large and complicated building projects, it is customary for the architect to recommend to the employer the appointment of consulting engineers, usually specialists in structural work and mechanical and electrical engineering services. The architect's knowledge and experience in these specialised areas will be normally insufficient to cover these functions effectively. Some integrated practices will contain these specialists within the practice and can then offer the services 'in house'.

The engineer prepares the necessary designs, specification and other relevant documents, obtains quotations for the work, and submits a report. He will subsequently supervise the work on the site under the overall control of the architect, who will retain responsibility for coordinating the work of the specialists. The engineer not only contributes his own special expertise, but also contributes to the combined work of the design team. Hence consultant engineers should be appointed to the team at the earliest stage. For example, the structural engineer must ensure structural efficiency and stability but, at the same time, with economy in mind, he will minimise avoidable obstruction by structural members and assist in producing a logical and systematic construction process.

Engineering services encompass methods of controlling the internal environment by means of heating, ventilating, air conditioning and lighting installations, and providing utilities such as electrical supplies, lifts and compressed air. The proportions of capital costs devoted to services vary widely with building design and function, but in general they account for between one-third and one-half of the total cost. The high cost of energy places on designers a duty to relate reasonable environmental standards, the thermal performance of the building fabric and energy demand of services to minimum energy consumption.

Services may be designed by independent consultant engineers, by services engineers within an integrated practice, or by engineers on a design and construct basis. Independent consulting engineers work with a number of architects and thus acquire a wide experience, although coordination will be more difficult. An integrated practice avoids this problem but 'in-house' engineers may lack authority. With small or straightforward contracts, engineering contractors may be engaged but a conflict may arise between their commercial and design activities, and it may be advisable to negotiate a separate design fee with the contractor.

## Other Consultants

Other consultants who may be engaged include landscape architects, interior designers and acoustic consultants. Landscape architects can be engaged on a wide range of activities from the landscaping of residential development, to environmental work around a power station or alongside a length of motorway. In like manner to an architect the landscape architect undertakes responsibility for all stages of the landscaping work from inception to completion, including cost control and contractual arrangements.

Interior design is the integration into a single harmonious concept of structure, dimensions, finishes and furnishings. In the post-war years the interior designer has been given many opportunities to experiment with new and novel ideas. These have resulted in the emergence of the landscape concept in deep office blocks, enclosed shopping precincts and new factory complexes. The requirements of sport and tourism offer further exciting challenges to the interior designer.

The acoustic consultant will advise the architect from his experience in acoustic measurement, calculation and prediction, coupled with his knowledge of sound and noise control design techniques. He can provide satisfactory noise control and good interior acoustical conditions by recommending materials and construction methods to control noise and sound to meet the architect's design requirements and standards. His services can be of particular value in the design of theatres, concert halls, conference centres, hospitals, television studios, lecture theatres and many other complex structures.

### Project Managers

On some very large projects such as the National Exhibition Centre in Birmingham and the Queen Elizabeth II Conference Centre in Westminster, a project manager is appointed by the employer to take overall control of the project from inception to occupation and to coordinate the work of all members of the building team. This procedure should secure maximum efficiency and enables the employer to obtain all information concerning the project from one person, who is thoroughly familiar with all aspects of the scheme. This arrangement will be examined in more detail in chapter 12 and is well described by Day (1994).

### The Contractor

The contractor is the person or firm who undertakes to complete a building project in accordance with the contract documents on behalf of the employer. He is accordingly one of the most important parties to a building contract and should have full control of all operations on site, including the work carried out by nominated subcontractors, with whom he has a direct contractual relationship by means of subcontracts. All instructions and payments to nominated subcontractors must come through the main contractor.

Under the conditions of contract (JCT80), the contractor is to proceed regularly and diligently with the works and to complete by a specified date. Failure to comply with this requirement may render the contractor liable for the payment of liquidated damages. The contractor is required to comply with all statutory requirements affecting the works and to give all notices that may be legally demanded. He must insure the building operations against fire and possible injury to persons or property. He has to attend to many matters of considerable complexity and has many obligations under the terms of the contract. He receives all his instructions through the architect but has dealings with other parties to the contract, such as the quantity surveyor, when he measures and values completed work. The contractor maintains close contact with the clerk of works in his supervision of the work and also the building control officer inspecting the work on behalf of the local authority.

The contractor employs personnel to take charge of work on the site and for large projects there can be a site agent, while on a small scheme a foreman will probably suffice. A foreman is a competent craft operative and must be able to control and coordinate the workforce and ensure that the work is performed satisfactorily and on time. The contractor often employs a quantity surveyor to safeguard his financial interests.

The range of contractors as regards size and categories of work performed is described in chapter 1, as are also the contracting associations.

## The Subcontractor

In chapter 1 the main categories of the specialist subcontractor were identified and described and they embrace crafts, structural work, building services, and decorative and other finishes. It is generally possible to achieve a cheaper and higher standard of workmanship by employing specialist subcontractors to undertake these classes of work. The employer, through his agent the architect, retains control of the subcontractors' work as they are normally appointed or selected by the architect.

The main contractor is, however, responsible for all operations on the site and subcontractors must look to him for instructions and payment. There is no contractual relationship between the employer and the subcontractor. For this reason alone it is imperative that a subcontract is entered into between the main contractor and the subcontractor to safeguard the interests of the main contractor. The principal provisions in a subcontract are as follows:

(1) To ensure that the main contractor secures full and proper cooperation from the subcontractor in carrying out the subcontract.
(2) To ensure that the subcontractor indemnifies the main contractor against the same obligations as those for which the main contractor is responsible to the employer under the terms of the main contract.
(3) To ensure that the subcontractor has full use of all general site facilities, although he is liable for any misuse or damage caused by his employees.
(4) To ensure that the subcontractor undertakes the necessary obligations as to time and progress.
(5) To ensure proper payment of the subcontractor by the main contractor.

### The Supplier

Building suppliers supply building materials and components to contractors and subcontractors. The architect may specify the use of certain materials to be obtained from nominated suppliers and the contractor is entitled to a cash discount of 5 per cent, provided he meets the contract requirements as to payment, but all other discounts pass to the employer. The quantity surveyor often needs to examine suppliers' invoices when checking prices in the basic schedule or against claims for recoverable costs of fluctuations, daywork and related matters.

### The Clerk of Works

The clerk of works is appointed by and acts as inspector on behalf of the employer, but he carries out his duties of detailed day to day inspection and issues verbal instructions only under the directions of the architect, who subsequently confirms them in writing in the form of an architect's instruction.

The clerk of works is usually a craft operative with wide experience of building work and his main function is to ensure that all work on the site complies with the contract documents and architect's instructions. The architect is not entitled to put implicit trust in a clerk of works; although he may delegate to him the supervision of the work, the architect has ultimate responsibility and should advisably decide the action to be taken in respect of any major matter.

The clerk of works usually records details of underground services, foundation depths and other work below ground or that which will be hidden by subsequent building operations. He agrees these details with the contractor's foreman or agent. He also keeps such other records as the architect may require. Cox and Hamilton (1994) have described typical duties and functions of the clerk of works.

### The Resident Engineer

On civil engineering contracts awarded to outside contractors, a resident engineer is usually appointed to supervise the work on site and to ensure that it complies in all respects with the contract documents. He is normally a qualified and experienced civil engineer and acts as the engineers' representative under the ICE Conditions of Contract. He will also keep extensive site records, certify payments to the contractor, check regularly on progress and ensure that health and safety requirements are satisfied (Ballantyne, 1986).

**The Building Control Officer**

Building control officers employed by local authorities administer the Building Regulations within their local government areas. These officers inspect the work on site to ensure that it complies with the current regulations and the approved plans. If the contractor carries out work that does not comply with the operative regulations or the approved plans, or fails to give the appropriate notices, he is placing himself in a very vulnerable position and may have to subsequently take down and rebuild the non-conforming work at his own expense. There is also provision in the regulations for the appointment of private inspectors.

## NATURE AND FUNCTIONS OF PROFESSIONAL AND ASSOCIATED BODIES

There are a number of professional bodies connected with the construction industry and these will be listed and described, together with their main functions. However, before considering individual professional bodies, it will probably be helpful, more particularly for students, to consider the place of professions in society.

Thompson (1968) described how the professional becomes his client's agent, acting on his behalf in matters of great moral, physical or financial importance, often in ways whose validity or purpose the layman has no means of judging for himself, and it becomes of vital importance that a client should have a reasonable assurance that it is prudent to surrender control of important sections of his affairs, and that his agent will be honest, reliable and incorruptible. Members of professional societies can only maintain the demand for their services by continually demonstrating that they are more expert and proficient than any private individual or any unqualified practitioner.

Thompson (1968) identified the task of a professional institution as to substitute its corporate reputation for the individual reputations of practitioners, as a basis of public trust. It has a double role to play where there can be some in-built conflict; to protect its members and to advance their interests, and to assume a public duty and strive to safeguard the public interest in relation to the professional activities of its members. Ethics and professional conduct are examined at length in chapter 16.

An excellent definition of a professional body was expounded by Wickendon, while President of the Institution of Electrical Engineers, and subsequently restated by the late Lord Butler in his gold medal address to the Centenary Conference of the Royal Institution of Chartered Surveyors in 1968. The various elements were analysed in the following admirable way:

'We must place first a Body of Knowledge (science) or of Art (skill) held as a common possession and to be extended by united effort.

Next is an Educational Process based on this body of knowledge and art, in ordering which the professional group has a recognised responsibility.

The third is a Standard of Professional Qualifications for admission to the professional group, based on character, training and proved competence.

Next follows a Standard of Conduct based on courtesy, honour and ethics, towards colleagues and the public.

Fifth, we may place more or less formal Recognition of Status by one's colleagues or by the State as a basis of good standing.

Finally there is usually the Organisation of the Professional Group, devoted to its common advancement and its social duty rather than the maintenance of an economic monopoly.'

In more recent times, eminent Past Presidents of the Royal Institution of Chartered Surveyors have expressed their own authoritative and well-considered views in their presidential addresses and the following represent short but pertinent extracts.

Wilson (1979) saw the Royal Institution of Chartered Surveyors 'as a body of people – a group of free minds – joined together by those tenuous threads of common qualification and near-common interests. . . . It is measured by the quality of its members and what they put into their profession, into their professional body and into society generally. . . . The Institution is not here to do for surveyors what they could and should do for themselves, nor is it here to strengthen the weak by weakening the strong. It is not here to create privilege without responsibility or merely to ensure that its members enjoy a reasonable and ever-improving standard of living.'

James (1980) saw in a world of changing values 'the need for the integrity, objectivity and impartiality of the professional ethic to become increasingly important. . . . The professional ethic is the code of principles of conduct to which the members of this profession subscribe. Those principles reflect the characteristic spirit or 'genius' of our profession, establishing fundamental tenets which determine the standards we should observe. Our code of professional principles is an amalgam of many virtues including competence, humanity, discretion, responsibility, integrity and impartiality. It denotes acceptance of a duty in its widest extent for the benefit not of the individual but of the community.'

Watkins (1981) took a very down-to-earth stance when he said 'But what nonsense it is to say that the journal is the only benefit of membership, even if it is coupled with the right to use the library, attend meetings and obtain advice on problems of practice and so on. The most important advantage is a recognition of professional status and the right to call oneself a chartered surveyor. This right is so difficult to quantify and evaluate that it is often taken for granted.'

Luff (1982) believed 'that the real test of a profession is the maintenance of an outward-looking view, with its first priority a strong commitment to the ethics of confidentiality, integrity and a high sense of public responsibility.' Further authoritative views on professional ethics and conduct are given in chapter 16.

The nature and scope of the principal professional bodies connected with the construction industry are now described.

### The Royal Institute of British Architects (RIBA)

This body was founded in 1834 and the Royal Charter was granted in 1837. Its objects were defined as 'the general advancement of civil architecture, and for promoting and facilitating the acquirement of the knowledge of the various arts and sciences connected therewith.' The long term objectives are to improve the general attitude of mind of the government and the public at large towards architectural, environmental and ecological appreciation and to promote the wisest use of technology and techniques.

Improvements in professional education in research and practice, based on the concept of a total design service, are aimed at influencing the architect's position and role in changing building technology and processes. As a learned society, of over 30 000 members in 1994, the Institute seeks through its journal, sessional programme, educational policy, and library and drawings collection to extend the body of knowledge upon which the practice of architecture is based.

### The Architects' Registration Council of the United Kingdom (ARCUK)

The architectural profession in the United Kingdom and Northern Ireland is controlled by the Architects (Registration) Acts of 1931, 1938 and 1969. The 1931 Act set up the Architects' Registration Council and charged it with three main duties: the maintenance and annual publication of the Register of Architects; the maintenance of correct standards of professional conduct; and the award of scholarships and maintenance grants to students in architecture who lack sufficient means to pursue their studies. For all practical purposes, and subject to one exception, admission to the register of architects is granted only by passing one of the examinations in architecture recognised by the Council. Under the 1938 Act it is illegal for anyone to carry on business as an architect unless they are registered with ARCUK.

### The Royal Institution of Chartered Surveyors (RICS)

The Royal Institution is the leading professional body concerned with land and property, its management and development. It comprises seven separate divisions, of which the quantity surveying division is the second largest. The Institute of Quantity Surveyors amalgamated with the Royal Institution in 1983, with considerable benefits to all the quantity surveyors concerned. The Institution was founded in 1868, received its Royal Charter in 1881 and its Royal status in 1946. Its corporate membership in 1994 was about 70 000.

The Institution is very much concerned with professional education, research and public relations. All applicants for membership are required to pass an assessment of professional competence (APC) and all members are required to keep up to date by undertaking an adequate programme of continuing professional development (CPD), as described in chapter 18. Chartered surveyors subscribe to a strict code of professional conduct and, depending on their position and class of work, may be required to conform to regulations with regard to professional indemnity insurance and members' accounts. These requirements are described in more detail in chapter 16.

The Institution's motto – *Est modus in rebus* (There is method in all things) is not without significance. An important feature is the considerable degree of autonomy of the divisions – for example, each has its own President, Secretariat, Divisional Council and committee and working party structure. The Institution also undertakes considerable work in the development of the profession overseas through the International Committee and other organisations and on impending legislation where the profession can make a valuable contribution. The Institution is becoming more involved with current major political issues of relevance to the profession and the needs of its clients than hitherto. It operates an extensive library information service at headquarters, and has strong branch and junior organisations throughout the country.

In addition there is a Construction Market Panel, being one of six such panels, which oversee the work of the various practice panels. RICS Business Services Ltd, a wholly owned subsidiary of the Institution, provides conferences, publications including the RICS Directory of Members and the official journal, BCIS and BMI, surveyors recruitment consultancy in conjunction with NB Selection Ltd (NBS), computer services, insurance services, an information centre and the RICS Westminster Centre.

### The Society of Surveying Technicians (SST)

The Society was formed in 1970 with the help and encouragement of the Institution to provide a recognised organisation for surveying support staff who, although not professionally qualified, are specialists in a specific branch of surveying. Although a completely independent body it has always worked in close cooperation with the Institution. The Society aimed to further the career development of its members and to create a qualification recognised by employers as proof of high technical competence in each branch of surveying. A partnership arrangement with the RICS was initiated in 1994, which enabled the SST to use the established Institution back up services. The later developments are described in chapter 18.

## The Society of Chief Quantity Surveyors in Local Government (SCQSLG)

The Society was formed in 1973 and by 1983 had about 250 members representing an influential proportion of local authority quantity surveyors. The objects of the Society are to promote discussion on professional, administrative, technical and other matters affecting the local authorities served by members of the Society and to afford help and advice to members in the execution of their duties. The Society also arranges for the interchange and collation of information concerning such quantity surveying and allied matters and to provide a means whereby the knowledge, experience and views of members of the Society can be made available to the body of membership and, where appropriate, to their employing authorities, professional institutions, central government and such bodies as may be approved and agreed from time to time by the Society. By 1995, changes in the structuring of local authority offices reduced the number of chartered quantity surveyors in the local government service.

## The Chartered Institute of Building (CIOB)

The Institute was founded in 1884 and obtained its Royal Charter in 1981. In 1994 its total membership exceeded 30 000. Its principal aims are to promote, for the public benefit, the science and practice of building, and the advancement of education and research. It operates appropriate standards of competence and professional conduct and provides a valuable technical information service and a good range of advisory services.

## The Chartered Institution of Building Services (CIBS)

The Institution was founded on the granting of its Royal Charter in 1976 from the former Institution of Heating and Ventilating Engineers (IHVE). The Institution aims to promote, for the benefit of the public in general, the art, science and practice of such engineering services as are associated with the built environment and with industrial processes. Members must have had training and education in building services engineering and a specialist knowledge of one or more branches. The total membership in 1994 exceeded 15 000.

## The Architects and Surveyors Institute (ASI)

The Institute was formed in 1989 from a merger of the former Construction Surveyors Institute and the Faculty of Architects and Surveyors to provide a professional home for surveyors and architects. Membership is obtained through examinations and experience requirements. Its members numbered around 5300 in 1996 and they are engaged in many areas of activity. The Institute publishes an informative journal bi-monthly.

### Incorporated Association of Architects and Surveyors (IAAS)

The Association was founded in 1925 to encourage and facilitate cooperation between architecture and surveying. Corporate membership is obtained through examination and experience requirements.

### The Landscape Institute

The Institute was founded in 1929 to promote the advancement of skills in the design of land and the maintenance of high standards. It ensures high standards of education and professional conduct.

### The Chartered Institute of Arbitrators (CIArb)

The Institute was founded in 1915 to promote the cause of arbitration in the settlement of disputes. It obtained a Royal Charter in 1979. It provides facilities for the education and training of arbitrators, and cooperates with other professional bodies in their appointment under the Arbitration Act 1950.

### The Institution of Civil Engineers (ICE)

The Institution was founded in 1818 and obtained its Royal Charter in 1828. It is primarily concerned with the advancement of the science and art of civil engineering and produces many publications. Its members play an important role in the majority of large civil engineering projects, both in the areas of design and contracting. Its corporate membership in 1994 exceeded 50 000.

### The Institution of Structural Engineers (ISE)

The Institution is a chartered engineering body obtaining its Royal Charter in 1934 and was formed to advance the science and art of structural engineering and furthering the education, training and competence of its members, and has strong research interests. Its corporate membership exceeded 12 000 in 1994.

### The Association of Consulting Engineers (ACE)

The Association exists to promote high professional standards and to safeguard the interests of consulting engineers. It is an independent, non-statutory body, yet its rules of professional conduct are observed not only by its members, but generally by British consulting engineers.

## The International Federation of Consulting Engineers (FIDIC)

The Federation was founded in 1913 by five national associations of independent consulting engineers within Europe, and since the last war has expanded substantially to embrace over thirty countries within its membership. Its objects are to disseminate information of interest to its members and promote their professional interests. Probably the best known of its publications is the FIDIC Conditions of Contract which are used internationally.

## The Association of Cost Engineers

This Association assumed considerable prominence in quantity surveying circles when it joined with the RICS and IQS in 1980 to form the Joint Documentation Board for Heavy Engineering Works. It is a very active professional association which aims to enhance cost management services in engineering.

## The Joint Contracts Tribunal (JCT)

This Tribunal comprises representatives from the Building Employers Confederation, the Royal Institute of British Architects and the Royal Institution of Chartered Surveyors, together with several associations representing local authorities. The Tribunal draws up standard forms of building contract, revises them periodically and issues practice notes stating the opinion of the Tribunal on matters that have given rise to difficulties of interpretation. The different standard forms of contract are examined in chapter 4.

## THE DESIGN PROCESS

The Royal Institute of British Architects has prepared a suggested pattern of procedure for architects in the preparation and implementation of building projects. The plan of work represents a sound and practical analysis of the design operations and has been applied successfully on many contracts. The plan is outlined in table 2.1 and shows the roles to be played by the various members of the design team throughout the design and construction period. Each of the various stages will now be examined.

Table 2.1 Plan of work for design team operation
Outline Plan of Work

| Stage | Purpose of work and decisions to be reached | Tasks to be done | People directly involved | Usual terminology |
|---|---|---|---|---|
| A. *Inception* | To prepare general outline of requirements and plan future action. | Set up client organisation for briefing. Consider requirements, appoint architect. | All client interests, architect. | *Briefing* |
| B. *Feasibility* | To provide the client with an appraisal and recommendation in order that he may determine the form in which the project is to proceed, ensuring that it is feasible, functionally, technically and financially. | Carry out studies of user requirements, site conditions, planning design, and cost, etc., as necessary to reach decisions. | Clients' representatives, architects, engineers, and QS according to nature of project. | |
| C. *Outline proposals* | To determine general approach to layout, design and construction in order to obtain authoritative approval of the client on the outline proposals and accompanying report. | Develop the brief further. Carry out studies on user requirements, technical problems, planning, design and costs, as necessary to reach decisions. | All client interests, architects, engineers, QS and specialists as required. | *Sketch plans* |
| D. *Scheme design* | To complete the brief and decide on particular proposals, including planning arrangement, appearance, constructional method, outline specification, and cost, and to obtain all approvals. | Final development of the brief, full design of the project by architect, preliminary design by engineers, preparation of cost plan and full explanatory report. Submission of proposals for all approvals. | All client interests, architects, engineers, QS and specialists and all statutory and other approving authorities. | |

*Brief should not be modified after this point*

| | | | |
|---|---|---|---|
| | | | *Working drawings* |
| E. Detail design | To obtain final decision on every matter related to design, specification, construction and cost. | Full design of every part and component of the building by collaboration of all concerned. Complete cost checking of designs. | Architects, QS, engineers and specialists, contractor (if appointed). |
| *Any further change in location, size, shape, or cost after this time will result in abortive work.* | | | |
| F. Production information | To prepare production information and make final detailed decisions to carry out work. | Preparation of final production information i.e. drawings, schedules and specifications. | Architects, engineers and specialists, contractor (if appointed). |
| G. Bills of quantities | To prepare and complete all information and arrangements for obtaining tender. | Preparation of bills of quantities and tender documents. | Architects, QS, contractor (if appointed). |
| H. Tender action | Action as recommended in paras. 7–14 inclusive of 'Selective Tendering'.* | Action as recommended in paras. 7–14 inclusive of 'Selective Tendering'.* | Architects, QS, engineers, contractor, client. |
| | | | *Site operations* |
| J. Project planning | Action in accordance with paras. 5–10 inclusive of 'Project Management'.* | Action in accordance with paras. 5–10 inclusive of 'Project Management'.* | Contractor, subcontractors. |
| K. Operations on site | Action in accordance with paras. 11–14 inclusive of 'Project Management'.* | Action in accordance with paras. 11–14 inclusive of 'Project Management'.* | Architects, engineers, contractors, subcontractors, QS, client. |
| L. Completion | Action in accordance with paras. 15–18 inclusive of 'Project Management'.* | Action in accordance with paras. 15–18 inclusive of 'Project Management'.* | Architects, engineers, contractor, QS, client. |
| M. Feedback | To analyse the mangement, construction and performance of the project. | Analysis of job records. Inspections of completed building. Studies of building in use. | Architect, engineers, QS, contractor, client. |

*Publication of National Joint Consultative Committee for Building. Reproduced by permission of RIBA Publications Ltd.

**Inception**

The difficulties inherent in establishing the employer's or client's brief are often substantial. The main problem is identifying the employer's needs and then reconciling these against a background of financial, technological and legislative constraints, the employer's personal requirements, and also a number of design variables including structural form, aesthetics and environment. By examining the constraints and design variables together, fundamental design decisions can be made judged against a further criterion of quality.

The architect cannot start his work until the employer decides what he wants, which he frequently does not know. The architect is then likely to have difficulty in extracting the essential information from him. Furthermore, it is necessary for the architect to include matters omitted by the employer and to relate the requirements to the conditions of the site and then solve the consequent design problems against the background of constraints previously outlined. It is important that the employer should be kept well informed of the various decisions as they are made.

The extent of the problem of communication will vary also with the type of employer and the nature of the project. There is a vast difference between the method of communication used with an experienced, sophisticated employer requiring a large chemical works and a young couple wanting to have their first house designed and erected. Many public authorities and some major private employers have considerable in-house professional skills within their organisations, while, on the other hand the design team may be faced with major projects commissioned by unskilled amateurs, often possessing considerable enthusiasm but little knowledge of the construction industry. Each new commission will therefore need dealing with individually and there cannot be one universally applicable approach at the brief stage. Where an architect is dealing direct with a board of directors and the senior executives of a large organisation, a chaotic situation can arise if they give a host of conflicting instructions and, in these circumstances, it would probably be desirable for them to appoint a single spokesman who can put their collective views to the architect.

The architect will require as much information as possible from the employer. Certain basic details are common to most projects, and these are now listed:

(1)  the nature, size and function of the proposed building;
(2)  the timescale and financial limits relating to the project;
(3)  information relating to ownership of the site, boundaries, restrictive covenants and other associated matters;
(4)  current position with regard to any planning application; and
(5)  other members of the design team to be appointed, with agreement as

to the service to be provided and basis of fees. A useful checklist for the selection of design consultants has been produced by Gray, Hughes and Bennett (1994).

The broad contents of an employer's brief may help to clarify this aspect. The example selected is the Queen's Medical Centre in Nottingham, where the brief was prepared by the Medical School Advisory Committee in June 1965. The complex was to comprise a large general district hospital of some 1400 beds to provide a substantial and integral part of Nottingham's future hospital needs, to be designed with undergraduate medical student teaching in mind; a medical school offering teaching in the full range of disciplines leading to the award of medical qualifications; and a nurses' training school to supply the nursing staff not only for the new hospital, but also for the other hospitals in Nottingham. The medical school should be sufficiently near the main University campus to maintain physical contact. The new building should be started as early as possible with a certain proportion available for use early in the programme, and the whole complex of some 186 000 m$^2$ should be completed in a continuous building operation. The complex and varied functions would generate considerable foot and trolley traffic within the building and the main aim would be to minimise the movement of this traffic by arranging the departments as compactly as possible (Constrado, 1973).

More of historic and personal interest to chartered surveyors was the employer's brief supplied to the five selected architects for the new headquarters building for the then Institution of Surveyors at 12 Great George Street, Westminster, in 1895. The schedule of accommodation included in the brief was accompanied by sketch plans 'appended only as a graphic illustration . . . of the instructions with regard to size of rooms, etc.'

The principal rooms required were a lecture hall, library, council room, reading room, two arbitration rooms served by a separate entrance, forestry museum and a 'good staircase', together with offices and caretaker's quarters – the type of accommodation that any professional body would require, although the provision of a forestry museum was the exception and reflected the interests of the land agents. The budget was a fairly modest one, £16 000 to £20 000 excluding fittings (Lever, 1981).

Gray, Hughes and Bennett (1994) have devised checklists relating to the statement of need and the client's brief and have advocated the use of standard briefing documents based on the architect's previous experience.

## Feasibility

The main objective at this stage is to examine all the technical, functional and financial aspects of the project and to advise the employer as to its feasibility. The first step would normally be a meeting of the design team to consider in detail the employer's requirements and to make their recommendations thereon for submission to the employer, and the subsequent establishment of the design brief. It will then be possible to decide the responsibilities of the individual members of the design team and to determine the programme for the design stage.

A detailed survey of the site and any adjoining buildings is undertaken and information obtained on the position, size and capacity of the various services (gas, water, electricity, sewers and telephones). Consultations will take place with Planning and Building Control Officers, Fire Authority and any other relevant bodies. Any restrictions on the use of the site, such as easements or restrictive convenants must be investigated and their effect on the proposed development determined.

It is common practice to visit modern buildings with similar uses to those envisaged by the employer, to investigate user requirements and facilities, particularly in the case of specialised buildings such as theatres and leisure complexes. An application is often submitted at this stage for outline planning permission to ensure that the project as envisaged will be permitted under planning regulations.

The architect will then compile a report based on the findings and recommendations of all the members of the design team. The report must effectively assess the overall feasibility of the scheme having regard to all relevant factors, so as to give the employer all the information he needs to make a decision as to whether or not to proceed with the scheme. The report will be supported by conceptual designs, usually in the form of diagrammatic plans and elevations and estimates of cost. In the case of an investment proposal, a developer's budget is also likely to be required to show the probable percentage return on the outlay.

## Outline Proposals

At this stage alternative schemes are prepared and compared in order to determine the general approach to the layout, design and construction. This involves very close collaboration between all members of the design team and the employer. Each member of the team has an important role to play – the architect will be particularly concerned with functional and aesthetic aspects, the structural engineer with different structural forms and the services engineer with alternative services layouts. Each of these will impinge on each other and may have a significant effect on the overall design. The quantity surveyor will maintain close contact with all the other professionals in order to determine the cost implications of the alternative proposals as they are prepared, and to subsequently ensure that the proposals keep within the approved budget and outline cost plan.

The architect is concerned with a number of general design principles in order to achieve a satisfactory building. For instance, the building should have character which is generally conveyed by the building itself and the site layout; the various parts of the building should be in suitable proportion to each other and to the whole structure and, at the same time, be in scale having regard to the size of the human figure. Materials, colours and texture should all be skilfully selected to blend in with the surrounding buildings and help to produce an attractive environment.

The architect is concerned not only with the size of rooms or work areas and their interrelation, but also the lighting, heating, thermal and acoustic aspects, which can be exceedingly important. Most occupants of buildings prefer daylight to artificial illumination, but unfortunately windows have a considerable effect on internal temperature through the problems of heat loss and solar gain. Noise has also been found to be a disturbing factor in a working environment. This arises from three main causes: the increasing use of machinery; rise in urban traffic volume; and use of light construction with continuous glazing.

It is important to identify the essential activities to be undertaken within the building and their optimum sequence. There are two main reasons for locating an activity in a particular place in the building and each has many dependent considerations. These are environmental, such as noise and daylight, and relationships between activities, which are mainly functional and circulatory. More sophisticated design procedures are now being used to an increasing extent, such as computer aided design (CAD), and these will help in locating the activities in their optimum juxtaposition within the available space, and provide the ability to change and modify the model of the building rapidly and so test alternatives.

**Scheme Design**

Having finalised the outline design proposals, the main purpose at this stage is to critically examine the wider issues of appearance, method of construction, outline specification and the preparation of a cost plan, showing the proposed distribution of costs over the elements or components of the building. The services and structural engineers will develop their work in sufficient detail for the architect to be able to apply for full planning permission.

As a result of design team meetings, all relevant design information is collated and a detailed report on the design and full cost implications can be supplied to the employer. The employer's approval is needed before the scheme is developed further, as the brief cannot be modified beyond this point without the possibility of incurring extra fees.

A wide range of design and constructional matters will be considered at this stage and the quantity surveyor will assist by providing cost estimates of the various alternatives, desirably incorporating both initial and future costs. Frequently a choice has to be made between a steel and a reinforced concrete frame with all the consequent ramifications. For example, one type of frame of a multi-storey building may show a lower cost but result in a longer construction period and subsequent reprogramming of the remainder of the works, and could have a significant influence on other constructional elements, such as floors and facades.

With industrial buildings, ground floors often require higher loading capacities than planned, up to 10 or even 15 per cent of floor space as office space, and other frequent requirements are the opportunity for expansion and improved weathertightness, thermal insulation, energy conservation and security.

Tall buildings on small inner city sites frequently require special treatment. A good example is the Arts Centre in Hong Kong, where the type of accommodation that is normally spread horizontally has been stacked up into a building of great complexity. The top seven stories of offices pay for the most impressive arts accommodation below. The architect, Tao Ho, regards the three essential elements in architecture as function, economy and aesthetics. He has successfully fused these by the traditional Chinese method of expressing the structural features as part of the decorative motifs of the building. The author was also very impressed by the skilful service arrangements and the fascinating use of colour.

With hospitals it is common for mechanical and electrical services to account for 40 per cent or more of the total cost of the building. This aspect was given special consideration in the design of the Queen's Medical Centre, Nottingham, where separate space for services is provided between usable floors; the service floor also forms the ceiling to the accommodation below. This system aids speedy construction and allows for future service modifications required for growth or changed uses within the building.

These examples serve to illustrate the wide range of matters to be considered at the scheme design stage.

Gray, Hughes and Bennett (1994) have described how the architect's design function is made difficult because the perfect solution is rarely obtainable; problems have to be identified as well as solved; subject value judgements are required; there is no simple scientific approach to solving design problems; and designers' work in the context of a need for action requiring project delivery objectives to be met. The authors have also produced a useful scheme design checklist embracing architect's work, engineering, mechanical services, electrical services, vertical transportation, briefs, cost and time budgets, method statement, and planning approvals.

**Detail Design**

The design will be developed in depth at this stage with the final decisions being taken in respect of all the individual components which together make up the complete building. All members of the design team will continue to work very closely together to secure an integrated and efficient scheme. The interchange of ideas and views of the various professionals will have contributed greatly to the final outcome. The quantity surveyor will be kept busily engaged on the preparation of estimates of cost for the alternatives proposed and will continue with cost checks on the cost plan to ensure that the employer's budget will not be exceeded.

The employer will be consulted with a detailed report and the various detail design documents, including the revised cost plan. This is a very important stage as when the employer accepts the detail design, he commits himself to the proposals and any changes that he requires subsequently will result in abortive work and almost certainly in additional cost.

## Production Information

The architect, engineers and any other specialists will now proceed with the preparation of the final drawings and schedules, and supporting specifications/specification notes. With specifications, there is extensive support from contractors for separate specification sections, covering materials and workmanship requirements, to be bound in the bills of quantities immediately preceding the corresponding bill section, and for the specification particulars to be of a more standardised nature than at present. It is good practice to convey information on components such as windows, doors, ironmongery and manholes, and also finishes in the form of schedules, as they provide a clear and concise way of conveying the architect's requirements to the quantity surveyor and later the contractor (Aqua Group, 1992).

There is a generally accepted minimum set of drawing types which should be available for a building project and details of these follow.

*Architect's drawings*   Site plan; general elevations; general sections; floor plans; selection of main construction details; special components.

*Structural engineer's drawings*   Below-ground drawings; floor plans; structural details, structural components.

*Services drawings*   Site drainage; building drainage; other services.

It is customary at this stage also to determine the preliminary tendering procedures and to prepare a list of potential main contractors, a questionnaire for issue to them, enquiries to subcontractors and suppliers, invitations to tender, information for tender documents, amendments to the JCT Form of Contract, procedures for selection of subcontractors and suppliers, and possibly preparation of advance orders to subcontractors and suppliers. Some of these activities will necessitate further consultation with the employer.

## Coordinated Project Information

Research by the Building Research Establishment (BRE) has shown that the biggest single cause of quality problems on building sites is unclear or missing project information. Another significant cause is uncoordinated design, and on occasions much of the time of site management can be devoted to searching for missing information or reconciling inconsistencies in the data supplied.

The crux of the problem is that for most building projects the total package of information provided to the contractor for tendering and construction is produced in a variety of offices of different disciplines.

To overcome these weaknesses, the Coordinating Committee for Project Information (CCPI) was formed on the recommendation of the Project Information Group, sponsored by the following four bodies: Association of Consulting Engineers (ACE), Building Employers Confederation (BCE), Royal Institute of British Architects (RIBA) and Royal Institution of Chartered Surveyors (RICS). Its brief was to clarify, simplify and coordinate the national conventions used in the preparation of project documentation (CCPI, 1987).

The following five documents were published either by CCPI or by the separate sponsoring bodies, during 1987 and 1988.

(1) Common Arrangement of Work Sections for Building Works.
(2) Project Specification – a code of procedure for building works.
(3) Production Drawings – a code of procedure for building works.
(4) Bills of Quantities – a code of procedure for building works.
(5) SMM7 (Standard Method of Measurement of Building Works: Seventh Edition).

It is, however, unlikely that any single discipline office will require all these documents. For example, SMM7 conforms to the Common Arrangement and so quantity surveyors using the Standard Method will not need the latter document. Similarly, users of the National Building Specification (NBS) and the National Engineering Specification (NES) will not require the Common Arrangement (Seeley, 1989). However, the industry has been slow to adopt CPI in its entirety.

Coordination between the project documents is achieved by using the specification clause numbers on the drawings and in the bill of quantities. This also reduces the amount of descriptive entries in the latter two documents to a minimum.

Latham (1994) pointed out that to achieve coordination in the documents available to the constructors on site seems basic common sense. The CIEC final report to Latham also recommended the use (with some modifications) of CPI in civil engineering. Latham considered that CPI was a technique which should have become normal practice years ago, and recommended that its use should be made part of the conditions of engagement of designers.

Truman (1989) has described the use of CPI on a Nottingham office block and the lessons to be learnt. The main advantage appeared to be in imposing a discipline upon the design team.

## Bills of Quantities

At this stage the bill of quantities and all other tendering documentation are finalised. It is important that all drawings, schedules and specification particulars should be checked and steps taken to ensure that all cross references are correct. The quantity surveyor when measuring the work will be continually checking all particulars, including cross references, and will prepare a query list for the architect covering the omission of necessary information, discrepancies and ambiguities. This provides the architect with the opportunity to clarify these points and amend the documentation prior to tender stage. It is also advisable to check the adequacy of prime cost and provisional sums, ensure that no information is outstanding from consultants, that the quantity surveyor has all the latest information and incorporates it in the bill of quantities, and that all consents have been received and conditions complied with.

The bill of quantities is designed primarily as a tendering document, but it also provides a valuable aid to the pricing of variations and computation of valuations for interim certificates. It provides a good basis for cost planning and, if prepared in annotated form, will help in the locational identification of the work. The bill of quantities normally comprises preliminaries, preambles or descriptions of materials and workmanship, and the measured works. The preliminaries define the scope and nature of the work, contain details of the contract conditions, list of drawings and any special instructions to the contractor on pricing. More detailed information concerning the compilation of preliminaries and preambles and the actual measurement of building work is contained in *Advanced Building Measurement* (Seeley, 1989).

Prime cost sums will be inserted in the bill of quantities for work to be carried out by nominated subcontractors and statutory authorities or for goods to be supplied by nominated suppliers, and for which estimates or tenders have usually been obtained. Provisional sums cover work for which details have not been finalised or for which costs are unknown at the time of preparing the bill.

## Tender Action

Tender procedures should follow closely the NJCC (1994a) recommendations and tenders should normally be obtained from a selected list of contractors, agreed previously with the employer. Large contracts require advertising in the *Official Journal of the European Communities*. The number of firms to be invited will depend on the size and type of contract and the NJCC code (1994a) recommended a maximum number of six tenderers.

Selected tenderers will normally be approached about four to six weeks before the tender documents are circulated, supplying them with adequate information about the project, and asking them whether they wish to tender.

Contractors must be given adequate time for tendering and the NJCC (1994a) recommended a minimum of four working weeks (20 working days). The tender period must be sufficient to enable the tenderers to obtain competitive quotations for the supply of materials and for the execution of works to be sublet, and must have regard to the size and complexity of the project. Selected tenderers will usually receive two unbound copies of the bill of quantities, drawings, two copies of the Form of Tender and suitably addressed and labelled envelopes for the return of the tender and priced bill of quantities (if required). The most advantageous tender should normally be accepted. The method of checking priced bills is dealt with in chapter 4.

## THE ROLE OF THE QUANTITY SURVEYOR

The quantity surveying profession in the United Kingdom has largely developed over the last century. It has now grown to such an extent that it forms the second largest division in the Royal Institution of Chartered Surveyors. As building work increased in volume and complexity, employers became dissatisfied with the method adopted for settling the cost of the work and recognised the advantage of employing an independent quantity surveyor, who would prepare an accurate bill of quantities to be priced by tendering contractors and who would measure and value any variations that might occur during the progress of the works.

Today, not only is the quantity surveyor generally regarded as indispensible on any major building contract, and often on civil and heavy engineering contracts as well, but he is also being appointed as project manager, on occasions, to take control of the project from inception to completion and to coordinate the work of the design team and the main contractor and subcontractors.

There has been a considerable change of emphasis in his role during the last two decades and the quantity surveyor does, in particular, offer an enlarged and more efficient service, as described in the latter part of this chapter and examined in more detail in chapter 19. An authoritative RICS report (1970) covered this aspect well, when it stated 'The quantity surveyor's role is to ensure that the resources of the construction industry are utilised to the best advantage  of society by providing, *inter alia*, the financial management for projects and a cost consultancy service to client and designer during the whole construction process.' Nisbet (1989) has provided a detailed account of the principal changes in quantity surveying techniques from 1936–86. A later RICS report, *QS 2000* (1991) updates the RICS Quantity Surveying Division's views, which are referred to in various parts of the book as appropriate.

## Preliminary Cost Advice

From the inception of a new building project, the quantity surveyor can give practical advice on the likely cost of the scheme – however complicated or unusual it may be. He can advise on the comparative costs of alternative layouts, materials, components and methods of construction. He can also assess how long it will take to build and can produce estimates of future maintenance and operating costs. Once the employer's brief is settled, the quantity surveyor can prepare a realistic budget and a cost plan showing the distribution of costs over the various elements.

## Cost Planning

Cost planning is a specialist technique used by the quantity surveyor. It aims to help all members of the design team to arrive jointly at practical and efficient designs for the project and to keep within the budget. Effective cost planning will help to ensure that, once a realistic estimate is agreed, everything that follows is in accordance with it, from the successful contractor's tender to the final project cost.

If the employer decides to change his plans and introduce variations, the quantity surveyor will assess the cost implications. Cost planning obtains improved economy standards and better value for money. Constant monitoring means that the risk of overspending can be seen at an early stage and prompt corrective action taken.

## Contractual Methods

No two projects are the same – each needing individual solutions. Hence a number of contractual procedures have evolved from competitive tenders, as the traditional method, to negotiated tenders with a single contractor and 'design and build' contracts, where the contractor undertakes the whole task of design and construction. These and other contractual procedures are examined in considerable detail in chapter 3. The quantity surveyor can advise on the best form of contract for a specific project.

## Tendering

Competitive tendering, often from a selected list of contractors, remains a common basis for construction contracts and bills of quantities are fundamental to this process. Bills of quantities translate the drawings, schedules and specification notes produced by designers into a document listing in detail all the component parts required for the project, to enable each contractor to calculate his tender prices on exactly the same basis as his competitors. During construction, they are a vital element in the preparation of valuations, pricing of variations and effective cost control.

With the increasing scale and complexity of building operations, it would be impossible for a contractor to price a medium to large sized project without a bill of quantities. In the absence of such a bill being prepared on behalf of the employer, each contractor will have to prepare his own quantities in the limited amount of time allowed for tendering. This places a heavy burden on each contractor and involves him in additional cost which must be spread over the contracts in which he is successful.

### Choice of Contractor

The quantity surveyor provides frank, sound and practical advice on contractor selection. Work shortage or 'overheating' in the local construction industry, industrial relations record, past performance in meeting standards and time targets and the quality of management – all these can affect not only the level of tenders but the chances of the project being successfully realised. The quantity surveyor will take these and many other factors into account in making recommendations to the employer on the make up of the selected list.

### Valuation of Construction Work

In most construction contracts, the contractor is paid monthly. The quantity surveyor will value the work carried out each month on the project and submit a recommendation for certified payment. He will also settle the final account and prepare financial statements for the employer. If required he can also prepare statements of expenditure for tax or accountancy purposes and assess the project's replacement value for insurance purposes.

### Increased Efficiency

Modern construction has become more complex, leaving little room for guesswork. Tenders, contracts and construction claims have grown increasingly sophisticated. An overriding consideration is always to ensure that the employer is receiving value for money and that the contractor is paid a fair and reasonable price for the work that he undertakes. The quantity surveyor can also make the work of everyone involved with construction a great deal easier and more efficient.

### The Role of the Quantity Surveyor in the Public and Private Sectors

In both the public and private sectors the quantity surveyor is an important member of the design team. As has been indicated, he is essentially a building economist, advising employers and architects on the probable costs of alternative designs. His advice enables design and construction at all stages to be controlled within predetermined limits of expenditure. He also advises on procedures for arranging building and engineering contracts. He prepares bills of quantities and, where appropriate, negotiates contracts with contractors; he also prepares forecasts of final costs and valuations for payments to the contractor as work proceeds. He is responsible for the measurement and valuation of variations in the work during the contract, and for the preparation and agreement of the contractor's final account.

The development of effective cost control techniques in building design has had a profound influence on public sector investment especially in the social services. It is believed that the United Kingdom is the only country in the world to have consistently raised its standard of public sector investment in the nineteen-seventies, whilst reducing its unit costs (at constant prices). Much of the credit for this is due to a willingness on the part of users and administrators to look afresh at their methods and procedures and to the ability of designers to find new solutions, but there is little doubt that the motive force for these developments was the creation by quantity surveyors of effective mechanisms of budgetary control. Important developments have been made by quantity surveyors in the fields of contractual policy, by the application of such techniques as serial tendering and early contractor selection and in the statistical measurement of price movement, all of which have played a large part in the success of public sector building programmes (RICS, 1973).

### The Role of the Quantity Surveyor in Contracting Organisations

In contracting organisations, the duties of the quantity surveyor will vary according to the size of the firm employing him – tending to be very wide in scope with the smaller companies, but rather more specialised with the larger firms.

The contractor's quantity surveyor's activities can include preparing bills of quantities for small projects; collecting information about the cost of various operations from which the contractor can prepare future estimates; preparing precise details of the materials required for the contracts in hand; compiling target figures so that operatives can be paid production bonuses; preparing interim costings so that the financial position of the contract can be ascertained as the work proceeds and appropriate action taken when necessary; planning contracts and preparing progress charts in conjunction with site management; making application to the architect for variation orders if drawings or site instructions vary the work; agreeing the value of variations and subcontractors' accounts; and comparing the costs of alternative methods of carrying out various operations so that the most economical can be adopted.

The quantity surveyor's role in a variety of situations is examined in more detail in chapters 11, 13 and 14.

# 3 Construction Procurement Systems

## CHOICE OF PROCUREMENT METHODS

### General Introduction

It is vital that clients make the correct choice of building procurement method in an increasingly complex situation, with a wide range of objective criteria and procurement systems, as described in this chapter. Quantity surveyors have an important role to perform in building procurement selection and need to market and develop their skills.

Morledge (1987) has aptly described how construction clients range from experienced clients who may have their own professional team and an effective procurement policy to the inexperienced client with little or no knowledge of procurement methods and who requires extensive professional advice. Furthermore, the needs of clients vary considerably with regard to certainty of price, cost limits, time requirements, complexity of design and many other factors.

Turner (1997) has provided a comprehensive guide to building procurement supported by fourteen diverse and valuable case studies, which examine the procurement selection processes, the decisions made and the subsequent review following completion of the projects.

### Procurement Criteria

Traditional procurement selection processes result in clients prioritising the basic criteria of time, cost and quality. This approach is too simplistic as these important criteria form only part of what the client needs to examine. The following questions indicate the range of matters which the client will generally need to consider to make the most appropriate choice of procurement system.

- Design input: does the client want to influence the design and, if so, to what extent?
- Client control: how hands-on does the client wish to be in the management of the project?
- Cost certainty: what level of cost certainty does the client require before signing the contract and on completion of the project?
- Risk taking: is the client prepared to accept the risk by direct management or does he wish to transfer it to third parties?

66

- Flexibility: to what extent is the client's brief likely to be changed during the execution of the project?
- Market conditions: how are market conditions likely to change during the course of the project with possible consequences for design or construction?
- Programme security: how crucial is the final completion date?
- Value for money: does the client want to contribute to and take benefit from value management and value engineering, and how will any resulting savings be shared? (Gillespie, 1994).

A discerning set of procurement criteria was established by NEDO (1985) and refined by the Business Round Table in 1995, and the latter has been used by Turner (1997) and is tabulated in figure 3.1. These criteria are:

- Timing
- Controllable variation
- Complexity
- Quality level
- Price certainty
- Competition
- Management
- Accountability
- Risk avoidance.

**Procurement Systems**

Many new procurement systems evolved during the 1980s and 1990s, giving greater choice and flexibility, and there are several ways of classifying them. The checklist provided by The Business Round Table (1995) provides a realistic set of options and is used in figure 3.1. The main requisites of each system are briefly described and they are examined in more detail later in the chapter.

(1) Lump sum: the client appoints consultants for design and cost control and subsequently selects a main contractor to carry out the work. There are two variants:
    (a) Sequential: contractors bid on completed design and cost documents.
    (b) Accelerated: contractor appointed early on the basis of partial information, by negotiation or in partial competition, often in a two stage process.
(2) Design and build: the client purchases the completed building from a contractor who undertakes both design and construction. There are three variants:
    (a) Direct: designer/contractor appointed after appraisal without competition.

(b) Competitive: documents prepared by consultants to permit several contractors to offer designs and prices in competition.

(c) Develop and construct: consultants appointed to design the building to a certain stage and then contractors complete and guarantee the design in competition, either using the client's consultants or their own designers.

(3) Fee Construction (management method): client appoints design and cost consultants and a contractor or consultant to manage the construction for a fee. Specialist contractors are appointed to undertake the construction work by negotiation or in competition. There are two variants:

(a) Management contracting: a management contractor takes some contractual risks in delivering the completed project at an agreed price and on time and employs specialist contractors as subcontractors. Client retains some time and price risks.

(b) Construction management: a professional firm is paid a fee to provide the management service and the specialist contractors enter into direct contracts with the client, who retains the time and price risks.

(4) Design and manage: client appoints a single firm to design and deliver the completed project, although specialist contractors are appointed to carry out the construction work by negotiation or in competition. There are two variants:

(a) Contractor: project design and manage firm takes a contractual risk in delivering the completed project to an agreed price and on time and employs specialist contractors and sometimes designers as subcontractors.

(b) Consultant: project designer/manager is employed as client's agent and specialist contractors enter into direct contracts with the client, who retains time and price risks.

### Selecting a Procurement System

Gillespie (1994) has outlined the following very effective procurement selection process:

(1) Following agreement of the procurement criteria/objectives, the client is asked to determine the relative importance of each and to rank them in order of merit. For example, cost certainty might be paramount and could receive a ranking of nine or ten on a scale of 1 to 10, while the balance of risk might be less important with a ranking of three to four.

(2) Each procurement option is then discussed to ensure that the managerial and contractual arrangements between the client and the rest of the team are fully understood. The principal working methods of each

route, with their advantages and disadvantages, are discussed. This normally results in the elimination of one or two procurement routes, leaving a shortlist of preferred options.

(3) The extent to which each shortlisted route meets the client's list of specific objectives is then considered. Each procurement route is discussed in turn and a weighting assigned to each objective according to how well that route meets the objective. Thus under construction management, cost certainty would be highly weighted and might score eight or nine, whereas under a JCT two-stage contract, it might only merit a five.

(4) The final task is to determine the most appropriate procurement route for the project. Taking each route and each item on the client's objectives list in turn, the importance ranking is multiplied by the weighting assigned to it for each shortlisted procurement option. After totalling the tallies, the best option is identified as the one with the highest score.

Figure 3.1 illustrates the procurement selection method produced by the Business Round Table Ltd (1995), refining an earlier approach by NEDO (1985), comprising a procurement path decision chart to help in discussing and choosing from the options available. The decision chart has procurement options (paths) and criteria (priorities) to be satisfied in choosing the path. From the sum of the perceptions of the decision maker(s) concerning the relative importance of the various criteria on any particular project, a procurement path is chosen.

The notes for guidance issued by The Business Round Table Ltd describe the method of using the chart. The multiple choice answers to each question outlined on the right hand side of the chart are considered and the answer which appears most relevant is identified and the appropriate dot on the chart is ringed. When all the questions have been considered the number of ringed dots in each column is totalled and the procurement paths with most rings should be worthy of further investigation or, better still, suitably weighted according to the relative importance of the project priorities. It is emphasised that procurement decisions should not be made solely on the basis of the questionnaire. It is intended as a primer for discussion with the principal adviser.

In addition, The Business Round Table Ltd gives some sound advice, accompanied by checklists, for selecting construction professionals.

The original NEDO approach was developed further by Skitmore and Marsden (1988) and incorporates a weighting system whereby numerical criteria are adopted in comparing the NEDO value criteria. This later development was used by Bennett and Grice (1990) in evaluating procurement options, in which the client with his advisors records the weight to be given to each of the project objectives on a priority scale of five for essential and one for not required. In addition utility factors are allocated

**As a construction industry customer, you can choose from a wide selection of procurement routes. Which one suits your business needs?**

To help you home in on a suitable procurement route, the following customer priority checklist has been prepared as a prompt for discussions with your principal adviser. Study the list of project priorities, A–I. Consider each in turn and decide which option (1–23) most accurately reflects your preference. Move along the line and note where each procurement route gets a positive score. When every priority has been marked, you may simply add the scores in each column to give comparative totals.

Alternatively, you may wish to refine further the initial broad assessment by giving extra weight to certain priorities. If getting a firm price, say, is measurably more important to you as a business customer than getting an early completion, you could weight the scores accordingly. Priority E might be worth a score of 3, for example, to Priority A's 1.

| | | |
|---|---|---|
| **A Timing** | How important is early completion to the success of your project? | |
| **B Controllable variation** | Do you foresee the need to alter the project in any way once it has begun on site, for example to update machinery layouts? | |
| **C Complexity** | Does your building (as distinct from what goes in it) need to be technically advanced or highly serviced? | |
| **D Quality level** | What level of quality do you seek in the design and workmanship? | |
| **E Price certainty** | Do you need to have a firm price for the project construction before you can commit it to proceed? | |
| **F Competition** | Do you need to choose your construction team by price competition? | |
| **G Management** | Can you manage separate consultancies and contractors, or do you want just one firm to be responsible after the briefing stage? | |
| **H Accountability** | Do you want direct professional accountability to you from the designers and cost consultants? | |
| **I Risk avoidance** | Do you want to pay someone to take the risk of cost and time slippage from you? | |

**Figure 3.1** Procurement Selection: customer priority checklist – identifying your priorities

| No. | Lump sum contracting | | | | Design and build | | Fee Construction | | Design and manage | |
|---|---|---|---|---|---|---|---|---|---|---|
| | Sequential | Accelerated | Direct | Competitive | Develop and construct | Management contracting | Construction management | Contractor project manager | Consultant project manager | |
| 1 | | • | • | | | • | • | • | • | Crucial |
| 2 | | • | • | • | • | • | • | • | • | Important |
| 3 | • | | | | | | | | | Not crucial |
| 4 | • | • | | | | • | • | • | • | Yes |
| 5 | | | • | • | • | | | | | No |
| 6 | • | • | | | | • | • | • | • | Yes |
| 7 | | • | • | • | • | • | • | • | | Moderately |
| 8 | • | • | • | • | • | • | • | • | • | No |
| 9 | | | • | • | | | | | | Basic |
| 10 | • | • | • | • | • | • | • | • | • | Good |
| 11 | • | • | | | | • | • | | | Prestige |
| 12 | • | | • | • | • | • | | • | | Yes |
| 13 | | • | | | | | • | | • | Target |
| 14 | • | | | • | • | • | • | • | • | Work contractors |
| 15 | • | | | | • | • | | • | | Wks & const. mgt. teams |
| 16 | | • | • | | | | | | | No |
| 17 | • | • | | | • | • | • | | | Separate firms |
| 18 | | | • | • | | | | • | • | One firm only |
| 19 | | | • | • | • | | | • | | No |
| 20 | • | • | | | | • | • | | • | Yes |
| 21 | | | | | | | • | | • | No |
| 22 | • | • | | | | • | | | | Share |
| 23 | | | • | • | • | | | • | | Yes |
| TOTALS | | | | | | | | | | |

*Source*: Thinking about Building, The Business Round Table Ltd (1995)

to each procurement system in the light of the project and the client. The utility figure multiplied by the criteria weighting gives the score for each question and system. The totals will indicate the relative suitability of each system and the one achieving the highest score will be the preferred option.

Morledge and Sharif (1994) postulate that to identify procurement systems internationally, not only will the generic functions of the system need to be isolated, but the inherent risks of each system must also be identified.

## FIRM PRICE OR LUMP SUM CONTRACTS

Firm price or lump sum contracts can be formulated with various methods of documentation. An important decision is whether or not the work should be quantified in a bill of quantities or in a form in which quantification follows later, such as with a bill of approximate quantities or a schedule of rates. Where no quantification takes place, the written documentation consists of drawings and a traditional type specification or a performance specification.

The inclusion of quantities reduces the contractors' estimating risks, ensures fairer competition, reduces contractors' tendering costs, and provides improved cost feedback and a basis for valuing variations.

### Bill of Quantities

There are distinct advantages in incorporating full and accurate bills of quantities in the contract documents for substantial building projects and it is a well recognised practice. The principal advantages stemming from their use are as follows:

(1) Bills avoid the need for all of the tendering contractors to measure the quantities themselves before preparing an estimate. If all tenderers have to measure the works there is wasteful duplication of effort and an increase in the contractor's overheads, which eventually has to be passed on to employers.
(2) Bills, prepared in accordance with the Standard Method of Measurement of Building Works, ensure that an adequate description of the work in a recognised format is given to all tendering contractors and therefore all tender on the same basis. The absence of bills leads to greater variability, increased risk in estimating and consequently more disputes.
(3) The detailed breakdown of the contract sum permits proper financial management of the contract.

The National Joint Consultative Committee for Building (NJCC) advocates the competitive tendering system based on firm bills of quantities as ensuring maximum efficiency and this has traditionally been regarded as the best method of placing contracts for building work. Hence many public authorities and private organisations insist upon this system being adopted wherever possible, and exceptional circumstances have normally to exist before authority is given for competition to be dispensed with. The system, however, has one disadvantage in that full documentation involving the preparation of accurate bills of quantities is of necessity a lengthy process.

### Approximate Quantities

While the best procedure is to provide accurate quantities, there are circumstances where approximate quantities are necessary. The following examples will serve to illustrate suitable applications:

(1) Where speed is of paramount importance and the general design has been formulated, it may be necessary to select a contractor before production drawings can be completed. In this situation, it is probable that sufficient design information is available to enable approximate quantities to be produced that can form the basis of a contract. This method was used very successfully in the Addenbrooke Hospital maternity unit from 1981 to 1983 using 'fast track' methods of contract preparation.
(2) With work below ground, the information is likely to be too imprecise for the preparation of accurate quantities. The perimeter of the building can be established, but the depth of foundations and extent of soft areas will be uncertain.
(3) Provisional quantities may be included in accurate bills of quantities to cover work that is uncertain in extent and that is subject to remeasurement when the work is carried out. Excavating in rock, the removal of underground obstructions and work on the site of buildings subject to demolition are in this category.

Where bills of approximate quantities are used for tendering the procedures will normally be the same as with accurate bills. A Standard Form of Contract has been prepared by the Joint Contracts Tribunal (JCT) for use with approximate quantities.

The main problem arises with postcontract administration, particularly with regard to cost control. It is usual to adopt one of the following three approaches:

(1) The work is remeasured at the end of the contract and this becomes the basis for the final account. A considerable amount of interim approximate measurement will be necessary.
(2) The work can be measured when the production drawings are complete, and the new bills substituted for the approximate bills. This is a better

method but is more time consuming, especially as there are likely to be further variations on the production drawings.

(3) On large projects where a resident quantity surveyor is engaged, the work can be measured as soon as it is executed. This will keep the measurement up to date but is of more restricted value in forecasting future expenditure (Aqua Group, 1990a).

## Schedule of Rates

*Nature and uses* The schedule can take one of a number of different forms and is best suited for maintenance and repair contracts, where the exact nature and extent of the works may not be known until the work is executed. They may be confined to decorations and other finishing works. The contractor is generally required to insert rates in the schedule for the listed items of work. Alternatively, the rates may be inserted by the employer and the contractor then enters a percentage which he requires adding to or deducting from each item or trade. In this way the employer obtains a full range of rates applicable to all the varying operations of work that are likely to be encountered, irrespective of the actual quantities that will be required.

The absence of any quantities makes a ready assessment and comparison of the tenders very difficult. An examination and comparison of the priced schedules will inevitably show wide variations in the rates inserted by competing contractors, which will have a considerable effect on the cost of the proposed work and make an effective comparison and evaluation of the tenders extremely difficult. It is therefore advisable to insert approximate quantities in order that the effect of the differences in rates can be determined. The only difficulty arising from this procedure is the possibility of claims being subsequently submitted by the successful contractor, because the actual quantities of various items of work vary substantially from the original approximate quantities. It should, however, be possible to so word the document that claims of this nature are inadmissible. Another alternative is for the quantity surveyor to price out the schedules against a notional set of quantities.

Another disadvantage encountered with the use of schedules of prices is the fact that the total cost of the works cannot be determined and some difficulty may be experienced on occasions in obtaining the necessary financial authorisation for the work. Thus to use schedules of prices effectively, it is most advisable that they shall be prepared in such a way that they will produce a lump sum price for purposes of comparison. Normally this will take the form of approximate quantities, whereby each item in the schedule has a quantity entered against it, which forms an approximate, albeit as accurate as possible, estimate of the quantity of work likely to be required in respect of each item. In this way lump sum prices will be computed that bear some relation to the actual quantity of work to be executed and will enable a realistic assessment and comparison of the various tenders to be made.

*National schedules*  A number of local authorities have developed their own schedules of rates for building maintenance work and there are several national schedules of which probably the widest used is the *National Schedule of Rates* published jointly by the Building Employers' Confederation (BEC) and the Society of Chief Quantity Surveyors in Local Government (SCQSLG).

A fundamental requirement for schedules of rates for general use is that they be soundly based. This was, in fact, the basic aim of the BEC/SCQSLG document, although the Association of Metropolitan Authorities (AMA) came to the conclusion that a common standard schedule of rates for maintenance is not a practical proposition and that schedules should be individually prepared to suit the needs of each local authority.

Most national schedules are based on the Standard Method of Measurement of Building Works (SMM), but it is generally accepted that the SMM is better suited to larger projects rather than maintenance work, and to renewal as distinct from repair work. However, the degree of usefulness will be linked to the quality of schedule descriptions.

Most national schedules are of the prepriced type and, in diverse geographical areas with different maintenance requirements, the inconsistencies could cause difficulties for users of the schedules. Familiarity will help to overcome this problem, but a greater difficulty is likely to arise in assessing the percentage adjustment to apply to the rates. Usually schedules allow different percentages to be applied to individual SMM work sections and this permits some measure of flexibility, but the effect of inflation varies within sections in respect of different elements of cost, and the value of prepriced schedules is influenced by the time interval between reviews, revisions and updating.

*Use on term contracts*  Well established schedules for use on term contracts, usually of 1 to 3 years duration, are the Property Services Agency (PSA) (1990) schedules of rates. These encompass new projects as well as maintenance work and, although the items are measured in accordance with the Standard Method (SMM), the descriptions are well detailed. General directions and preambles are incorporated and the rates include preliminaries. However, rates are 'all-in' without any form of subdivision and this restricts the opportunities for analysis. A monthly maintenance index is available for use with the schedules for an annual charge, but the value of the index is partially offset by the relatively lengthy periods between new editions of the schedule, and the problems that can arise because of specification changes and alterations to maintenance techniques that can occur in the intervening period.

The PSA schedules are not restricted to large buildings or complex properties as they are used in connection with many houses for services personnel, but housing maintenance is not their prime function and local authorities are therefore unlikely to use them for building maintenance work to any significant extent. Hence any prospective user can best assess the usefulness of a national schedule by comparing the descriptions with his own maintenance requirements. Where these are compatible, pricing decisions are simplified and more realistic tenders are obtainable. The JCT issued a Standard Form of Measured Term Contract in 1989.

## Specifications

Small contracts are often let on the basis of drawings and specification and the contractor usually submits to the employer a fixed tender sum for the work, covering its execution and normally the maintenance for a specified period after completion. Bills of quantities are not considered necessary as the works will not be subject to remeasurement. Contracts of this type encompass such projects as small ancillary buildings, small extensions to existing buildings, works of decoration and demolition, construction of playing fields and the like. Hence, in the absence of any variations to the project, the employer will know the cost of the work from the outset. He will remain immune from increased costs that occur during the course of the contract.

The contractor is made liable for all the risks and uncertainties associated with the works and he must allow for them in his tender price. In periods of keen competition contractors will tend to price the risk elements on the low side as this may largely decide who is successful. On occasions this type of contract has been used for works that were extremely uncertain in extent and in these circumstances this represents an unfair and unsatisfactory contractual arrangement. Some of the more reputable contractors may even decline to tender. Where the successful tenderer finds subsequently that he has grossly under-estimated the cost of some of the uncertain parts of the work, he may well be tempted to skimp some of the project work in order to offset the under-estimate, which can only cause the employer dissatisfaction and highlight the need for increased supervision.

The following difficulties, as identified by the National Joint Consultative Committee for Building (NJCC), are likely to arise with competitive tendering for building works on drawings and specification alone:

(1)  in effectively comparing and evaluating tenders when each contractor has to prepare his own analysis of the information included in the tender documents;
(2)  in accurately evaluating work in progress for payment on account;
(3)  in accurately valuing variations; and
(4)  in maintaining proper financial management of the contract.

Where a contract is based on a specification and drawings, the specification will serve three main purposes.

(1) It will be a tendering document and must be prepared in a way that will help the estimator to price the work. This normally involves the separation of trades with the needs of subcontractors in mind.
(2) It will, on acceptance of the tender, become a contract document and must therefore be free of ambiguities.
(3) It will be a management document to inform the contractor of exactly what work is physically to be performed on site and to assist the architect and quantity surveyor in their cost control functions (Aqua Group, 1990a).

Specifications for small building projects normally follow a logical sequence of sections, each covering matters relating to building work in general and to the specific project in particular. The first section often covers such matters as the parties to the contract and their contractual relationship, the contract period, payments, insurances, delays, insolvency and other related matters. The second section seeks to ensure that materials and workmanship achieve an acceptable standard of quality, while the third section describes the work to be done and, in the case of existing buildings, needs to provide sufficient information to locate the works described.

In recent years some designers and engineers have substituted performance specifications for the traditional type. The contractor is then able to choose the materials and design the project subject to satisfying the prescribed performance requirements. Where competitive tenders are received the tendering contractors will almost inevitably submit different proposals which will require evaluation. Hence there are difficulties in drafting such a specification. If it is too tightly prescribed the contractors will have little opportunity to exercise initiative, while conversely a very open-ended specification will result in a wide range of solutions which will be difficult to compare. It is also necessary to take economic criteria into account when drafting a performance specification, otherwise the tenders may be too high and some difficult negotiations will be required if the scheme is not to be abandoned.

Some form of cost control of the work is needed to safeguard the employer's interests. Probably the only feasible approach is to separate the various work sections into the smallest convenient elements or functions, to enable the price for each to be identifiable. A substantial variation to a particular section can then be identified subsequently. For example, with water services, it would be desirable to separate hot water services, cold water services and equipment. It would also be possible to quantify the successful contractor's solution to the design and to incorporate the quantities in the contract document to assist with cost control. If this procedure is to be adopted, it should be included as a condition of tender to avoid any subsequent objections from the contractor (Aqua Group, 1990a).


## COST REIMBURSEMENT CONTRACTS

In a cost reimbursement contract the employer (client) pays the contractor the actual cost of the work plus a management fee which will include the contractor's overhead charges, supervision costs and profit. The management fee may be calculated in one of three different ways which will now be described.

### Prime Cost plus Percentage Contracts

This type of contract provides for the management fee payable to the contractor to be calculated as a percentage of the actual or allowable total cost of the building. This form of contract permits an early starting date as the only matter requiring agreement between the employer and the contractor is the percentage to be applied in respect of the contractor's overheads and profit. It is accordingly relatively simple to operate and was used extensively during the Second World War for war damage repair work and some defence installations, and was extensively abused on occasions.

It is a generally unsatisfactory contractual arrangement as higher costs also entail higher fees and there is accordingly no incentive for efficiency and economy. The use of this form of contract should therefore be confined to situations where the full nature and extent of the work are undetermined and urgent completion of the project is required, resulting in an emergency situation. Even then every care should be taken to safeguard the employer's interests by employing a reputable contractor and arranging effective supervision of the work. The main deficiency is that an unscrupulous contractor could increase his profit by delaying the completion of the works. No incentive exists for the contractor to complete the works as quickly as possible or to try to reduce costs. Furthermore, the fee will fluctuate proportionately to any prime cost fluctuations but these will not necessarily bear any relation to any changes in the actual costs of management.

The method of contractor selection generally follows the orthodox approach of selective tendering based on certain predetermined criteria. Where it is difficult to assess the quality and reputation of firms under consideration, they may be invited for interview as a means of determining their capabilities. The only element of competition relates to the percentage addition which can vary from one contractor to another.

A typical percentage fee contract might contain an addition of 100 per cent on the actual cost of wages, fares and allowances paid by the contractor to foreman, operatives and staff (other than clerical, administrative and visiting staff) for time wholly spent on the works, together with amounts paid in respect of such wages for national insurance, graduated pensions, holidays with pay, employer's contributions, employer's liability and workmen's compensation insurance; an addition of 20 per cent on the actual cost of materials used upon the works after the deduction of all trade, cash and other discounts and rebates; an addition of 5 per cent on the actual cost of any subcontractors' accounts in connection with the works and any payments made by the employer, and an addition of 10 per cent on the actual cost of any mechanical plant used on the site upon the works.

## Prime Cost plus Fixed Fee Contracts

In this form of contract the sum paid to the contractor will be the actual cost incurred in the execution of the works plus a fixed lump sum, which has previously been agreed upon and does not fluctuate with the final cost of the project. No real incentive exists for the contractor to secure efficient working arrangements on the site, although it is to his advantage to earn the fixed fee as quickly as possible and so release his resources for other work. This type of contract has advantages over the prime cost plus percentage contract from the employer's standpoint.

In order to establish a realistic figure for the fixed fee, it is necessary to be able to assess with reasonable accuracy the likely amount of the prime cost at the tender stage, otherwise the employer may have to revert to a prime cost plus percentage contract with its inherent disadvantages. The Aqua Group (1990a) emphasised the importance of preparing a document showing the estimated cost of the project in as much detail as possible so that the work is clearly defined and also the basis on which the fixed fee is calculated.

The estimated prime cost should be subdivided between the work to be carried out by the main contractor, subcontractors and nominated subcontractors, as the management expenses attributable to the main contractor's own labour could be higher than that of subcontractors. As a further refinement it may be feasible to subdivide the estimate into trades. Where it is decided that the fixed fee shall be settled in competition, this will also form the basis for contractor selection and the detailed estimate should be sent to selected contractors with the other tendering documents. When the contractor has been selected a form of contract will be signed, usually based on the JCT Standard Form of Prime Cost Contract (1992).

*The JCT Standard Form of Prime Cost Contract* (1992) is most commonly used for works of alteration and addition or refurbishment, where it is relatively easy to describe the general scope of the proposed work but very difficult to set it out in precise detail. It may also be applicable where the work is to be carried out in buildings that are partially or completely occupied, and the disruptive effect of the work on the occupier may be an important aspect of the contract and one that it would be very difficult to describe and price. The Aqua Group, (1990a) has suggested that there may be circumstances when it could be advisable to vary the arrangements in the JCT Standard Form as, for example, where joinery will probably be supplied from the main contractor's own joiner's shop, when it could help to calculate the prime cost for this work differently from the prime cost of work on site. Fewer problems may arise if small tools and consumables are included in the fee rather than in the prime cost.

The method of calculation of the prime cost is described in the Standard Form of Contract. Fuller details are contained in an Agreement between the Royal Institution of Chartered Surveyors and the Building Employers Confederation. This agreement prescribes the percentages to be applied for various items on a trade by trade basis, delineating those matters that are deemed to be within the fixed fee and those that are chargeable to the contract. It is important to identify the overhead element which will be paid for at cost, such as site staff and site facilities including huts, canteens, plant and scaffolding, and to be satisfied that the use of these facilities is in a reasonable relationship to the amount of work being undertaken.

### Prime Cost plus Fluctuating Fee Contracts

In this form of contract the contractor is paid the actual cost of the work plus a fee, with the amount of the fee being determined by reference to the allowable cost by some form of sliding scale. Thus the lower the final cost of the works (prime cost), the greater will be the value of the fee that the contractor receives. An incentive then exists for the contractor to carry out the work as quickly and cheaply as possible, and it does constitute the most efficient of the three types of prime cost contract that have been described (Seeley, 1993).

### Control of Prime Cost Contracts

It is essential that an effective method of checking the prime cost is implemented, as a much greater measure of control and supervision of the work will be required as compared with fixed price contracts. The employers would be well advised to employ a quantity surveyor to undertake the following duties:

(1) Supervision of the delivery of materials on to the site, their retention and use and checking of labour force and the activities performed.
(2) Checking of all records relating to the contractor's payments and transactions and certifying of amounts due to subcontractors and suppliers.

(3) Constant monitoring to ensure that no money is wasted by the use of unnecessary or unsuitable plant or in the lack of coordination of services or discontinuity of the work.

One of the first actions on a prime cost contract is to verify who is to be the full-time foreman, to be paid as part of the prime cost, and which personnel are in a supervisory capacity and are therefore covered by the management fee. Contractors often pay labour more than the recognised rate and this should ideally be discussed at an early stage, and possibly be included as part of the tender. The amount of labour employed on the contract is usually obtained from the certified weekly timesheets. It is advisable to agree the extent of normal overtime to be worked and to require the contractor to seek approval before working any additional overtime (Aqua Group, 1990a).

Competitive quotations should be obtained for materials as far as practicable. In cases of expediency, the quantity surveyor should take a fair and reasonable view, but should always check to ensure that materials are purchased within the current market price. A check is needed to ensure that the materials invoiced equate with the materials delivered, and this is often done by inspecting delivery tickets. In many cases a physical check on site provides the most effective procedure. Credits should be allowed for surplus materials, usable materials from existing buildings and materials supplied by the employer (Aqua Group, 1990a).

The conditions of contract should state how plant is to be charged, and the two most commonly used plant hire schedules are the *RICS Schedule of Basic Plant Charges* and the *Federation of Civil Engineering Contractors Daywork Schedule*. They are both intended for use with dayworks but they can be suitably adjusted for use with cost reimbursement contracts. Where the plant is required for a long period on site it may be cheaper to purchase it, in which case some limit should be placed upon the period of hire, possibly restricted to 80 per cent of the capital cost.

On all cost reimbursement contracts there will be a proportion of work that is sublet on a measured basis to a subcontractor. Where the proportion of sublet work is greater than that estimated when the fee was originally fixed, it may be necessary to vary the fee. As this needs establishing before the contract is signed, the contractor's intentions regarding subletting should be determined before he is finally selected (Aqua Group, 1990a).

The definition of prime cost is crucial, as generally what is not specifically included in the definition is automatically deemed to be included in the fee. This needs emphasising to the smaller contractors who may not be used to working on this method. Most prime cost contracts contain a clause requiring the contractor to make good defects at his own expense. Where they are made good during the progress of the work then the cost must be separated from that of the normal work.

Valuations for cost reimbursement contracts are normally prepared on the basis of wages paid and invoices submitted by the contractor each month. The retention may be a percentage of the value of the work done or a proportion, possibly 25 per cent, of the fixed fee. There is always a considerable time lag between the work being carried out and invoices being submitted, and for cost control purposes it is necessary to reconcile the actual prime cost with the estimated prime cost (Aqua Group, 1990a).

## TARGET COST CONTRACTS

Target cost contracts are used on occasions to encourage the contractor to execute the work as cheaply as possible. A basic fee is generally quoted as a percentage of an agreed target estimate usually obtained from a priced bill of quantities. The target estimate may be adjusted for variations in quantity and design and fluctuations in the cost of labour and materials. The actual fee paid to the contractor is determined by increasing or reducing the basic fee by an agreed percentage of the saving or excess between the actual cost and the adjusted target estimate. In some cases a bonus or penalty based on the completion time may also be applied (Seeley, 1993).

Hence actual prime costs are recorded and a fee agreed for management services as in cost reimbursement contracts. The actual amount paid to the contractor depends on the difference between the target price and the actual prime cost. In practice various methods have been used for computing this sum. An alternative method that has been used is to pay the contractor the prime cost plus the agree fee and for the difference between target price and prime cost, whether a saving or an extra, to be shared between the employer (client) and the contractor in agreed proportions. Yet another method is to pay either the target price or the prime cost plus the agreed fee whichever is the lower. This latter form of contract combines the characteristics of both the fixed price and cost reimbursement contracts.

Fluctuations in fee due to differences between target and actual costs operate as a bonus to the contractor if his management is efficient or as a penalty if it is inefficient. The benefits to be obtained by the employer from this contractual arrangement are mainly dependent on the target price being agreed at a realistic value, as there will be a great incentive for the contractor to increase the estimated price as much as possible in the first instance. It is essential that the employer obtains expert advice in evaluating this price. It may be negotiated with the contract or contractor or established in competition. Target cost contracts should not be entered into lightly as they are expensive to manage, requiring accurate measurement and careful costing on the employer's behalf (Aqua Group, 1990a).

## EARLY CONTRACTOR SELECTION

At this stage it may be helpful to the reader to consider the principal methods used in early contractor selection, in order to set the total scene, even although they are not strictly contractual arrangements. There are two main reasons for early selection:

(1) When the contractor can make a technical contribution to the pre-contract process.
(2) When the employer wishes to start work on site before all production information is available and this is generally dictated by a very tight programme.

The contractor may be able to make a technical contribution to design through the use of a proprietary system of construction, by assisting with the solution of complex site problems or the preparation of the design to suit particular specialist plant.

On occasions a subsidy or grant may be directly linked to a date for starting work on site. With annual accounting, projects can sometimes be sanctioned only if a start can be made before the end of the financial year. Another alternative is where substantial additional income will accrue from an earlier completion. In all these cases it is necessary to make a careful assessment to determine whether the anticipated savings are likely to exceed any additional costs involved.

### Two Stage Tendering

Two stage tendering is of value where early contractor selection is required but it is not feasible to negotiate with a single contractor without any form of competition. It is advisable that the tendering process should conform to the NJCC Code of Procedure (1994b). The first stage involves the competitive selection of the contractor, while the second stage embraces the determination of the contract price based on pricing data obtained from the first stage. Sufficient information will be supplied to tendering contractors to enable them to establish the basis on which the final price will be determined. Thus the offers of selected contractors may be considered in the light of such factors as management and plant capacity, labour rates and overheads, and they may even be asked to price a nominal bill. In some instances the first stage is preceded by a preliminary stage in which the selected contractors are interviewed to determine the extent of their resources and the contribution they can make. It can, however, only be a subjective assessment leading to the first stage tender.

In the second stage after the design has been prepared, with the contractor participating in the design team, the contract price is determined. At this stage it is very important that the determination of the actual price is closely related to the quantity and specification of the work to be done, and that the cost control mechanism for use during construction is established. The documentation for the initial selection can take various forms but three quite popular approaches, as identified by the Aqua Group (1990a) are now described.

The *general estimating method* is most suitable for large contracts of a general nature, where the contractor is making no particular design contribution. The tendering documents normally include:

(1) a preliminaries bill;
(2) the contractor's allowances for head office overheads and profit relating to his own work and that of his own subcontractors and nominated subcontractors;
(3) the basis of labour rates;
(4) some means whereby the productivity of labour can be assessed;
(5) the method for dealing with subcontractors' work; and
(6) the method for dealing with materials, including waste allowances.

The quantity surveyor has to make an objective evaluation of the various tenders, weighting items as necessary to obtain a common denominator. Having selected a contractor, the second stage of the process is implemented to determine the contract price. This is desirably done as speedily as possible but has to be geared to the development of the design. A bill of quantities will normally be prepared incorporating any amendments in measurement method dictated by the particular project.

The second method is the *approximate quantities method*, forming an alternative to the first method and used in similar circumstances, but establishes competitive prices for measured rates in place of labour constants and materials prices. The first stage documentation includes an approximate bill of quantities. After a contractor has been selected and the design completed, the approximate quantities can be confirmed and the contract price determined. It is important that the initial quantities are reasonably close to the 'designed' quantities so that the original level of contractor's overheads, plant and profit can stand without renegotiation or adjustment.

The third method is the *cost plan/proprietary design method* which is normally used when the contractor has an important contribution to make to the design through the use of a proprietary system. The object is to select the contractor whose proprietary system best meets the requirements of the architectural scheme and to assess the offers received in terms of best value for money. Preliminary discussions will be needed with the various contractors to establish that their systems can be adapted to accommodate the architect's requirements as delineated on the sketch plans.

In preparing the tendering documents it is usual to identify the areas of work under six main headings:

(1) work to be carried out in a traditional manner, such as the external services and works, and probably the substructure;
(2) work included in the proprietary system;
(3) nominated or approved subcontract work;
(4) fixing or assembly of supplier's work in the superstructure;
(5) preliminaries and site management; and
(6) head office overheads and profit.

The tendering documents should provide for flexibility, although the architect will wish to have maximum control over those parts of the project that are not an integral part of the proprietary system. The following documentation is often used in these circumstances:

(1) Bill of quantities or approximate quantities for the external services and works.
(2) Bill of quantities or approximate quantities for elements unconnected with the selected proprietary system, such as foundations.
(3) Performance specification for the work covered by the system. Provision should be made for a breakdown of the price, for control purposes after the contractor has been selected. Where quantities are included they are normally on a functional basis, for example, areas of floor related to superimposed load.
(4) The remaining work will largely be covered by subcontractors who will either be nominated or agreed and approved between the architect and the main contractor. The details of this work normally take the form of the cost plan with a specification describing what is intended for each element.
(5) Specific items of builder's work in the superstructure, such as fixing of joinery fittings, which will either be covered by a schedule of rates or approximate all-in quantities.
(6) Site management costs and head office overheads and profit dealt with as in the general estimating method (Aqua Group, 1990a).

When the offers are received they will not be on a strictly comparable basis, making for problems in assessment. Evaluating the part covered by the performance specification needs careful consideration to determine which contractor is giving best value for money. Financial and aesthetic aspects must be considered separately.

Following the selection of the contractor, a considerable period will be required for the refinement of the design, to ensure that the architectural requirements are satisfactorily integrated with the proprietary system. Concurrently, negotiations on the price will be proceeding and finally a firm contract price will be determined on the basis of the finalised drawings and other documents. These documents will incorporate bills of quantities and performance specifications, with such prime cost (PC) sums as are necessary.

The main advantages to be gained from two stage tendering are:

(1) Early contractor selection accompanied by a quicker start to and completion of the contract.
(2) The contractor's detailed pricing policy is known after the first stage, following the receipt of competitive offers, and this can be used in determining the rates at the second stage. Knowledge of the contractor's pricing methods is of considerable value when negotiating future items in the contract.
(3) There are benefits at the design stage with the availability of the expertise and experience of the contractor and his organisation.
(4) Construction may start before the design is complete, although there are dangers inherent in this procedure.

Useful guidelines for two stage selective tendering have been published by the NJCC (1994b).


## CONTINUITY CONTRACTS

Continuity contracts are concerned with securing continuity in a programme of building work, covering both tendering and construction. The three main categories of contract in this area are serial contracts, continuation contracts and term contracts and each of these is now examined.

### Serial Contracts

In many programmes of building work as, for example, in the provision of primary schools, there is an element of continuity. Because of the continuity, the building resources, including associated professional services, may be used more economically if all the work is carried out by the same contractor, rather than by employing a different one for each site. This should result in lower costs, better value for money and improved relationships between the contractor, employer and the design team.

Serial contracting has been broadly defined as an arrangement whereby a series of contracts is let to a single contractor, and this may also be referred to as serial tendering. Furthermore, in a serial contract the approximate extent of the series is known when the offers are obtained, and the individual projects tend to be of a similar order of size. The serial tender is a standing offer to carry out a series of projects on the basis of pricing information contained in the competitive tendering documents, which is generally a master bill of quantities, incorporating most of the items that are likely to occur in the projects involved.

The series often consists of around three to fifteen projects. The number is usually known at the time of the original tender, although additional projects may be added by agreement at a later date. Final designs will not be available for all the projects at the tendering stage, but the employer (client) and his architect will know their probable characteristics, which will be reflected in the tendering documents prepared by the quantity surveyor. Exact quantities will be prepared as the design for each project is finalised, and these form the basis of each contract which will be priced from the rates contained in the master bill. Because of the similarity in the projects, the same type of contractor's management is suitable for them all. As much of the detailing will be similar, it is a relatively straightforward task to produce for tendering a master bill that is likely to embrace a high proportion of the work items that will eventually be carried out.

Having negotiated the individual contracts on the basis of the master bill, work thereafter will proceed very much as for a normal contract. In the initial stages, however, before formal tendering takes place, it is advisable to take great care in the selection of tendering contractors. An error of judgement in selection will have much more serious repercussions than when dealing with one-off contracts. It is therefore very important to verify the financial and physical resources of the contractor. An insolvency, covering, as it would, many contracts would be a serious matter. Before the formal tendering it is common practice to interview possible tenderers and then prepare the short list on the basis of the information obtained (Aqua Group, 1990a).

### Continuation Contracts

A continuation contract differs from a serial contract in that it is an *ad hoc* arrangement to take advantage of a particular situation. For example, where a housing contract is proceeding on a particular site, another site may become available with similar housing requirements. In this case there will have been no standing offer to undertake further work, and the original tendering documents were prepared for the single project. However, they can provide a good basis for a continuation contract.

Alternatively, it is possible to make provision for continuation contracts in the tendering documents for the original project. There is, however, no firm commitment and the continuation contract may not materialise. In either case the contractor has to price the tender documents for the original contract on the basis of securing that contract alone. The continuation contract, if and when it arises, is then dealt with separately.

There is a wide variation in the size of housing contracts. One site may be suitable for 120 houses, while another, with dwellings of similar design, will accommodate 800. Moreover, planning permission for housing in urban locations may not always be readily obtainable, and even when obtained can be subject to a variety of conditions. The transfer of residents from sites scheduled for redevelopment may depend on the completion of other contracts, which makes site acquisition and handing over ready for building rather uncertain. For these reasons housing rarely lends itself to serial contracting but there are many examples of the use of continuation contracts for this purpose.

In continuation contracts the tendering documents will be closely related to the documentation for the original project on which the continuation is to be based. The contract sum is then negotiated on the basis of the original contract but with the following two substantial adjustments:

(1) the contractor will require allowance to be made for increases in the costs of labour and materials because of the time difference between the two tenders; and
(2) the employer (client) will expect to receive the benefits of continuity arising from increased productivity and further savings stemming from the same contractor doing similar work for the same architect and employer.

These factors are difficult to assess and it is essential that the contractor discloses to the quantity surveyor the details of build-up of the tender for the original contract. This matter should be discussed at an early stage, because if the contractor is not prepared to cooperate in the release of this pricing information, it may not be worthwhile to proceed with a continuation contract.

Where it is envisaged that several continuation contracts may arise from one original contract, the contractor should be offered an incentive to make cost reductions. For example, the contractor might be given the full benefit of any reduction in cost he makes, even if it involves a change in specification, but not performance, on the first continuation contract, with the proviso that the full amount of the cost reduction is passed on to the employer in the succeeding continuation contracts. This is most likely to occur where the contractor is involved in a considerable amount of design work and where the detailed specification is closely related to his production requirements (Aqua Group, 1990a).

### Term Contracts

Term contracts differ from both serial and continuation contracts in that they envisage a contractor doing certain work for a period or term. The contractor signs a contract to do all the work requested within agreed parameters over a period such as 18 to 24 months. The tendering document will usually be a schedule of rates. Orders for the work are issued from time to time, the work is measured and the contractor paid accordingly. Term contracts are ideal for maintenance and repair work on large housing estates and the like, and were used extensively by the Property Services Agency (PSA). They provide a satisfactory method of controlling the work with a measure of accountability and a straightforward procurement procedure for the individual project.

The JCT produced the Standard Form of Measured Term Contract in 1989 for clients with programmes of regular maintenance and minor works. The client (employer) is required to list the properties and type of work to be covered by the contract and the term for which it is to run, and provide an estimate of the total value of the contract and minimum and maximum value of any one order. Payment is made on the basis of the SCQSLG National Schedule of Rates, or an alternative priced schedule, and the contractor quotes a single percentage adjustment to the base document.

## JOINT VENTURE CONTRACTS

The services' installations of modern buildings with complex environmental services can approach 50 per cent of the total contract value. In these circumstances, mechanical and electrical contractors take on an increasingly important role in determining the method of working and programming of the works. In traditional forms of contracting where the services contractor is a nominated, specified or domestic subcontractor to the building contractor, a potential source of enmity has been created and a communication barrier often exists between the specialist installer and the design consultant.

In the approach often adopted by the Department of Health and Social Security, the building contractor is asked to form a joint venture with a mechanical and electrical contractor to tender and carry out the work as a joint main contractor. Thus the services work is given equal recognition with the building works and the installer is given an opportunity to make a contribution to the overall control and running of the contract, and a direct link to the design consultant.

Joint venture tendering has been adopted in the Department of Health and Social Security since the mid nineteen-seventies. Selection of contractors has been by means of an initial invitation to building contractors asking them if they wish to tender on a joint venture tendering basis and, if so, to name their partners (mechanical and electrical contractors). The named partners may not always be on the Authority's approved list and the applicants then have to be scrutinised.

The only amendment to the JCT Standard Form of Building Contract was to define the contractor in the Articles of Agreement as the named parties, who are jointly and severally liable for all the obligations of the contractor contained in the Agreement. The Regional Health Authority particularly emphasised the legal liability of each party to perform the entire contract and all contractors are advised to consider arrangements for their mutual protection. The JCT introduced a guidance note on Joint Venture Tendering for Contracts in the UK in 1985, which was updated in 1994.

The joint venture could comprise one of the following:

(1) A building contractor, mechanical contractor and electrical contractor.
(2) A building contractor and a combined mechanical and electrical contractor.
(3) A building contractor having its own mechanical and electrical engineering divisions.

In the light of experience, joint venture tendering is being restricted to projects in excess of £1m, and even then only when the services element forms a substantial part of the whole. Subsequently, mechanical and electrical contractors were given the opportunity of naming partners as well as the building contractor. These developments stem from the large proportion of services work in hospital contracts and the desire to obtain competitive tenders on a selective basis but, at the same time, to secure closely integrated construction arrangements subject to satisfactory safeguards. Haley (1994) produced a useful guide to joint venture tendering, which examines the principal reasons for and considerations in establishing a joint venture, the legal structures and competition aspects. The reasons are listed as limitation of risk, pooling of risk, exploiting opportunities and harmonisation of the whole operation.

## NEGOTIATED CONTRACTS

As a general rule negotiation of a contract with a single contractor should take place only if it can be shown to result in positive advantages to the employer (client). There are a number of situations in which negotiations may be beneficial to the employer and some of the more common instances are now listed:

(1) The employer has a business relationship with the contractor.
(2) The employer finds it difficult, or even impossible, to finance the project in any other way.
(3) The employer has let an initial contract in competition and then another project of similar design comes on programme.
(4) In some geographical areas where there may only be one contractor available to do the work.
(5) A certain contractor is the only one available with either the expertise or the special plant required to carry out the project.

(6) At times when the construction industry is grossly over-stretched and negotiation offers the best approach.
(7) Where a rapid start is required, as for example when the original contractor has gone into liquidation (Aqua Group, 1990a).

The two principal methods of negotiation are:

(1) Using the competitive rates obtained for similar work under similar conditions in another contract; but there are many inherent problems in adjusting the existing rates to provide a basis of pricing the new work.
(2) An agreed assessment of the estimated cost, to which will be added an agreed percentage for head office overheads and profit, which can be subsequently documented in a normal bill of quantities.

There are certain essential features which are required if the negotiation is to proceed satisfactorily. These include equality of the negotiator for each party, parity of information, agreement as to the basis of negotiation and an approximate apportionment of cost between suitable heads, such as site management, contractor's own labour, direct materials, plant, contractor's own subcontractors, nominated subcontractors, nominated suppliers, provisional sums and contingencies, and head office overheads and profit.

**Advantages of Negotiation**

A number of advantages can stem from a decision to select a contractor and to negotiate a contract sum with him. The contractor can be brought in at an early stage as a member of the design team, so that full advantage can be taken of his ability and experience. He can take an active role in the cost planning process, which should help towards producing a better design solution at lower cost and with possibly a shorter completion time. Further benefits may accrue if subcontractors are brought in at the same time. Quantity surveyors can benefit as, through the closer working relationships with the contractor, they will learn more about the practical organisation and management of contracts and the costing of contracts and pricing of tender documents.

With the contractor appointed, agreement can be reached on the form of the bill of quantities which will be of greatest use to the contractor in programming, progressing and cost controlling the project. Where bills are produced by computer, it is relatively easy to produce bills in any alternative form very quickly.

**Contractor Selection**

The selection of the contractor can be carried out in a number of different ways and the following methods are employed fairly frequently:

(1) appointment of a contractor who the employer and design team believe can carry out good work at a fair price;
(2) appointment of a contractor already employed on a similar project, and pricing the bill of quantities at the same rates as in the existing contract but subject to adjustment where necessary;
(3) selection of a contractor following the interview of a number of suitable contractors, agreement having been reached on the percentage to be added to the estimated net cost for overheads and profit; and
(4) selection of a contractor following the submission in competition by a number of contractors, whose ability to execute the contract efficiently has been established, of a document which discloses in considerable detail the method to be employed in pricing the bill of quantities when it becomes available.

**Procedural Aspects**

There will have to be complete and free disclosure of all information relevant to the negotiation, and considerable skill by all concerned in building up unit rates from pricing expertise and site cost feedback. The estimator may be able to assess only the probable average time and cost of each item, but he should certainly know the price at which the work cannot be undertaken. In this connection the CIOB (1983) Code of Estimating Practice provides useful guidelines.

A competitive element can be introduced into the selection of the contractor. Contractors invited to compete should ideally be of similar calibre, so that it can be assumed that each will be able to purchase materials and place work with subcontractors on equally favourable terms. It will then only be necessary for competitive quotations to be obtained at the appropriate time for materials and work to be sublet. It will therefore be possible to concentrate on the work that would normally be carried out by the contractor employing his own labour.

Contractors' estimating methods vary widely and so the tender documents should be presented in a way that permits maximum flexibility. The information falls under three main heads:

(1) Articles of agreement and conditions of contract, including the normal preambles and preliminaries clauses.
(2) Information relating to the build-up of effective labour rates, productivity factors and examples demonstrating how these should be applied in calculating unit rates. All the major items of work which the

contractor is expected to undertake will be included, with the approximate quantity of each stated.

(3) Information on the pricing of insurances, attendances and the like, and the percentage each contractor will require to be added to the estimated net prices for overheads and profit.

Some quantity surveyors point to the weaknesses in the negotiating procedure, such as the length of time required for pricing and negotiation and that there is rarely any guarantee that a lower price will be obtained than by the normal competitive tendering procedure. It may be argued that the allowance for business risk is a matter of opinion, the anticipated profit is based on hope, and off-site overheads are dictated by the efficiency of the construction organisation. Hence the probability of negotiating a contractor's margin equal to or less than that prevailing in the competitive market is unlikely, and the reason why this section started with a specific proviso.

Boyce (1993) has produced a useful guide on contract negotiation from the legal aspects through to the conduct of the negotiation process and the criteria for success.

## MANAGEMENT CONTRACTING

The management contract is a system whereby a main contractor is appointed, either by negotiation or in competition, and works closely with the team of professionals. All physical construction is undertaken by subcontractors (works contractors) selected in competition. The management contractor provides common services to the subcontractors such as welfare facilities, any plant and equipment that is not confined to one subtrade, and sufficient management both on and off the site to undertake the planning and management, coordination and control of the project. He is paid a fee for his services and, in addition, the cost of his on-site management, common services and the cost of all the work undertaken by subcontractors. In 1987 the JCT published the Standard Form of Management Contract between the employer and the management contractor for the construction of the building project, which is not a lump sum contract. A useful reference book is that of Powell-Smith and Sims (1988).

### Uses

The management contract, which emanated from the United States, is most appropriate to large, complex projects, such as the £200m Terminal 4 complex at Heathrow, which exhibit particular problems that militate against the employment of fixed price contract procedures. Typical examples are:

(1) Projects for which complicated machinery and/or computer equipment are to be installed concurrently with the building works.
(2) Projects for which the design process will of necessity continue throughout most of the construction period.
(3) Projects on which construction problems are such that it is necessary or desirable that the design and management team includes a suitably experienced building contractor appointed on such a basis that his interests are largely synonymous with those of the employer's professional consultants.

The JCT Practice Note MC/1 also includes the use of an independent design team, the need for early completion and maximum price competition for the building works.

## Procedures

The procedures to be adopted in management contracting usually incorporate the following activities and requirements.

(1) The management contractor is precluded from carrying out any of the physical works using directly employed labour. His role is primarily that of planner, manager and organiser.
(2) The works are divided into packages agreed by the professional team and the management contractor as being most appropriate for the particular project. Competitive tenders are normally invited for each package from tenderers selected by the professional team (usually architect or contract administrator, quantity surveyor, and structural, mechanical and electrical engineers), the employer and the management contractor. Thus the architect is precluded from obtaining single quotations from his own selected subcontractors, and the management contractor cannot invite tenders from specialist contracting or subcontracting firms within his own Group. However, large contracting firms may tender for subcontract packages as, for example, the substructure or structural frame and, if successful, will operate as subcontractors. The packages sent to potential subcontractors (works contractors) may include, in addition to the permanent works, such temporary works as temporary lighting and power, temporary plumbing, scaffolding and canteen operation.
(3) The management contractor provides from his own resources:
    (a) Site supervisory, technical and administrative staff to run the contract.
    (b) Those facilities to be shared by the subcontractors where they are not included in any of the agreed subcontract packages, such as the normal preliminaries items.
(4) The management contractor is normally paid by monthly instalments computed on the following basis:

(a) The amounts due to be paid by him to subcontractors which have been valued by the quantity surveyor in conjunction with the management contractor and have been certified by the architect. Such payments will be nett, that is all cash discounts will be deducted.

(b) The nett cost to him of providing site staff and shared facilities. The nett cost will be in accordance with a set of definitions of prime cost included in the contract documentation.

(c) A management fee which may be in two parts:

(i) a precommencement management fee, which will be a lump sum including the cost of any staff involved in precommencement activities of the management contractor; and

(ii) a construction management fee which is a percentage of the nett costs to the management contractor, including payments to be made by him to the subcontractors. The percentage is usually in the order of 4 to 5 per cent. It will be appreciated that the management contractor takes very little risk and his financial outlay is comparatively low, although he will be liable for liquidated and ascertained damages if completion is delayed due to his mismanagement.

## Special Role of the Management Contractor

It is contended with some justification that because the management contractor is not at risk in the way that a building contractor would be and has no means of increasing his profit margin, his attitude to the project will be similar to that of the professional team. For example, he will be concerned with keeping the cost of the works within the project budget price, reporting to the employer on possible extras and also in dealing with the subcontractors in regard to such matters as claims for loss and expense and the settlement of accounts. It is probable that future commissions will depend to a large extent on the quality of past performance in the same way as the professionals. Hence large contractors involved in management contracting normally set up separate management contracting divisions, as it could be detrimental to their interests to have claims-conscious staff involved in management contracting.

The management contractor attends all design and progress meetings and it is good policy for a representative of the employer also to be in attendance. The management contractor will be able to report, among other matters, on the dates by which he will require design information and on any information that is already late. The presence of the employer will put pressure on the designers to produce the designs and details by the required times.

### Benefits to the Employer

One of the benefits of management contracting is that the works may be commenced on site at a very early stage in the design process, without the employer foregoing the benefits of competitive tendering, as the sub-contract packages are sent out to tender progressively as the design information becomes available. Management contracting also lends itself very well to cost budgeting and the employer will be able to see exactly where extras have occurred and can request explanations. Apart from monitoring costs and possible trends, the budget reports to the employer should include cash flows. Computerisation of these activities enables the information to be produced quickly and in the required amount of detail.

The JCT Standard Form of Management Contract states that one of the primary roles of the quantity surveyor is to ensure that the prime cost covering the subcontract packages, site management and other costs, is controlled. The quantity surveyor also prepares a contract cost plan which is to be agreed by the management contractor. The Standard Form also standardises the operative procedures.

### CONSTRUCTION MANAGEMENT CONTRACTS

Bennett (1986) defined 'Construction Management' as a consultant responsibility for identifying all the roles needed to undertake any one project, selecting teams to perform these roles and establishing the coordination needed to ensure that the project organisation as a whole works efficiently to meet agreed objectives. In this approach to procurement, the actual construction work is carried out by specialist works contractors who enter into direct contracts with the client.

Hughes (1991) considered that the construction management approach was most suited to various combinations of the following circumstances:

- the client is familiar with construction processes and techniques and knows some or all of the professional team
- the risks associated with the project are dominated by timeliness and cost
- the project is technically complex, involving diverse technologies and subsystems
- the client needs to retain the right to make minor variations to requirements as the project proceeds
- the nature of the project is such that it is realistic to separate professional responsibility for its design from professional responsibility for its management
- the client requires an early start on site
- the cost to the client needs to be competitive, but the control of cost in

terms of securing value for money is more important than simply securing the least possible cost.

Thus the construction work is carried out by 'works contractors' engaged directly by the client, and hence the client assumes the contractual position of the main contractor. Since most clients do not have sufficient expertise to manage the works contractors, they usually employ a construction management firm, on a fee basis, to do this on their behalf. The firm could be a contracting organisation or a professional consultant. In general the organisations using this system are large ones with rolling programmes, considerable experience of similar projects and often some in-house expertise.

Construction management has been a common method of procurement in North America for many years past. In more recent times, very large projects have been undertaken in the UK using the construction management approach, including Broadgate, Canary Wharf, Stansted Airport and Sainsbury's major stores; furthermore, the Terminal 5 complex at Heathrow is to be executed on this basis, using the New Engineering Contract (NEC). In addition, Continental contracting methods tend to operate on a similar basis, and their influence in the UK has been accentuated since the formation of the single market (Seeley, 1995a).

A Construction Management Forum Report (1991) concluded that construction management is likely to replace management contracting for future major contracts, and this has already been evidenced at Heathrow. There are distinct similarities between construction management and management contracting, but one major difference is that with construction management all work packages are treated as direct contracts between the client and the various package contractors, as illustrated in figure 3.2. The construction manager is appointed in a similar manner to the professional consultants with similar liability to the client. This procedure avoids some of the drawbacks of management contracting, which can prove to be more confrontational and expensive, and carry a greater degree of risk for the client, works contractors and management contractor. Views differ as to the stage at which the construction manager is to be appointed, either before, simultaneously with or after the appointment of the design team (Herbert, 1991).

## DESIGN AND BUILD CONTRACTS

A design and build contract, sometimes referred to as a package deal, is a contractual arrangement whereby the contractor offers to design and build a construction project for a sum inclusive of both the design and construction costs. Contractual arrangements vary considerably, ranging from projects where the contractor uses his own professional design staff

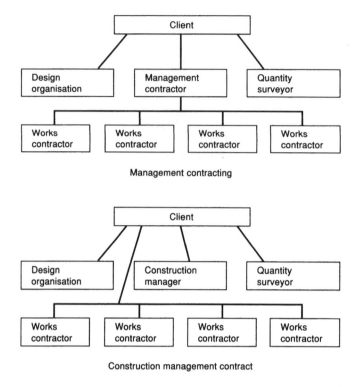

Management contracting

Construction management contract

**Figure 3.2**   *Contractual arrangements in management contracting and construction management contracts*

and undertakes both complete design and construction, to projects where the contractor, specialising in a certain form of construction, offers to provide a full service based on preliminary sketch plans provided by the employer's architect. Design and build contracts can be on a fixed price or cost reimbursement basis, competitive or negotiated. Where they include the total financing of the project they are generally referred to as turnkey contracts (Aqua Group, 1990a).

**Uses**

Design and build contracts are not suitable for all projects. As, for example, where architectural quality is of paramount importance the employer will probably wish to choose the architect independently or by means of an architectural competition, and would not want to be tied to a single contractor. Refurbishment work rarely lends itself to this type of arrangement, while employers requiring purpose made buildings will generally prefer an independent design team. With large or complex projects there are a variety of ways of introducing contractors at an early stage to work with the design team and these tend to militate against the use of design and build.

There are, however, standard building systems developed by contractors where the use of this procedure could be beneficial. Housing is another area where there is scope for its use and many housing associations used this approach in the 1990s. Where a contractor's proprietary system can be used without detriment to the employer's requirements, there can be economic advantages in the use of a modified form of design and build, preferably incorporating choice of layout, finishings and external works.

### Selection of Contractor and Evaluation of Submission

The selection of a design and build contractor should be based on a brief of the employer's requirements. The brief should ideally be prepared by independent professional advisers and costed by them, so that the contractors are tendering on a brief that is within the employer's budget. It is costly for design and build contractors to tender in competition, as each contractor will have to produce a design to meet the brief and a price for construction. Where this process is taken to excess at the tendering stage it will result in an uneconomic use of resources. Hence most contractors are not prepared to proceed beyond outline sketch design and an indicative price (Aqua Group, 1990a).

The evaluation of contractors' tenders is complicated as each contractor is likely to interpret the brief in a different way. Hence considerable adjustments are needed to reduce them to a common basis for purposes of comparison. The adjustments can be fairly straightforward such as adjusting different finishes to the more complicated task of adjusting different roof types. Furthermore, the evaluation can be very subjective where aesthetic aspects are important. Hence the employer's professional advisers, preferably an architect and quantity surveyor, are constrained in the amount of advice they can give to the employer on both financial and aesthetic aspects. In practice one of the main problems can be differences of opinions between the employer and the contractor as to what is to be provided.

### Contractual Aspects

The JCT Standard Form of Building Contract with Contractor's Design is the most suitable basis for the contract where the contractor designs the project. Other contractual arrangements apply where the employer's organisation undertakes the design work. It is important to ensure that the employer's professional advisers have adequate powers to monitor the work and secure effective performance of the contract. After selection of the contractor, a considerable period normally elapses when the contractor's designers are refining the design and incorporating the specific requirements of the employer where these differ from the brief or have been misinterpreted. The specification and performance requirements of the project will then be finalised and the final contract price negotiated.

The responsibility for implementing the design lies with the contractor's management and design personnel. The employer's professional team will monitor the work and ensure that the specification and performance requirements are met. Valuations for payments on account will be made by the employer's quantity surveyor and he will value any variations requested by the employer and negotiate them with the contractor (Aqua Group, 1990a).

The design and build process is becoming more sophisticated and adaptable. As the designer takes all the main decisions affecting cost, time and quality, it seems illogical to isolate him from those erecting the building. A properly integrated design and build organisation can operate on a project team basis, with those possessing different but complementary skills getting to know and respect each other. There are distinct advantages in enabling the contractor to use his management skills and experience in the precontract period to ensure that design and performance are more closely coordinated and better related to time and cost. Economy and efficiency should flow from the continuity of joint experience. However, it may result in some constraints being placed upon the client in developing his requirements, variations may prove expensive and the quality of work may suffer. Readers requiring more information about design and build are referred to Janssens (1991) and a useful client's guide (Nisbet, 1990).

## DESIGN AND MANAGE CONTRACTS

Turner (1997) has aptly described how the 'design and management' system combines some of the characteristics of 'design and build' with those of 'management'. A single firm is appointed, following a selection process that may include some degree of competition on price, although this is not usually the main selection criterion. The client will need adequate in-house skills or obtain appropriate professional services in order to formulate his requirements and carry out his responsibilities under the contract.

The main components of the system are:

- establishing the need to build
- determining the client's requirements
- selecting and inviting tenderers to bid
- the contractor or contractors preparing their proposals for management, design, time and cost

- evaluation and acceptance of a tender which becomes a contract
- management, design and construction of the works.

In this system the client enters into a contract with a design and manage contractor and possibly a scope designer, while the design and manage contractor, in addition to his contract with the client, will be in contact with consultants for design and/or cost consultancy services and with works contractors who may number as many as 60 to 100 organisations. Alternatively, the client may enter into a contract with a design and manage consultant who will be selected from one of the building professions and who will enter into similar contractual arrangements with a design and/or cost consultancy and numerous works contractors. These systems offer advantages when factors such as timing, controllable variation, complexity, quality level and competition are significant.

## DEVELOP AND CONSTRUCT CONTRACTS

Turner (1997) has described how in this type of contract consultants design the building to a partial stage, often called 'scope design', and competitive tenders are obtained from contractors who develop and complete the design and then construct the building. The system offers maximum flexibility as the 'develop' part can begin from anywhere between RIBA Plan of Work stage C up to any point starting before the end of stage E. Develop and construct requires analysis/creativity of a design team, which may be chosen in competition, and then acceptance/implementation by a construction team which may also be chosen in competition. This system offers advantages where timing, good quality level, price certainty, competition, professional responsibility and risk avoidance are considered important.

## BRITISH PROPERTY FEDERATION SYSTEM

In late 1983, the British Property Federation launched on an unconsulted industry a Manual which describes in detail the way in which the BPF intended projects under its members' control to be managed. The key man is the client's representative who takes control of the project and could come from any of the relevant professions – it seems that he could have power without responsibility and it is essential that he should act impartially. Subservient to him is a design leader responsible for design and cost.

The contractor tenders for the work on the basis of drawings and specification, with bills of quantities replaced by a priced schedule of activities, whereby the contractor will be paid as each activity is completed, and the contractor will be encouraged by financial rewards to reduce costs. The contractor will choose the subcontractors, as there will be no nominations, and he will also be responsible for a proportion of the design, which could prove burdensome.

The scheme received a mixed reception from the industry and it is evident that the effectiveness of the system will be largely dependent on the drawings and specification being clear and comprehensive. Inexperienced clients could be faced with many problems, although the main aim is to reduce both the time and cost of construction contracts.

# 4   Contract Arrangements

The general procedure adopted by most quantity surveying offices up to the stage of inviting tenders was described in chapter 2. The present chapter examines tendering procedures in more detail and then describes the normal arrangements for the evaluation of tenders. This is followed by an examination of general contractual arrangements, and the nature and form of the various standard forms of contract and subcontract.

## TENDERING PROCEDURES

It is advisable for the tendering procedures to follow the guidelines prepared by the National Joint Consultative Committee for Building (1994a), unless there are very good reasons for departing from these recommendations. The NJCC Code takes into account the positive recommendations contained in a NEDO report (1975) and 'Action on the Banwell Report' (1967).

It is most desirable where a contractor is to be selected by competitive tendering that a short list of suitable tenderers should be drawn up either from the employer's approved list of contractors or from an *ad hoc* list of contractors of established skill, integrity, responsibility and proven competence for work of the character and size contemplated. The suggested maximum number of contractors is six. For contracts less than £50 000 in value, it is common practice to invite tenders from less than the six contractors recommended by the NJCC Code (1994a). Typical numbers of tenderers are 1 to 3 for contracts not exceeding £1000, 3 for contracts exceeding £1000 and not exceeding £10 000, and 3 to 5 for contracts exceeding £10 000 and not exceeding £50 000. Most public authorities require alternative quotations wherever possible, and a single tender for contracts not exceeding £1000 would be permissible only when it is in the authority's interest and it is not considered reasonably practical to obtain alternative quotations.

On contracts where it is desired to secure the early involvement of the contractor before the development of the design is completed, two stage tendering procedures may be adopted. This procedure is described in chapter 3 and should conform to the NJCC Code of Procedure for Two Stage Selective Tendering (1994b).

The cost of preparing tenders is high and hence the tender list should be kept as small as practicable to reduce abortive tendering costs. Specialist engineering contracts involve particularly high tendering costs and this should be borne in mind when preparing the tender list. In all cases it is wise to append one or two further names to replace any firms on the initial list that do not accept the preliminary invitation.

When selecting the short list the following are some of the principal matters to be considered:

(1) the firm's financial standing and record;
(2) whether the firm has had recent experience of building at the required rate of completion over a comparable contract period;
(3) the firm's general experience and reputation in the area of work;
(4) whether the management structure of the firm is adequate for the type of contract envisaged; and
(5) whether the firm will have adequate capacity at the relevant time (NJCC, 1994a).

Approved lists should be reviewed periodically, extracting firms whose performance has been unsatisfactory and allowing the introduction of suitable additional firms. If the compilation of the list has been performed in a sound and efficient manner, it should be possible to entrust any named contractor with a project. The final choice will then merely consist of accepting the lowest tenderer. Only very exceptional circumstances would justify a departure from this procedure. Approved lists may be categorised by value in ranges such as under £20 000, £20 000 to £100 000, £100 000 to £300 000, £300 000 to £1m, £1m to £5m and over £5m, and by type such as new buildings (traditional), new buildings (CLASP or other system) and alterations. The NJCC has produced application forms/questionnaires for admission to select lists of contractors (for one particular contract) and approved lists of contractors.

## Preliminary Enquiry

In order that contractors may be able to decide whether to tender, each firm should be sent, and should reply promptly to, a preliminary invitation to tender in the form illustrated in Appendix A of the NJCC Code (1994a). This procedure should eliminate subsequent withdrawals and the submission of cover prices (prices that are sufficiently high to be well above the lowest tender). The letter states that the acceptance will imply the submission of a wholly *bona fide* tender in accordance with the principles prescribed in the Code, and that tenderers are not to divulge the tender price to any person or body before the time for submission of tenders. Once the contract has been let the employer will supply all tenderers with a list of the tender prices. The inability of a contractor to accept the invitation will not prejudice his opportunities for tendering for further work, neither will the inclusion in the preliminary list guarantee the receipt of a formal invitation to tender for the works. The preliminary enquiry letter will contain adequate particulars of the project (parties, location, general description of work, approximate cost range, nominated subcontractors for major contracts, form of contract and amendments, contract period and anticipated dates, approximate date for despatch of tender documents, tender period, guarantee requirements and any particular conditions). A suitable format for this letter is shown in Appendix A of the NJCC Code of Procedure for Single Stage Tendering (1994a), and appendix A at the back of this book.

It is suggested that an appropriate time period between the preliminary enquiry and the despatch of tender documents should be 4 to 6 weeks, although certain situations might warrant periods as long as 3 months. Once a contractor has notified his initial agreement to tender, such acceptance should normally be honoured. All contractors should be notified of their position shortly after the latest date for the acceptance of the preliminary invitations (NJCC, 1994a).

## Tender Arrangements

All tender documents should be sent to the tenderers on the day stated in the preliminary invitation. Where a bill of quantities is supplied, it will be accompanied by 1:100 or 1:50 scale drawings and probably supporting component details. Where no bill is provided, each contractor must be supplied with a complete set of drawings from which to prepare his estimate. With alteration works clear and well-annotated drawings are essential and the best practice is to provide one set of drawings depicting the existing arrangements and another showing the proposed alterations, as combined drawings are rarely satisfactory. Where there is no bill of quantities, a full specification will be supplied to all tenderers and this constitutes a contract document, whereas a bill of quantities will contain all necessary specification notes, usually in the form of preambles to works sections, and no specification will be needed. The remaining contract documents consist of the conditions of contract, which normally comprise the appropriate JCT standard form, and the form of tender.

The formal invitation to tender can with advantage follow the format contained in the NJCC Code (1994a), which is illustrated in appendix B at the back of this book. It is customary to send two copies of the bill of quantities, drawings and form of tender, and addressed envelope for the return of the tender. The closing date for the return of the documents is also given.

The form of tender suggested by the NJCC (1994a) is shown in appendix C and contains the contractor's offer and the contract period, which is usually inserted by the architect to avoid problems in the assessment of tenders resulting from differing completion dates. Contract periods are likely to range from about 4 months for contracts estimated not to exceed £75 000 up to 18 months for contracts in the £1.75m to £2m range. This form also prescribes the method for dealing with errors in pricing or arithmetic in the priced bill of quantities discovered before acceptance of the offer. In practice, many public authorities and other organisations have their own form of tender with various changes of wording. Reference may, for example, be made to compliance with the general conditions of the Fair Wages Resolution passed by the House of Commons on 14 October 1946 and an assurance that the tenderer has not adjusted the amount of the tender in accordance with any agreement or arrangement with any other person, nor has communicated it to any other person before the closing date. In the case of public authority contracts, the contractor may be required to procure the issue of a performance bond or other guarantee from an insurance company or other financial institution. These bonds or guarantees are generally limited to 10 per cent of the contract price, and are in the nature of a guarantee on the part of the bondsman that the contractor will perform his contractual obligations. Contractors may also be required to name a surety for tenders in excess of say £25 000. Contractors are generally opposed to bonds particularly in selective tendering where the contractors chosen should be financially sound. Bonds cost money and in the case of bank sureties can restrict credit facilities. The NJCC (1986) guidance note on performance bonds briefly explains their nature and operation and gives a relevant example. Performance bonds are dealt with in more detail later in the chapter. Some forms of tender include a schedule of percentage additions on the prime cost of daywork calculated in accordance with the RICS and BEC *Definition of Prime Cost of Daywork carried out under a Building Contract*. Instructions to tenderers often state that no undertaking will be given that the lowest or any tender will be accepted.

The National Joint Consultative Committee for Building (NJCC, 1994a) recognises that the terms of the standard forms of building contract are not mandatory and that the employer and his professional advisers may amend them at their discretion. Nevertheless the Committee believes that alterations to the standard forms impede the move towards greater standardisation of building procedures. There was, however, considerable dissatisfaction with the JCT80 Standard Form of Building Contract and an alternative form was produced by the Association of Consultant Architects (ACA) in 1982 to further confuse the situation. The NJCC believed that any changes in the Standard Form should be kept to an absolute minimum, and then only with serious prior consideration and drafting by a person competent to ensure that all consequential alterations to other clauses are made.

It is further recommended that the tenderer's attention should be specifically drawn to any alterations in the preliminary invitation to tender and, where appropriate, with the reasons for the changes. The contractor then has the opportunity to consider the implications of the amendments prior to acceptance of the invitation to tender and so reduce the possibility of queries at the tender stage which could result in an extension of the tender period (NJCC, 1994a).

In the public sector, projects exceeding a certain threshold value (excluding VAT and the value of nominated subcontracts), are required to be advertised in the *Official Journal of the European Economic Community* before tenders are invited.

## The Needs of the Contractor

The time allowed for the preparation of tenders will be influenced by the size and complexity of the project. A period of four weeks is common, but this will need to be increased to possibly six weeks for large contracts exceeding £1 m estimated value, complex projects, smaller projects without quantities and in other special circumstances. Tender periods must be sufficient to allow tenderers to familiarise themselves with the documents and sites, obtain quotations, price the preliminaries and unit rates and carry out the adjudication process, which may involve adjusting preliminaries, unit rates or the summary by an agreed percentage (C10B, 1983).

Harrison (1981) described how the decision as to the final tender price has to be made once the estimate has been completed and the risks assessed. Three major points generally need considering:

(1) What is the probable cost of carrying out the works and to what extent could this vary according to circumstances that can be foreseen, but the likelihood of which cannot be assessed with any accuracy – namely, what is the extent of the risk?
(2) What is the minimum price at which the contract is likely to be of benefit to the company?
(3) At what price level is the firm:
     (i) certain to get the work?
     (ii) likely to get the work?

The contractor also has to consider what contribution the contract should be expected to make towards the payment of overhead costs. This is usually derived from an assessment of the anticipated average ratio of turnover to overheads and requires continual monitoring. Hence the contractor's task of building up the tender price is fraught with difficulties and the design team should do all they can to assist him to carry out efficient tendering on the basis of the maximum amount of sound information and the reducing of contractual risks to a minimum. For example, the contractor needs to know in advance the approximate cash demands imposed by the tender programme in order to produce a cash flow chart; thus the establishment of a realistic programme at the outset by the contractor is a necessity, based on the information supplied by the design team.

Shash (1993) identified the factors which contractors consider when formulating tender bids by means of a structured questionnaire and these were as follows: degree of difficulty of work; risk owing to the nature of the work; current workload; need for the work; contract conditions; anticipated cost of liquidated damages; owner/client identity; and past profit in similar work. This helps to emphasise the problems and risk involved in the bidding process. Smith (1995) has described the use of bidding models in the preparation of tendering bids. In the mid 1990s, electronic tendering was moving closer.

## Qualified Tenders

To ensure fair competitive tendering, it is essential that the tenders submitted by each contractor shall be based on identical tender documentation and that tenderers should not vary the common basis by qualifying their tenders. Where a tenderer considers that any of the tender documents are deficient in any way and require clarification or contain unacceptable alterations to the standard form of building contract not listed in the preliminary invitation to tender, he should as soon as possible notify the issuing authority or architect, with a copy to the quantity surveyor. If it is decided to amend the documents, all tenderers should be informed and the tendering period extended if necessary (NJCC, 1994a). Pretender meetings, preferably in the form of a group meeting of all tenderers, to clarify points of doubt and uncertainty, as suggested for civil engineering contracts, can prove invaluable (ICE, 1983).

A tenderer who submits a qualified tender should be given the opportunity to withdraw the qualifications without amending the tender. Failure to comply with this request will result in his tender being rejected. For example, a tender invited on a 15 month contract period and submitted on the basis of a 17 month period will be rejected, as it fails to meet the stipulated requirements and if considered would give this particular tenderer an unfair advantage over his competitors.

However, with civil engineering contracts it is fairly common for the tender documents to expressly permit the submission of alternative designs with their prices. Furthermore, as long ago as 1964, the Banwell Committee, when considering the placing and management of both building and civil engineering contracts, expressed the view that 'if a firm has the initiative to produce a novel and possibly better technical solution, fully documented, to a problem, we cannot see why this should be ignored or disclosed to rival tenderers. Such alternatives should be considered on their merits.' In this type of situation it would obviously be wise and prudent practice for the tenderer concerned to advise the professional team in advance and to seek its cooperation and confirmation that such an approach will be acceptable. It is evident that alternative offers require the closest scrutiny and there is the possibility of any design liability devolving upon the tenderer, whereby the burden may ultimately exceed the benefit. The whole aspect of alternative designs and prices is an exceedingly complex one and hence for this reason is normally avoided.

### Contractor's Tendering Prices and Costs

An interesting survey of the cost effects of alternative tendering procedures was undertaken by the RICS Essex Branch (Smith, 1981). The quantity surveyors in the survey from Kent, Essex, Hertfordshire and East London considered tenders based on negotiated bills of quantities to be on average 6 per cent above the comparable prices based on selected competitive bills of quantities, as compared with a 1 per cent addition quoted by contractors from the same location. The difference will probably be smaller in practice as quantity surveyors tend to base their estimates on one of the lowest three or four in competition, but not the lowest tenderer.

Many contractors questioned expressed the view that they would be prepared to tender on a plan and specification up to a value of £500 000 in 1981 for straightforward projects. It was interesting to note that the total cost of each contractor preparing his own quantities from plans and specification for a selected list of the size recommended by the NJCC was approximately the same as the cost of a single quantity surveyor preparing a detailed bill of quantities plus the contractors' tendering costs. The additional risk of the contractors preparing builders' quantities is reflected in their price. Although contractors stated that varying costs were incurred in submitting tenders of different types, only 41 per cent specifically reflected the different cost in their tenders.

Main contractors were asked to indicate the ratios of unsuccessful to successful bids and the average was 6.5:1 in normal market conditions. Contractors interviewed frequently stated that small plan and specification projects failed to proceed after the client received the tender. Two thirds of the main contracts handled by quantity surveyors were based on competitive firm or approximate bills of quantities and a further 14 per cent on competitive plans and specification. In 1995 a significant proportion of contracts were let on a design and build or management contract.

## EXAMINATION AND EVALUATION OF TENDERS

### Opening and Action on Receipt of Tenders

The tenders must not be opened before the prescribed day and time, but as soon as possible thereafter, and the procedure varies considerably between different organisations. Most public authorities have delegated this task to designated officers, who may vary according to the size of the contract. It is advisable to arrange for a group of persons to be present at the opening to avoid any suggestion of an irregularity occurring during the proceedings. A convenient group for overseeing the opening of tenders in the private sector will be the employer, architect and quantity surveyor.

The first step after opening the tenders is to tabulate them, starting with the lowest, and listing the tender sums and the contract periods where these are not stipulated in the tender documents. If any tender is qualified in any way the qualification should be recorded. The tabulated list of tenders should then be signed by at least two of the persons present at the meeting. Photocopies of this schedule and the completed forms of tender will be circulated to appropriate persons concerned with the project.

Where a tender is received after the stipulated date and time, it is generally advisable to reject it since there is a possibility that the tenderer may be aware of the amounts of some of the other tenders before submitting his own. The matter should be reported to the employer who will make the final decision. The NJCC recommendation is not to admit tenders received after the prescribed time.

The lowest tenderer should be asked to submit his priced bill of quantities as soon as possible, and normally within four working days, unless it has already been submitted with the tender. The NJCC (1994a) recommends that all but the three lowest tenderers should be informed immediately that their tenders have been unsuccessful, as it may have considerable impact on their current and future tendering plans. In order to avoid loss of time in the event of the lowest tenderer withdrawing his offer, the second and third lowest tenderers should be informed that although their tenders were not the most favourable received, they will be approached again if it is decided to give further consideration to their offers. They should subsequently be notified as soon as a decision has been taken on the acceptance of a tender. Once the contract has been let every tenderer should be supplied with a list of the firms who tendered and the tender prices.

### Examination of Priced Bills

The examination of the priced bill(s) of quantities supporting a tender should be undertaken by the quantity surveyor who must treat the document as confidential. No details of a tenderer's pricing shall be disclosed to any person, other than the architect or other appropriate consultant, without the express permission of the tenderer. The object of examining priced bills is to detect any errors that may have occurred in the computation of the tender, or any anomalies that could cause problems at the postcontract stage.

All the entries in a bill of quantities must be checked arithmetically as errors can occur in a variety of ways. Billed items may unintentionally have been left unpriced, there may be errors in item extensions (multiplication of quantities by the unit rates), page totals, transfers of page totals to collections or summaries, or in the General Summary or even in the transfer from there to the Form of Tender. Fortunately, electronic calculators have replaced the manual processes formerly used for this work.

Pricing errors can also occur, such as the insertion of a superficial rate for a linear item or the inclusion of different rates for identical items in separate sections of the bill. This latter type of error is most likely to occur when there are repetitive items in separate bills as with housing contracts, where there may be a separate bill for each house type. Even decimal points can be misplaced with serious consequences where large quantities are involved. Probably the best method of recording the errors is to enter them on double cash paper identified by item references.

At the same time the quantity surveyor will be concerned with the general form of the pricing. For instance it could be that very few of the items in the Preliminaries Bill have been priced and presumably their cost has been spread over measured rates. This practice can cause problems later should adjustment of the Preliminaries become necessary. The earlier work sections in the bill, such as Groundwork, Concretework and Masonry, which will be mostly completed during the early months of the contract, could contain highly priced major items, counterbalanced by low priced items in the Surface Finishes section of the bill. The contractor is thus able to obtain higher payments than those to which he is really entitled during the early part of the contract and so improve his cash flow position at the employer's expense, while keeping his total tender sum the same. It also places a greater risk upon the employer should the contractor go into liquidation during the execution of the contract.

These matters do pose problems for the quantity surveyor as it will be appreciated that the contractor is entitled to adopt whatever pricing strategy he considers most appropriate. If, however, the quantity surveyor is genuinely concerned about the method adopted in pricing the bill and its possible consequences for the employer, he can recommend that the tender is not accepted. Wide differences can often occur in pricing levels of competing tenderers, and although the totals for work sections can vary by as much as 10 to 15 per cent, the difference in tenders is unlikely to exceed 2 to 5 per cent. Work sectional price differences in the order of 25 per cent would certainly justify consideration. Apart from pricing strategy, the anticipated method of working, use of plant, materials purchasing sources and other related matters will all have their effect on the unit rates.

The contract may include a list of basic materials which the contractor is requested to price for subsequent use in connection with any daywork or for the evaluation of price fluctuations under the traditional method, as opposed to the NEDO price adjustment formula which is examined in chapter 5. The quantity surveyor will examine the prices to satisfy himself that they conform to the current market prices delivered to the site in full loads. In case of doubt the contractor can be requested to submit recent quotations. Daywork should ideally be quantified and incorporated in the bill in such a way that it will affect the tender figure and thus form part of the competitive bid.

In the case of tenders based on approximate quantities, the rates for items where the quantities could conceivably be subject to a significant increase justify careful scrutiny. In like manner, the schedule of principal rates used in the preparation of a tender based on drawings and specification deserves careful study.

### Adjustment of Priced Bills

The NJCC Code of Procedure for Single Stage Selective Tendering (1994a) recommends two alternative methods of dealing with errors that have been found in priced bills, and these are now listed:

(1) The tenderer should be given details of the errors and afforded an opportunity to confirm or withdraw his offer. If the tenderer withdraws, the priced bill of the second lowest tenderer should be examined and, if necessary, this tenderer will be given a similar opportunity. Where the tenderer confirms his offer, an endorsement should be added to the priced bill indicating that all rates or prices (excluding preliminary items, contingencies, prime cost and provisional sums) inserted by the tenderer are to be considered as reduced or increased in the same proportion as the corrected total of priced items exceeds or falls short of such items. This endorsement should be signed by both parties to the contract.
(2) The tenderer should be given an opportunity of confirming his offer or of amending it to correct genuine errors. Should he elect to amend his offer and the revised tender is no longer the lowest, the offer of the firm now lowest in competition should be examined. If the tenderer elects not to amend his offer, an endorsement will be required as described in method (1). If the tenderer does amend his tender figure, and possibly certain of the rates in his bill, he should either be allowed access to his original tender to insert the correct details and to initial them or be required to confirm all the alterations in a letter. If in the latter case his revised tender is eventually accepted, the letter should be conjoined with the acceptance and the amended tender figure and the rates in it substituted for those in the original tender.

Summing up, there are two alternative approaches put forward by the NJCC for dealing with genuine errors – either for the contractor to confirm or withdraw, or to confirm or correct. Whichever alternative is adopted, it needs to be communicated to the tenderers in the preliminary enquiry for invitation to tender, the formal invitation to tender and the form of tender itself. In practice the first alternative is generally adopted as there is often difficulty in deciding exactly what are genuine errors and the fact that the contractor is allowed up to four working days to produce his priced bill of quantities, except where it is already supplied with the tender, which tends to be the exception.

Where the contractor decides to confirm his tender, the errors must be adjusted by the quantity surveyor in agreement with the contractor, so that arithmetical and obvious pricing errors are clearly and neatly corrected and the corrections carried through to page totals, collections, summaries and finally to the General Summary. A sum equal to the nett amount of the aggregated errors may then be inserted on the General Summary immediately before or after the final total as an addition or deduction. This will result in the adjustment of the corrected bill total to the amount inserted on the Form of Tender. It will be noted that in the original document the bill total may be rounded off before being inserted in the Form of Tender. The average net addition or deduction to be applied to all rates and prices in the bill is then calculated as a percentage of the value of the contractor's work (Tender figure less prime cost and provisional sums and contingencies). The NJCC Code (1994a) also excludes preliminaries, but the general practice is to include them as the preliminaries items are calculated and inserted in the bill by the contractor in a similar manner to the measured rates and can also be the subject of arithmetical errors. The following example and that given as table 4.1 will serve to illustrate the main procedures.

*Typical adjusted pricing error*

| | | | £ p |
|---|---|---|---|
| C *In situ* concrete foundations | | | |
| (21 N/mm$^2$ − 20 aggregate), | | | *11 268.00* |
| poured on or against earth. | 180 m$^3$ | *62.60* | ~~*11 196.00*~~ |

Note: Page totals, Collections and Summaries
        will be individually corrected by hand
        in the same way.

Table 4.1    Correction of errors in priced bill of quantities

---

*GENERAL SUMMARY*

| | £ p |
|---|---|
| Preliminaries | *83 767.00* |
| Substructure | *65 231.00* |
| | *417 174.46* |
| Superstructure | ~~*417 274.46*~~ |
| | *141 390.54* |
| Drainage and External Works | ~~*131 590.54*~~ |
| Prime Cost and Provisional Sums and Dayworks | *100 050.00* |
| | *807 613.00* |
| TOTAL CARRIED TO FORM OF TENDER | ~~*797 913.00*~~ |
| *Deduction to adjust for errors* | *9 700.00* |
| *Total* | *£ 797 913.00* |

---

The General Summary (unamended) shows a total of £797 913.00, which includes a sum of £100 050.00 for prime cost and provisional sums and dayworks.

Hence the percentage adjustment is calculated as follows:

$$\frac{9\,700}{(797\,913\,-\,100\,050)} \times 100 = 1.39 \text{ per cent (reduction)}$$

The contract copies of the priced bill should then be endorsed with a statement that all rates and prices (other than prime cost and provisional sums and dayworks) are to be considered as reduced by 1.39 per cent.

### Negotiation of and Reporting on Tenders

The quantity surveyor will report to the architect and the employer on completing his examination of the tenders. Where a tender is free of errors, or the tenderer is prepared to stand by his tender in spite of error(s), or a tender after amendment, where this is permitted, is still the lowest, this should be recommended to the employer for acceptance in the absence of any major point of concern to the quantity surveyor.

Should the tender under consideration exceed the employer's budget the NJCC (1994a) recommended procedure is for a reduced price to be negotiated with the tenderer. The basis of negotiations and any agreements made should be fully documented. Only when these negotiations fail should discussions be commenced with the next lowest tenderer, and if these also fail, then similar action may be taken with the third lowest tenderer. In the event of all these negotiations failing, new tenders may be invited. Sound cost planning throughout the design process, as described in chapter 8, should avoid the need for these protracted and possibly abortive negotiations.

The most common approach when the lowest tender exceeds the employer's budget is to make amendments to the proposals which are normally incorporated in an addendum bill, which will probably contain omissions and the substitution of some cheaper items, and thus bring the revised tender sum within the employer's budget. It is important that the scope of the works is not changed so significantly that the original tenders no longer form a valid basis upon which to make the award, otherwise all tenderers should be given the opportunity to submit a new tender for the amended project. In conducting these negotiations, the guiding principle must be to ensure that the confidentiality and fairness of tendering is preserved, and that no one tenderer is given an unfair advantage over others (ICE, 1983).

The quantity surveyor's report will comment on the general pricing levels and any unusual aspects and possible future matters of concern identified

Table 4.2   Report on tenders

---

*Shops and Offices, Howarth Street, Oakbury, Suffolk*

1. *Tender Llist*
   Eight tenders were received as follows:                          £
   Johnson Brothers                                         1 497 320
   Buildwell Ltd                                            1 506 226
   Smith and Brown Ltd                                      1 518 190
   W. T. Robson                                             1 525 780
   Baker and Thompson Ltd                                   1 531 670
   Europa Builders                                          1 553 840
   Hadley, Hudson and Company                               1 569 432
   Blackworth (Contractors) Ltd                             1 581 560

2. *Tender Levels*
   The lowest tender is only 0.59 per cent below the second lowest tender and the total range over all eight tenders is 5.63 per cent, indicating extremely keen tendering in a highly competitive situation. The lowest tender is within the final cost plan prepared on 18 January 1996 and totalling £1 520 000.

3. *Pricing*
   The general level of pricing was consistent and realistic with no significant anomalies. There were arithmetical errors amounting to an addition of 1.39 per cent of the value of the contractor's work but Johnson Brothers elected to confirm their initial tender sum, and the rates and prices will be adjusted accordingly.
   Johnson Brothers are keen to undertake further work in this area and have a good reputation for sound workmanship and completion on time.

4. *Recommendation*
   We recommend that the contract be awarded to Johnson Brothers in the sum of £1 497 320.

Lewis and Lloyd
Chartered Quantity Surveyors
28 South Street
Congleworth
Essex
12 June 1996

---

in the priced documents submitted by the lowest tenderer, the extent of any errors or inconsistencies in the pricing and the action that has been taken, details of any qualifications to the tender, the likely cost of the project compared with the budget, and a recommendation as to acceptance or otherwise. A typical tender report is contained in table 4.2.

## PLACING OF CONTRACTS

### Precontract Meeting

It is good practice for the architect to arrange a meeting with the contractor, at which the employer, quantity surveyor and other consultants should also desirably be present. The main aim is to settle all outstanding matters and to clarify any points of doubt or obscurity. In particular, the dates for possession of the site and completion of the work should be confirmed or agreed. The contractor is then requested to prepare his master programme for the execution of the works, preferably supported by a method statement. The design team should establish a programme for the supply of working details, schedules and other necessary documents to establish a sound working framework against which progress can be periodically monitored.

The architect should supply the contractor with all available information about nominated subcontractors and suppliers and their likely programmes. The architect will seek assurances from the contractor that he has no objections to the engagement of any of the nominated persons. The contractor may wish to tender for certain nominated work and the decision of the architect is required. Since there are controls on subletting under the standard forms, it is desirable that the contractor should give details of the firms that he wishes to use. Finally the details in the appendix to the conditions of contract should be confirmed and all insurance arrangements agreed.

### Signing the Contract

As soon as all preliminary matters have been resolved, arrangements should be made for the signing of the contract documents, which normally consist of the Articles of Agreement, drawings and bill of quantities. Ideally the formal acceptance of the tender and signing of the contract should be carried out within two months of the receipt of the offer. The contractor should not be left in a state of uncertainty any longer than is absolutely necessary and it must always be borne in mind that the tenderer has the option of withdrawing his offer at any time before acceptance.

The quantity surveyor often plays a major role in assembling the contract documents and preparing them for signing because of his familiarity with them and the fact that he drafts the Preliminaries. The Articles of Agreement require completing to provide details of the employer, contractor, architect/supervising officer and quantity surveyor, a brief statement of the nature and location of the works, the numbers of the contract drawings and the contract sum (in words and figures). It will also be necessary to make a number of deletions and amendments to the Conditions of Contract and to complete the appendix to the Conditions. These deletions, amendments and insertions will have been notified to all tendering contractors and no further changes should be made without the prior agreement of the contractor. All deletions and amendments must be initialled by the parties at the time of signing the contract documents.

There are also two supplements to the JCT Standard Form of Building Contract (1980) which require consideration. One is bound in with the JCT Form dealing with Value Added Tax and the other covers sectional completion of the works and is optional. The sectional completion supplement is used where the works are to be completed in phased sections, and can be used only where tenderers are notified that the employer requires the works to be performed in this way and that he will take possession on the practical completion of each section.

The priced bill of quantities will contain the amended rates and prices and/or the percentage to be applied to the measured rates agreed between the contractor and the quantity surveyor. An endorsement is normally made at the back of the bill indicating that 'These are the Contract Bills referred to in the Contract' and signed by the parties. In like manner the drawings will be stamped to show that they are the contract drawings and the ones from which the bills of quantities have been prepared.

Some local authorities and other public organisations prefer to execute their contracts under seal, although since 1960 these bodies have been able to enter into contracts under the signature of an authorised official. The principal difference between the two methods related to the period of the contractor's liability for latent defects, of 12 years under seal and 6 years when under hand, but even this was removed by the Latent Damage Act 1986, which provided for 3 years from the date of actual or ascertainable knowledge of the existence of the cause of action accompanied by an overriding 'long stop' of 15 years (Wallace, 1995). The employer may require the contractor to provide a bond for the due performance of the contract and, if so, action will be required from the contractor.

### Issue of Documents

Immediately after the signing of the contract, the architect must supply the contractor, free of charge, with:

 (i)   one copy certified on behalf of the employer of the contract conditions;
 (ii)  two further copies of the contract drawings;
 (iii) two copies of the unpriced bills of quantities.

If the bills have not been annotated, the architect must also provide the contractor with two copies of the specification or other descriptive schedules.

The contractor shall provide the employer with two copies of his master programme for the execution of the works. Any further copies of the priced bill that may be required will be produced by photocopying, which avoids the risk of any transcribing errors. With the priced bill now available, the quantity surveyor can reassess the cost plan, update it and so provide an effective basis for the cost control of the work as it is executed. Some employers may require a forecast of their expenditure for cash flow purposes. This entails an analysis of the contract sum set against the contractor's programme and an assessment of the effect of possible variations, claims and fluctuations (Turner, 1983).

## GENERAL CONTRACTUAL ARRANGEMENTS

### Nature and Form of Contracts

It is probably helpful to start by briefly outlining the general characteristics of contracts. A simple contract consists of an agreement entered into by two or more parties, whereby one of the parties undertakes to do something in return for something to be undertaken by the other. A contract has been defined as an agreement that directly creates and contemplates an obligation. The word is derived from the Latin *contractum*, meaning drawn together. Where the contract terms are set out in writing in a document, which the parties subsequently sign, then both parties are bound by these terms even if they do not read them. Thus by setting down the terms of a contract in writing one secures the double advantage of affording evidence and avoiding disputes. The law relating to contracts imposes on each party to a contract a legal obligation to perform or observe the terms of the contract, and gives to the other party the right to enforce the fulfilment of these terms or to claim damages in respect of the loss sustained in consequence of the breach of contract (Seeley, 1993).

### Nature of Building Contracts

The construction industry in this country approaches one million people who work for varying periods on many different sites and the resultant buildings are often erected on land owned by others. The majority of employers know what quality of buildings to expect and understand in general terms the implications of the contracts into which they enter.

The Standard Form of Contract sets out to establish a series of relationships which apply on building sites for most types of contract, and enables people working on different projects to carry out their activities in as standardised form as practical. In setting up these relationships the Standard Form covers the method of ordering work by the employer, dealing with delays and defaults on both sides, arrangement of insurances and calculation of the final account, together with stage payments and variations in cost. It is generally accepted that buildings are less than perfect – hence the operation of the defects liability period and the preparation of certificates for making good defects, coupled with taking over the project at practical completion with an anticipated list of outstanding or irregular matters. This emphasises the difference between a one-off product produced on land owned by the purchaser and a product manufactured in the factory of a vendor and delivered as a completed article to the purchaser and which in general can be inspected by the purchaser before he receives it.

Under the Standard Form, the contractor undertakes to carry out the work under prescribed conditions and in accordance with the contract documents previously described. Should there be discrepancy between the Form of Tender and the Conditions of Contract, provided that the contract has been signed by the relevant parties, it is probable that the contract conditions will prevail. In the Standard Form with quantities, it will be seen that the Bill of Quantities performs several functions, namely:

(1) a precise measure of the work to be completed for the contract sum in terms of both quantity and quality;
(2) a basis for the measurement and valuation of variations; and
(3) a means of incorporating the necessary specification information as part of a contract document.

The parties to the contract are the contractor and the employer (client). The architect, quantity surveyor, engineers and other consultants are not parties to the contract. Each has their own terms of employment with the employer, usually on a standard form issued by the appropriate professional body, and at individually agreed fees. The architect is given extensive agency powers under the Standard Form (Fellows, 1995).

Most contracts entered into between building contractors and their employers are entire contracts, whereby the agreement is for a specific project to be undertaken by the contractor for an agreed sum which does not become due until the work is complete. In these circumstances the contractor is not entitled to any payment if he abandons the work prior to completion, and will be liable in damages for breach of contract. Where the work is abandoned at the request of the employer, or results from circumstances that were clearly foreseen when the contract was entered into and provided for in its terms, then the contractor will be entitled to payment on a *quantum meruit* basis, that is, he will be paid as much as he has earned. However, contractors are rarely willing to enter into contracts, other than the very smallest, unless provision is made for interim payments to them as the work proceeds. For this reason the JCT Standard Form of Building Contract (1980) provides for the issue of interim certificates at various stages of the work (Seeley, 1993).

It is usual for the contract further to provide that only a proportion of the sum due on the issue of a certificate shall be paid to the contractor. In this way the employer retains a sum, known as retention money, which will operate as an insurance against any defects that may arise in the work. The contract does, however, remain an entire contract, and the contractor is not entitled to demand payment in full until the work is satisfactorily completed, the defects liability period expired and the certificate of completion of making good defects issued.

That works must be completed to the reasonable satisfaction of the architect does not give him the right to demand an exceptionally high standard of quality throughout the works, in the absence of a prior express agreement. Otherwise the employer, acting in liaison with the architect, might be able to postpone indefinitely his liability to pay for the works. The employer and the architect are normally entitled to expect only a standard of work that would be regarded as reasonable by competent men with considerable experience in the class of work covered by the particular contract. The detailed requirements of the specification particulars, whether in a separate document or inserted in the bill of quantities, will have an important bearing on these matters (Seeley, 1993).

## Operational Aspects of Building Contracts

Local authority standard contract administration procedures are often formulated to lay down a series of guidelines to ensure efficient operation and uniform practice. For example, they may provide for projects not exceeding £20 000 estimated value to be placed on an order incorporating the local authority's own abridged general conditions of contract, while larger contracts require a formal contract based on a Standard Form. In addition, projects in excess of £100 000 estimated value might require a performance bond for 10 per cent of the contract value.

*Liquidated and ascertained damages* are often assessed at £2.50 per £1000 estimated contract value for projects exceeding £10 000 estimated value. *Retention percentages* can vary considerably. The percentages prescribed in the JCT Standard Form (1980) are 5 per cent reducing to 3 per cent for contracts estimated at £0.5m or more.

*Contingency sums* can show considerable variations, but a common approach is to insert 2.5 per cent for all contracts other than for alterations, and 5 per cent for alteration contracts. The minimum amount of third party insurance liability is likely to be in the order of £1m, irrespective of the value of the works to be executed.

The limiting value for contracts based on drawings and specification adopted on some public sector contracts is £40 000, but even this figure is often subject to variation for individual contracts.

Certain procedures may be advocated to help in avoiding claims, of which the following are typical examples:

(1) The extent of architect's instructions issued during the contract should be strictly limited.

(2) The flow of information to the contractor must be adequate to avoid delays on site.
(3) Nominated suppliers and subcontractors must be nominated sufficiently early to ensure that the progress of the works is not impeded.
(4) Direct contracts let by the employer shall not lie on the critical path.

## Refurbishment and Alteration Work

The RICS (1982) has described how the prime aim of documentation, prepared by the quantity surveyor on refurbishment contracts, is to provide the contractor with an unambiguous definition of the work to be carried out in a manner that is readily priceable and that can also be issued for the administration of the work. The type of project will influence significantly the documentation to be used.

Whatever type of documentation is prepared, the following conditions should apply:

(1) Unless it is in the form of a specification, it should contain such measurements and/or quantities as are necessary for pricing purposes.
(2) The quantity surveyor should be responsible for the accuracy of the quantities supplied.
(3) It should not only state what work is required but also where this work is required to be done.
(4) It should be specific as to the work to be carried out and avoid vague clauses, such as 'overhaul and repair roofs'.
(5) A bill of quantities, where provided, should be in elemental form, but in the case of specialist trades that are traditionally sublet, it should enable the main contractor to obtain prices from subcontractors readily and easily.

On refurbishment projects, drawings should always be provided of both the existing and proposed buildings, with the latter annotated with room, door and window numbers. The other documentation can be:

(1) A quantified specification which is generally ideal when there are extensive minor alterations.
(2) A priceable schedule which can be particularly useful where there is a large degree of repetition of similar items, such as occurs in multiple housing projects.
(3) Elemental bills of quantities which would be the most satisfactory documentation for a 'one-off' project involving fairly extensive alterations.
(4) A combination of the methods in (1) to (3).

## STANDARD FORMS OF BUILDING AND CIVIL ENGINEERING CONTRACTS

Three standard forms of contract cover the bulk of construction work carried out in the United Kingdom. These are the JCT Standard Form of Building Contract (1980) intended primarily for building works; the ICE Conditions of Contract (1991) used primarily for civil engineering works; and PSA/1 (1994) produced by central government for use in both public and private sectors and for both building and civil engineering work. In addition, the ICE has produced conditions of contract for design and construct and for minor works. There has been a limited amount of cross-fertilisation with the successive editions, but attempts to weld them into a common form, as recommended by Banwell (1964), have not met with any success. The Joint Contracts Tribunal (JCT) has produced Local Authority and Private Editions of the Standard Form, each for use with and without quantities or approximate quantities. In addition standard forms have been produced for use with prime cost, contractor's design and management, and measured term contracts and there is a JCT Agreement for Minor Building Works, and an Intermediate Form of Contract (IFC), being a form intermediate between the standard form and minor works. In addition there is a DoE (1991a) form of contract for building, civil engineering, mechanical and electrical small works.

In 1991, the ICE adopted a completely new approach to engineering contracts with three key objectives, namely flexibility, clarity and simplicity, and the promotion of good management. The New Engineering Contract (NEC) can be used on any engineering, building or construction project, in any country and on any scale. It comprises a core contract and six main options encompassing a conventional contract with activity schedule; conventional contract with bill of quantities; target contract with activity schedule; target contract with bill of quantities; cost reimbursable contract; and management contract; with a variety of secondary options for use where necessary, to allow the employer to choose the version most appropriate to his needs. Latham (1994) in his report 'Constructing the Team' strongly advocated the wider use of NEC.

It is proposed to give a brief outline of the main provisions of the JCT Standard Form of Building Contract, with appropriate clause references in brackets, but for more detailed commentaries readers are referred to the books by Fellows (1995), Turner (1994), Chappell (1993), Powell-Smith and Sims (1990) and Price (1994). A useful explanatory publication on the ICE Conditions of Contract has been produced by Eggleston (1993). For the sake of uniformity throughout the book, capital letters have not been used for such terms as employer, contractor, architect and works, as is the practice in the Standard Form.

## Contractor's Obligations

Under the JCT Standard Form the contractor is required to carry out and complete the works in accordance with the contract documents and to use materials and workmanship of the specified quality and standard and they shall be to the reasonable satisfaction of the architect (2.1). It should be noted that nothing contained in the contract bills shall override or modify anything contained in the Articles of Agreement, Conditions and Appendix (2.2). The contractor shall provide the architect with two copies of his master programme for the execution of the works (a network will show the effects of delays far more clearly than a bar chart) (5.3). The contractor has an overriding obligation to comply with all statutory requirements, including the regulations of statutory undertakers, and pay and indemnify the employer in respect of all statutory fees and charges legally demandable (6.1). He is responsible for setting out the works based on information supplied by the architect (7).

The contractor is to keep a competent person-in-charge on the site, generally interpreted as throughout normal working hours, as agent on his behalf (10). He must have the written consent of the architect to sublet any part of the works, and such consent must not be unreasonably withheld (19.2).

The contractor is liable for and must indemnify the employer against any expense, liability, loss, claim or proceedings at statute or common law for personal injury or death due to the carrying out of the works (20.1). A similar obligation is also imposed in respect of real or personal property, except damage at the sole risk of the employer, and damage due to any negligence, omission or default by the contractor or persons for whom he is responsible (20.2). He must take out and maintain insurances and require subcontractors to do likewise in respect of the liabilities placed upon them for personal injuries, or damage to real or personal property arising from the works (21.1). If there is provision in the Appendix requiring insurance against damage to other properties caused by the execution of the works, the contractor is to maintain insurances in the joint names of the employer and contractor (21.2).

Possession of the site is to be given to the contractor on the date of possession, and he must regularly and diligently proceed with the execution of the works and complete (practical completion) on or before the completion date (23.1). If the contractor fails to complete the works by the completion date, subject to extension of time awards, the architect must so certify and the contractor will be liable for the payment of liquidated and ascertained damages for the period between the completion date and the date of practical completion (24). Once it is reasonably apparent that the progress of the works is being or is likely to be delayed, the contractor must inform the architect in writing stating the circumstances causing the delay and noting the relevant events he considers applicable, in furtherance of a claim for extension of time (25.2). Relevant events are listed in the Standard Form and range from *force majeure* to compliance with architect's instructions (AIs), delay in the supply of necessary information and delay on the part of nominated subcontractors or nominated suppliers, which the contractor has taken all practicable steps to avoid or reduce (25.4). Many of the relevant events may also give rise to a claim by the contractor for loss and expense (26.1).

The contractor may by registered post or recorded delivery serve notice of determination of his employment under the contract to either the employer or the architect, for one of the reasons listed in the Conditions, such as the employer failing to honour a certificate after due notice (28.1). All fossils and antiquities discovered on the site during the execution of the works are the property of the employer, and the contractor must endeavour not to disturb the object and to preserve it and to notify the architect or clerk of works (34.1).

### Architect's Duties

The architect can issue written instructions to the contractor with which he must comply, except where a variation is involved to which the contractor has objected to the architect in writing (4.1). The architect or his representative are entitled at all reasonable times to have access to the site and all workshops, including those of subcontractors, where items for the project are being made (11). The employer is entitled to appoint a clerk of works who acts solely as an inspector on behalf of the employer but under the directions of the architect (12).

The architect can issue variations altering or modifying the design, quality or quantity of the works as described in the contract drawings and bills, or altering, extending or omitting obligations or restrictions imposed by the employer in the contract bills regarding certain matters listed in the Conditions (13.1)

As soon as the architect considers that practical completion of the works has been achieved, he must issue a certificate to that effect (17.1). Defects, shrinkages or other faults which appear within the defects liability period and are due to materials and workmanship not in accordance with the contract must be:

(1) specified on a schedule of defects by the architect, to be delivered as an architect's instruction to the contractor not later than 14 days after the expiration of the defects liability period; and

(2) made good by the contractor at his own cost (unless subject to an architect's instruction to the contrary) and within a reasonable time (17.2).

Despite the latter provisions, such defects, including damage caused by frost prior to practical completion, may be the subject of an architect's instruction for their making good – the contractor to comply within a reasonable time and normally at his own cost (17.3). This clause enables individual defects, usually of a significant nature, to be made good before the issue of the full defects list.

When the defects specified in the architect's instructions and/or the schedule of defects have, in the architect's opinion, been made good, he must issue a certificate of completion of making good defects (17.4). This certificate is a prerequisite for the issue of the final certificate and release of the balance of retention (Fellows, 1995).

Prior to practical completion of the works, the employer may take possession of any part or parts of the project provided that he has the consent of the contractor to do so. This will entail the architect certifying its estimated value, and practical completion is deemed to have occurred with its implications for the certificate of completion of making good defects, fire insurance and liability for payment of liquidated damages (18.1). The RICS (1995) has analysed the problems associated with practical completion.

The architect shall issue interim certificates stating the amount due to the contractor from the employer, usually at monthly intervals (30.1). A statement of the contractor's and nominated subcontractor's retention must be prepared by the architect, or the quantity surveyor under his instruction, on the date of each interim certificate. The statement must be issued to the employer, the contractor and all relevant nominated subcontractors (30.5). All items to be included in the adjustment of the contract sum are listed in clause 30.6.

The architect shall issue a final certificate and inform each nominated subcontractor of its date of issue. The date of issue will be as soon as possible but before the expiry of 2 months from the latest of:

(1)  end of the defects liability period, or
(2)  completion of making good of defects, or
(3)  date of architect sending to the contractor, copies of the ascertainment and statement under clause 30.6.1.2.

The final certificate must state:

(1)  the sum of the amounts already stated as due in interim certificates, and
(2)  the contract sum adjusted as necessary in accordance with clause 30.6.2.

and the difference between the two sums is expressed as a balance due in either direction and constituting a debt from the 28th day after the date of the final certificate (30.8).

Subject to the provisions regarding legal action or arbitration being instigated prior to the issue of the final certificate or within 28 days of its issue (and excepting instances involving fraud), the final certificate shall, in any proceedings due to the contract, be conclusive evidence that:

(1) work, goods and materials, stipulated to be to the architect's satisfaction, so comply, and
(2) all computations in determining the amount of the final account are correct, except for any mistakes whether arithmetical or due to incorrect inclusion or exclusion of items (30.9).

### Determination by the Employer

The employer's right, expressly without prejudice to his other possible rights and remedies, become operative in the event of default prior to practical completion by the contractor in any or all of:

(1) wholly suspending execution of the works without reasonable cause;
(2) failure to proceed regularly and diligently with the works;
(3) refusal or persistent failure to remove defective items which materially affect the works and regarding which the architect has given him written notice to move;
(4) failure to comply with clause 19.1.1 or 19.2.2 (assignment and subcontracts).

The architect may give written notice specifying the default to the contractor by registered post or recorded delivery. If the contractor continues the default for 14 days from receipt of the notice, the employer may within 10 days, by registered post or recorded delivery, serve notice of determination of the contractor's employment upon the contractor (27.2).

If the contractor goes into liquidation except for the purposes of reconstruction of the firm or an amalgamation, the employment of the contractor is automatically determined (27.3). Where the employment of the contractor is determined, the employer may employ another contractor to complete the works who may use all the items (temporary buildings, plant and materials) on or adjacent to the works and intended for use on them (27.6).

## NOMINATED SUBCONTRACTORS AND SUPPLIERS

### Contract Provisions

Virtually all the provisions on the nomination of subcontractors are set out in clause 35 of the JCT Standard Form and in the various subcontract documents listed in table 4.3, which are quite complex. Attention is also drawn to certain provisions elsewhere, mainly in clause 30, which deal with the inclusion of amounts in respect of nominated subcontract work in interim certificates and the penultimate certificate for final payment in respect of nominated subcontractors. A brief outline only is given. For fuller information on the JCT contract provisions for nominated subcontractors and suppliers, readers are referred to Fellows (1995) and Price (1994).

Nominated subcontractors (NS/Cs) are defined as persons whose final selection and approval for the supply and fixing of any materials or goods or the execution of any work has been reserved to the architect. Such reservation to the architect can be effected either by the use of a prime cost sum or by the naming of a sole subcontractor in the contract documents so that the contractor is, therefore, bound to enter into a subcontract with that person. There is, however, provision for the contractor to choose a domestic subcontractor from a list of three or more subcontractors.

### Procedure for Nomination of a Subcontractor

The nomination of a subcontractor is effected in accordance with clauses 35.4 to 35.9 inclusive and the documents listed in table 4.3 are used for this purpose.

Table 4.3    Nominated subcontract documents

| Name of document | Identification term |
|---|---|
| The Standard Form of Nominated Subcontract Tender 1991 edition comprising: | NSC/T |
| Part 1: The Architect's Invitation to Tender to a Subcontractor | – Part 1 |
| Part 2: Tender by a Subcontractor | – Part 2 |
| Part 3: Particular Conditions (to be agreed by a Contractor and Subcontractor nominated under 35.6) | – Part 3 |
| The Standard Form of Articles of Nominated Subcontract Agreement between a Contractor and a Nominated Subcontractor, 1991 edition | Agreement NSC/A |
| The Standard Conditions of Nominated Subcontract 1991 edition, incorporated by reference into Agreement NSC/A | Conditions NSC/C |
| The Standard Form of Employer/Nominated Subcontractor Agreement, 1991 edition | Agreement NSC/W |
| The Standard Form of Nominated Instruction for a Subcontractor | Nomination NSC/N |

The contractor has a right to reasonable objection to a proposed nominated subcontractor (NS/C), to be made in writing within 7 working days of receipt of NSC/N (35.6). Where the contractor makes a reasonable objection to a nomination, the architect may:

(a) issue further instructions to remove the objection (35.7) or
(b) cancel the instruction of nomination and issue an instruction which either omits the work or nominates another subcontractor (35.6).

The architect must use NSC/N to issue architect's instructions (AIs) of nomination to be sent to the contractor, accompanied by NSC/T Part 1 (completed by the architect), NSC/T Part 2 (completed and signed by the NS/C and by or on behalf of the employer), and a copy of the numbered tender documents. NSC/W is then completed and entered into by the employer and NS/C (35.6).

The architect must send a copy of the instruction to the NS/C together with a copy of the appendix to the main contract. On receipt of the instruction, the contractor must:

(1) agree with the NS/C and complete NSC/T Part 3 – on completion both the contractor and NS/C must sign the document
(2) execute NSC/A with the NS/C and then send a copy of the completed documents (NSC/T Part 3 and NSC/A) to the architect.

### Payment of a Nominated Subcontractor

Upon the issue of each interim certificate, the architect must:

(1) direct the contractor as to the amounts of any (interim or final) payments to NS/Cs included, as computed by the architect in accordance with the relevant provisions of NSC/C. This will usually be in the form of a schedule attached to the contractor's copy of the interim certificate and
(2) inform each NS/C of any payments so directed.

The contractor must make the payments to the NS/Cs in accordance with the relevant NSC/C provisions, namely within 17 days of the date of issue of the interim certificate, less $2\frac{1}{2}$ per cent cash discount.

Prior to the issue of the second and subsequent interim certificates, and the final certificate, the contractor must provide the architect with reasonable proof that payment to any NS/Cs has been discharged, usually by producing a receipt detailing the amounts involved (35.13).

If the contractor fails to produce reasonable proof of discharge of payment to a NS/C, the architect must issue a certificate to that effect, stating the amount involved, and issue a copy of that certificate to the appropriate NS/C. The employer must pay the amounts due direct to the NS/C, provided the employer can set them off against amounts due from him to the contractor (35.13).

Further detailed provisions as to direct payment of NS/Cs are contained in clause 35.13. Other JCT provisions related to NS/Cs include failure to complete NS/C works (35.13); practical completion of NS/C works (35.16); and early final payment of NS/Cs (35.17–19).

### Position of Employer and Contractor in Relation to Nominated Subcontractor

Nothing contained in the Conditions shall render the employer in any way liable to any NS/C, except as provided in NSC/W (35.20). In like manner, a contractor is not responsible to the employer in respect of any NS/C works, although he still has general obligations under the Contract relating to the quality and standard of workmanship, materials and goods (2.1).

### Other Matters Related to Nominated Subcontract Works

Clause 35.24 details the circumstances where renomination is necessary and the procedure to be adopted by the contractor and architect. The contractor must not determine any nominated subcontract without an appropriate AI (35.25).

Fellows (1995) has emphasised the following important criteria relating to nominated subcontract works:

- If the contractor determines the employment of a NS/C without the relevant AI, he is in breach of contract.
- If the architect does not renominate as required, the employer and contractor should make a separate agreement regarding what is to be done and the contractor executes the work. Payment is at an agreed price or *quantum meruit* (as much as he has earned).
- If a NS/C repudiates the subcontract, the contractor may seek to recover from the NS/C only damages he has incurred directly resulting from the repudiation.
- Any work to be executed by a NS/C is still the responsibility of the contractor and he must ensure that it is carried out satisfactorily by the nominee. If the contractor fails to do this he is in breach of his contract with the employer.

Additionally:

- A NS/C is still a subcontractor of the contractor.
- The NS/C is merely nominated by the architect.
- The contractor is responsible for executing all work under the contract.

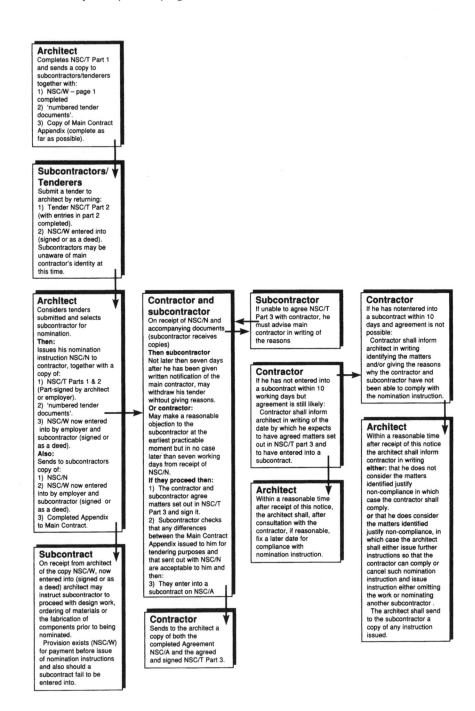

**Figure 4.1**  *Subcontractor nomination procedures (Tate, 1992)*

- The architect is responsible for all design work as far as the contractor is concerned.
- The contractor is entitled to only $2\frac{1}{2}$ per cent cash discount for prompt payment to NS/Cs ($\frac{1}{39}$th to add to nett accounts).

Figure 4.1 summarises the subcontractor nomination procedures.

*Employer's option list of subcontractors*  The Standard Form recognises the practice of some employers who set out in detail in the main contract documents certain work for pricing in full by the contractor, but which the employer requires executing by a subcontractor selected by the contractor from a list supplied by the employer (clause 19). Provided the list contains not less than three subcontractors who are able to carry out the prescribed priced work and the contractor has free selection, then the subcontractor so engaged is termed a domestic subcontractor. Rules are framed to cover the situation where prior to a binding subcontract becoming operative, the number on the list falls below three.

*Nominated suppliers*  The requirements are set out in clause 36, whereby a supplier is nominated or deemed to be nominated if:

(i) a prime cost sum is included in the contract bills or in an architect's instruction relating to expenditure of a provisional sum and the supplier is named in the bills or instructions or subsequently named by the architect; or
(ii) in an instruction relating to expenditure of a provisional sum or a variation, the architect specifies materials or goods which can be purchased only from one supplier.

The parties are permitted to make other arrangements where a nominated supplier (NSup) is not prepared to include in his sale contract the terms set out in clause 36.4, which include quality, standard, delivery and ownership. The contractor is entitled to only 5 per cent cash discount for prompt payment ($\frac{1}{19}$th to add to nett accounts) and all other discounts must be passed to the employer.

## PERFORMANCE BONDS

### Introduction

Great concern has been expressed by contractors and members of the design team about the adverse effect of performance bonds and particularly the use of on-demand bonds. Hence in 1994 the Secretary of State for the Environment set up a working party to prepare guidance on the use of performance bonds in government construction contracts. As a first step the working group asked the Construction Industry Council (CIC) for its views on the text of Central Unit on Procurement (CUP) guidance notes on the use of bonds and guarantees in government contracts (1994).

In December 1994 the Association of British Insurers launched a model performance bond which was intended to become a standard guarantee for all construction industry clients and contractors, to protect employers from any losses resulting from a contractor going out of business. There were also worries that the Appeal Court ruling in *Trafalgar House Construction (Regions) Ltd v General Surety and Guarantee Company Ltd, 1994* could result in simple performance bonds being treated as onerous on-demand bonds.

The Latham Report (1994) recommended that the DoE guidance on bonds should be formulated within the following principles:

1. They should be drafted in comprehensible and modern language.
2. They should not be on-demand and unconditional, but should only operate in clearly defined circumstances.
3. If the circumstances/conditions provided for in the bond are fulfilled, the beneficiary should be able to obtain prompt payment without recourse to litigation.
4. They should have a clear end date.

The Secretary of State made it clear that in the construction sector, unconditional on-demand bonds are not acceptable within government work, and the DoE agreed to build Latham's four principles into its guidance. The NJCC in guidance note 2 (1986) recommended that performance bonds in the UK should not be of the 'on-demand' or 'unconditional' type, and hence this has long been established good practice in the construction industry.

### Nature and Use of Bonds

The CUP guidance notes (1994) defined a bond as a legally enforceable financial guarantee given by a third party (the guarantor) to a purchaser (the client) to guarantee the obligations of a supplier of goods, works or services (the contractor) under a contract. The guarantor agrees to pay the client a sum of money if the contractor defaults on its obligations. The purpose of requiring a bond is to help the client to meet the extra expenses to remedy the default and/or complete the contract.

Bonds are generally provided by the financial market, either by a bank or a surety company. The contractor and the guarantor will seek to establish the terms and conditions under which the bond can be called. Clients for their part will wish to know that the guarantor issuing the bond is a sound, reliable and responsible corporate body and to be satisfied that if the need arises to call the bond for payment, the guarantor will comply promptly (Central Unit on Procurement (CUP), 1994).

A guiding principle of procurement best practice should be not to place a contract with a contractor if there is reasonable doubt about his ability to meet the terms and conditions of the contract satisfactorily. Such doubts can arise with regard to the adequacy of the contractor's management and technical resources to deliver on time and to the required standard, or where available information indicates that the contractor may have inadequate financial resources. Hence bonds cannot be a substitute for considered judgements about the capabilities and financial resources of available contractors and the degree of risk involved, and they have the serious disadvantage that they eat into the contractor's credit limit and can thus act as a break on development of even the most efficient firms.

### Different Types of Performance Bond

A performance bond, where required, is usually provided on the award of the contract for an agreed percentage of the total contract value (normally 10 per cent). These bonds generally have an expiry date linked to some event in the contract, so that time slippage is allowed for automatically. If the value or duration of the contract increases, then the bond may require amendment.

Unconditional on-demand bonds are grossly unfair and should not be used in construction contracts. In like manner, conditional on-demand performance bonds place a substantial burden on a contractor and, if used, should be confined to high risk and/or high value projects where the costs and/or other consequences of default by the contractor are high and even then only after careful consideration, including appropriate professional and legal advice (CUP, 1994).

Conditional on-default performance bonds are mainly provided by surety companies and are usually called only when there is a serious breach by the contractor of the agreed terms and conditions of the contract, including bankruptcy of the contractor. CUP (1994) believed that properly expressed this type of performance bond provides a third party guarantee that the contractor will not default from a contract into which it has freely entered. They can be used where there are identifiable risks of default by the contractor, subject to a value for money assessment.

## COLLATERAL WARRANTIES

### Introduction

It seems reasonable that an owner or user of a building will wish to claim against the builder or designer if a major problem arises. Claims in tort are unlikely to be successful and hence the owner or user may require an enforceable contractual commitment. Probably the developer is in the strongest position to ask for warranties before he appoints the contractor and the members of the professional team, such as the architect, quantity surveyor and structural engineer, and this list could be extended to encompass nominated subcontractors who make an important specialist input (Latchmore, 1992).

### NJCC Guidance on Collateral Warranties

In common with many in the construction industry, the NJCC has been concerned at the large growth in demands for collateral warranties in recent years and the 'dangers posed by *ad hoc* bespoke forms of warranty, which are demanded by a variety of beneficiaries. Hence the NJCC guidance notes (1992) give valuable help to those in the construction industry who are asked to give collateral warranties and duty of care undertakings, be they professional consultants, contractors or subcontractors.

The NJCC guidance notes describe in lay terms the nature of collateral warranties and why they are requested, covering recent legal developments, the impact of full repairing leases and contracts for sale as seen, and the question of design by subcontractors. They also give detailed advice about limitations on the use of warranties, who benefits from them, dangers of common form warranties, appropriate and onerous terms of warranties, assignment, insurance, joint and several liability, how long liability lasts, and administration.

The NJCC emphasises that collateral warranties increase the cost of building, because of the substantial costs incurred by the industry in negotiating one off forms of warranty, the cost of maintaining insurance cover and the fact that those asked to give warranties will often price the additional liabilities they are asked to accept. The NJCC believes that the ultimate answer to the problem may lie in the development of a form of latent defects insurance which is now becoming available in the UK.

### CoWa/F Warranty Agreement

To overcome the problems resulting from the use of disparate warranty agreements, in 1990 the BPF, ACE, RIBA and RICS published a warranty agreement, *CoWa/F*, for use where a warranty is required by a company providing finance for a development. The warranty is given on behalf of the architect, engineer or surveyor in favour of the financier and provides for the following:

• The consultant warrants that it has exercised and will continue to exercise reasonable skill and care in the performance of its duties to the client. It is a condition of the warranty that under certain limited circumstances the financier may give instructions to the consultant, in which case the obligations of the consultant to the financier are no greater than those to the client.

- The consultant also warrants not to specify certain materials, such as high alumina cement, wood wool slabs in permanent formwork, calcium chloride in reinforced concrete, asbestos products, and naturally occurring aggregates which do not comply with BS 882: 1983 or BS 8110: 1985.
- The warranty makes it clear that the financier with limited exceptions has no authority to issue instructions to the consultant.
- The warranty provides that where the financial agreement between the financier and the consultant is terminated, the consultant, subject to receiving written notice from the financier, will accept instructions from the financier or his appointee in accordance with the original terms of appointment of the client.
- There is a rather complex arrangement whereby the consultant may give the financier not less than 21 days notice in writing of an intention to exercise any right to terminate the consultancy agreement with the client or to discontinue the performance of any duties under the agreement.
- The warranty provides for the copyright in all drawings, reports, specifications, bills of quantities, calculations and other similar documents to remain vested in the consultant. However the financier or his appointee are given a licence to copy and use the documents and to reproduce the designs and use them for any extension of the development.
- The consultant undertakes to maintain professional indemnity insurance for a specified period of time and minimum amount of cover.

Knowles (1990) advises consultants to avoid entering into warranties or duty of care agreements wherever possible. If there is no choice then it seems reasonable for the consultant to insist upon the use of *CoWa/F* and to refuse to sign warranties which contain onerous conditions.

Readers requiring more detailed information about the legal principles, provisions, uses and practical implications of collateral warranties, with a commentary on the *CoWa/F* agreement and references to relevant statutes and cases, are referred to Fearon (1990).

## AMERICAN BIDDING AND CONTRACT DOCUMENTATION

American construction tendering and contractual practices vary considerably from those operating in the United Kingdom and a brief outline of the normal American approach could be helpful to readers. In the United States an employer requiring construction work performed will publicly or privately seek contractors to perform this work by inviting them to bid on the project or to submit a proposal for doing the work. Upon selection of a contractor by the employer, a contract is signed by the two parties and work proceeds in accordance with its terms. The bidding documents will be accompanied by general conditions, supplementary conditions, drawings and specification, but there is unlikely to be any bill of quantities.

The invitation to bid gives a general description of the project and its location, states where contract documents may be obtained, establishes the deadline for receipt of bids and provides other general instructions. Instructions to bidders may be included in the invitation to bid or be presented in a separate document. Typical items that can be included are:

(1) *Bid bond* – this requires a surety which guarantees that the contractor will accept the contract if it is offered. A typical bid bond would be 5 to 20 per cent of the bid price. Should the contractor fail to accept the contract after bidding, the employer will be reimbursed for the difference between the contractor's bid and the next higher responsive bid.

(2) *Performance bond* – this is a guarantee that the contractor will actually complete the contract. It is normally for 100 per cent of the contract price. Should the contractor default, the surety must take action to complete the contract.

(3) *Payment bond* – this guarantees payment of labour and suppliers of the contractor. It will be in the range of 40 to 50 per cent of the contract price. Failure of the contractor to meet these obligations requires surety assumption of the debts.

(4) *Other items* – these include a variety of matters ranging from details of contract forms to be used, method of obtaining drawings and specifications, procedure for administering modifications prior to bid opening date, prebidding conference and procedure for selecting contractor, to any special requirements (Neil, 1982).

The employer may also request a list of equipment available to the contractor which will be used on the project. This will assist in assessing the contractor's capability to perform the work.

Trade or professional associations such as the Associated General Contractors of America (AGC) and the American Institute of Architects (AIA) have formulated standard sets of general conditions that are available for use by employers. Construction contract documents comprise two packages which describe the structures to be built.

(1) Drawings which provide dimensional information on the project.

(2) Specifications which establish quality, performance and methods to be employed. These can take three forms – proprietary (such as specific makes and models), descriptive (such as descriptions of materials to be used), and performance (in terms of what an item must do) (Neil, 1982).

# 5 Contract Administration

This chapter deals with some important aspects of contract administration, including the issue of architect's instructions, supervision, coordination, programming and progressing, valuations and payments.

## RIGHTS AND DUTIES OF PARTIES

The extent of the risks to be borne by all parties involved in a construction project (employer, contractor, subcontractors, architect, quantity surveyor and engineers) is of great importance. There are two main elements – the probability of the occurrence of an event and the consequences that are likely to flow from it. Desirably, the party who is best able to control an event that has an element of risk attached to it should bear the risk. There are, however, many matters that are outside the direct control of any particular party, such as delay in the completion of the works owing to exceptionally adverse weather conditions, and it is essential that the contract makes it clear as to who will be responsible in the event of this happening. In this particular case JCT80, clause 25.4.2, includes this as a 'relevant event' to be considered as one of the reasons for granting extension of time. Some of the principal rights and duties of the various parties will now be considered.

### Employer

The employer's first duty at the commencement of a construction project is to give the contractor possession of the site on the date stipulated in the contract. He cooperates with the design team to ensure the smooth and effective operation of the contract, and will appoint an architect and other consultants, and replace them with successors in the event of death, resignation or dismissal. The employer should at all times avoid hindering the contractor's progress, particularly by ordering significant extras or alterations to the works, which disrupt the general sequence and progress and create changed economic conditions.

In general, the employer, apart from signing the contract and paying for the work as executed, has a very limited part to play and his representation at site meetings renders the management of the contract more difficult. Although the employer is named in the contract as insured in the event of damage to the works during construction, he does not take an active part in any insurance settlement and his entitlement will be limited to the percentage included for professional fees. He will be pressing for a quick settlement and an early start to the reinstatement work.

The employer and the contractor agree the rate of liquidated damages for delay in completion on signing the contract, and this will avoid any possible disputes later as to the actual loss incurred. Where the architect certifies that the contractor has failed to complete the works by the completion date or extended completion date prevailing at the time, the employer can commence deducting damages. If a further extension of time is granted, the amount of damages will be corrected. The ability to extend the completion date exists primarily to protect the employer's rights. Where an act or default of the employer or architect prevents the contractor from achieving the completion date, in the absence of the provision for extension of time the contractor's obligation would be to complete the contract is a reasonable time. The employer's rights to claim damages for major defects occurring after final completion are dealt with in the next section.

### Contractor

The contractor can expect to be given the site on the date of possession and to retain it until completion. Delay in handing over the site would nullify the employer's right to liquidated damages and the contractor's obligation in respect of completion would change to completion within a reasonable time.

The contractor assumes on occasions that as he is subject to visits by the employer, architect, clerk of works, building control officer and others, he is freed of any obligation to supervise the work and that he is relieved of any liability for faulty work that has been approved. However, this is not the case and the contractor has a supervisory duty; under clause 2.1 of the JCT Conditions of Contract, he is required to carry out and complete the works with materials and workmanship of the specified quality and standards and where applicable to the reasonable satisfaction of the architect. The contractor thus has a duty to adequately supervise the work to ensure that all the work undertaken, including that which is sublet, is done satisfactorily. In the case of *Anns v. London Borough of Merton* (1977), a building control officer of the local authority approved, for the purpose of building regulations, foundations of a block of flats which were defective. The House of Lords was unanimous in ruling that approval by the building control officer did not render the contractor immune from liability. The contractor had a duty to comply with the building regulations and failure to do this made him liable.

The Defective Premises Act 1972 imposes greater obligations on contractors and others in the building team than are contained in the Standard Form (JCT80). The Act prescribes that 'any person taking on work for or in connection with the provision of a dwelling ... owes a duty to see that the work he takes on is done in a workmanlike manner, with proper materials ... and so as to be fit for the purpose required ...' Houses and flats built under the National House-Building Council scheme are exempt from the provisions of this statute as the contract used provides express warranties in the same terms as the ones implied by the statute.

It is in the best interest of all the parties that the contractor should be as free as possible to execute building or civil engineering works in the way that he wishes. His preference for a particular design and method of carrying out temporary works, and his proposals for a particular sequence of work may differ from those of the architect or engineer, but it is advisable for him to be permitted to proceed in the way that he considers best (ICE, 1986). Furthermore, the architect derives no power from the JCT Form to dictate either the method or sequence of working.

The contractor is obliged to complete the works even if they have proved to be far more costly to him than he envisaged at the tender stage. The building Standard Form provides for handing over the building to the employer on 'practical completion', while the civil engineering form uses the term 'substantial completion' (ICE, 1991). Practical completion can be interpreted as a state of completion that enables the employer to occupy the building and carry on his activities without inconvenience.

If a contractor does his work badly, so that it is likely to give rise to trouble thereafter, and then covers up the bad work so that it is not discovered for some years, then he cannot rely on the normal 6 year limit in the Limitation Act 1939 as a bar to the claim. The defect has been concealed by 'fraud', as the term is used in this context (*Archer v. Moss, 1971*).

When a contractor is excavating in land, particularly in urban areas, he will be aware of the possibility of his excavation damaging pipes, cables and other apparatus belonging to one of the statutory authorities. Before excavating he should make enquiries of all the authorities concerned as to the position of their services and, if there is a possibility of damaging a service, work with great care and closely with the authority. Where services have to be moved, it is customary for the appropriate authorities to insist that they move the services, and they do have substantial legal powers regarding the laying, maintenance and protection of their services.

The following two legal cases illustrate some of the problems that can arise in practice. In the High Court case of *Post Office v. Mears Construction Ltd 1979*, the Council sent the Post Office a drawing of a new bridge and highway improvements for the services to be marked on it. The Post Office inserted a disclaimer on the drawing 'the information given is compiled from records and is believed to be correct. There may, however, be departures from the course(s) and depth(s) shown; there may also be items of Post Office plant of which no record is held.' A cable was shown by the plan to be under the footpath but it was subsequently struck by an excavator in the grass verge. The Court decided that the Post Office's disclaimer as to the accuracy of the plan absolved them from any duty of care that might have existed. In another case before the Court of Appeal

(*Post Office v. Hampshire County Council, 1979*) council workmen damaged a telephone cable in a grass verge after being informed verbally by Post Office employees that it was under the carriageway. The Post Office were not reimbursed the cost of repairs despite it being awarded, because of the negligence of their employees.

**Subcontractors**

It is now quite common for subcontractors to carry out a substantial part of the work on many building projects. The main contractor faces considerable difficulties in correlating and coordinating the activities of a number of subcontractors, each intent on undertaking as profitably as possible his own section of the works. The main contractor's problems are often compounded by a lack of detailed knowledge of subcontract operations and also, in many cases, by an insufficiently systematic approach to subcontract management. Furthermore, the main contractor can incur heavy losses through delay and disruption if subcontract work is permitted to proceed without adequate control.

Viewing it from the subcontractor's position, the JCT main contract and subcontract have been prepared in such a way as to reduce the subcontractor's risk of not receiving proper payment. The subcontractor can look to the employer for payment if the contractor for any reason, except his bankruptcy or liquidation, fails to discharge properly his duty to pay. See chapter 4 and Price (1994) for more detailed information on direct payment by the employer to the nominated subcontractor.

Each calendar month the nominated subcontractor will receive payment of almost the total value of the work properly executed during the previous month, within 17 days of the date of issue of each certificate. If the valuation is accurate, the retention low, the progress equal to or better than planned, and the rates profitable, the subcontractor may receive even more cash than he has actually spent. The subcontractor should therefore take full advantage of these favourable recycling terms by ensuring:

(1) prompt measurement and agreement of quantities;
(2) quick settlement of new or revised rates and prices;
(3) advance face-to-face discussions on contentious matters;
(4) regular production of supporting data, records, returns and invoices; and
(5) accuracy of the valuation submitted to be included in the amount stated as due in an interim certificate.

If the subcontractor still feels aggrieved by any amount substantially undercertified or not directed by the architect, then the subcontractor can have recourse to arbitration, and in a cash crisis the right to suspend work exists in NSC/C, 35 days after the issue of the interim certificate in dispute. Legal advice should, however, be taken before following either of these courses of action. In the reverse direction, the employer has the right to sue a subcontractor direct for defective work, following the decision of the House of Lords in *Junior Books v. Veitchi Co. Ltd* (1982).

Considerable subcontract labour is now employed on projects which can cause the main contractor problems. The main interest of labour-only subcontractors is often to complete as much priced work as possible in the shortest time, thus providing the highest income. This can lead to the larger and more straightforward elements of their work being given greatest priority, while leaving the smaller or less financially rewarding tasks in abeyance. For example, walls in rooms are plastered, yet reveals are left untouched. Long, straight brick walls are built, but special sill bricks are left unlaid. This type of approach often prevents or hinders subsequent activities and jeopardises contract programmes.

Many labour-only subcontractors fail to provide permanent on-site supervision to ensure the correct implementation of their work. This often results from the submission of low, keenly competitive prices which do not provide for this measure of control. The lack of supervision can lead to construction errors, poor quality control and poor site safety. The main contractor needs to prescribe sound contractual arrangements and effective site control to safeguard his own interests.

Detailed information on all aspects of the operation of subcontracts under the JCT Standard Forms of Contract is provided by Price (1994).

## Architect

The architect must ensure that his first brief is sufficiently comprehensive to enable him to prepare sketches and an adequate specification, which will enable his interpretation not only to meet the employer's requirements but also the environmental demands of the proposed site, as well as taking into account the employer's budgetary and time constraints.

The design of any project is of the utmost importance both on practical and legal grounds, as poor design can result in technical problems on site and possibly delay in completion. The legal liability of the architect for inadequate or faulty design is found in common law and normally in express terms and sometimes in implied terms in contracts. For many years it was held that the obligation of a professional man to the employer or client lay solely in contract, and this was confirmed in court rulings in 1957, 1960 and 1971. However, in *Esso v. Marden* (1976), the Court of Appeal decided that every time a designer enters into a contract he also assumes a liability in tort. Although the designer's first obligation is likely to be towards the employer to provide a design that is suitable for its purpose, he may also be liable to others for damages arising out of the malfunction or non-feasibility of his design.

Furthermore, design responsibility in construction projects is widening, with an increasing use being made of specialist contractors carrying out the design of their own work. Indeed the majority of contracts involve some element of design by the contractor and/or by the nominated subcontractors or suppliers. For example, it is quite common for the supplier of steel reinforcement in a reinforced concrete framed building to design all the reinforced concrete work, and this example can be extended to include mechanical and electrical installations designed by the respective specialist subcontractors, or their preparation of working drawings for approval by a consultant services engineer. The employment of contractors, subcontractors or suppliers to undertake design work can create problems for both parties to the contract, in settling liability in the event of faulty design.

Under the contract between the architect and the employer, the architect has an express or implied obligation to perform his services, including design aspects, with that standard of care that would be exercised by an architect of ordinary skill. A definition of the required standard of care was given in the House of Lords in *Bolam v. Friern Hospital Management Committee* (1957) – 'a man need not possess the highest expert skill; it is well established law that it is sufficient if he exercises the ordinary skill of an ordinary competent man exercising that particular art.' Courts are unwilling to hold a professional man to be negligent in the absence of expert evidence being given to the courts by a similar professional man to show that what has been done does not accord with standard practice.

Where an architect or engineer uses new or novel techniques in his design without being expressly requested to do so by the employer, he places himself in some jeopardy if the design is defective. In these circumstances he will have difficulty in proving that he exercised the necessary standard of care in his design.

The architect is under a continuing duty to check that his design will operate effectively in practice and to rectify any errors that may emerge. This was confirmed in *Brickfield Properties Ltd v. Newton* (1971) and more recently in *London Borough of Merton v. Lowe and Another* (1981) in the following terms: 'I am now satisfied that the architect's duty of design is a continuing one, and it seems to me that the subsequent discovery of a defect in the design, initially and justifiably thought to have been suitable, reactivated or revived the architect's duty in relation to design and imposed upon him the duty to take such steps as were necessary to correct the results of that initially defective design.'

The JCT contract is quite specific on the architect's duty to supply the contractor with all necessary information at the appropriate time. It requires that requests be made by the contractor 'neither unreasonably distant from nor unreasonably close to the date', when the information is required. Designers who lack foresight and initiative in the preparation and supply of information and who shelter behind this condition, seem reluctant to acknowledge that the contractor requires a great deal of information before he can identify which particular detail is outstanding and also that substantial lead times are required for the measurement of quantities, ordering, manufacture and delivery of materials, as well as for work scheduling and communications to the rest of the building team. The ideal and most economic way to build can be achieved only if all information is available before work starts and for no changes to be made during construction. The first objective underlies the basic philosophy of SMM7.

As with other professional men, the architect is expected to keep abreast of developments in the law, including relevant recent legislation and court decisions, and will be liable for damages for negligence to the employer if he does not do so (*B. L. Holdings v. Wood, 1979*). In particular, he will be expected to have a detailed and expert knowledge of the appropriate Standard Forms of Contract.

Architects, as well as contractors, have an obligation to ensure that buildings are erected in accordance with contract requirements, and the employer normally expects the architect to supervise all the work contained in the project. The RIBA Conditions of Engagement make provision for the architect to advise on the need for independent consultants and to be responsible for the direction and integration of their work, but not for the detailed design, inspection and performance of the work entrusted to them.

In the case of *Leicester Board of Guardians v. Trollope* (1911) the architect was under a duty to supervise the work under the contract. The clerk of works corruptly allowed the contractor to omit the damp-proof course from the ground floor and helped him to conceal the fact, and subsequently there was an attack of dry rot. When sued the architect claimed that the clerk of works, who had been appointed by the employer, was unfit for the job and it was his negligence and corruption that had caused the faulty work to be approved. The court, however, held the architect liable for his failure to supervise properly.

An architect when certifying the value of goods supplied or work done for an interim or final certificate under a building contract is liable in contract and in tort to his employer for over-certification, and is liable in tort to the contractor for under-certification. The House of Lords in *Sutcliffe v. Thackrah* (1974) established that the architect was not entitled to special dispensation as a quasi-arbitrator when issuing a final certificate.

**Other Consultants**

Quantity surveyors, structural engineers and other consultants engaged by the employer in connection with a building project, all have separate contractual relationships with the employer. They all undertake to carry out their work in a professional and workmanlike manner to the accepted standards of their professions. If failure to meet these standards results in problems, causing additional expense or delays, they will be in breach of their contracts and may be liable for damages. Apart from contractual obligations there exists a duty in tort to take reasonable care, as described earlier in the chapter with regard to architects. Thus the quantity surveyor will be expected to display a reasonable level of professional competence and care in all his duties concerned with the measurement and cost control of projects.

## ARCHITECT'S INSTRUCTIONS

No matter how well the project has been planned, it will almost inevitably be necessary from time to time for the architect to issue further drawings, details and instructions, which are collectively termed 'architect's instructions' (AIs). Under clause 4.1.1 of the JCT Conditions of Contract the contractor is required to comply expeditiously with all architect's instructions which the architect is expressly empowered by the Condition to issue. The only exception is where a variation is issued to which the contractor has objected in writing to the architect. An examination of the Conditions of Contract reveals a large number of matters that can validly be the subject of architect's instructions, indicating their important role in the operation of a building contract, and these are listed in table 5.1.

The procedure for the issue of architect's instructions is set out in clause 4 of the Conditions of Contract and this is now summarised.

(1) An instruction issued by the architect shall be in writing.
(2) An oral instruction is not operative unless confirmed in writing by the contractor or architect within seven days. If confirmed by the contractor within the stated period and the architect does not dissent, then it is deemed to be an architect's instruction.
(3) The instruction is effective from the date of issue of the architect's instruction or the expiration of the seven-day period referred to in (2).
(4) If neither the architect nor the contractor confirms any oral instructions, but the contractor still carries out the work, then the architect may at any time up to the issue of the final certificate confirm the oral instructions in writing. These instructions shall then be effective from the date when they were given orally.

Clause 12 provides that if the clerk of works issues any directions to the contractor, these are effective only if they relate to matters about which the architect is empowered to issue instructions, and provided that they are confirmed by the architect within two working days.

All instructions from the architect to the contractor should be issued or confirmed on a standard form. A typical local authority form is shown in table 5.2, but there are other formats issued by the RIBA and many architects' offices. They all contain the same basic information and offer sev-

Table 5.1   Matters for which the architect can issue instructions

| Clause | Subject |
| --- | --- |
| 2.3 | Discrepancies or divergences between documents |
| 2.4.1 | Discrepancies or divergences in performance related work |
| 6.1.3 | Divergence between statutory and contract requirements |
| 6.1.6 | Contractor's correction of divergence between statutory requirements and contractor's statement |
| 7 | Errors arising from contractor's inaccurate setting out |
| 8.3 | Opening up work for inspection or testing of materials or goods |
| 8.4.1 | Removal of work, materials or goods not in accordance with the contract |
| 8.4.3 | Variations in consequence of AI to remove items not in accordance with contract |
| 8.4.4 | Open up and test for non-compliance of items with the contract |
| 8.5 | Work not executed in a proper and workmanlike manner |
| 8.6 | Exclusion of persons from the works |
| 13.2 | Variations |
| 13.3.1 | Expenditure of provisional sums included in the contract bills |
| 13.3.2 | Expenditure of provisional sums included in a nominated subcontract |
| 17.2 | Making good defects, shrinkages or other faults |
| 17.3 | Schedule of defects |
| 23.2 | Postponement of work |
| 32.2 | Protection of work on outbreak of hostilities |
| 33.1.2 | Removal of debris and war-damaged work and execution of protective work |
| 34.2 | Action in respect of antiquities |
| 35.5.2 | Revised nominated subcontractor arrangements |
| 35.8 | Action in the event of the contractor being unable to reach agreement with a proposed nominated subcontractor |
| 35.18 | Nomination of the substituted subcontractor in the event of the original subcontractor failing to rectify defects |
| 35.24.6 | Renomination in the event of a nominated subcontractor determining the subcontract |
| 36.2 | Nomination of supplier |
| 42.8 | Expenditure of provisional sum in contract bills for performance of specified work |

eral distinct advantages in use. They reduce the amount of work involved in the issue of instructions and are immediately recognised by all the parties involved. Some offices produce them in distinctive colours for ease of identification.

It is good practice for all verbal instructions given by the architect on site visits and directions given by the clerk of works to be recorded in a duplicate book or on a duplicate pad of site orders, which can be signed at the time and subsequently confirmed by a written instruction prepared on the accepted format. Instructions should always be clear and precise

Table 5.2  Typical architect's instruction

---

CITY OF BLANKMOUTH
DEPARTMENT OF ARCHITECTURE
City Hall, Marion Way, Blankmouth BL1 3DG

K. J. Stevenson BA RIBA FRTPI
City Architect

ARCHITECT'S INSTRUCTION

| Project Nr and title | Instruction Nr |
|---|---|
| LS8: Broadway Leisure Complex | 8 |

---

To (Main Contractor)
Johnboy and Smithson Ltd
18 Grange Road, Blankmouth BL4 2ES

Under the terms of the Contract, I issue the following instruction(s). Where applicable, the contract sum will be adjusted in accordance with the terms of the relevant Condition.

| | | for office use: | |
|---|---|---|---|
| | | Omit £ | Add £ |
| 1. | *Aluminium windows*<br>*Omit* PC sum of £27 300 (BQ item 98D) for supply of aluminium windows. | | |
| | *Add*  Place order for the supply of aluminium windows with Bright Windows Ltd, Hillside, Newtown, in accordance with their quotation ref: J/621/P dated 16 February 1996 in the sum of £26 460, including 5% cash discount (copy attached). | | |
| 2. | Cut out one-brick infill panel 1.65 m x 2.60 m high between reinforced concrete portal frames, and extend and make good floor finish of PVC tiles in position shown on Drawing LS8/16 revision D (copy attached). (Note: work to be recorded as daywork). | | |
| | Date 15 May 1996    . . . . . . . . . . . . . . . .<br>City Architect | | |

| | | | | | |
|---|---|---|---|---|---|
| Contractor | ☐ | Quantity Surveyor | ☐ | Heating Engineer | ☐ |
| City Treasurer | ☐ | Clerk of Works | ☐ | Electrical Engineer | ☐ |
| Admin. (accounts) | ☐ | Architect's file | ☐ | Structural Engineer | ☐ |

Table 5.3   Quantity surveyor's responsibilities under the contract

| Clause | Subject |
| --- | --- |
| 5.1 | Possible custody of contract drawings and B/Q |
| 13.4.1 | Value variations under clause 13.4 rules, unless otherwise agreed by contractor and employer |
| 13.6 | Allow contractor to be present and take notes and measurements for valuing variations |
| 26.1 & 26.4.1 | Ascertain amount of contractor's loss/expense for delays at architect's option make interim valuations |
| 30.1.2 | Make interim valuations |
| 30.5.2.1 | Statement of retentions – at architect's option |
| 30.6 | Final adjustment of contract sum |
| 34.3.1 | Ascertain amount of contractor's loss/expense in respect of antiquities – at architect's option |
| 38.4.3 & 39.5.3 | Agreement with contractor of amount of fluctuations |
| 40.5 | Agree with contractor any alterations of formula fluctuations recovery methods |

and, where drawings are revised, specific reference should be made to the revision. Instructions issued by consultants should be passed to the architect for confirmation in the form of architect's instructions.

Copies of architect's instructions should be distributed to the main contractor (two copies), clerk of works, quantity surveyor and consultant(s) (where involved). It is advisable for the architect to emphasise at the initial site meeting that no adjustment will be made to the contract sum unless the matter is covered by an architect's instruction issued in accordance with the terms of contract.

The quantity surveyor's responsibilities under the contract are shown in table 5.3.

## PROGRESS AND SITE MEETINGS AND RELATED MATTERS

### Preliminary Arrangements

The architect, together with his consultants, is responsible for ensuring that the necessary statutory and other approvals have been obtained before the contract is let. In addition to the supply of the necessary drawings and schedules, nominated subcontractors' and suppliers' quotations will be needed to enable the contractor to place his orders. It is good practice for the architect to arrange with the employer for the advance ordering of materials or components that are in short supply or equipment with a lengthy manufacturing period. The employer should give full and clear instructions and prompt decisions on all matters referred to him.

Immediately prior to the award of the contract, the architect and the contractor should meet to satisfy themselves that no misunderstandings exist between them as to the full nature and extent of the project. Before starting work on the site the contractor has to finalise his construction programme and assemble plant, materials and labour, and the architect needs to ensure that arrangements are made for systematic and effective communication between the members of the building team. It would not be in the best interests of the employer to start construction before these arrangements are complete.

## Initial Briefing Meeting

As soon as practicable after the contract has been placed, the architect should arrange a meeting of all concerned, probably at his own office or that of the contractor, so that they may be made known to each other and to decide the procedure to be adopted at subsequent meetings. The meeting should be attended by the architect, who usually chairs it, the quantity surveyor and other consultants, the clerk of works if appointed, representatives of the contractor, and possibly principal nominated subcontractors and suppliers. At this meeting the architect should supply the contractor with all available information relating to the site and works, and factors affecting the carrying out of the works. These factors could include access to the site, space available, any restrictions on the method or time of working, building lines, buried services, site investigation, protection of the works, unfixed materials, adjoining buildings, and safety and welfare. The contractor's initial programme should also be considered along with such other matters as subcontracts, method of communication and insurances.

Many problems on building contracts are caused by delay in the issue of instructions by the architect relating to nominated subcontractors and suppliers and difficulties in liaison between the contractor and subcontractors. The position concerning all relevant subcontract sums in the bill of quantities should be clarified. It is important that all nominations are made in adequate time for the subcontract work to be phased into the contractor's programme without causing disruption (Greenstreet, 1994).

## Contractor's Programming and Associated Arrangements

Immediately after the award of the contract, the contractor should finalise his programme for the construction of the works by trades and/or work sections. This should be followed by a work statement showing the latest dates on which specified items of information are required for the implementation of the works programme. The contractor will appoint supervisory staff and start to assemble labour, materials and plant. The contractor should discuss with the architect the location of site huts, storage compounds, plant, service supply points and safety and welfare arrangements. The contractor should also inform the architect if he wishes to sublet any parts of the contract in order to obtain the architect's agreement.

On or about the date agreed for possession of the site and the start of construction work, the architect normally arranges a site meeting for the following purposes:

(1) to consider the contractor's refined programme for the works;
(2) to agree the contractor's programme for the supply of further information by the architect and his professional consultants; and
(3) to confirm all arrangements for communications, including the procedure with regard to architect's instructions (Aqua Group, 1990b).

**Communication Processes**

Clear and agreed lines of communication between all members of the building team and strict adherence to them is vital to the successful management of a building contract. In particular, procedures for the following activities should be established before construction starts.

*(1) Architect's site meetings*

The architect should arrange regular site meetings to identify the information required and to arrange consequential action. The frequency of these meetings will be influenced by the complexity of the project. All meetings should be preceded by the circulation of an agenda and decisions are recorded in minutes prepared by the architect. A typical agenda is shown in appendix D. The normal frequency and attendance at meetings, the distribution and agreement of minutes, and the responsibility for implementing instructions arising from them, should be agreed between the architect and contractor and confirmed in correspondence. There will be occasions when the architect will require the presence of subcontractors at these meetings. The Aqua Group (1990b) have suggested subdividing site meetings into two separate parts to make the best use of the time available, for example (1) the method of carrying out the work and (2) to monitor progress and performance.

*(2) Distribution of correspondence, drawings and other information*

The distribution of correspondence, instructions, orders and drawings should be agreed between the architect, quantity surveyor and other consultants, and the contractor. All drawings prepared by, or requirements or instructions of, consultants and nominated subcontractors and suppliers, must be issued through the architect, for he is responsible for ensuring that all instructions for work comply with the terms of the contract. All drawings and other information required by subcontractors or suppliers to enable them to plan and execute their sections of the work must be issued to them by the contractor, for he is responsible for ensuring that their work is carried out in accordance with the terms of the main contract and the subcontract.

### (3) Architect's instructions

Arrangements for the issuing of architect's instructions and the powers of the clerk of works to issue directions to the contractor normally form part of the terms of the building contract. All instructions should be given or confirmed in writing and should be in strict accordance with the terms of contract, as described earlier in this chapter.

### (4) Daywork sheets and records

The architect, quantity surveyor and contractor should agree upon arrangements for:

  (i) The signing of daywork sheets by the architect's representative. To avoid difficulties in interpretation, daywork sheets should be completed clearly and in adequate detail.
 (ii) The examination by the quantity surveyor of daywork sheets, and of operatives' time sheets and materials invoices, if the contract sum is to be adjusted to allow for fluctuations in the rates of wages and prices of materials by the orthodox method.
(iii) The authorisation of overtime.

### (5) Visits to site by the architect, quantity surveyor and other consultants

The building contract normally provides for reasonable access to the works by the architect and his representatives. It is good practice to agree the arrangements with the contractor and confirm them in writing. The quantity surveyor will need to visit the site periodically to measure the work.

### SUPERVISION OF WORK ON SITE

The main object of site supervision is to ensure that the employer's requirements as defined in the contract documents are correctly interpreted and that any problems are satisfactorily resolved. Responsibility for supervision is shared between the contractor and the architect as described earlier in the chapter. The nature and extent of the supervision will depend on the size and complexity of the works. For example, on a small project a contractor can usually rely on a competent general foreman, while on larger contracts a number of foremen may be engaged working under a site agent. The architect may be able to supervise a small project by periodic personal visits, while large or complex projects will require at least one clerk of works (Aqua Group, 1990b).

The architect's site supervision generally falls into three categories.

## (1) Routine Site Inspections

The architect when carrying out such inspections should inform the clerk of works and the contractor's representative of his visit, and they normally accompany him around the works. Any instructions will be given to the person in charge of the works and will subsequently be confirmed in writing. The architect will be concerned with health and safety aspects as well as constructional issues and a checklist is often used to avoid items being overlooked.

## (2) Records and Reports

The architect normally receives weekly reports from the clerk of works covering such matters as:

   (i) number of men employed in the various trades;
  (ii) weather conditions and details of time lost;
 (iii) principal deliveries of materials and any shortages;
  (iv) plant on site;
   (v) details of drawings and other information required;
  (vi) general progress compared with the programme;
 (vii) any other matters affecting the operation of the contract; and
(viii) reference numbers of architect's instructions received.

The weekly report is a valuable means of keeping the architect well informed of progress on the site and will provide a useful source of reference should disputes arise subsequently. The weekly report is normally compiled from the diary kept by the clerk of works. A copy of the contractor's programme should be kept in the clerk of works' office with actual progress checked weekly against the programme. The clerk of works also keeps records of any departures from the working drawings and this applies particularly where work, such as foundations and drains, is subsequently covered up, and this information is also of considerable value to the quantity surveyor. Progress photographs may also be taken periodically of the work.

## (3) Samples and Testing

The architect may require samples of various materials and components to be submitted for approval as a check on quality. He may also require sample panels of materials such as facing bricks to be erected on site. Regular laboratory tests of basic materials such as concrete are often required, with test reports submitted by the contractor to the architect or structural engineer (Aqua Group, 1990b).

## COORDINATION OF MECHANICAL AND ELECTRICAL ENGINEERING SERVICES

If a construction project does not run to time, delays most frequently occur in those areas about which the main contractor has least knowledge. A NEDO report in 1975 noted that two-thirds of the projects surveyed finished behind programme and it was suggested that the performance of major subcontractors was a critical factor. The combined values of prime cost sums for engineering services in building projects vary between 30 and 80 per cent of the whole contract sum. The responsibility for coordinating this work rests with the main contractor who is responsible also for overall performance.

Apart from the differences in attitude and communication problems, a major problem associated with the design/construction interface is the accepted practice for the nomination of services subcontractors by the architect. Nominated subcontractors are seldom appointed at the same time as the main contractor. They may be appointed before the main contractor, in which case any planning of their work upon which they have based their estimates cannot have been taken into consideration in the main contractor's planning policy for the contract. Any discrepancies between the nominated subcontractors' and the main contractor's plans are likely to be a source of coordination problems throughout the contract. Alternatively, the nominated subcontractor may not be appointed until work has started on the site with the result that errors or omissions in the work may arise because the main contractor is not fully aware of the nominated subcontractor's requirements (Barton *et al.*, 1978).

Successful services coordination entails the orderly organisation of mechanical and electrical systems from inception to completion to ensure that:

 (i)   no conflict occurs between the various services during installation;
 (ii)  there is no physical interference between service elements and the building structure or finishes; and
 (iii) a correct installation sequence is established and followed by all subcontractors.

When the latest dates for the issue of nominations and design information are finalised, the following matters require careful consideration:

(i)   the time required by subcontractors to prepare their shop drawings;
(ii)  the period required by subcontractors to obtain materials, components and equipment; and
(iii) the time required by the main contractor to implement the entire coordination procedure.

In particular the starting and finishing times of activities become control points which must be reasonably achievable.

## CONTRACTOR'S PROGRAMMING AND PROGRESSING

The following procedure provides a satisfactory approach:

(1) The contractor is responsible for ensuring that the programme of construction he has prepared is implemented. This necessitates effective planning of operations on the site to ensure that labour and plant are used at optimum efficiency, and that wastage of materials is minimised. To achieve this entails planning several weeks ahead and a short term programme specifying each day's work by each gang, with alternatives for use if weather or other factors make changes necessary. These site programmes will identify potential causes of delay and progress chasing should then become preventive rather than remedial.
(2) The contractor should ensure that his own work and that of subcontractors is coordinated in accordance with his programme and site planning. Hence the contractor should arrange and chair regular meetings with the subcontractors to review progress, to confirm and where necessary coordinate their detailed programmes of work, and to initiate any action required by the architect and his consultants, in connection with subcontractors' work. It is good practice for these meetings to precede the architect's site meetings, at which their proceedings and decisions can be reported.
(3) The architect is responsible for ensuring that all relevant information is supplied to the contractor in accordance with the contractor's programme agreed at the start of the project.

Where any changes to the programme are required by the architect or contractor, all members of the building team should be notified promptly, so that other related programmes can be suitably adjusted.

The CIOB (1980) has highlighted how the contractor's construction programme is an important common reference. It shows how the contractor has interpreted the detail of the contract documents and other supporting information and sets out the whole as a statement of intent for building. As the key programme for the construction work its content is of great significance to all parties charged with handing over to the employer a building that is well fitted for its intended use, within the contract time and with optimum economy. Throughout construction the programme is a common reference for predicting, quantifying and communicating performance as well as providing a basis for compiling records.

The simplest form of construction programme is a written list of activities, dated and arranged in the order in which they will be carried out. However, in the majority of projects more sophisticated graphic programming techniques are used to show the relationship of individual activities to one another and collectively to a common timescale. The five main techniques used are bar charts, elemental trend analysis (line of balance), network diagrams, partly linked bar charts and precedence network diagrams. These processes are now briefly described and compared but readers are referred to the publications by the CIOB (1980b), Cooke (1992) and Pilcher (1992) for more detailed explanations, illustrations and applications.

*Bar charts* are widely used and easy to understand, with bar lines representing the time period allocated to each operation, and the relationship between the start and finish of each activity is clearly seen. The timescale, usually related to a calendar, is shown horizontally and the activities listed vertically. It is relatively simple to convert a draft programme into a bar chart. They are, however, incapable of showing in detail a large number of interrelated activities and dependencies, or whether one activity is delaying or about to disrupt another. Hence they have substantial limitations and do not permit a high degree of control.

*Elemental trend analysis* or line of balance is not so easily understood as a bar chart but it highlights the importance of activity completion, production rates and relationships between selected activities. It also has a horizontal timescale and calendar, with cumulative output shown vertically. Bar lines representing trades or operations are inclined at different slopes to indicate the rate of working. It is particularly suitable for repetitive work and strict trade sequencing and permits a high degree of control.

*Network diagrams* concentrate on the logical relationships between activities and their relevance to project completion. The earliest and latest event times, critical path and activities with spare time (float) are all identified. This technique shows up potential problems and possible ways of shortening the contract period, and is well suited for the coordination and control of complex non-repetitive projects. The preparation, presentation and assimilation require greater effort and large projects need subdividing into sections. Quantity surveyors find them valuable in the assessment of contractors' claims.

*Precedence network diagrams* resemble a collection of linked boxes with visual emphasis on activity descriptions. Activities can be grouped by trade and location and there is a strong emphasis on construction sequence. The technique is ideal for large complex projects, but reference to calendar time is only indirect and skill is required in both presentation and interpretation.

*Partly linked bar charts* retain the visual benefits of bar charts with added emphasis on dependencies. This takes the form of vertical links between the completion of one activity and the start of another. The increased emphasis on coordination and construction sequence allows the technique to be used for more complex projects than a bar chart. The float concept is generally missing and there is a limit to the amount of linking that is possible.

Figure 5.1 shows a *bar chart programme* for a factory project which also includes a cumulative value forecast. This is very useful to both the contractor and the client and his professional advisers as an indication of how work will progress, the value of work and the payments to be made. Adjustments will however need to be made for deduction of retention and monthly payments as distinct from the four weekly values shown, possibly by inserting five week periods periodically to balance weeks and months.

## INTERIM CERTIFICATES AND PAYMENTS

### General Procedure

Under the JCT Form of Contract (clause 30) the architect is required to issue interim certificates at the intervals stated in the appendix to the conditions, normally one month, and this can continue up to the settlement of the final account and the issue of the final certificate. As soon as the final accounts of all the nominated subcontractors have been agreed, and in any event not less than 28 days before the issue of the final certificate, the architect must issue an interim certificate which includes the final amounts due to all nominated subcontractors.

The contractor relies on regular and fairly assessed payments to conduct his business. The employer could face problems in the event of over-certification by the architect if it coincided with the contractor becoming insolvent. The contract further provides for the employer to pay the contractor any monies due under a certificate within 14 days of its date of issue, and the contractor must, in his turn, pay any monies due to nominated subcontractors within 17 days of the same date.

The contractual basis of the certificate is the valuation of the work properly executed and the architect must satisfy himself as to the acceptable quality of the work before certifying payment by the employer. It is good practice for the architect to keep the quantity surveyor informed of any defective or improperly executed work, as at the valuation stage the quantity surveyor's

Right-side annotations: "20 week Project period" ; "Project value £164000"

| OPERATION | BUDGET | 1 | 2 | 3 | 4 | 5 | 6 | 7 | 8 | 9 | 10 | 11 | 12 | 13 | 14 | 15 | 16 | 17 | 18 | 19 | 20 |
|---|---|---|---|---|---|---|---|---|---|---|---|---|---|---|---|---|---|---|---|---|---|
| Establish | 5000 | 5 | | | | | | | | | | | | | | | | | | | |
| Piling | 10000 | | 10 | | | | | | | | | | | | | | | | | | |
| Caps & G. Beams | 21000 | | | 7 | 7 | 7 | | | | | | | | | | | | | | | |
| Drainage | 6000 | | 2 | 2 | 2 | | | | | | | | | | | | | | | | |
| Erect frame | 20000 | | | | | | 10 | 10 | | | | | | | | | | | | | |
| Roof cladding | 14000 | | | | | | | | 7 | 7 | | | | | | | | | | | |
| Ext. brickwork | 30000 | | | | | | | | | 6 | 6 | 6 | 6 | 6 | | | | | | | |
| Floor slab | 12000 | | | | | | | | | | | | | 4 | 4 | 4 | | | | | |
| Internal services | 20000 | | | | | | | | | | | | | | | 5 | 5 | 5 | 5 | | |
| External works | 16000 | | | | | | | | | | | | | | | | | 4 | 4 | 4 | 4 |
| Preliminaries | 10000 | 0.5 | 0.5 | 0.5 | 0.5 | 0.5 | 0.5 | 0.5 | 0.5 | 0.5 | 0.5 | 0.5 | 0.5 | 0.5 | 0.5 | 0.5 | 0.5 | 0.5 | 0.5 | 0.5 | 0.5 |
| Weekly value (£'000) | | 5.5 | 12.5 | 9.5 | 9.5 | 7.5 | 10.5 | 10.5 | 7.5 | 13.5 | 6.5 | 6.5 | 6.5 | 10.5 | 4.5 | 9.5 | 5.5 | 9.5 | 9.5 | 4.5 | 4.5 |
| Cumulative weekly value | | 5.5 | 18.0 | 27.5 | 37.0 | 44.5 | 55.0 | 65.5 | 73.0 | 86.5 | 93.0 | 99.5 | 106.5 | 116.5 | 121.5 | 130.5 | 136.0 | 145.5 | 155.0 | 159.5 | 164.0 |
| Monthly value | | | | | 37.0 | | | | 73.0 | | | | 106.0 | | | | 136.0 | | | | 164.0 |
| Valuation periods | | | | | 1 | | | | 2 | | | 3 | | | | 4 | | | | 5 | |

Time in weeks

**Figure 5.1** *Factory project – bar chart programme – cumulative value forecast*

prime concern is one of quantity not quality. The contract (30.10) specifically provides that the certificates are not conclusive evidence that any works, materials or goods included are in accordance with the contract.

When issuing a certificate the architect must direct the contractor as to the amounts included for nominated subcontractors and must notify each nominated subcontractor of the amount included. Each interim certificate shall be accompanied by a statement stating the amount of retention held on the main contractor's work and on the work of each nominated sub-contractor. Copies of this statement are issued by the architect to the employer, the contractor and each nominated subcontractor. These provisions are illustrated in appendices G, H and J, using the standard RIBA forms.

## Components of Interim Certificates

The general guidelines for the determination of the items to be included in interim certificates are detailed in clause 30 of the Conditions of Contract. It is usual for the quantity surveyor to prepare valuations for certificates and a typical example is included later in this chapter. The gross value to be included in interim certificates is detailed under two headings which are now summarised. The retention provisions were described in chapter 4, and are further elaborated under interim valuations, later in this chapter.

*(a) Items subject to retention*

(1) The total value of the contractor's work as properly executed, including the appropriate proportion of preliminaries, any variations, and, where applicable, adjustments in respect of fluctuations where the price adjustment formula applies (30.2.1.1).
(2) The total value of materials and goods delivered to or adjacent to the works provided they are reasonably, properly and not prematurely delivered and are adequately protected (30.2.1.2).
(3) The total value of materials and goods off site where authorised by the architect (30.2.1.3).
(4) The total value of nominated subcontractors' works, materials and goods (30.2.1.4).
(5) The contractor's profit on the amounts included in respect of nominated subcontractors (30.2.1.5).

*(b) Items not subject to retention*

(1) Amounts due to the contractor under the terms of the contract in respect of statutory fees or charges (6.2), errors in levels and setting out (7), opening up for inspection and testing (8.3), royalties (9.2), making good defects where the architect authorises payment (17.2

and 17.3), insurance against excepted risks (21.2.3) and non-insurance against all risks or specified perils by the employer (22B and 22C).
(2) Amounts due to the contractor by way of reimbursement for loss and expense arising from matters materially affecting the regular progress of the works (26.1) or from the discovery of antiquities (34.3).
(3) Final payments to nominated subcontractors (35.17).
(4) Payments to the contractor in respect of fluctuations, other than those recovered under the price adjustment formula (38 and 39).
(5) Amounts properly payable to a nominated subcontractor relating to statutory fees and charges, making good defects where the architect authorises payment, and fluctuations calculated under the orthodox method (NSC/C 4.17.2).

If under the orthodox fluctuations arrangements, there are any sums payable by the contractor to the employer or by a nominated subcontractor to the contractor, then these monies must be deducted from the amounts due in an interim certificate which are not subject to retention.

The inclusion of off-site materials or goods in the gross value of an interim certificate is at the discretion of the architect. The conditions prescribed in clause 30.3, to safeguard the employer's interest, must be met before the architect authorises payment.

The gross valuation for an interim certificate consists of the total amounts subject to retention, plus the total amounts not subject to retention, less any amount allowable to the employer in respect of fluctuations as described. From the gross valuation, retention and the total amounts stated as due in previous interim certificates will be deducted to arrive at the sum due to the contractor at this stage.

## Payments to Nominated Subcontractors

Prior to the issue of each interim certificate, the contractor must provide the architect with reasonable proof that he has made the relevant payments due to nominated subcontractors under previous certificates. If reasonable proof of payment cannot be provided, the architect must issue a certificate to that effect, stating the amount involved, with a copy to the nominated subcontractor(s). The payment to the contractor under the next interim certificate will be reduced by the amount by which the contractor has defaulted in his payment to the nominated subcontractor(s) and the employer will pay the subcontractor(s) direct.

There is no obligation on the employer to make direct payments to nominated subcontractors in excess of the amount due to the contractor. Where there is more than one nominated subcontractor involved and the monies due to the contractor are insufficient to meet the total of the direct payments, then the employer will apportion the money available either on a *pro rata* or some other fair and reasonable basis (Aqua Group, 1990b).

## LIQUIDATED DAMAGES

The Standard Forms of Contract provide for the payment of liquidated and ascertained damages for delay in completion beyond the completion date inserted in the contract or a substituted date following the grant of extension of time by the architect. A sum inserted for liquidated damages to be enforceable must be a genuine pre-estimate of damages. Whether the sum is a penalty or is liquidated damages will be largely influenced by the terms and inherent circumstances. If it is held to be a penalty, the sum is unenforceable as a measure of damages and the employer would have to revert to his remedy of common law damages. In many building and civil engineering contracts the liquidated damages are not a genuine pre-estimate of damages to be suffered by the employer, and are often related to those included in previous contracts of a similar nature.

The Society of Chief Quantity Surveyors in Local Government set up a working party to investigate the procedures adopted for the assessment of liquidated damages on local authority building contracts in 1981 and its revised report was published in 1993.

The precedent set by the courts for a valid assessment of damages was summarised as follows:

(1) If the parties make a genuine attempt to pre-estimate the loss likely to be suffered, the sum stated will be liquidated damages and not a penalty, irrespective of actual loss.
(2) The sum will be a penalty if the amount is extravagant having regard to the greatest possible loss that could be caused by the breach.

The form the loss might take was illustrated by the SCQSLG (1993) by taking extracts from *Halsbury's Laws of England* (Fourth Edition):

(a) The measure of damages for failure by the contractor to complete a building . . . will include . . . any loss of rent of the building, or any loss of use of the building or, in appropriate circumstances, of business profit which may accrue to the employer in consequence of any delay . . .'
(b) 'In certain cases the measure of damages may be the loss of interest on the cost of the contract works and of the land on which they are constructed.'

The legal requirement thus suggests five alternative losses to be considered:

(i) loss of rent (does not necessarily represent economic rents and considered inappropriate);
(ii) loss of use of building;

(iii)  loss of business profit (can be appropriate in quasi commercial ventures);
(iv)  loss of interest on the cost of land; and
 (v)  loss of interest on the cost of the contract works.

The earlier SCQSLG report (1981) suggested that category (v) appeared to be the only satisfactory basis for assessment in local authority building contracts. However, the later report (1993) suggested that changes in the climate of financial accountability with the differing needs of local authority clients could make it necessary in the future to consider all five categories when assessing likely losses. The report also emphasised that the contract figure for liquidated damages where reasonable and agreed by the contractor is not adjustable and should be expressed as a weekly or daily figure.

The SCQSLG working party recommended that the following approach provided a reasonable basis for the assessment of liquidated damages, provided the project is funded from resources already held by the local authority.

(1)  Loss of interest on the cost of the contract works, based on the assumption that 80 per cent of the monies is paid at theoretical completion and that annual interest payable is 10 per cent. This amounts to 0.15385 per cent of the contract sum per week but will need monitoring to take account of significant fluctuations in interest rates.

(2)  Professional fees and charges (architect, quantity surveyor and clerk of works) for post-contract services based on scale fees but adjusted to reflect current level of fee bids received by an authority. Consideration should also be given to all consultants likely to be affected by the delay.

(3)  Further costs such as costs of a temporary nature awaiting completion of the building, for example, temporary housing, caravans and mobile libraries, and client departmental costs (such as housing managers, social service workers, education supply officers and treasurers). These are best assessed on an *ad hoc* basis.

(4)  Fluctuations, where applicable, normally calculated by reference to the index numbers relating to the valuation period.

In the case of housing contracts some local authorities state damages at a rate per dwelling per week and a direct division of the damages total per week by the number of units will give a reasonable figure per unit per week. A further subdivision into units, garages and external works can be applied if required. In the case of modernisation contracts where staff time per unit is greater, the professional cost could reasonably be doubled.

Table 5.4 shows a typical calculation of liquidated and ascertained damages prior to tender, based upon an example contained in the SCQSLG report (1993).

Table 5.4  Typical calculation of liquidated and ascertained damages prior to tender

---

| | | | |
|---|---|---|---|
| **Project** | LIBRARY | | |
| **RIBA Category** 3 | | **Interest Rate** | 10% |
| **RICS Scale** 37 Class B | | | |
| | | **£** | Item |
| **Estimated Contract Sum** | | 1 200 000 | *(A)* |
| **Fees** (taken from fee scales or fee bids) | | 180 000 | *(B)* |
| **Total** *(A + B)* | | 1 380 000 | *(C)* |
| **Contract Period** (weeks) | | 50 weeks | *(D)* |

| | £ |
|---|---|
| **1. Interest on Capital:** (Loss factor for 10% from Annexe B) 0.15385% of *(C)* | 2 123 |

**2. Professional Costs**     £

Architect: RIBA Category 3 New Work

$25\% \times 6.25\% \times \dfrac{(A)}{(D)}$         375

**Quantity Surveyor:** RICS Scale 37
Post Contract Alternative II (para. 6.2)

$£1500 + \dfrac{(0.4\% \times £900\,000)}{(D)}$         102

**Clerk of Works:** $1\frac{1}{2}\% \times \dfrac{(A)}{(D)}$         360

**Structural, Mechanical and Electrical Engineers**
(to be assessed as appropriate)
                                £        837

**3. Further Applicable Costs:**
Rent of existing leased building         450

**4. Fluctuations**                     N/A

|  |  | £ |
|---|---|---|
| | **TOTAL PER WEEK** | £ 3410 |
| or | **TOTAL PER DAY** | £  487 |

**Calculated by** ......................     **Date** .......................

---

*Source*: SCQLG – Liquidated and ascertained damages on building contracts

## INTERIM VALUATIONS

### General Procedure

Prior to the architect issuing an interim certificate it is usually necessary for the quantity surveyor to prepare an interim valuation. The main items to be considered when valuing the work are:

(1) measured work, including preliminaries;
(2) value of variations and extras;
(3) value of nominated subcontractors' and suppliers' work;
(4) fluctuations, where applicable;
(5) unfixed materials on site;
(6) retention monies; and
(7) previous payments.

A valuation for an interim certificate should be as accurate as is reasonably possible. The contractor is entitled to the total value of the work properly executed, less a specified percentage for retention. If the valuation is depressed, it has the effect of increasing the retention which must be contrary to the terms and spirit of the contract. The contractor may have a number of contracts in hand and an aggregation of unnecessary and improper sums retained could place an excessive strain on his capital and ability to pay his debts. On the other hand, the employer must be protected against overpayment, as in the event of insolvency of the contractor, any excess payments cannot be recovered and the employer is likely to be faced with additional costs in employing another contractor to perform the work already certified.

The amount of measurement to be undertaken depends on the nature and complexity of the works and the stage they have reached. A common practice is for the quantity surveyor and the contractor's representative to meet at regular intervals for the purpose of measuring the work, although it is normally the prime responsibility of the quantity surveyor to make the valuation. Other meetings may be necessary for the purpose of measuring substructural and other work which will subsequently be covered up. Appendix E shows a completed quantity surveyor's Valuation on the standard form issued by the RICS.

### Valuation of Preliminaries

The preliminaries section of a bill of quantities gives the contractor the opportunity to price items of an organisational and general nature which affect the cost of works but which are not restricted to any particular works section. Each of three methods is used on occasions to value the items contained in the preliminaries section of a bill of quantities, and these are now described and evaluated.

*(1) Based on the amount of the contract period that has elapsed*

For example, assuming the total value of the preliminaries bill to be £125 000 on a 60 week contract, the amount included in an interim valuation will be at the rate of

$$\frac{£125\,000}{60} = £2083 \text{ per week}$$

*(2) Based on the value of work done*

For example, assuming the total value of the contract to be £2 100 000 and the total value of preliminaries to be £157 500, the amount included in an interim valuation is at the rate of

$$\frac{£157\,000}{£2\,100\,000} \times 100 = 7.5 \text{ per cent}$$

Under both of these methods, any provisional sums included in the preliminaries must be deducted prior to the calculation. Neither method is entirely satisfactory as they present an oversimplified approach. For example, the first method fails to take account of the fact that the work may not be on target. Neither method deals effectively with lump sums inserted in the preliminaries relating to items that will be executed early in the contract resulting in under-valuation, while other items may be included prematurely.

*(3) Based on valuation by individual assessment*

Each item is assessed individually and this permits a much more realistic and accurate approach to be adopted to suit the particular circumstances as shown in the following examples.

  (i)  Certain items may be required at the outset and can accordingly be valued in full, such as the provision of temporary site fencing, construction of temporary roads, and the erection of hoardings, storage compounds and site huts.

  (ii)  Items valued in full or in part according to their usage, such as scaffolding and plant.

  (iii)  Items valued in proportion to the works executed, such as wages of general foreman and cost of telephone calls and insurances.

  (iv)  Items valued on completion such as cleaning and drying out.

    The actual method adopted will be influenced by the way in which the preliminaries have been priced. Estimators use different pricing methods and these vary from a single lump sum to pricing all preliminaries items.

The third method would be unsuitable in the former case and the first and second methods in the latter. The quantity surveyor must use his skill and judgement in assessing a fair and reasonable value in the prevailing circumstances.

**Remeasurement and Valuation of Provisional Work**

An accurate bill of quantities (contract bills) should not require remeasurement, except where provisional quantities and sums have been inserted. However, where errors are discovered in the bill of quantities they shall be corrected by remeasurement and are deemed to be a variation issued by the architect (13.1.1). Approximate quantities will need remeasuring in order to value the work as executed, although these are much less prevalent where bills of quantities are measured in accordance with SMM7.

The extent to which measurements are taken on the site will depend on the nature of the work. For instance substructural and drainage works will normally require only the depths of excavation measured on site and the quantities can be taken off at a later stage in the office. The quantities of builder's work in connection with specialist trades for such items as holes and chases in walls and floors will be taken from site measurements.

**Valuation of Measured Work**

The interim valuation of measured work may be carried out in the following ways or a combination of them.

(1)  Items of work done may be individually extracted from the bill of quantities, where they are readily identifiable as being complete, as illustrated in the following example.

|  |  | £ |
|---|---|---|
| Items 32a and b | Hardcore bed | 1062.00 |
| Items 35d and e | Reinforced concrete bed | 2976.00 |
| Items 39c–f | Fabric reinforcement to ditto | 1549.00 |
| Items 48b–e | Horizontal damp-proof courses | 683.00 |

(2)  Approximate quantities may be measured for the work as executed and priced at billed rates. This method is ideally suited for such items as concrete and brickwork in the early stages of the contract, as indicated in the following example.

|  |  |  | £ |
|---|---|---|---|
| One-and-a-half brick wall | 186 m² | @ £82.00 | 15 252.00 |
| Half brick skin of hollow wall | 1280 m² | @ £28.20 | 36 096.00 |
| Ditto in facework | 1170 m² | @ £43.30 | 50 661.00 |

Forming cavity and wall ties          1210 m² @ £ 1.90    2 299.00
Concrete block partition, 100 thick    540 m² @ £17.80    9 612.00

(3) It may be practicable to value the work as an estimated percentage of the total value of a work section. This procedure is most likely to be adopted where the work consists of a large number of small items, such as the plumbing and painting work shown in the next example.

Plumbing installation          80% x £31 680       £25 344
Painting and Decorating        20% x £29 920       £ 5 984

(4) Predetermined stages of completion may be assessed as a percentage of the value of the total work in the contract. This method is particularly useful on work containing many similar items, such as housing schemes. The values for the different house types will be assessed separately. A schedule is prepared for each house type showing each work category with its monetary value. The extent of completion of each listed work category is indicated in the column provided for each monthly valuation, and there may be as many as 25 or more separate categories.

As an alternative to regular monthly certificates, the contract may provide for *stage payments*, whereby payments are made when certain defined stages in the construction work have been completed. A typical breakdown of a house type contract sum for use in stage payments in interim valuations is shown in table 5.5, although in practice the schedule would show a range of house types.

Table 5.5  Typical breakdown of house type contract sum for stage payments

| Stages | Breakdown of contract value for house type A |
|---|---|
| | £ |
| 1. Substructure | 2 665 |
| 2. External walls | 6 017 |
| 3. Roof, internal walls and partitions, first fixings | 7 221 |
| 4. Plumbing, electrical work, internal finishings and second fixings | 7 869 |
| 5. Painting and decorating | 3 042 |
| 6. External works | 3 128 |
| 7. Preliminaries | 4 258 |
| Total | £ 34 200 |

## Valuation of Variations and Extras

The architect may issue instructions from time to time encompassing variations. The contractor may also be involved in direct loss or expense arising from actions of the employer. The quantity surveyor will assess the value of these items which should be included in interim valuations. The valuation of variations involves the omission of work as originally included in the bill of quantities and the addition of the work as executed in its place. Each variation should be considered separately and their net values carried forward to a general summary. The total value of omissions and the total value of additions are computed to arrive at a net balance which is added to or deducted from the valuation, and this aspect will be covered in more detail in chapter 6.

## Valuation of Nominated Subcontractors' and Suppliers' Work

Nominated subcontractors' and suppliers' accounts are included in interim valuations as and when the work has been executed or materials supplied. The nominated firm will submit accounts to the contractor who will pass them to the quantity surveyor for assessment. The accounts often include work executed to date in accordance with the original quotation, extra works, materials on site and possibly fluctuations. All accounts are checked before they can be included in a valuation for interim payment, and this aspect is illustrated in chapter 6.

The architect should be notified of the specific amounts included in an interim valuation in respect of each nominated subcontractor's account, and the subcontractors notified accordingly. Proof that the contractor has paid the nominated subcontractors on previous interim certificates is required by the architect. In the event of default by the contractor, the employer will pay the subcontractors direct and these payments will be set off against the next payment to the contractor.

### Fluctuations

A fluctuation clause may be inserted in a building contract which provides for the adjustment of the nett increase or decrease in the cost of labour and/or materials after the date of tender. Where this is operative the fluctuations shall be assessed and included in each interim valuation. The assessment of the value of fluctuations is progressive and the cumulative total added to each valuation. The two methods used are assessment by analysis (the orthodox method) and application of the building price adjustment formula.

*Assessment by analysis*  This method of assessment is based on information supplied by the contractor in respect of the amounts and values of labour and materials used on the works.

Assessment of increased labour costs is based on an examination of the contractor's timesheets to determine the number of hours worked. The net increase in wage rates is obtained from information published by the National Joint Council for the Building Industry. Labour fluctuations are assessed in the manner shown in the following example.

| Craft operatives | £ |
|---|---|
| 960 hrs @ 3p (nett increase per hour) | 28.80 |
| 1580 hrs @ 6p (nett increase per hour) | 94.80 |
| Labourers | |
| 1290 hrs @ 2p (nett increase per hour) | 25.80 |
| 2740 hrs @ 5p (nett increase per hour) | 137.00 |

The fluctuations in the cost of materials are based on a schedule of basic rates of the principal materials agreed at the time of tendering. This schedule is incorporated in the bill of quantities and forms the basis of prices from which fluctuations can be calculated. The list should contain such commonly used materials as sand, gravel, cement, common bricks, facing bricks and other materials that the contractor considers necessary. The contractor will produce invoices for the materials used and from these the quantity surveyor will be able to extract the relevant quantities and prices in order to calculate the fluctuations, as shown in the following example relating to facing bricks. Assume that:

(i)  The basic price of facing bricks is £325.00 per thousand as shown on the basic price list.
(ii)  The number of bricks required is 120 000. This is calculated from the drawings or quantities with allowance made for waste.
(iii)  The invoices submitted show the following particulars.
     Invoices from Newtown Brick Co. Ltd

   Invoice ref: 1721 45 000 facing bricks @ £323.50/1000
   Invoice ref: 1934 25 000 facing bricks @ £326.60/1000
   Invoice ref: 2262 25 000 facing bricks @ £328.70/1000
   Invoice ref: 2485 25 000 facing bricks @ £331.00/1000

From this information the fluctuations can be calculated.

| Material | Quantity | Basic price/1000 £ | Invoice price/1000 £ | Net increase £ | Net decrease £ | Total increase £ | Total decrease £ |
|---|---|---|---|---|---|---|---|
| Facing bricks | 45 000 | 325.00 | 323.50 | – | 1.50 | – | 67.50 |
| Ditto | 25 000 | 325.00 | 326.60 | 1.60 | – | 40.00 | – |
| Ditto | 25 000 | 325.00 | 328.70 | 3.70 | – | 92.50 | – |
| Ditto | 25 000 | 325.00 | 331.00 | 6.00 | – | 150.00 | – |
| | | | | | | £282.50 | £67.50 |

The net increase in the cost of facing bricks is therefore £215.00.

Only those prices on the basic price list are to be taken into account for the purpose of calculating the fluctuations in prices of materials. The net totals of price fluctuations for each material are taken to a general summary and the total net increase or decrease is added to or deducted from the valuation.

*Building formula price adjustments*   This price adjustment formula method was introduced in 1977 by the Property Services Agency as an alternative to the orthodox laborious method of assessment by analysis. It is a much quicker and easier method showing a reasonable degree of accuracy and results in earlier payments to the contractor. The method is based on monthly indices published by HMSO in monthly bulletins. The Central Statistical Office provides the information for materials indices and the Department of the Environment (DoE) for labour indices. A description of the indices and a guide to their application and procedure, based on the 1990 series of indices, have been produced by the DoE (1995).

The work is divided into work categories or work groups (combinations of work categories into weighted work groups), each of which has an index listed monthly. A schedule is prepared before the contract is signed to show the work categories/groups that are being used. A base month is established at the time the tenders are submitted. The base index for any work category/group is the index shown on the list for the base month. Each work category/group is calculated separately using the following formula

$$C = \frac{V(I_v - I_0)}{I_0}$$

where $C$ = the amount of the price adjustment for the work category/group to be paid to or received from the contractor

$V$ = the value of work executed in the work category/group during the valuation period

$I_v$ = the index number for the work category/group at the mid-point of the month during which the valuation occurred

$I_0$ = work category/group index for the base month.

For example, assuming the base indices are as follows:

        work category 1        202.1
        work category 2        197.9
        work category 3        200.5

Category indices for the valuation period are:

        work category 1        201.5
        work category 2        200.2
        work category 3        202.6

Value of work executed during the valuation period:

        work category 1        £12 500
        work category 2        £20 800
        work category 3         £7 600

The price adjustments are then calculated as follows:

work category 1: $C = 12\,500\,\dfrac{(201.5\,-\,202.1)}{202.1} = £37.11$ (decrease)

work category 2: $C = 20\,800\,\dfrac{(200.2\,-\,197.9)}{197.9} = £241.74$ (increase)

work category 3: $C = 7\,600\,\dfrac{(202.6\,-\,200.5)}{200.5} = £79.60$ (increase)

The price adjustment for each category is taken to a general summary and the net increase or decrease is added to or deducted from the valuation.

### Unfixed Materials

The value of unfixed materials is also included in the interim valuation. This value is usually based on a list prepared by the general foreman or clerk of works which can be checked on the site. The materials should be of the specified quality, be insured and properly protected and not be brought on to the site prematurely. The value of certain off-site materials may also be claimed providing the following conditions are fulfilled:

(i)  they are intended for inclusion in the works;
(ii)  they are complete and satisfactory in every respect and are adequately insured;
(iii)  they have been set apart and are clearly and visibly marked and referenced to identify where they are stored, the person to whose

order they are held and that their destination is the works; and
(iv)  where such materials are ordered from a supplier or subcontractor,
the subcontract expressly provides that the materials shall pass un-
conditionally to the contractor.

## Retention

Provision is usually made in building contracts for the retention by the
employer of a proportion of the monies due to the contractor on an interim
certificate, as described in chapter 4. Retention is generally expressed as a
percentage agreed between the parties. The object of the retention is to
create an incentive for the prompt and satisfactory performance of the work
and to provide security for the employer against default by the contractor.

The retention percentage under the JCT Contract (30.4) is 5 per cent
unless a lower rate is agreed at the tender stage between the parties and
specified in the appendix. Where the contract sum is estimated to be
more than £500 000 it is recommended that the retention percentage should
be not more than 3 per cent, and hence most contracts are at the reduced
rate. If no percentage is stated it will automatically be 5 per cent. The
agreed percentage retention applies only to interim certificates before practical completion. In the case of interim certificates issued after practical
completion, the employer may retain only half the retention percentage in
respect of the additional amount certified (30.4.1.3).

Appendix F shows a completed statement of retention and nominated
subcontractors' values, based on the standard form issued by the RICS.

## Release of Retention

The contractor is entitled to the release of part of the retention monies
when the following events occur:

(1)  sectional completion (18.1.2);
(2)  final payment of a nominated subcontractor before practical completion (30.2.3.2);
(3)  practical completion (30.4.1.3); and
(4)  residue on issue of final certificate (30.8).

## Release of Retention arising out of Sectional Completion (18.1.2)

The employer may by agreement take possession of a section of the works
before practical completion. Half the retention on the value of the relevant part is released within 14 days of the employer taking possession of
that section. The other half of the retention is released when the defects
have been made good at the end of the defects liability period. The total
value of the relevant part is, for the purpose of the contract, to be reduced

by the estimated value of any nominated subcontractor's work for which final payment has already been made (30.2.3.2). The following example illustrates the procedure.

The value of a relevant part of the contract to which sectional completion applies is £104 000, and the value of a reinforced concrete frame for which a final payment has been made is £20 576.

|  | £ |
| --- | --- |
| value of relevant part | 104 000.00 |
| *less* value of reinforced concrete framework (paid in full) | 20 576.00 |
|  | £83 424.00 |

The percentage of retention to be released is one-half of 5 per cent (2.5 per cent). Hence the retention to be released is 2.5 per cent of £83 424 = £2085.60.

It is good practice for the released amounts to be recorded on an architect's certificate, thus ensuring that they will be incorporated in the next valuation.

### Release of Retention by virtue of the Final Payment of a Nominated Subcontractor before Practical Completion

Where an architect wishes to secure final payment of a nominated subcontractor before the main contract is completed, he may release the retention held against the completed subcontract work. The subcontractor must satisfactorily indemnify the contractor against latent defects. The gross retention will be reduced by the amount of retention released.

### Typical Interim Valuation

An interim valuation is produced in table 5.6 covering the later stages of an office block contract to show the general format and the more usual contents. The adjustment of prime cost sums for nominated subcontractors and suppliers and of preliminaries items is considered in greater detail in chapter 6.

Another interim valuation and the supporting statement of retention and of nominated subcontractors' values on the standard RICS forms are provided in Appendices E and F.

Table 5.6   Typical interim valuation

---

OFFICE BLOCK, JACKSON STREET, BROADWELL

Valuation Nr. 14: 15 May 1996

| | £    p | £    p |
|---|---:|---:|
| * Bill Nr. 1 Preliminaries | | See subsequent entry |
| Bill Nr. 2 Substructure | | 108 909.00 |
| Bill Nr. 3 Superstructure | | |
| Excavation and Earthworks | 73 901.00 | |
| Concrete work | 92 504.00 | |
| Brickwork and Blockwork | 89 803.00 | |
| Woodwork 86A–91D   45 493.00 | | |
| 94A–95F   14 280.00 | | |
| 99A–101E   3 707.00 | 63 480.00 | |
| Metalwork | 16 789.00 | |
| Plumbing Installation | 64 963.00 | |
| Electrical Installation | 14 198.00 | |
| Finishings   85% of £52 256 | 44 418.00 | |
| Glazing | 16 256.00 | |
| Painting   25% of £40 864 | 10 216.00 | 486 528.00 |
| Bill Nr. 4 External Works | | |
| Drainage | 20 160.00 | |
| Fencing   75% of £23 392 | 17 544.00 | |
| Pavings   60% of £12 026 | 7 216.00 | 44 920.00 |
| | | 640 357.00 |
| * Preliminaries (priced items analysed) | | |
| Cost related items 2.9% of £640 357 | 18 570.00 | |
| Time related items 14 × £3 344 | 46 816.00 | |
| (total cost spread over contract | | |
| period of 15 months) | | |
| Lump sums | 10 032.00 | 75 418.00 |
| | | 715 775.00 |
| Variations Nrs 1–27 | | 17 936.00 |
| Nominated subcontractors: | | |
| Frames Construction – floor and roof beams | 23 200.00 | |
| Hot Air Ltd            – heating installation | 52 000.00 | |
| Electrics Ltd          – electrical work | 59 200.00 | |
| | 134 400.00 | |
| Profit                        5% | 6 720.00 | |
| Attendances – floors and roof beams   1 120 | | |
| – heating installation   1 440 | | |
| – electrical work   800 | 3 360.00 | 144 480.00 |

| | | | |
|---|---|---|---|
| Nominated suppliers: | | | |
| Ironmongery | | 11 520.00 | |
| Profit | 5% | 576.00 | 12 096.00 |
| Statutory bodies: | | | |
| Severn Trent Water – water main connection | | 896.00 | |
| MEB           – electric main connection | | 800.00 | |
| Northern Council  – sewer connection | | 1 920.00 | |
| | | 3 616.00 | |
| Profit | 5% | 181.00 | 3 797.00 |
| Materials on site: | | | |
| Paint and wall tiles | | | 4 480.00 |
| Fluctuations, as agreed | | | 39 747.00 |
| | | | 938 311.00 |
| Retention | | 938 331.00 | |
| *less* amounts subject to nil retention | | | |
| Statutory bodies | 3 616 | | |
| Fluctuations | 39 747 | 43 363.00 | |
| *less* retention @ 3% of | | 894 948.00 | 26 848.00 |
| Total of Valuation Nr. 14 | | | 911 463.00 |
| *less* total of certificates 1–13 | | | 840 538.00 |
| Total amount due | | | £70 925.00 |

# 6 Variations and Final Accounts

This chapter is concerned with the adjustment of preliminaries and prime cost and provisional sums, the valuation of daywork, variations and extras, the preparation of financial statements for the employer and final account procedure.

## ADJUSTMENT OF PRELIMINARIES

The first bill in a bill of quantities normally contains a list of contract clause headings and detailed requirements relating to contractor's obligations and general facilities to be provided as listed in clauses A40–A44 inclusive of the Standard Method of Measurement of Building Works (SMM7, 1988). The preparation of a preliminaries bill is detailed by the author in *Advanced Building Measurement* (1989).

A contractor often prices his on-site and contract costs in the preliminaries and these may amount to 5 to 20 per cent of the contract sum. As described in chapter 5, the contractor frequently incurs very high costs in setting up his site organisation and the recovery by monthly proportional costs, as a quick and simple method of assessment, will be far less than the actual cost expended in the early stages of the contract. For example, typical on-site costs priced in the preliminaries which could be included in the first valuation are shown in table 6.1.

As described in chapter 5, preliminaries items may be cost related, time related, lump sum or a combination of these categories. To illustrate this a relatively simple example is taken of a number of commonly occurring preliminaries items on a factory contract with a contract value of £1 310 000 and a contract period of 12 months commencing on 1 May 1996. Table 6.2 shows the likely cost patterns over the contract period for twelve selected items.

Many of the items have a time-related element within them, but they may also peak at the start of the contract and sometimes also at the end. For example, the supply of water for the works includes the cost of temporary plumbing in the first month's payment and after that equal monthly payments for the supply of water which is often calculated on the basis of 0.25 per cent of the contract value. Temporary roads and access entail heavy constructional costs at the outset, much smaller monthly costs of maintenance over the intervening period and increased expenditure in the last month of the contract for breaking up and removal of the temporary

Table 6.1    Typical on-site costs to be included in Valuation Nr. 1

|  | £ |
|---|---|
| Bond | 8 000 |
| Insurance | 4 000 |
| Site accommodation (erection only) | 700 |
| Equipment (purchased direct for site use) | 600 |
| Fencing/hoarding (purchased direct for site use) | 1 100 |
| Electrical installation to site accommodation | 650 |
| Water installation for site use and Water Authority charges | 350 |
| Temporary electrical services | 250 |
| Temporary installation (two telephones) | 300 |
| Access road (part only recoverable in this valuation) | 1 000 |
| Site notice boards | 500 |
| Site signs | 150 |
| Transportation of plant and equipment (part only) | 300 |
| Total to be included in Valuation Nr. 1 | £17 900 |

works. Temporary accommodation also involves heavier expenditure during the first and last months of the contract for transport and erection, and dismantling and transport respectively. The temporary telephones entail higher expenditure at the beginning of the contract for installation followed by an even cost throughout the remainder of the contract covering telephone rentals and charges for telephone calls. The major cost in removal of rubbish and cleaning the works occurs at the end of the contract, although some cost is involved each month in keeping the site reasonably clean and tidy and preventing the excessive accumulation of debris.

The incidence of scaffolding costs is determined by the timing and extent of the permanent works that they serve, while the cost of protective measures in inclement weather is confined to the winter months when conditions are normally at their worst. Insurance premiums constitute a lump sum which is payable at the start of the contract.

A breakdown of this type assists in making realistic assessments of the cost of preliminaries at any particular stage of the contract and provides the basis for the type of subdivision shown in the interim valuation in table 5.6. Table 6.2 shows clearly the significant levels of costs occurring in the first month of the contract (£17 425) representing over 15 per cent of the total of preliminaries. If calculated on a time-related basis the payment would amount to only £9429 or 8.33 per cent of the total value of the preliminaries.

Table 6.2 Breakdown of preliminaries items over contract period

| Item | Total | | | | | | Months | | | | | | |
|---|---|---|---|---|---|---|---|---|---|---|---|---|---|
| | | 1 | 2 | 3 | 4 | 5 | 6 | 7 | 8 | 9 | 10 | 11 | 12 |
| | £ | £ | £ | £ | £ | £ | £ | £ | £ | £ | £ | £ | £ |
| Site staff | 39 840 | 3 320 | 3 320 | 3 320 | 3 320 | 3 320 | 3 320 | 3 320 | 3 320 | 3 320 | 3 320 | 3 320 | 3 320 |
| Plant, tools and vehicles | 20 520 | 1 710 | 1 710 | 1 710 | 1 710 | 1 710 | 1 710 | 1 710 | 1 710 | 1 710 | 1 710 | 1 710 | 1 710 |
| Scaffolding | 13 200 | – | – | – | 500 | 900 | 2 200 | 3 200 | 3 200 | 1 800 | 600 | 400 | 400 |
| Site administration and security | 3 600 | 300 | 300 | 300 | 300 | 300 | 300 | 300 | 300 | 300 | 300 | 300 | 300 |
| Water for the works | 3 300 | 550 | 250 | 250 | 250 | 250 | 250 | 250 | 250 | 250 | 250 | 250 | 250 |
| Protecting the works from inclement weather | 1 650 | – | – | – | – | – | – | – | 550 | 550 | 550 | – | – |
| Lighting and power for the works | 6 000 | 500 | 500 | 500 | 500 | 500 | 500 | 500 | 500 | 500 | 500 | 500 | 500 |
| Temporary roads and access | 7 400 | 5 200 | 175 | 175 | 175 | 175 | 175 | 175 | 175 | 175 | 175 | 175 | 450 |
| Temporary accommodation | 6 350 | 750 | 500 | 500 | 500 | 500 | 500 | 500 | 500 | 500 | 500 | 500 | 600 |
| Temporary telephones | 1 620 | 245 | 125 | 125 | 125 | 125 | 125 | 125 | 125 | 125 | 125 | 125 | 125 |
| Remove rubbish and clean the works | 5 150 | 350 | 350 | 350 | 350 | 350 | 350 | 350 | 350 | 350 | 350 | 350 | 1 300 |
| Insurances | 4 500 | 4 500 | – | – | – | – | – | – | – | – | – | – | – |
| Monthly totals | | 17 425 | 7 230 | 7 230 | 7 730 | 8 130 | 9 430 | 10 430 | 10 980 | 9 580 | 8 380 | 7 630 | 8 955 |
| Cumulative totals | £113 130 | 17 425 | 24 655 | 31 885 | 39 615 | 47 745 | 57 175 | 67 605 | 78 585 | 88 165 | 96 545 | 104 175 | 113 130 |

## ADJUSTMENT OF PRIME COST AND PROVISIONAL SUMS

### Prime Cost Sums

These are included in the contract to cover specialist work and the supply of specific materials, which entail the employment of nominated subcontractors or nominated suppliers. For example, nominated subcontractors are usually appointed to carry out such work as electrical installations, heating and ventilation systems and the like. Nominated suppliers may be appointed to supply specific materials or goods, such as facing bricks, sanitary appliances and ironmongery, which are subsequently fixed or built into the work by the main contractor. A prime cost sum is a sum provided for work or services to be executed by a nominated subcontractor or for materials or goods to be obtained from a nominated supplier. Such sum shall be deemed to be exclusive of any profit required by the general contractor and provision shall be made for the addition thereof.

By way of contrast provisional sums are included in the contract bills to cover the cost of work to be executed by statutory authorities who are empowered to carry out their functions by virtue of Acts of Parliament, regulations and byelaws. Examples of these categories of work are gas, electricity, water and sewer connections. Note that the term 'contract bills' is used in the 1980 JCT Form of Contract to describe the operative bill of quantities.

### Provisional Sums

These are included in the contract for work of which the full extent and character cannot be determined precisely at the time the bill of quantities is prepared. For example, the existence of a stone-lined underground storage tank may be known when new basement details are being prepared and measured, but the full nature and extent of the work required in connection with the underground tank in the construction of the basement is unlikely to be fully determinable. The cost of performing the necessary work is therefore included in the contract as a provisional sum. Under the JCT Standard Form the architect has the power to authorise work contained in provisional sums to be executed either by the main contractor or by a nominated subcontractor (13.3).

SMM7 (1988) provides for two categories of provisional sum:

1. Provisional sum for *defined work* is a sum provided for work which is not completely designed but for which the following information is provided – (a) nature and construction of work; (b) statement of how and where the work is to be fixed to the building and what other work is to be fixed thereto; (c) quantity or quantities which indicate scope and extent of work; (d) any specific limitations.

2. Provisional sum for *undefined work* is a sum provided for work where the information listed in 1(a)–(d) cannot be given.

## Contingency Sum

This is a sum included in a contract to cover the cost of work or expenses not contemplated or implied in the contract sum. These monies are to be expended at the discretion of the architect, and where no unforeseen circumstances arise this figure constitutes a saving to the employer. This provides an opportunity for the employer to secure the execution of essential additional work within the contract sum. For example, underground structures and services may be discovered which were not previously known to exist. Work must be carried out on them before the contract works can be continued and the extra cost of this work must be met by the employer.

## Contractual Provisions

The amounts included in the contract bills for prime cost and provisional sums are deducted and the actual amounts contained in the certified accounts of the nominated firms or the value of the main contractor's additional work added to the variation account, and the contract sum is adjusted accordingly.

Under the 1980 JCT Standard Form of Building Contract, the architect is enabled to authorise the expenditure of prime cost and provisional sums by virtue of the following clauses:

(i)   clause 13.4 relating to the expenditure of provisional sums included in the contract bills and valued by the quantity surveyor;

(ii)  clause 35.1.4 provides for the nomination of subcontractors for the supply and fixing of any materials or goods or the execution of work for which prime cost sums have been included in the contract bills; and

(iii) clause 36.1.1 provides for the nomination of suppliers of materials or goods for which prime cost sums have been included in the contract bills.

## Discounts

These are the amounts by which quoted costs or prices may be reduced when the contractor is paying accounts and are usually dependent on certain specified conditions being met. Discounts are not obligatory unless there is prior agreement as, for example, with cash discounts under the JCT Form of Contract. There are two types of discount allowed by suppliers and subcontractors.

### (i) Trade discounts

These discounts are allowed off the standard prices of certain materials and goods by suppliers for the benefit of regular customers. Under the

JCT Contract, the benefits of these discounts pass to the employer. The amount of discount often varies considerably between different suppliers.

*(ii) Cash discounts*

This type of discount is given by suppliers or subcontractors as an incentive for prompt payment. Prime cost sums are deemed to include a cash discount of 2.5 per cent in respect of subcontractors and 5 per cent where suppliers are involved. With subcontractors the cash discount of 2.5 per cent applies only where the subcontractor is paid by the main contractor within the prescribed period; normally within 7 days of the date of issue of the architect's certificate. In the case of suppliers, the cash discount of 5 per cent is allowed provided that payment is made within 30 days of the end of the month during which delivery of the materials or goods is made (36.4.4).

On occasions subcontractors and suppliers quote nett prices (exclusive of cash discount) or prices inclusive of discount but at the wrong percentage. Under the JCT Standard Form the contractor has the right to the appropriate discount and the employer has the right not to pay more than this. The use of the JCT Standard Form of Nominated Subcontract NSC/C should ensure that the cash discount included in a subcontractor's tender is correct. However, in cases where the JCT nomination procedure for a subcontractor is not operated, problems may arise in dealing with cash discounts.

The easiest solution where quotations or tenders are nett or include the wrong percentage is for the architect to request the subcontractor or supplier to revise the quotation to include the discount and a satisfactory response is usually forthcoming. Where the architect instructs the contractor to place an order against a quotation that includes an incorrect discount or no discount at all, the contractor has a duty to advise the architect and to seek instructions. Failure by the contractor to seek such instructions could be held to constitute a waiver by the contractor of his rights under clauses 35 and 36. Upon receiving a request for instructions, the architect may vary the contract by instructing the contractor to place an order with the firm concerned and to allow the appropriate cash discount as an extra. The adjustments will take the following form:

(i) nett addition of 1/39 to an amount subject to 2.5 per cent cash discount;
(ii) nett addition of 1/19 to an amount subject to 5 per cent cash discount; and
(iii) nett addition of 1/39 to an amount which includes 2.5 per cent cash discount but was subject to 5 per cent cash discount.

Some examples will serve to illustrate the practical application of these adjustments.

(1) A nominated subcontractor rendered an invoice for £2990 for supplying and laying PVC floor tiling, the invoice being marked 'nett monthly account.' The invoice should have included 2.5 per cent cash discount and to correct the error it is necessary to add 1/39, as the percentage is deductable from the gross or inclusive total.

|  | £ |
|---|---|
| Amount of invoice | 2990.00 |
| *Add* 1/39 (2.5 per cent cash discount) | 76.67 |
| Payment due to contractor | £3066.67 |

The sum of £3066.67 will be paid by the employer to the contractor, who will retain the 2.5 per cent cash discount (£76.67) and pay the balance of £2990 to the subcontractor.

(2) A nominated supplier submitted a nett invoice of £10 270 for the supply and delivery of metal windows. The invoice should have included a cash discount of 5 per cent and an addition of 1/19 is required to correct the error.

|  | £ |
|---|---|
| Amount of invoice | 10 270.00 |
| *Add* 1/19 (5 per cent cash discount) | 540.53 |
| Payment due to contractor | £10 810.53 |

(3) A nominated supplier of ironmongery invoiced the goods supplied at £6148.24, with the invoice showing a cash discount of 2.5 per cent instead of 5 per cent. The 2.5 per cent should be deducted and the 5 per cent substituted by an addition of 1/19 to the nett amount.

|  | £ |
|---|---|
| Amount of invoice | 6148.24 |
| *Deduct* 2.5 per cent | 153.71 |
|  | 5994.53 |
| *Add* 1/19 | 315.50 |
| Payment due to contractor | £6310.03 |

A check can be made by taking 5 per cent of £6310.03, which equals £315.50.

Clause 4.17 of the JCT Nominated Subcontract NSC/C provides for the addition of 1/39 to fluctuations and certain other costs so that the main contractor can deduct the discount without reducing the original amounts. Statutory bodies, such as water, gas and electricity boards, do not normally give cash discounts and this is formally recognised in the JCT Standard Form (6.3). In these instances the tenderer is usually informed that he will be deemed to have made allowance in his addition for profit for any extra amount he may require to offset in the absence of cash discount.

### Prime Cost Sums for Materials

It is often convenient for the architect to request that a prime cost sum be inserted in a bill of quantities to cover the supply of specific materials or components. The actual choice by the employer of the items to be used on the project can be deferred and, at the same time, all contractors will be tendering on the same basis. SMM7 (1988) has, however, tightened up on this procedure by requiring that the bill of quantities shall, for example, fully describe the bricks or blocks to be used and to draw attention to any limitations on laying so that the work can be properly priced. This is particularly so where the brick or block is the subject of a prime cost sum for nominated supply where it is possible for different types of brick or block with identical purchasing prices to have widely differing laying costs. It is recommended in the SMM7 Measurement Code (1988) that where the supply is the subject of a prime cost sum or a prime cost rate per thousand, the assumptions to be made by the tenderers should be clearly stated. Hence, there are positive advantages from all points of view in selecting the product to be used at the time of finalising the bill of quantities.

Where a product has been selected, it is good policy for a quotation to be obtained for entry in the bill, to avoid the need for all contractors individually to have to obtain quotations. There are other factors that deserve consideration at this stage.

(i) Where a prime cost price is given for a specific material it should be extended to provide prices for all necessary associated materials. For example, with facing bricks, prime cost prices may be required for standard bricks, bullnosed bricks, splayed bricks and the like. Similarly with roofing tiles prime cost prices may be needed for plain tiles, eaves tiles, tile-and-a-half tiles, ridge tiles, hip tiles and valley tiles.

(ii) Prices of materials should preferably be 'delivered site'. Where quotations are 'ex works', it is necessary to make a suitable addition to cover transport or to inform the contractor of the location from which they are to be collected.

(iii) Where the contract makes provision for prime cost prices to allow a specified cash discount, the estimate must be checked to ensure that this has been done. If the discount is not clearly stated then confirmation should be obtained and a revised estimate requested if necessary.

### Main Contractor's Profit

Prime cost sums are deemed to exclude the main contractor's profit and so provision is made in the bill of quantities for a separate addition which is normally calculated as a percentage of the prime cost sum. The tenderer inserts the percentage that he requires. When the contract sum is adjusted by the omission of each prime cost sum and the substitution of the actual cost involved, the addition for profit will be adjusted on a *pro rata* basis, by applying the same percentage to the actual cost as that inserted in the bill against the initial prime cost sum.

Where a profit addition is given as a lump sum, without any indication of the method of calculation, this sum will also be adjusted *pro rata* to the actual cost. In the event of the employer paying a subcontractor direct, the main contractor is still entitled to the profit addition, as it represents a management fee for arranging and taking responsibility for subcontract work or for arranging and checking the supply of materials and goods.

### Attendances

The main contractor attends upon nominated subcontractors, as he is responsible for their work and for the satisfactory integration of it with his own work. He has to provide facilities to subcontractors so that their work may proceed uninterrupted. Tenderers are accordingly directed in the tender documents to allow for all such attendances in their tender sums, and they should be given as much information as possible about the nominated subcontractors' work.

Where bills of quantities are provided, separate items are included for 'general attendances' and 'special attendances'. The item for general attendance is intended to be an indication of the facilities that are normally available to subcontractors where they are provided by the main contractor to meet his own needs. Under SMM7, clause A42.1–16, general attendance is deemed to include the use of the contractor's temporary roads, pavings and paths, standing scaffolding and standing power operated hoisting plant, the provision of temporary lighting and water supplies, clearing away of rubbish, provision of space for the subcontractor's own offices and for the storage of his plant and materials and the use of messrooms, sanitary accommodation and welfare facilities provided by the contractor.

Any items of special attendance on nominated subcontractors that are required are separately listed and described in the bill. The object is to enable proper provision to be made for significant costs beyond those envisaged in the definition of general attendance. For example, all special scaffolding or scaffolding additional to the contractor's standard scaffolding or standing scaffolding required to be altered or retained required for use by subcontractors must be described in adequate detail. Typical examples of special attendances are as follows:

(i) Special scaffolding or scaffolding additional to the contractor's standing scaffolding.

(ii) The provision of temporary access roads and hardstandings in connection with structural steelwork, precast concrete components, piling, heavy items of plant and the like.

(iii) Unloading, distributing, hoisting and placing in position, giving in the case of significant items the weight and/or size.

(iv) The provision of covered storage and accommodation including lighting and power thereto.

(v) Power supplies giving the maximum load.

(vi) Maintenance of specific temperature or humidity levels.

The prices included in tenders for attendances are normally regarded as fixed sums and so, unlike the profit additions, are not usually adjustable *pro rata* to the actual cost of the subcontract works. This is because the amount and type of attendance work or services is unlikely to vary significantly with variations in the actual cost of the subcontract work compared with the prime cost sum. However, where the subcontract works vary substantially from what was originally intended and thereby change the nature and extent of the attendances, then some adjustment of the attendance amounts is required. On occasions contractors insert percentages against attendance items of prime cost sums in the same way as for profit items. Nevertheless, the amounts shall still be treated as lump sums and the quantity surveyor should obtain the contractor's agreement to this and to the deletion of the percentage rates, prior to the signing of the contract.

### Practical Applications of the Adjustment of Prime Cost Sums

Some examples covering the adjustment of prime cost sums for subcontractors' work, materials and goods and statutory bodies' services follow and these illustrate the normal approach.

|  |  | Omissions £ p |  | Additions £ p |
|---|---|---|---|---|
| *Structural steel frame* |  |  |  |  |
| Prime cost sum as item 189A |  | 85 000.00 |  |  |
| Profit as item 189B | 5% | 4 250.00 |  |  |
| General attendance as item 189C |  | 2 300.00 |  |  |
| Special attendances as item 189D |  | 2 600.00 |  |  |
| Trentlock Steel Fabrication Ltd, |  |  |  |  |
| approved account (dated 24/5/96) |  |  |  | 81 290.00 |
| Profit *pro rata* item 189B |  |  | 5% | 4 064.50 |
| General attendance as item 189C |  |  |  | 2 300.00 |
| Special attendances as item 189D |  |  |  | 2 600.00 |
|  |  | 94 150.00 |  | 90 254.50 |
| *less* additions |  | 90 254.50 |  |  |
| Nett omission carried to Summary |  | £ 3 895.50 |  |  |

| *Electrical installation* | | *Omissions* | | *Additions* |
|---|---|---|---|---|
| Prime cost sum as item 189E | | 52 000.00 | | |
| Profit as item 189F | 5% | 2 600.00 | | |
| General attendance as item 189G | | 1 300.00 | | |
| Providing hardstanding for sub-contractor's own offices and storage hut as item 189H | | 880.00 | | |
| Providing power, maximum load 7.5 kVA as item 189J | | 190.00 | | |
| Jameson Electrical Ltd, approved account (dated 21/6/96) | | | | 54 590.00 |
| Profit *pro rata* item 189F | | | 5% | 5 729.50 |
| General attendance as item 189G | | | | 1 300.00 |
| Special attendances: | | | | |
|   hardstanding as item 189H | | | | 880.00 |
|   providing power as item 189J | | | | 190.00 |
| | | 56 970.00 | | 62 689.50 |
| *less* omissions | | | | 56 970.00 |
| Nett addition carried to Summary | | | | £ 5 719.50 |

| *Ironmongery* | | | | |
|---|---|---|---|---|
| Prime cost sum as item 190A | | 11 000.00 | | |
| Profit as item 190B | 5% | 550.00 | | |
| Modern Ironmongery Ltd, approved invoice (dated 13/6/96) | | | | 11 254.00 |
| Profit *pro rata* item 190B | | | 5% | 562.70 |
| | | 11 550.00 | | 11 816.70 |
| *less* omissions | | | | 11 550.00 |
| Nett addition carried to Summary | | | | £ 266.70 |

*Note*: It will be noted that no provision is made for attendance on nominated suppliers. Unloading, storing, hoisting and returning packing materials are deemed to be included in the items of fixing.

| *Sanitary Appliances* | | | | |
|---|---|---|---|---|
| Prime cost sum as item 190C | | 16 000.00 | | |
| Profit as item 190D | 5% | 800.00 | | |
| Murphy Sanitation Ltd, approved invoice (dated 4/6/96) | | | | 15 651.20 |
| Profit *pro rata* item 190D | | | 5% | 782.56 |
| | | 16 800.00 | | 16 433.76 |
| *less* additions | | 16 433.76 | | |
| Nett omission carried to Summary | | £ 366.24 | | |

|  | | Omissions | | Additions |
| --- | --- | --- | --- | --- |
| *Water main connection* | | | | |
| Prime cost sum as item 190E | | 560.00 | | |
| Profit as item 190F | 5% | 28.00 | | |
| General attendance as item 190G | | — — — — | | |
| Severn Trent Water approved account (dated 19/4/96) | | | | 580.16 |
| Profit *pro rata* item 190F | | | 5% | 29.01 |
| General attendance as item 190G | | | | — — — — |
| | | 588.00 | | 609.17 |
| *less* omissions | | | | 588.00 |
| Nett addition carried to Summary | | | | £ 21.17 |

## Procedure where a Contractor carries out Work covered by a Prime Cost Sum

The main contractor may be permitted to submit a tender for work included in a prime cost sum in the circumstances outlined in chapter 4. If his tender is accepted, the question arises as to whether he is to be paid the profit addition included in the contract sum. Where the tender for the subcontract works is submitted in competition, then the main contractor is entitled to the profit addition, but otherwise it is deemed to be included in the contractor's price. The contractor should be informed of this at the tender invitation stage. A main contractor cannot attend upon himself and hence any amount included in the contract sum for general attendance by the main contractor will be deducted, but prices inserted for special attendances will stand.

## DAYWORKS

### General Procedure

Daywork is a method of payment for work based on the prime cost of all labour, materials and plant used in carrying out the work, normally with a percentage addition to the total cost of each of the three groups of items for overheads and profit. From time to time, particularly when carrying out work in connection with or adjacent to existing buildings, it is necessary to undertake certain activities which by their very nature cannot be measured and valued at billed rates. Typical examples are the removal of an unexpected obstruction in excavations and the alteration in size of a door or window opening after it has been formed or partially formed.

The JCT Standard Form (13.5) prescribes rules for the valuation of additional or substituted work that cannot properly be valued by measurement. The normal method of evaluation is to adopt the prime cost of the work calculated in accordance with the RICS *Definition of Prime Cost of Daywork carried out under a Building Contract* (1975), together with percentage additions to each section of the prime cost at the rates inserted by the contractor in the contract bills. In the case of electrical and heating and ventilating works, there are separate prime cost of daywork definitions published by the RICS (1980a/1981). The old method of the tenderer pricing provisional hours of labour and plant and quantities of materials is now little used. The BCIS (1994) produced calculations of standard hourly base rates for labour which are periodically updated and show the changes since 1984.

Vouchers, commonly termed daywork sheets, detailing the time spent daily upon the work, the workmen's names and the plant and materials used, shall be submitted for verification to the architect or his authorised representative, usually the clerk of works, not later than the end of the week following that in which the work has been executed. It should be emphasised that these only represent a record of work done that constitutes a variation to the contract and they do not necessarily indicate that the work will be valued at daywork rates in preference to using measured rates. Furthermore, the verifying signature of the architect or his representative relates only to the amount of labour, materials and plant used and not to any prices inserted. The valuing is undertaken by the quantity surveyor and he may reduce the quantities recorded if he considers them to be excessive.

The rates and prices used in the valuation of daywork shall be those current at the time the work is carried out, and not those operative at the date of tender. Labour rates are based on standard wage rates, regardless of whether or not these are the actual rates paid. Where the contractor pays operatives wages above the basic rates, he is deemed to have included the difference in the addition for profit. Plant costs can be calculated on the basis of the BCIS *Schedule of Basic Plant Charges* (1990), but the rates quoted in this schedule will need updating. The PSA (1990) published a useful schedule containing over 21 000 items of building work, which will also need adjusting. The main weaknesses with dayworks are the difficulty of checking the time spent on such works, how long they should have taken and the absence of competition.

### Preparation of Daywork Vouchers

Gerrity (1980) prepared a set of useful guidelines in the preparation of daywork vouchers (daywork sheets) in order to avoid subsequent problems, and they comprise, in essence, a code of good practice.

(1) Each daywork sheet must relate to a specific architect's instruction, preferably to a variation order or site works order, and the reference numbers of these must be stated.

(2) Daywork vouchers should be submitted by the main contractor to the architect or clerk of works for signature. Nominated subcontractors should first submit their vouchers to the main contractor who will then include his own attendances and submit new vouchers to the architect or his representative.

(3) Daywork vouchers must be submitted for verification not later than the end of the week following that in which the work has been done. This means that daywork records must be prepared weekly. Subcontractors must be informed of the procedure and of the day on which their own vouchers must be submitted to the main contractor.

(4) It is essential that daywork vouchers should properly describe the work done. Many daywork sheets are prepared in such a way that only minimal space is allowed for the description of the work. Since the daywork account may not be valued for some months, exact, clear and concise descriptions of the work are essential, even to the extent of supplementing the record with sketches or photographs of the work done. There may also be other work proceeding at the same time to be paid for on a measured basis, and it will be necessary to distinguish between the separate activities.

(5) Plant and materials must be recorded as well as labour. It is, therefore, important to record the description and type of plant used, the cost of any small quantities of materials specially ordered and any extra handling costs in addition to the labour required. When plant on site is being used uneconomically instead of bringing the more appropriate plant on to the site, the agreement of the architect should be obtained as special rates may apply. In accounting for these items it is necessary to observe the provisions of the *Definition of Prime Cost of Daywork.*

(6) Most architects require the recording of operatives' names as well as the time spent on daywork, as prescribed in JCT80 (13.5.4). Subcontractors should be instructed to prepare their daywork accounts in a similar format.

(7) Wage rates for labourers, craft operatives and foremen differ and, therefore, the status of the operative must be shown in addition to his name. Time must be allowed for setting out, covering up finished work, clearing away rubbish and making good, and these costs may be higher than normal.

(8) Daywork vouchers must be produced at least in triplicate, one copy for the clerk of works, one to the contractor's surveyor and one to the site manager. These should be carefully filed in sequence, together with the relevant instructions relating to them.

(9) Once a voucher is signed as a true record of work done, this may not be the end of the matter and so all relevant drawings, sketches, instructions and delivery notes must be retained or passed to the contractor's surveyor dealing with the particular contract.

(10) Gerrity (1980) emphasises how it is easy to forget the obvious when preparing daywork sheets. In particular, cognisance should be taken of any scaffolding, shoring, special sections of formwork with only one possible use, extra long barrow runs, protection of adjacent finished work, access problems, difficulties in working in confined space, sharpening tools for breaking concrete, disorganisation of following trades, and other significant matters.

A typical daywork sheet is illustrated in table 6.3.

## Cost of Daywork

Quantity surveyors often contend that work executed on a daywork basis is more costly than similar work undertaken at billed rates, principally because there is no incentive for the contractor to execute the work expeditiously. However, the nature of the work is such that in the majority of cases it is likely to take longer to perform than seemingly similar billed items. Many contractors argue that even although they insert seemingly high percentages in respect of dayworks, these rarely cover the actual costs incurred. They further state that they would prefer to have as little work as possible undertaken on this basis. Admittedly, the work normally executed on a daywork basis is generally far from straightforward and is likely to require increased supervision and cause considerable dislocation of the organising and programming of the works, apart from the incidental additional costs previously described.

## Daywork Accounts

The daywork account is built up from the daywork sheets which are normally submitted on a weekly basis. The quantity surveyor will check the daywork sheets against the architect's instructions, and he will list any of them that will be superseded by measurement and those covered by work in the bill of quantities. He will also check them for any inconsistencies and will record any details that appear excessive for subsequent discussion with the contractor's representative, normally his surveyor.

The resulting particulars, after verification, can then be tabulated in a daywork account of the form illustrated in table 6.4, where the percentage additions inserted by the contractor in the daywork schedule in the bill of quantities have been applied. The daywork account will subsequently be incorporated into a variation account.

Table 6.3 Typical daywork sheet

*Contractor* Johnson and James Ltd   *Contract* Hay's Wharf, Warehouses   *A.I.* nr. 27 (12/6/96)

*Work commenced* 8 July 1996   *Work completed* 12 July 1996   *Daywork sheet* nr. 7   *Date* 15 July 1996

*Description of work* Demolition of Stores marked 'X' on drawing HWW/6/2B and making good boundary wall

### LABOUR

| Name | Trade | M | Tu | W | Th | F | Sa | Su | Hrs | Rate | £ | p |
|------|-------|---|----|---|----|---|----|----|-----|------|---|---|
|      |       |   |    |   |    |   |    |    |     | £ |   |   |
| J. Thompson | B/layer | 8 | 8 | 8 | 8 | 8 | | | 40 | 5.50 | 220 | 00 |
| S. Dicks | Labourer | 8 | 8 | 8 | 8 | 8 | | | 40 | 4.70 | 188 | 00 |
| T. Jackson | Labourer | 8 | 8 | 8 | 8 | 8 | | | 40 | 4.70 | 188 | 00 |
| R. Smith | Labourer | 8 | 8 | 8 | 8 | 8 | | | 40 | 4.70 | 188 | 00 |
| B. Stocking | Labourer | 8 | 8 | 8 | 8 | 8 | | | 40 | 4.70 | 188 | 00 |
| G. Appleby | Labourer | – | 8 | 8 | 8 | 4 | | | 28 | 4.70 | 131 | 60 |
| P. Jinks | M/operator | – | 8 | 8 | 8 | 2 | | | 26 | 5.50 | 143 | 00 |
| R. Jones | Ganger | – | 8 | 8 | 8 | 5 | | | 29 | 5.50 | 159 | 50 |
| | | | | | | | | | | | £ 1406 | 10 |

### MATERIALS

| Description | Qty | Rate | £ | p |
|-------------|-----|------|---|---|
| | | £ | | |
| Class B engineering bricks | 1800 | 390.00 per M | 702.00 | |
| Portland cement | 1 tonne | 77.00 | 77.00 | |
| Sand | $3\frac{1}{2}$ tonnes | 11.50 | 40.25 | |
| | | | £ 819.25 | |

### PLANT

| | | | £ | p |
|---|---|---|---|---|
| JCB 3C excavator | 26 | 11.10 | 288 | 60 |
| 2 tonne dumper | 26 | 3.30 | 85 | 80 |
| Concrete mixer | 20 | 0.90 | 18 | 00 |
| | | | £ 392 | 40 |

Site agent's signature ..................................   The information on this sheet has been verified (excluding rates and prices)

Architect/architect's representative's signature ...............

Table 6.4   Daywork account

| | £ | p |
|---|---|---|

*Hay's Wharf Warehouses*

*A.I. nr. 27* (issued 12 June 1996)

*Executed 12 July 1996 Daywork sheet nr. 7*

*Description*
Demolition of Stores marked 'X' on drawing
HWW/6/2B and making good boundary wall

| | | £ p |
|---|---|---|
| *Labour* | | |
| 188 hours labourer | 4.70 | 883 60 |
| 40 hours craft operative | 5.50 | 220 00 |
| 26 hours machine operator | 5.50 | 143 00 |
| 29 hours ganger | 5.50 | 159 50 |
| | | 1406 10 |
| Percentage addition | 110% | 1546 71 |
| | | 2952. 81 |
| | | |
| *Materials* | | |
| 1800 class B engineering bricks | 390.00 per M | 702 00 |
| 1 tonne, Portland cement | 77.00 | 77 00 |
| $3\frac{1}{2}$ tonnes, sand | 11.50 | 40 25 |
| | | 819 25 |
| Percentage addition | 15% | 122 89 |
| | | 942. 14 |
| *Plant* | | |
| 26 hours JCB 3C excavator | 11.10 | 288 60 |
| 26 hours 2 tonne dumper | 3.30 | 85 80 |
| 20 hours concrete mixer | 0.90 | 18 00 |
| | | 392 40 |
| Percentage addition | 15% | 58 86 |
| | | 451. 26 |
| Total to Summary | | £4346. 21 |

## VARIATIONS AND EXTRAS

With a lump sum contract, the contract sum may be adjusted under various heads, the most common of which are:

(i)   approximate quantities (considered in chapter 5);
(ii)  variations;
(iii) extras permitted by the contract;
(iv)  prime cost and provisional sums (detailed earlier in this chapter); and
(v)   fluctuations (considered in chapter 5).

### Origins of Variations and Extras

(a) *Variations*  A variation is defined in the JCT Standard Form (13.1.1) as 'the alteration or modification of the design, quality or quantity of the Works'. This includes the addition, omission or substitution of any work or the alteration of the kind or standard of any materials or goods to be used in the Works or the removal from the site of any work executed or materials or goods brought thereon by the contractor for the purposes of the Works other than materials or goods which are not in accordance with the contract. For example, the position of an internal partition may be changed resulting in altered lengths and/or heights, or the material may be changed from brickwork to blockwork.

A variation may originate from a number of causes specified in the JCT Standard Form.

(i)   A discrepancy or divergence between the contract drawings, contract bills, architect's instructions, or any drawings or documents issued by the architect (2.3).
(ii)  Compliance with statutory requirements (6.1.2).
(iii) Alterations or modifications of the design, quality or quantity of the works (13.1.1).
(iv)  Restoration work in making good damage resulting from fire or other causes (22B.3.5).
(v)   Instructions relating to the finding of antiquities or other objects of interest or value on the site (34.2).

(b) *Extras*  The contract sum may be adjusted because of extras that are permitted under the JCT Standard Form.

(i)   Divergence between statutory requirements and architect's instructions (6.1.2).
(ii)  Cost of opening up works or testing of materials or goods which are subsequently found to be satisfactory (8.3).

(iii) Any infringement of royalties and patent rights in complying with architect's instructions (9.2).

(iv) Disturbance to the regular progress of the works and/or for any direct loss and/or expense, where it is not reimbursable under any other contract clause (13.5). Disturbance costs of this kind are usually considered under clause 26, with clause 13.5 being used only as a last resort.

(v) Costs of making good defects, shrinkage or other faults where the architect, at his discretion, so directs in exceptional circumstances (17.2 and 17.3).

(vi) Payment for insurances by the contractor upon default of the employer to insure the works where the employer is responsible (22B.2).

(vii) Any direct loss and/or expense arising out of any matters affecting the regular progress of the works for the reasons specified (26.1).

(viii) Any direct loss and/or damage arising out of the determination of the contract upon default by the employer (28A).

(ix) Fluctuations in the basic prices of labour and materials (39 and 40).

(x) Any direct loss and/or expense in dealing with antiquities and other objects of interest or value found on the site (34.3.1).

**Ordering of Variations**

Modification to the works must be the subject of variations orders from the architect to the contractor if the latter is to receive payment for any additional cost involved. A variation order constitutes an instruction from the architect requesting the contractor to substitute new work or materials for those originally incorporated in the contract or to make such other changes as may be specified. It is extremely unlikely that the project will proceed exactly as originally conceived by the architect for a wide variety of reasons. The issue of variation orders ensures that the quantity surveyors representing the employer and the contractor are kept informed of amendments to the scheme and are given the opportunity to determine their monetary effect. Moreover, it is advisable for the architect to ascertain the likely cost effect of a proposed variation from the quantity surveyor before he issues the variation order.

If a contractor carries out a variation involving additional expense without receiving written confirmation from the architect, he may be unable to recover the extra costs. Where instructions are given orally, these should be confirmed in writing by the contractor to the architect within seven days. The architect has the opportunity during seven days following receipt of the confirmation to dissent, otherwise the instruction takes immediate effect. Failing this, the architect may confirm at any later time prior to the issue of the final certificate (4.3). The architect can, however, subsequently sanction any unauthorised variation that a contractor has previously carried out (13.2). Any directions given by a clerk of works become operative only if confirmed in writing by the architect within two working days (12). The quantity surveyor will need to determine the status of each variation before assessing the financial consequences of it.

Variation orders are often issued on printed forms in triplicate, one copy being sent to the contractor, another to the quantity surveyor and the third copy is retained by the architect. Alternatively, the architect may forward a letter to the contractor detailing the variation to the works and sending a copy to the quantity surveyor. The forms have the advantage of maintaining a uniform, complete, orderly and distinctive approach. The architect's instruction containing a variation must give the title of the contract, name of contractor, date and serial number of the instruction and full details of the variation accompanied by appropriate drawing(s) where necessary, and be signed by the architect. In the case of instructions involving the adjustment of prime cost sums, particulars must be given of the estimate to be accepted, stating the date of the quotation and its reference number.

It is customary to include in the architect's and quantity surveyor's copies of the variation orders, the approximate cost of the variation and, if required, a summary of the balance remaining in the contingency sum. It would not be advisable to include this information on the contractor's copy as he would almost certainly regard the cost as a minimum with consequent inhibiting effects on subsequent negotiations. Where the variation entails work with no counterpart in the contract, a firm estimate will be required from the contractor, which will require close scrutiny by the quantity surveyor prior to acceptance.

The Department of the Environment, Audit Inspectorate, (1983) has identified the need for defining a framework within which the design team can issue variation instructions without prior reference to the public sector employer. Discretion should accordingly be afforded to the design team to instruct variations that are:

(i)  unavoidable and essential for the contract to proceed;
(ii)  of a minor, day-to-day nature, with minor cost implications; or
(iii)  urgently required.

Otherwise it is recommended that all variations that are significant in terms of design or cost should receive the prior approval of the employer. The design team should identify the cost of the proposed variation, the existing budget allowance, the nature of the variation and the justification for it.

## Measurement of Variations

The quantity surveyor will almost certainly receive periodically throughout the contract revised drawings incorporating variations to the works. Each drawing should be stamped with the date of receipt and a large red

'V' inserted in a predetermined position on the drawing, often the bottom right hand corner, to distinguish it from the original drawings from which the billed quantities were taken off.

Variations to foundations occur in most contracts and, at or before the time of the valuation for the first certificate, it is good practice for the quantity surveyor to arrange to meet the contractor's representative on the site to discuss the procedure to be adopted with regard to variations. At this stage it should be arranged for the clerk of works and the foreman to keep certain records of the work as executed and for the contractor to make application for the issue of variation orders as and when they become necessary. For example, the clerk of works will be requested to keep clear and accurate records of works that will subsequently be covered up and to agree his notes and dimensions with the contractor's foreman. Works in this category include the depth of foundations, position of steps in foundation trench bottoms and thickness of hardcore.

As soon as there is sufficient work available to constitute a day's measurement, the quantity surveyor should arrange an appointment with the contractor's representative, normally a surveyor, on the site. If an early start is made with this work, there is a much better chance of completing the variation account within a reasonable time after the completion of the works. Another advantage of the early measurement and valuation of variations is that the financial effect of the larger and more important variations can be settled and, in consequence, more accurate values can be incorporated in respect of them in valuations for interim certificates.

When visiting the site for the purpose of measuring variations, the quantity surveyor will have the appropriate variation orders and supporting documents and possibly a list of items for which the contractor is applying for variation orders. Each of the items will be systematically taken in turn, measuring various items of work as executed, but leaving the omissions to be dealt with in the office. The question as to whether additional work should be measured on the site or taken from drawings in the office is largely dependent on the particular circumstances. For example, a dormer that is fully illustrated on 1:20 scale drawings and 1:5 details, can be more readily and accurately measured from the drawings. Whereas if the same item had been built from sketchy particulars, then measurement on site would achieve a more satisfactory solution. A combination of the two methods may prove to be the best approach in some situations, as is often the case with drainage work. In both cases the approach to measurement will follow the rules laid down in SMM7, and, for example, the dimensions of trench excavation and earthwork support may vary from the actual quantities of work carried out on the site.

The omissions will often be extracted from the quantity surveyor's original dimensions, unless they constitute entire items or sections in the bill of quantities, in which case reference can be made to the appropriate billed items. The quantity surveyor should be prepared to make the original dimensions available for inspection by the contractor's representative.

The quantity surveyor will be well advised to consider the form in which the variation account will eventually be presented when proceeding with the measurement of variations, in order to avoid unnecessary problems occurring at a later stage. Each set of variation measurements is best recorded under a heading that will suitably describe the nature and scope of the work. It must also be made very clear as to which dimensions relate to additions and which to omissions. It is usual practice to insert the words 'add' or 'omit', suitably underlined, at the top of each page and where a change takes place from additions to omissions or vice versa. It is customary to use dimensions books when measuring on site and dimensions sheets in the office.

Variation dimensions should be provided with appropriate subheadings and explanatory notes in order to effectively explain the nature and scope of each set of dimensions and how the measurements have been taken. The quantity surveyor cannot be too careful in preparing these particulars as he never knows when he may be called upon to explain the dimensions to the contractor, employer or architect. Furthermore, there is also a possibility that a variation account could become the subject of court proceedings or an arbitration.

It is not always possible to secure the optimum arrangement for the variation dimensions at the time of measurement. For this reason it is important for the quantity surveyor to carefully read through the variation dimensions, prior to handing them over to a technical surveyor (previously described as a technician) or other member of staff responsible for their processing. The quantity surveyor should enter the variation item number on each page, preferably in coloured ink, making the changes from one item to another and from additions to omissions and vice versa abundantly clear. It is also good practice to prepare an index or list of items covered by the variation measurements with their respective item numbers.

As the number of variations increases, the quantity surveyor will be able to decide on a suitable method of grouping them. For example, he might find one variation order authorising the increase in capacity of a cold water storage cistern from 227 to 327 litres, another variation might provide for the installation of two additional washbasins, another variation order omits a shower and yet another order requires the provision of an additional draw off point. Each of these items will be measured separately, but the quantity surveyor will probably decide to group all four items together under a heading of variations to internal plumbing. Although these items are dealt with under several separate sections in the Standard Method.

All estimates supplied by specialists and approved by the architect should be passed to the quantity surveyor for checking. These are best kept together in a suitable folder for ease of reference.

Specimen dimension book entries are shown in table 6.5, which are subsequently worked up into a variation account sheet in table 6.6.

**Valuing Variations**

Valuations of variations are generally made by the quantity surveyor in accordance with the following rules.

(1) Where the work is of similar character and executed under similar conditions and with no significant change in the total quantity to that in the main contract, then the measured quantities of the variation will be priced at rates contained in the contract bills.
(2) Where the work is of similar character to that described in the contract bills but is not executed under similar conditions or where there is a significant change in total quantity, then it shall be valued on the basis of the rates and prices in the contract bills but with a fair allowance being made for the differences in conditions and/or quantity.
(3) Where, however, the work is not of similar character to that contemplated under the terms of the contract, then a fair valuation will be made by the quantity surveyor for agreement with the contractor.
(4) Where it is impossible by the normal methods of measurement or price build-up to fairly represent the cost of the work carried out, then, unless otherwise agreed, the work shall be assessed at daywork rates at the prices ruling at the date the work is actually carried out, and including any percentage additions inserted by the contractor in the contract bills. Daywork is normally valued at the prime cost of such work calculated in accordance with the RICS *Definition of Prime Cost of Daywork carried out under a Building Contract* together with percentage additions to each section of the prime cost as described earlier in this chapter.

Omissions are valued at the rates entered in the bill of quantities but if, in consequence of these omissions, the conditions under which the remainder of the work is to be executed are substantially varied, then the same principles are to apply as in the case of additions that are not entirely in conformity with the original billed items. For example, if the area of coloured profiled steel sheeting to be used on the roof of a building was reduced from 550 $m^2$ to 340 $m^2$, the actual terms of the original quotation would have to be examined and the billed rate per square metre suitably adjusted to take account of the higher unit price for the smaller quantity of material.

In the case of significant variations, where it is not possible to use billed rates as a means of computing a price for the varied work, an architect may request the contractor to supply him with a lump sum quotation for the substituted work. In this situation the architect would be well advised to ask the quantity surveyor to prepare an approximate estimate of nett extra cost, allowing for any omitted work, and then to proceed only if the contractor's price bears a reasonable relationship to the quantity survey-or's estimate.

With measured work, there are no problems where the character of the work and the conditions under which it will be carried out are similar to those as originally envisaged when the contract bills were priced. Where, however, these conditions do not apply then the contract conditions re-quire the work to be valued at fair rates and prices (13.5.1.3). The unit rates are then normally built up from the prime cost of the necessary materials, labour requirements valued at 'all in' labour rates, appropriate allowances for plant and additions for overheads and profit. In between the two extremes are varied items where the work is of similar character, but is not executed under similar conditions and/or there are significant changes of quantity, and the rates need varying accordingly. In some cases comparable billed rates can be analysed and new rates synthesised on the basis of the analyses as is so amply demonstrated in estimating textbooks, although this procedure can be applied to only a few cases in practice. Skill and careful judgement are required to assess fair and realistic prices for the substituted items. It will immediately be appreciated that the exca-vation of a trench for a pipe not exceeding 200 mm nominal size, which is deemed to include earthwork support, consolidation of trench bottom, trimming excavations, filling in, compaction and disposal of surplus exca-vated materials, average 3 m deep will not be double the price for a trench average 1.50 m deep. *Pro rata* rates cannot always be applied and frequently a much more thorough scrutiny of all the relevant factors is required.

Clause 13.5.3.3 of the JCT Standard Form requires that in the valuation of additional, omitted or substituted work, allowance, where appropriate, is to be made for any addition to or reduction of preliminaries items of the type referred to in SMM7 (Preliminaries/General Conditions). For ex-ample, where an extension of time has been granted by the architect because of variations, then those items of preliminaries with a time-re-lated content may be proportionately increased, but cost-related and lump sum items remain unchanged. Other factors, apart from extension of time, can also justify variations in preliminaries. For instance, if plant is kept on the site longer than would otherwise have occurred, while varied work is being carried out, then the actual additional expense needs to be assessed with the assistance of supporting data from the contractor. Adjustments can only be made by determining the amount by which individual items of preliminaries have been affected by variations.

**Example of Variation**

Table 6.5 shows a variation dimensions book entry covering the substitu-tion of a 75 mm concrete block partition for a 100 mm stud partition, although in practice the entries would be handwritten.

Table 6.5    Variation dimensions book entries

---

A.I. Nr. 7    *Substitution of 75 mm concrete block partition for 100 mm stud partition in canteen*

---

| OMIT | | | |
|---|---|---|---|
| 22/ | 3.45 | | 100 × 38 sn. swd. in ptn. |
| | | | membrs. |
| 2/ | 11.00 | | 100 × 50 do. |
| 2/ | 11.00 | | Gypsum lath pla. bd. 10th. to BS 1230, |
| | 3.50 | | fxd. to swd. studs & |
| | | | scrimmg. jts. |
| ADD | | | |
| | 11.00 | | 75 conc. blk. ptn. in |
| | 3.50 | | type B lightwt. agg. blks., |
| | | | size 400 × 215 to BS2028, w. |
| | | | keyed fin., b. & j. in g.m. |
| | | | (1:1:6). |
| 2/ | 11.00 | | Lightwt. gyp. pla. to |
| | 3.50 | | BS1911 Pt.2 in 2 cts. 10th. |
| | | | o/a on blk. walls w. bondg. |
| | | | pla. backg. & final ct. |
| | | | of fin. pla., w. trowld. fin. |

## Variation Accounts

The normal rules for the preparation of a bill of quantities will be followed in working up the variation account. Each variation will normally incorporate a bill of omissions and a bill of additions. The descriptions of the items that appeared in the original bill of quantities can be abbreviated considerably and in many instances the insertion of the billed item reference, such as 49D-F, will suffice. Where, however, entirely new items are encountered, then comprehensive descriptions must be provided in the variation account as these will be needed to build-up the new rates. Many of these items will be billed direct from the dimensions, but some processing may be necessary. Each separate sheet should be labelled with the particular variation or architect's instruction number with clear identification of omissions and additions. The insertion of dimensions book references is also helpful, such as 2/73 (page 73 of dimensions book nr. 2).

Once the variation account has been completed, the quantity surveyor will proceed to price the component items. Expenditure against prime cost and provisional sums will be readily determined by reference to the approved invoices for the work carried out against them. Much of the remainder of the pricing will be undertaken by reference to and analysis of billed rates, including the correction of any errors in the descriptions or quantities of the contract bills and the inclusion of any daywork. When

Table 6.6   Variation account sheet

---

*A.I. Nr. 7*

*Substitution of 75 mm concrete block partition*
*for 100 mm stud partition in canteen*

OMISSIONS

| | | | | |
|---|---|---|---|---|
| A | 100 × 38 sawn softwood in partition members. | 76 | m | |
| B | 100 × 50 sawn softwood in partition members. | 22 | m | |
| C | Gypsum lath plasterboard 10 thick to BS 1230 fixed to softwood studs and scrimming joints. | 77 | m$^2$ | |
| | TOTAL OMISSIONS CARRIED TO SUMMARY | | | £ |

ADDITIONS

| | | | | |
|---|---|---|---|---|
| D | 75 concrete block partition in type B lightweight aggregate blocks, size 400 × 215 to BS2028, with keyed finish, bedded and jointed in gauged mortar (1:1:6). | 39 | m$^2$ | |
| E | Lightweight gypsum plaster to BS1911 Part 2 in 2 coats 10 thick overall on block walls with bonding plaster backing and final coat of finishing plaster, with trowelled finish. | 77 | m$^2$ | |
| | TOTAL ADDITIONS CARRIED TO SUMMARY | | | £ |

---

the pricing is complete, the account will be extended and cast and each variation will have separate totals for omissions and additions, as shown in table 6.6. Alternatively, the variations can be billed on sheets with double pricing columns; the first set used for omissions and the second for additions. The total amounts for each variation, or possibly groups of variations or work sections, will subsequently be carried to a summary in the form shown in table 6.7. The prices inserted in the omission and addition columns in the summary will be totalled and the difference added to or deducted from the contract sum. Abridged running schedules of variations are often prepared showing the net addition or omission in respect of each variation and the cumulative addition or reduction.

The opportunity may be afforded in the bill of quantities for certain items, such as insurances, to be priced on a percentage basis. These items are normally proportional to the total cost of the work and it is, therefore, reasonable to apply a similar adjustment to the total of the variation account. There may also be variations in the prices of some preliminaries items, as described earlier in the chapter, where significant changes to the works or the authorised contract period have occurred.

Table 6.7   Summary of variation account

| | | Omissions | Additions |
|---|---|---|---|
| *Elderly Persons' Home, Bishopsgate* | | | *May 1996* |
| SUMMARY OF VARIATIONS | | | |
| *Item nr.* | *Nature of work* | *Omissions* | *Additions* |
| | | £       p | £       p |
| 1. | Foundations | 2 171 20 | 2 334 96 |
| 2. | Roof coverings | 1 371 28 | 1 241 68 |
| 3. | Metal windows | 3 259 04 | 3 204 92 |
| 4. | Floor finishes | 973 96 | 894 94 |
| 5. | Ironmongery | 358 84 | 378 68 |
| 6. | Heating and hot water services | 5 139 20 | 5 227 08 |
| 7. | Electrical work | 1 905 24 | 1 929 96 |
| 8. | Sanitary appliances | 1 130 87 | 1 050 48 |
| 9. | Reduction in sanitary accommodation | 827 68 | 434 70 |
| 10. | Drainage | 252 88 | 289 75 |
| 11. | Access road | 636 69 | 594 83 |
| | | £18 026 88 | £17 581 98 |
| | Price adjustments | | |
| | Labour | – – | 448 92 |
| | Materials | 60 68 | 1 139 24 |
| | | £18 087 56 | £19 170 14 |
| | | | 18 087 56 |
| | Net addition carried to Statement | | £ 1 082 58 |
| | STATEMENT | | |
| | Contract sum | | 558 603 00 |
| | Net addition for variations | | 1 082 58 |
| | Adjusted contract sum | | £559 685 58 |

The quantity surveyor then submits a copy of the variation account to the contractor for comments. It is usual to arrange a meeting between the quantity surveyor and the contractor's surveyor to settle any disputes concerning quantities or prices. Following the settlement of these differences, it may be necessary to make some adjustments to the variation account, which can then be finalised and signed by the contractor as evidence of his agreement to it. Where the contract has been undertaken for a public authority, the variation account and summary is generally submitted to the authority for examination by the appropriate officers. With private employers, a simplified statement showing the principal variations and their values may be sufficient.

Note: the normal vertical rulings have been omitted from tables 6.6 and 6.7 for greater clarity.

# FINANCIAL STATEMENTS

## Monitoring of Construction Costs

At the outset of a project it is advisable to determine the employer's requirements as to the form of cost control that he requires and the extent of the authority that he is prepared to delegate to the design team. Cost control encompasses all the measures necessary to ensure that the authorised cost is not exceeded and these measures are examined in some detail in chapter 8.

The quantity surveyor should keep the employer informed of the financial position of the contract at agreed regular intervals, often monthly. The employer is concerned not only with the current cost position but also the likely pattern of future payments over the contract period and the probable total cost, in order that he may have sufficient funds available to honour the architect's certificates as they are issued. In the event of substantial additional expenditure being forecast, the employer may raise further capital and/or request savings to be made on the project to keep construction costs within the budget. It is generally accepted that the contingency sum cannot be used for design alterations without the prior approval of the employer.

As described earlier in the chapter, it is advisable to consider the probable cost effects of all architect's instructions before they are issued. The likely expenditure against prime cost and provisional sums, and fluctuations where applicable, and the evaluation of possible variations and contractor's claims should be undertaken periodically by the quantity surveyor to permit the continuing updating of the forecast final cost. A further refinement is for the quantity surveyor to prepare a graph showing the probable monthly nett total costs of the works over the contract period. This is normally based on the contractor's master programme with the work priced from the contract bills and allowance made for the allocation of preliminaries items, as described earlier in this chapter, and the retaining and release of retention as appropriate.

As the work proceeds, the quantity surveyor will need to assess the adequacy of the balance of the contingency sum on a monthly basis, and the unexpended balance should be included in the financial statements or reports to the employer, in consultation with the architect. The format, frequency and detail of these statements is a matter of professional judgement in the light of prevailing circumstances, and having regard to the wishes of the employer. The reporting arrangements should, in any event, be consistent with the procedure for the approval of variations. The Department of the Environment (1983) recommended that the financial implications of any claims should be reported only upon settlement with the contractor, in order not to prejudice the employer's position. The primary objectives of the report are to provide a clear build-up of the latest forecast

Table 6.8   Financial statement

| | £ |
|---|---|
| Waterby Jones and Partners | Contract: Office block |
| Chartered Quantity Surveyors | Broadway |
| 15 Shady Walk | Ref: 0/15/96 |
| Cabletown | Financial statement nr. 12 |

| | £ |
|---|---|
| Contract sum | 2 586 420 |
| *less* contingencies | 65 000 |
| | 2 521 420 |
| Variations (as attached schedule) | 22 080 |
| Claims (as attached schedule) | 9 215 |
| | 2 552 715 |
| *add* allowance for contingencies for remainder of contract | 6 500 |
| Estimated final cost (excluding fluctuations) | 2 559 215 |
| *add* allowance for fluctuations | 62 500 |
| Estimated total final cost | £2 621 715 |

Date: 31 May 1996                Signed . . . . . . . . . . . . . . . . . . . . . . . . . .

final cost and a ready reconciliation with previous reports. A typical financial statement is illustrated in table 6.8, showing the estimated total final cost of a contract. The variations and claims would be detailed in supporting schedules.

## FINAL ACCOUNT PROCEDURE

### Contractual Provisions

The responsibilities of the contractor and quantity surveyor in connection with the final account are detailed in the JCT Standard Form (30.6.1).

(1) Not later than 6 months after practical completion of the works, the contractor should send to the architect or, when instructed, the quantity surveyor all the documents necessary for preparing the final account.
(2) Not later than 3 months after receipt of the documents, the quantity surveyor shall prepare the final account (statement of all adjustments to be made to the contract sum).
(3) When the final account has been completed the architect shall send a copy to the contractor and the relevant extract to each nominated subcontractor.

Clause 30.6.2 tabulates all the matters that shall be dealt with in the final account in order to adjust the contract sum in accordance with the conditions.

*Sums to be deducted:*

(1)  Prime cost sums and amounts in respect of named subcontractors and associated contractor's profit.
(2)  Provisional sums and value of work for which approximate quantities are included in contract bills.
(3)  Variations that are omissions including, where appropriate, omissions of other works carried out under changed conditions as a result of variations.
(4)  Amounts allowable to the employer under the fluctuations clauses.
(5)  Any other amount that is required by the contract to be deducted from the contract sum.

*Sums to be added:*

(1)  The total amounts of nominated subcontracts finally adjusted in accordance with the relevant subcontract conditions.
(2)  Where the contractor has tendered for work that was to have been performed by a nominated subcontractor and his tender has been accepted, the amount of the tender suitably adjusted.
(3)  Any amounts due to nominated suppliers, including cash discounts of 5 per cent, but excluding Value Added Tax (VAT).
(4)  The contractor's profit on the amounts referred to in (1), (2) and (3).
(5)  Any amounts payable by the employer relating to statutory fees and charges, opening up and testing, royalties and patent rights, and insurances under clause 21.2.3.
(6)  Additions in respect of valuations including, where appropriate, additions for other works carried out under changed conditions as a result of variations.
(7)  The value of work carried out against provisional sums or approximate quantities included in the contract bills.
(8)  Any amounts payable by the employer to the contractor by way of reimbursement for direct loss and/or expense arising from matters materially affecting the regular progress of the works or from the finding of antiquities.
(9)  Any amount expended by the contractor as a result of loss or damage by fire or other perils where the risks are insured by the employer and the contractor is entitled to reimbursement.
(10) Any amount payable to the contractor under the fluctuations clauses.
(11) Any other amount that is required by the contract to be added to the contract sum.

### Final Account Preparation Procedure

A quantity surveyor prepares the final account in the manner that is best suited for the particular project, with the original contract sum as the starting point. The bulk of the final account will consist of measured work priced at the original billed rates. If the contractor has reason to doubt the accuracy of any of the original billed items, he can make a request to the quantity surveyor for the work concerned to be measured on site. The adjustment of the contract sum in the final account normally falls under the following main heads, although the quantity surveyor must have regard to all the matters listed in the JCT Standard Form, as tabulated in the previous section of this chapter.

  (i) variations;
 (ii) remeasurement of approximate quantities in the contract bills;
(iii) nominated subcontractors' accounts;
(iv) nominated suppliers' accounts;
 (v) loss and expense caused by disturbance of the regular progress of the works; and
(vi) fluctuations in the rates of labour and prices of materials, where applicable.

All relevant items must be shown separately in the final account and the nett amount of each variation and amount due to each nominated subcontractor and nominated supplier listed. When preparing the final account the quantity surveyor should give the contractor's surveyor or other representative the opportunity to be present when measurements and details are taken or recorded, so that the document is prepared in full liaison with the contractor.

The draft final account is a useful mechanism in maintaining cost control of the contract, if it is commenced as building work begins and is updated at monthly intervals, while the details and physical conditions are still fresh in the minds of those involved. Hence, remeasurement of work, variations, daywork and claims should all be dealt with as early as possible. From the draft account a report on known and anticipated expenditure can be prepared and submitted to the architect, probably at monthly intervals, as a financial forecast of the probable final cost in the manner outlined earlier in the chapter. This procedure also enables the account to be finalised fairly soon after the building work is complete and ensures that the interim payments to the contractor represent a realistic assessment of the value of work performed.

Delays in settlement of the final account represent additional cost to the contractor and in the majority of cases the employer is anxious to know his ultimate financial commitment. The architect and the quantity surveyor have a contractual responsibility to keep to the date stipulated in the contract for completion of the account and the contractor should provide every assistance in the prompt provision of subcontractors' and suppliers' accounts, agreement of measurements and prices, and the supply of all necessary supporting data.

**Preparation of Final Account Summary**

A worked example will serve to illustrate a common method of approach to this particular task, based on the contract information supplied, and should help examination candidates faced with problems of this kind.

CONTRACT BILLS SUMMARY

|  |  | £ |
|---|---:|---:|
| Preliminaries |  | 34 320 |
| Measured work |  | 419 673 |
| (including £29 000 approximately |  |  |
| measured foundations) |  |  |
|  | £ |  |
| Prime cost and provisional sums | 123 000 |  |
| Profit | 6 150 |  |
| Attendance | 3 200 | 132 350 |
| Contingencies |  | 20 000 |
|  |  | 606 343 |
| *less* adjustment for contractor's pricing |  |  |
| errors (1.5% of £422 873) |  | 6 343 |
| (measured work and attendance) |  |  |
| Contract sum | | £600 000 |

Further Details

|  | £ |
|---|---:|
| Nominated subcontractors' and suppliers' invoices |  |
| (supplied and agreed) | 124 660 |
| Variation account:  measured additions at billed rates | 6 490 |
| daywork | 1 300 |
| measured omissions at billed rates | 830 |
| Error in contract bills (incorrect quantity) add |  |
| 168 m² @ £3.00 | 504 |
| Remeasured work in foundations at billed rates | 33 540 |
| Increased costs under clause 39 | 25 410 |
| Previous payments | 603 580 |
| Liquidated damages for non-completion (10 weeks) £500 per week | |

*See table 6.9 for solution.*

Table 6.9    Final account summary

|  | OMIT £ | ADD £ |
|---|---|---|
| Contract sum |  | 600 000.00 |
| *less* contingencies |  | 20 000.00 |
|  |  | £ 580 000.00 |
| *Add* error in contract bills | 504.00 |  |
| −1.5% | 7.56 | 496.44 |
| *Omit* provisional foundations | 29 000.00 |  |
| −1.5% | 435.00 | 28 565.00 |
| *Add* foundations as remeasured | 33 540.00 |  |
| −1.5% | 503.10 | 33 036.90 |
| *Omit* prime cost and provisional sums |  | 132 350.00 |
| *Add* subcontractors' and suppliers' invoices | 124 660.00 |  |
| profit *pro rata* $\dfrac{6\,150}{123\,000} \times 124\,660$ | 6 233.00 |  |
| attendance | 3 200.00 | 134 093.00 |
| Variation account: *omit* | 830.00 |  |
| −1.5% | 12.45 | 817.55 |
| *add* | 6 490.00 |  |
| −1.5% | 97.35 |  |
|  | 6 392.65 |  |
| + daywork | 1 300.00 | 7 692.65 |
|  | 161 732.55 | 755 318.99 |
|  |  | 161 732.55 |
|  |  | 593 586.44 |
| *Add* fluctuations |  | 25 410.00 |
| Adjusted contract sum |  | 618 996.44 |
| *less* previous payments |  | 603 580.00 |
| Balance due |  | £ 15 416.44 |

From this balance the sum of £5000 may be deducted
as liquidated and ascertained damages, provided the
requirements of clause 24 have been complied with.

## FINAL CERTIFICATE

The architect is required to issue the final certificate not later than two months after the end of the defects liability period or after the completion of making good defects or of the date on which the architect sent a copy of the ascertainment and statement of adjustments to the contractor, whichever is the latest.

The amount of the final certificate is the difference between the total of the final account and the amounts previously stated as due under interim certificates. This wording safeguards the contractor's rights to any money previously certified that the employer may not have paid. Not less than 28 days before the issue of the final certificate the architect has to issue an interim certificate which includes the final amounts due to all nominated subcontractors (30.7). Hence the final certificate will not include any further payments to nominated subcontractors, but the architect must notify them of the date of issue of the final certificate.

The final certificate provides conclusive evidence that the quality of materials and standard of workmanship are to the reasonable satisfaction of the architect, as decided in Colbart Ltd v Kumar (1992) and in the Court of Appeal in Crown Estates Commissioners v John Mowlem & Co Ltd (1994), and that all the terms of the contract with regard to the adjustment of the contract sum have been implemented (30.9.1). A completed final certificate on a standard RIBA form is illustrated in appendix K.

# 7 Claims and Insolvencies

## GENERAL BACKGROUND TO CLAIMS

The construction industry covers a complex field of activity involving many operative skills and conditions which vary considerably from one project to another. Site and climatic conditions, market conditions, project characteristics and available resources are some of the variables, each of which can have a significant effect on the operation of the contract. In the 1990–95 recession contract prices were pared to the bone with the result that even relatively minor changes to projects frequently gave rise to the submission of claims, which often caused substantial increases in the eventual costs, as evidenced in the Channel Tunnel and Canary Wharf.

Most construction contracts make provision for these complexities and uncertainties by the inclusion of clauses permitting the contractor to claim for loss or expense resulting from specific contingencies. The standard forms of contract attempt to clarify the contractual requirements and remove any ambiguities as far as possible. In the absence of these provisions, contractors would have to include in their tenders for many more uncertainties than they do now, which would result in a significant increase in tender figures. However, under the standard forms of contract, the employer will only have to meet the cost of such contingencies if they arise and have been duly verified.

The term 'claim' as used in this context is a request by the contractor for recompense for some loss or expense that he has suffered, or an attempt to avoid the requirement to pay liquidated and ascertained damages. It is in this light that claims should be viewed by both sides of the industry. Frivolous claims by contractors to redress the effects of inefficiency or profit shortfall are unlikely to receive sympathetic consideration by the architect or employer. It has been argued that the term 'claim' should be used only in respect of fundamental breaches of the contract and that the remainder are contractual entitlements. A claim to be successful must be well prepared, based on the appropriate contract clauses and founded on facts that are clearly recorded, presented and provable.

The well-organised contractor will be able to recognise the occurrence of events that are likely to result in his ultimately suffering loss or expense. A contemporaneous record of such events should be sent to the architect as a basis of the claim. The late submission of claims by a contractor will inevitably result in difficulties, since neither the employer nor the architect will have had the opportunity to check the factors giving rise to the additional expense, at or about the time of their occurrence. A contractor can so easily fail to obtain reimbursement of monies to which he is entitled because of the late submission of a claim or notification of his intention to submit one.

The potential loss and expense must be clearly identified, quantified and valued. Additionally, other parties to the contract must be convinced that they are valid claims and that the integral parts are claimable and correctly valued.

Robinson (1977) aptly described how, depending on the quality of the records, a good, indifferent or bad claim is produced. Most poorly produced claims are the result of a hasty, last minute analysis of sketchy and incomplete records. The best prepared claims come from the managements of contractors who appreciate that loss and expense situations are likely to arise on contracts, and accordingly establish procedures to identify and record all relevant background information and data, in order that an accurate and well-founded evaluation can be made promptly. The contractor should ideally have an efficient organisation which is continually looking ahead to identify and diagnose possible future problems. Claims settlement invariably becomes protracted and difficult where records are poor or non-existent.

The contractor must apply in writing for the issue of instructions, details, drawings, levels, or for the nomination of subcontractors as appropriate. He must also give written notice to the architect of any cause of delay in the progress of the work and written application in respect of any claims that he is contemplating making in respect of variations or loss and/or expense. He must take positive steps to ensure that architect's instructions are issued in writing or verbal instructions confirmed within seven days. The contractor should ensure that the various certificates required under the contract are issued by the architect, particularly in respect of practical completion, completion of making good defects, final certificate and extensions of time, to prevent any unnecessary problems arising in the future. It is always a better policy to avoid disputes rather than being involved in their settlement.

Burke (1976) described how some architects appear to interpret any letter from a contractor asking for details or information as a prelude to a claim; it is, however, a contractual duty of the contractor to apply in writing for such details if they are not supplied to him. By so doing, he may remove the necessity for making a claim, if he receives the information promptly.

The design team should recognise that most contractors will be looking for opportunities of submitting claims, whether fully justified or otherwise, especially in periods of economic depression, and should not take offence when the proper notice is given. It is in the interests of all parties to the contract to deal with extensions of time and evaluation of monetary entitlement at the time when the relevant events occur and the pertinent facts are still fresh in everyone's mind. Furthermore, their satisfactory resolution has the effect of keeping the contractor on target both for money and for time. Another advantage is that a contractor tends to keep his best management and employees upon those contracts that he knows have effective completion dates and cost targets and where claims will be dealt with expeditiously (Trickey, 1979).

Hughes (1983) advised that whatever the merits or demerits, the architect, engineer or quantity surveyor would be wise to encourage the contractor to keep him informed of anything that is happening or has happened involving the possibility of additional expense. In some cases work in hand may be concerned or affected and the architect or engineer may be able to take remedial action, where the fault rests with either of them or the employer, or avoiding or mitigating action, where the cause is one for which the employer has accepted the risk.

It is the responsibility of the contractor to formulate his claim in detail and to furnish the evidence on which the claim is based. A properly supported claim will be carefully examined to establish whether the facts are properly founded, whether the matters submitted equate to the circumstances provided for in the relevant contract clauses and whether the amount claimed can be justified (Davies *et al.*, 1980). Trickey (1983) summarised the main actions of the quantity surveyor, acting under the responsibility delegated to him by the architect, on receipt of a claim from the contractor, as determining:

(i)   which clauses of the contract apply and how they are to be interpreted?;
(ii)  what elements of cost are involved?; and
(iii) what is the monetary entitlement?

In most cases when assessing a claim the quantity surveyor will require the master programme, often in the form of a bar chart, a method statement showing in general terms how the contractor intends to carry out the work, and a detailed breakdown of the costs of preliminaries (Trickey, 1979).

Simmonds (1979) has set the scene in emphasising that contracting is a high risk business. Therefore the contractor at the time of tendering has the right to be able to plan and expect to proceed with his work in an orderly manner. The employer, on the other hand, has the right to expect experience and competence from the contractor. The contractor should not attempt to make the employer pay for his mistakes and the employer must not expect the contractor to bear the cost of errors or changes made by him or the design team. A contractor, when pricing a tender, is asked to take risks. When the contract is let and underway and the employer or his agents prevent the contractor from carrying out his contractual obligations properly and effectively, then he will probably find it necessary to make a claim.

## TYPES OF CLAIMS

### Broad Categories of Claims

*(1) Contractual claims*

These are claims that are founded on specific clauses within the terms of the contract. This type of claim will be considered in more detail later in this chapter, with particular reference to the principal standard forms of contract.

*(2) Ex-contractual claims*

These claims are not based on clauses within the terms of a contract, although the basis of the claim may be circumstances that have arisen out of the project and have resulted in loss or expense to the contractor. On occasions a sympathetic employer has settled an excontractual claim (a term disliked by lawyers), because the contract was finished on time and the contractor suffered exceptional misfortune.

However, in general these claims are unlikely to succeed as there is no contractual obligation for payment and any payments made are in the nature of *ex-gratia* payments (act of grace). Typical examples are where late deliveries of materials by a supplier on a firm price contract resulted in substantial price increases for the materials, or where difficulty was experienced by the contractor in recruiting adequate labour and he was obliged to pay high additional costs to attract them.

*(3) Common law claims*

Many of the clauses in the standard forms are stated to be 'without prejudice to any other rights and remedies.' Such rights are established by taking action through the courts for damages for breach of contract, tort, repudiation, implied terms and other related matters.

### Contractual Claims

Contractual claims may be classed as either negative or positive in character, although these classifications do not have universal recognition.

*(1) Negative claims*

Negative claims are those where the contractor seeks to avoid a payment, such as liquidated and ascertained damages. A claim for liquidated damages is made by the employer against the contractor for alleged breach of contract, in that the contractor has not completed the works within the agreed contract period. In order to mitigate this claim by the employer, a negative claim may be submitted by the contractor.

Claims within this classification include those seeking an extension of the contract period, which will in consequence prevent payment by the contractor of some or all of the liquidated damages. It is necessary to consider the appropriate contract clauses on which such claims are based and the main contractual provisions are now detailed.

(i) *The JCT Standard Form of Building Contract (1980, as amended)*
Clause 25 deals with extension of time. The contractor is required to give written notice to the architect of the material circumstances, including the cause or causes of the delay, upon it becoming reasonably apparent that the progress of the works is being or is likely to be delayed for one or more of the relevant events listed in this clause. If not in the notice, the contractor shall give in writing, as soon as practicable thereafter, particulars of the expected effects and an estimate of the extent of the expected delay in the completion of the works.

If the architect is of the opinion that any of the events stated by the contractor is a relevant event and that the completion of the works is likely to be delayed beyond the completion date, he shall in writing to the contractor give an extension of time by fixing such later date as he considers to be fair and reasonable. Examples of relevant events are listed in clauses 25.4.1–16 and each of these will now be considered.

25.4.1: *Force majeure*, being an act of God or man-made events beyond the control of the parties.
25.4.2: Exceptionally adverse weather conditions as distinct from normal adverse weather which is mainly predictable and assumed to be allowed for in the programme.
25.4.3: Loss or damage from specified perils, such as floods.
25.4.4: Civil commotion, strike or lock-out at the site, subcontractor's or supplier's premises, or transport directly connected with the works.
25.4.5.1: Compliance with architect's instructions encompassing discrepancies and divergences (2.3), variations (13.2), provisional sums (13.3), postponement (23.2), antiquities (34) nominated subcontractors (35) and nominated suppliers (36).
25.4.5.2: Inspection and testing where items comply with the contract (8.3).
25.4.6: Non-receipt by the contractor in due or sufficient time of necessary instructions, drawings, details, or levels from the architect for which the contractor specifically applied in writing. The programme is of particular significance in this connection, especially if supported by a schedule of key dates for the release of information (Fellows, 1995).
25.4.7: Delay by nominated subcontractors or nominated suppliers

which the contractor has taken all practicable steps to avoid or reduce. The contractor's obligation is to minimise delays as far as possible, not to make up time (Fellows, 1995).

25.4.8.1: Delay or failure to execute work by artists and tradesmen. These persons are engaged direct by the employer and are thus outside the scope of the contract and the contractor's control.

25.4.8.2: Delay in or failure to supply goods and materials which the employer is obliged under the contract to supply.

25.4.9: By the Government of the United Kingdom exercising any statutory power after the base date and thereby adversely affecting the contractor procuring necessary labour, goods, fuel and energy essential for the proper carrying out of the works.

25.4.10.1: The contractor's inability to secure adequate labour for the proper carrying out of the works. This must be beyond the contractor's control and not reasonably foreseeable at the base date.

25.4.10.2: A similar problem as covered by 25.4.10.1 but for goods and materials. The contractor must exercise reasonable foresight and endeavour as far as practicable to procure the specified items.

25.4.11: Delay in execution or failure by a local authority or statutory undertaker in pursuance of a statutory obligation.

25.4.12: Failure by the employer to give, in due time, access to the site through property in his possession and control in accordance with the contract bills and/or drawings, provided any required notice for such access has been given by the contractor to the architect, or the architect and contractor have themselves agreed an access provision through the employer's property. Access includes both ingress to and egress from the site and is particularly relevant to conversion projects.

25.4.13: Employer's deferment of giving possession of the site to the contractor where the appendix states that clause 23.1.2 applies.

25.4.14: Execution of work covered by an approximate quantity in the contract bills where such an approximate quantity is not a reasonably accurate forecast of the work actually required.

25.4.15: Delay due to a change in statutory requirements occurring after the base date which necessitates alteration/modification to any performance specified work.

25.4.16: Threatened or actual acts of terrorism or actions of authorities in dealing therewith.

(ii) *The ICE Conditions (sixth edition, 1991)*
Clause 44 is the relevant clause, whereby any claim for extension of time that the contractor wishes to make shall be submitted within 28 days after the cause of delay has arisen, or as soon as is reasonable thereafter with full and detailed particulars.

(iii) *PSA/1: General Conditions of Contract for Building and Civil Engineering Works (1994)*

This form is intended for use in both public and private sectors for UK building and civil engineering works. The date for completion and extensions of time are dealt with in clauses 34 and 36 respectively. Any extension of time is dependent on notice being given to the project manager (PM) with the grounds for the request stated.

An extension of time clause has the advantage to the employer of keeping liquidated and ascertained damages alive, in that by establishing a new completion date the clause also sets up a new date from which damages accrue. Such extensions of the contract period have the immediate effect that the contractor will not be liable for liquidated damages for the period of extension, but the causes for the extension may also be reasons for which the contractor can recover cost under a positive claim.

### (2) Positive claims

It is this type of claim that attracts most attention. Such a claim, if successful, can result in an addition to the contract sum and consequently more money being paid to the contractor. Before examining the initiation, origination, preparation and presentation of such claims, it is desirable to consider the relevant clauses of the conditions of contract. The following provisions relate to the JCT Standard Form.

6.1: Any liability in respect of statutory requirements that are at variance with the contract documents. Particular attention should be paid to safety regulations and fire regulations, which are becoming increasingly stringent consequent upon a number of disasters which have occurred with large loss of life. This area is a possible source of conflict and the contractor needs to proceed with caution to avoid becoming involved in abortive work.

8.3: The cost of opening up for inspection any work covered up or the testing of materials or goods and the cost of making good, unless the inspection or test shows that the work, materials or goods are not in accordance with the contract. It is not difficult to envisage the circumstances in which this type of situation can occur. A clerk of works is temporarily away from the site and returns to find a drain trench filled in and he had not inspected the drain earlier; he becomes suspicious and requests the contractor to open up the trench.

9.2: Liability in respect of any infringement or alleged infringement of patent rights, such as proprietary techniques, or royalties.

13.5.6: Any direct loss or expense that is not reimbursable by virtue of the valuation of a variation.

17.2 and 17.3: The cost of making good defects that appear within the defects liability period, these costs being at the discretion of the architect. Premature failure may result from faulty design, poor construction or both. The causes are not always readily identifiable and there may be divided responsibility.

22.C.1.1 and 2: The cost of insurance premiums in respect of existing structures, particularly where the contract encompasses extensions to existing buildings, upon default of the employer to insure the works.

26.1: Any direct loss and/or expense resulting from disturbance of the regular progress of the works, caused by one of the matters listed in 26.2. It will be observed that some of the matters listed in the eight subclauses have previously been mentioned. The matters include delay by the architect in issuing instructions, discrepancies or divergences between contract drawings and/or contract bills, default by the employer in respect of direct employees or supply of materials and goods, postponement of work and failure by the employer to give access to the site when required.

28.2: Direct loss or damage arising out of the determination of the contract by the contractor upon the default of the employer. The action by the contractor could result from such factors as excessive interference by the employer, non-payment for work executed and excessive delays in the issue of instructions.

34.3: Loss or expense in dealing with antiquities found on the site. For example, a site may contain items of special historic interest and work may be suspended on part(s) of the site while archaeologists carry out investigations. This occurred in Central London following the discovery of Roman remains.

37: Fluctuations, where price fluctuation provisions are included.

## ADDITIONAL COST FOR LOSS OR EXPENSE

This is a major area for claims and the appropriate contract clauses will first be examined, followed by an analysis of their main implications.

### (i) JCT Standard Form of Building Contract (1980)

Under clause 26.1, the contractor must claim within a reasonable time of the loss being incurred or the regular progress of the works being delayed. As soon as any of the relevant facts are established, the contractor must make written application to the architect without delay, otherwise the claim may subsequently be regarded as inadmissible. The claim will normally only be in principle at this stage, being more in the nature of a notification of intention to claim pending the assembly of more detailed information and costs.

The basis of a claim under clause 13.5.6 will be direct loss and/or expense that would not be reimbursed under 13.5.1–5, which define the rules for the valuation of variations. Typical examples of claims in this category are as follows:

(1) disruption in the progress of the works causing uneconomic working of labour, plant and materials as a result of variation(s);
(2) standing and waiting time of plant;
(3) the cost of returning resources to site in the case of a variation issued late in the contract;
(4) part load delivery charges where the bulk of the particular materials or goods has already been delivered; and
(5) the cost of a prolonged contract period as a direct result of extensive variations, and for which an extension of time should be granted under clause 25.

Clause 26.1 is linked with clause 25. It is important that the contractor when requesting an extension of time under clause 25, as described under negative claims, for one of the matters covered by clause 26, states that reimbursement will be sought under clause 26. It will in consequence constitute a two-pronged claim. The basis of the claim in this case is direct loss and or expense for which the contractor would not be reimbursed under any other clause, by reason of the regular progress of the works being affected by the matters stated.

It should be emphasised that the late issue of instructions must be viewed in the light of the completion date and not the contractor's programmed completion, which may be earlier. The request for these instructions must neither be unreasonably distant from nor unreasonably close to the date by which they were required (26.2.1) – that is neither too early nor too late. The type of claim that is most likely to be based on this clause is one for the cost of a prolonged contract.

As these clauses refer to direct loss and or expense, the meaning and intent of the wording deserves attention. Firstly the costs claimed must be directly attributable to the clause identified, which is one recognised by the contract. 'Loss' is incurred where the contractor will not recover what he could reasonably have expected to receive as a direct result of the disruption; it is not merely a contributory factor. 'Expense' is where the contractor has had to increase his expected expenditure on an item of work to produce the same result, again as a direct result of the disruption.

Hughes (1983) defined 'disruption' as costs or loss and expense arising from uneconomic working or loss of productivity of labour and plant owing to works not being carried out in their planned or logical sequences. Additional cost may be incurred through overtime or weekend working or the provision of additional formwork, in order to maintain a required level of output or rate of progress, consequent upon architect's instructions that disrupt or delay the regular progress of the work.

### (ii) ICE Conditions of Contract (1991)

Most civil engineering contracts differ extensively in character from building projects and usually involve a much greater proportion of work below ground. Under clause 12, a claim may be submitted as a result of adverse physical conditions or artificial obstructions encountered during the execution of the works, and which could not reasonably have been foreseen by an experienced contractor. This clause has wide ranging implications and the final decision on the validity of any claim rests with the engineer who, in the opinion of the author, does sometimes adopt a rather harsh and uncompromising stance. There is provision for reference to arbitration but in practice the contractor rarely uses this option for a variety of reasons, including the high cost and the possible adverse effect on future invitations to tender.

A claim submitted under the provisions of clause 12 may be for additional payment to cover the cost of dealing with the conditions or obstruction(s). Clause 12(b) prescribes that this shall include 'a reasonable percentage addition in respect of profit, and any extension of time to which the contractor may be entitled.'

The type of adverse physical conditions that can occur may range from running sand to a landslide, while artificial obstructions could encompass old tips and foundations to unrecorded mine workings and underground services. During a conference at the former Trent Polytechnic, an attempt was made to define 'running sand'. Some felt it could be water-bound sand or sand that would not stand up unsupported. It was, however, generally accepted that the term applies to sand whose conditions change after being worked (Seeley, 1978).

Clause 52(4) gives a contractor an opportunity to claim for additional payment under other clauses of the contract. It is up to the contractor to give notice in writing to the engineer as soon as reasonably possible and in any event within 28 days after the occurrence of the event that gives rise to the claim. In respect of the pricing of variations, the engineer under clause 52(1) and (2) is given authority to determine certain prices. If the contractor intends to claim a higher price than that determined, he must under clause 52(4)(a) notify the engineer within 28 days.

The time limitation is all important. After giving notice that he intends to claim under this clause, the contractor must send to the engineer an interim account giving full and detailed particulars of the amount claimed to date and the grounds upon which the claim is based. The contractor shall update this information when the full facts and extent of costs are known.

### (iii) PSA/1: General Conditions of Contract for Building and Civil Engineering Works (1994)

Clause 46 of these conditions gives a contractor the opportunity to claim for any expense properly incurred by reason of compliance with any of the project manager's instructions and related matters resulting in prolongation or disruption. The amount of such expense is to be determined by the quantity surveyor. The contractor must give notice to the project manager immediately upon becoming aware that the regular progress of the works or any part of them has been or is likely to be disrupted or prolonged and the circumstances causing this. The contractor must provide full details of all expenses within 56 days of being incurred.

### Further Implications and Applications

Direct loss and/or expense often results from disturbance of the regular progress of the works and this may arise because of lack of instructions, drawings, details or levels; the opening up for inspection or testing if found unnecessary; discrepancy between contract drawings and contract bills; delay on the part of artists or tradesmen employed by the employer; and architect's instructions issued regarding postponement of any work under the contract.

If the contractor follows the correct procedure with regard to notification and the architect is of the opinion that the contractor has been involved in direct loss and/or expense, then the architect will ascertain or, more probably, instruct the quantity surveyor to determine the amount of such loss or expense. This ascertained loss is added to the contract sum and included in an interim certificate if one is issued after the date of ascertainment. Such payment will not be subject to retention.

The specified probable causes of interruption can also become grounds for extension of time (clause 25), as they are under the control of the employer or architect. There are also matters for which extension of time may be granted which would not necessarily give grounds for a monetary claim, such as *force majeure*; exceptionally inclement weather; fire damage; strikes and lock-outs; and delay in securing labour or materials.

The timing of requests by the contractor to the architect for instructions is significant and it must be realistic. For example, it would not be reasonable for a contractor to request a colour schedule in the sixth week of a contract for painting work to be executed in the forty-third week and then to regard the non-provision as a basis for a claim. Accurate daily or weekly reports to the architect, together with an agreed programme of work, preferably supported by a critical path network and a work method statement, should provide a realistic forecast of when information is reasonably required. This can be further refined by the supply by the contractor of key dates when specific details and information will be required.

Contractors' claims embrace a wide range of matters and the following list serves to illustrate their diversity.

(1) Cost of bringing back to the site, plant and operatives whose work was finished and their subsequent transfer to other sites.
(2) Cost of extra hire periods of plant.
(3) Cost of delay in the non-use of plant.
(4) Extra cost of removing unwanted or unused plant.
(5) Cost of waste of materials, such as cement and lime, through deterioration caused by long delays. The contractor must, however, provide adequate storage facilities and use the materials in the proper sequence.
(6) Cost of uneconomic use of plant and labour when diverted to other work, to avoid waiting or standing time.
(7) Cost of extra power, lighting and watching.
(8) Cost of extra dewatering, as a result of the extension of underground working.
(9) Cost of ready mixed concrete instead of site mixed concrete, after the removal of the batching plant.
(10) Cost of labour operations as against mechanised operations, after the removal of plant (Wood, 1978).

Changes in the architect's requirements may have significant financial effects. For example, changing 150 mm reinforced concrete walls to 300 mm walls will affect formwork prices considerably, although on occasions quantity surveyors have attempted to apply the same billed rates. Formwork to a 150 mm wall might, for instance, use 12 mm plywood with $50 \times 50$ mm softwood soldiers at 450 mm centres and Acrow propping at 900 mm centres. The 300 mm wall will require much stronger formwork, possibly 19 mm plywood and much stronger soldiers and closer propping.

Claims may be made under clause 26 if the regular progress of the work is materially affected, and this can happen without needing to extend the contract period. It may be that the varied work is not a critical item and can be undertaken in float time on the critical path network. It is disturbance that costs money even although the work in its entirety is not delayed. Such claims may revolve around acceleration with the need for increased productivity of labour or plant, the necessity of providing additional supervision and other related matters which are not time-related.

The tender for a project is based upon work being carried out in an orderly sequence with the most efficient use of labour and plant. A disturbance to flow through lack of details could conceivably increase the cost of the work by as much as 50 per cent, as the rate of output could revert from the peak rate to the much lower rate prevailing at the start of the contract. It must be borne in mind that other factors can also reduce productivity rates, such as exceptionally inclement weather, go-slow tactics by labour, poor supervision, inefficient subcontractors and poor or insufficient labour.

In order to assess a claim the quantity surveyor will often have to make extensive investigations. For instance, when faced with a claim concerning the additional cost of executing work brought about by delay in the supply of certain drawings, he would need to:

(1) ascertain the period between the latest date when drawings were required and the date at which they were actually supplied to the contractor;
(2) ascertain on which work the men were employed who would have otherwise have been engaged on the work detailed in the missing drawings;
(3) determine whether they were transferred to this latter work immediately the drawings were supplied; and
(4) ascertain the overall programme and labour position at the time the arrival of the drawings was being awaited.

The claim may be based on additional cost due to operatives standing idle or to indirect problems such as the reprogramming of the works with consequent inconvenience and loss of efficiency.

Another possible source of claims is where several contractors are working on the same site at the same time, as on a power station site where building and civil engineering, mechanical and electrical plant contracts are in operation, with a distinct possibility of interference between them, despite the careful wording of the contracts and excellent coordination arrangements on site. The superstructure contractor may be delayed while certain plant is being installed, then he has to leave access gaps in the structure and subsequently to close them at a much later date, possibly beyond the end of his contract period.

When evaluating loss and expense, the quantity surveyor has always to contrast what has happened with what would have happened had not the delay or disruption occurred. Almost inevitably an element of conjecture is involved but the quantity surveyor has throughout a professional duty to assess the matter in a fair and impartial manner and to exercise due professional skill and judgement. The building industry in the United Kingdom is generally believed to be potentially the most efficient in the world, but it does suffer from the large number of disputes that arise, relating principally to direct loss and expense.

Regular progress meetings, if conducted in an efficient manner, can help to secure improved working and reduced risk of disruption and claims, and a concerted effort by the employer's professional advisers to ensure that they do not cause delay to the works by matters within their control.

## ORIGINATION OF CLAIMS

### On Site during Course of Works

It is important that the occurrence that results in a situation where the contractor suffers loss or expense is identified soon after the event. The verification of the supporting information and facts, which are so important to the success of the claim, is very difficult in retrospect.

Members of the contractor's management team are likely to experience a variety of situations where certain aspects are not entirely satisfactory. The following examples will serve to illustrate some of these problems:

(1) the site surveyor observes an operation proceeding on site but cannot clearly determine how it is going to be reimbursed;
(2) the contracts manager experiences difficulty in obtaining information from the architect despite a number of approaches; and
(3) the managing director is unhappy about the profitability of the project.

In each of these circumstances immediate investigation is warranted, to identify the cause and to determine whether it provides a realistic basis for a claim and, if so, to give notice of the claim.

In a contractor's organisation information regarding the actual cost of a contract compared with its estimated cost or value is often produced far too late. As a result any investigation into the cause of the loss is in retrospect, and hence the information is more difficult to establish, collate and record. A detailed cost study should be carried out on the more important elements of the work. One approach is to relate actual cost to budgeted cost from the estimator's records on carefully selected elements where this form of comparison can be done readily, albeit approximately, and regularly on site. Alternatively, a properly integrated bonus scheme, to achieve the optimum level of productivity, will provide much of the information needed.

An accurate register of drawings is vital with their suffixes, date of receipt and the main changes introduced. Site minutes and notes of all meetings must be carefully filed after being scrutinised by the contractor to check their accuracy. Any doubtful aspects should be queried in writing and raised at the next meeting. Site agents and foremen should keep comprehensive diaries as these may subsequently prove invaluable in supporting claims and even more so at an arbitration.

Where it is evident that the contractor will sustain a loss that can form the subject of a claim, full records of all pertinent facts should be kept, notice given to the architect, and the contractor may possibly consult with the quantity surveyor on how to evaluate the loss. Here again an effective costing system is vital.

**Periodic Checks by the Management Team of Contract Particulars**

In addition to the occurrence of events during the course of the works that may give rise to a claim, it should be the duty of one of the management team, probably the contractor's quantity surveyor, to check if any fundamental changes have occurred between the contract documentation and the works themselves. Some examples will serve to illustrate this aspect:

  (i)  the site area and accesses, which may be more restricted than was indicated in the contract documents;
 (ii)  the foundations, as the details are frequently incomplete at the design stage;
(iii)  the proportions and quantities of each type of work in the contract bills – it is just possible that 500 000 m³ of excavation has been incorrectly billed as 50 000 m³; and
 (iv)  the nature and character of the works, to determine whether any significant change has occurred between the project as construed from the contract documents and that subsequently required on the site.

A check should also be made of the effect of variations on:

 (i)  the contract programme and period; and
(ii)  the smooth running of the project.

When checking any such fundamental differences or on-site problems, it is important that all relevant information regarding the assumptions made by the estimator at the time of tender is made available. For example, the length of time that scaffolding and plant should be on the site, and the assumptions made about winter and summer working should be disclosed.

The contractor must be on the look-out for any discrepancies or divergences between any of the contract documents, such as substantial differences between scaled and figured dimensions on drawings, incorrect concrete mixes and areas of preparing walls for plastering. Inconsistencies sometimes occur between the quantities of excavation on the one hand and fill and disposal on or off site on the other. Changes in design do not always find their way through to the contract drawings and contract bills.

The character of varied work may be so different from the original that the billed rates are no longer applicable, as for example a decision to incorporate a chapel in a new school project. A substantial variation in the conditions under which the work is to be carried out will also necessitate varied rates such as the substitution of restricted access to the site for free access, shaped work in lieu of straight work or the need for manual work in place of mechanical work.

## PREPARATION OF CLAIMS

Claims are submitted principally to cover the extra cost and/or expense resulting from disruption of the work or prolongation of the contract. In either case a large amount of supporting information is needed in order to prepare a sound and logical claim, as described earlier in this chapter. One useful approach is for the contractor to require site agents and foremen to insert daily comments against a numbered list of topics to avoid significant omissions. It is also wise to keep two sets of important records in different offices to guard against possible loss in the event of fire.

In compiling a claim, a contractor may need to refer to any of the following documents:

(1) correspondence (seems straightforward at first sight but it can have important implications);
(2) approved minutes of site meetings (can contain instructions, variations and additional requirements);
(3) architect's instructions (could be the most important single item);
(4) clerk of works' directions (normally require written confirmation by architect within 2 working days);
(5) contract and working drawings and other contract documents (identify divergences and inconsistencies between them);
(6) labour allocation sheets (showing location, tasks and standing time);
(7) correspondence with and claims from subcontractors and suppliers (may indicate additional requirements for main contractor or causes of delays);
(8) site diary (must contain accurate and comprehensive entries and will often highlight problems);
(9) daily weather reports (rainfall records for the previous 5 years are useful – a CIOB analysis (1981) shows rain occupying 4 to 7 per cent of daytime hours with considerable monthly variations; topographical features can cause variations within a few km);
(10) receipt of drawings schedule (identification of revisions to drawings as compared with those on which tender was based);
(11) progress photographs, dated by the photographer (can show lack of progress and identify disruptions as basis for prolongation claims);
(12) site level details (basis for earthwork quantities);
(13) effect of artists' and tradesmen's work (under auspices of employer and can be disruptive);
(14) photographs and report detailing condition of site at date of possession (could show obstructions and part only of site available);
(15) records showing time period between date of tender and date of possession, or order to start work (could show delay at outset);
(16) build up of tender (particularly allocations for preliminaries, site and general overheads and profit);
(17) extension of time claims and allowances certified by architect;

(18) materials schedule (quantities received and delivery dates);

(19) invoice lists (additional costs under fluctuations clause, where appropriate);

(20) schedule of anticipated plant output;

(21) plant records (show standing time and number of times transported to and from site, with reasons, and plant output on key activities);

(22) scaffolding records (showing time in use and whether reassembly was necessary);

(23) authorised daywork schedule (covering varied work which cannot be valued at billed rates);

(24) programmes and progress charts (showing contractor's anticipated programme and actual performance);

(25) borehole logs (actual soil conditions may be different from those which the contractor could reasonably have anticipated);

(26) work method statement (identify extent to which disruption has occurred and its effect);

(27) variation data sheets (nature and effect of variations);

(28) interim applications, certificates and payments (amounts and pattern of payments – cashflow aspects);

(29) cost of head office overheads each month;

(30) cost and value of work executed each month on the project;

(31) cost and value of work executed each month for all projects; and

(32) profit or loss made by the company for each accounting period.

It is relatively easy to find correspondence on the files that confirms a major instruction, but it rarely amplifies the probable consequences which are so important when preparing a claim. For example, the following consequences might result from an architect's instruction:

(1) work in several areas will be disrupted;

(2) labour and plant will have to be transferred to other areas;

(3) new materials will have to be ordered and replaced materials either scrapped or transported to another site;

(4) other related work will be more costly;

(5) delay will be caused and extension of time will be necessary; and

(6) reimbursement of loss and expense will be sought (Robinson, 1977).

One of the most important documents is the labour allocation sheet showing who is working where and what they are doing, with space for entries relating to disruption. The quantity surveyor and site management need to know and trace all periods of work that are unproductive. These must be shown on allocation sheets together with the reasons for stoppages. They may result from the issue of architect's instructions or shortcomings within the contractor's own organisation. A knowledge of both is vital to management.

Another important record is a variation data sheet which asks a whole range of pertinent questions, including the following:

(1) Was there any stoppage of work by operatives?
(2) Were any operatives redirected?
(3) Was there any loss of output of plant?
(4) Was any plant redirected?
(5) Were any subcontractors affected or disrupted?
(6) Did the variation result in the order of new materials?
(7) Were any materials made surplus?
(8) How much extra staff time was incurred? (Robinson, 1977).

The details from a variation data sheet enable a comprehensive cost assessment of the loss situation to be made, and the effects of a variation to be priced out in bill form. To secure official agreement it is advisable to request the clerk of works or site architect to examine the variation data sheet and subsequently to sign it.

Hence one of the main criteria in establishing the validity of claims is good, accurate records. Probably the next most important step is to inform the architect and quantity surveyor that claim situations are arising. A major problem can be the confidentiality aspect of much of the cost information which the contractor guards jealously and some of which may be needed to satisfy the architect or quantity surveyor of the validity of the claim.

Every claim should be produced as if it is to become evidence in court (which it may do) and should be carefully detailed and presented, preferably in a bound cover. An untidy and carelessly prepared claim is unlikely to receive very serious consideration.

The claim could conveniently be broken down into the following logical sequence:

(1) Contract particulars – details of the site (as contained in the preliminaries) and details of the contract (as contained in the articles of agreement and appendix).
(2) Claim particulars – a summary of the bases or heads of claim, stating all facts and details, together with full particulars of the specific contract clauses on which the claim is based.
(3) Evaluation of the claim – a summary of the contractor's financial loss and/or expense.
(4) Appendices – a section that collates all the back-up information described in (2) and (3).

Often the claim is based on a reasoned argument of loss rather than the proof of the loss. The cause of this may stem primarily from the variability in the quality and preciseness of the contractor's procedures. As a result of negotiation and compromise the less well prepared claims are frequently amicably settled or withdrawn, but it is far better for the contractor to start from a sound and realistically based claim. It is also beneficial for the contractor to reach a reasonable settlement with the employer's advisers, by means of a well presented and carefully considered claim, than to resort to costly arbitration.

To sum up, in preparing a claim the first essentials are for the contractor to determine the extent of his obligations under the contract and then to obtain details of the matters that hindered or prevented him from executing the work with expedition and economy. It then remains for the facts to be stated with the utmost precision and clarity and to calculate the amount of the additional expense incurred. It is bad policy for the contractor to submit inflated claims using the argument that he does not expect to receive the whole of his entitlement. It is far better if realistic claims are submitted and that a truly professional attitude is adopted by all the disciplines concerned.

## DISRUPTION OF THE WORKS RESULTING FROM VARIATIONS

The issue of certain types of variation order can have a significant effect on the efficiency of a project, and its resultant productivity and profitability. For example, a variation for an extension to a building under construction which impinges upon the contractor's main access to the site, will have a disruptive effect on the movement of operatives, plant and materials on and off the site. Also a variation changing demountable internal partitions to concrete blockwork is likely to affect finishing trades and the overall site planning. Proving that such disruption falls within the term 'direct loss and/or expense' can be difficult. Although there are some fundamental principles that are important in the preparation and assessment of claims, each case must normally be treated on its merits. Powell-Smith (1990) has examined the problems and likely outcome related to claims for variations, supported by relevant cases.

### Programme

A realistic, properly prepared and detailed programme is a vital aid to proving disruption of the programme, particularly when it identifies critical activities.

### Resource Allocation

As an aid to proving the costs of disruption it is advisable for the contractor to show, in a simple form on the programme, the resources he has planned to use throughout its various stages; for example, the number of operatives to be employed per day or per week, and the amount and type of plant expected to be used. These can then be compared with the records of the actual resources used on the works and the cost differential established. The clerk of works normally keeps his own records which can be used as an independent check. The type of information described may also be incorporated by the contractor in a method statement.

### Effect of Variations

The disruption or adverse effect of variations should be suitably recorded at the time the variation orders are issued. These are best recorded on a standardised form containing the type of questions posed earlier in the chapter for insertion in a variation data sheet. The information required in response to the questions will be inserted by site staff, as a basis for a claim to be submitted by management staff.

### Acceleration

Acceleration entails increasing the rate of the constructional work to meet the contract completion date. This can be achieved by increasing the number of operatives on site, the length of working hours or the amount of plant. If the issue of an architect's instruction has caused delay it may be in the best interests of all concerned to make up the lost time by acceleration. It would, however, be unwise for the contractor to expend money in this way without first obtaining the architect's agreement and his confirmation that he is prepared to meet the cost.

Since the alternative to this kind of disruption claim is likely to be a request from the contractor for an extension of time under clause 25 of the JCT Conditions, with an associated request for recompense under clause 26 (a prolongation claim), the architect will usually decide to accept the cost of acceleration, as this is likely to cost less than the other available alternatives and will ensure that the building is still completed on time. The cost of such an acceleration claim is normally obtained by using programme comparisons, as previously described. Revised programmes, but incorporating the same details, should be supplied by the contractor to the architect.

### Recovery of the Cost of Variations

Contractors have in the past often cooperated with quantity surveyors in applying *pro rata* rates as a ready means of valuing varied work. This procedure should, however, be applied only when the work is of similar character and executed under similar conditions to the work priced in the contract bills. In practice there are few variations that can be made to the contract works without altering either the character and/or the conditions under which the work will be performed, and in these circumstances it is necessary to assess fair rates and prices.

### Final Accounts

Some quantity surveyors appear to treat a final account as a technical operation rather than a matter of judgement. In some cases a project is subject to redesign to such an extent that it bears little resemblance to the original scheme. The quantity surveyor may decide to abandon the preparation of a bill of variations and to remeasure the whole of the work at billed rates. This is tantamount to an admission that the entire contract is varied and it is then patently incorrect to use a bill of quantities in the same way as a schedule of rates, and new rates and prices should be negotiated to arrive at a fair valuation of the work (Burke, 1976).

### Prolongation Claims

These claims may be presented as a result of extensive variations and be associated with a claim for extension of time under clause 25, or as a result of interference with the regular progress of the work under clause 26. Prolongation claims will be based on the direct loss and/or expense incurred by the contractor and resulting from the extension of the contract period. The various elements likely to appear in this type of claim are now separately considered. Delays which qualify for additional payment normally relate to activities which appear on the critical path.

### Preliminaries

The value of preliminaries may be increased because of a number of the components being directly related to the contract period. A claim under this head should preferably be based on actual expense, and not merely to *pro rata* the prices inserted against preliminaries items. The contractor should record the actual expense incurred during the period of the extension. Another approach is to subdivide the preliminaries items into the three categories of lump sums related to specific events, time based and value based in the manner described in chapter 6. The main expenses involved are likely to include the following:

(1) salaries of site staff;
(2) cost of plant and scaffolding retained on site;
(3) temporary lighting;
(4) offices and stores retained on the site and their upkeep;
(5) safety measures;
(6) protection of the works;
(7) insurance premiums;

(8) telephone costs;
(9) electricity costs; and
(10) rates on site buildings.

## Time of Year when Work is Carried Out

The extension of a contract from summer into winter may lead to extra costs as a result of reduced working hours, stoppages through bad weather and other associated additional costs. Carefully kept site records will record these additional costs which can then be compared with the programme and resource allocation.

## Extended Attendances on Nominated Subcontractors

When the contractor prices general and special attendances on nominated subcontractors these may, in certain circumstances, relate to the period of time that the nominated subcontractor is on the site. When as a result of variations or extensions of time these attendances are extended, the contractor should record the actual costs incurred.

## Overheads

Claims should ideally reflect actual costs or expenses that are provable and this principle should, as far as practicable, be applied to site overheads. In the case of general overheads, it can be argued that the mere extension of a contract will not automatically result in increased costs. If it has been necessary to increase the number of office staff employed as a direct result of the prolongation then the cost of such an increase would be claimable, based on the actual additional cost incurred. Otherwise, it could be argued that the overheads are the same but that their incidence is spread over a longer period.

Overheads are a budgeted amount calculated on the anticipated year's expenditure on staff, offices, equipment, stationery and the like, and are related to a budgeted year's turnover. On this basis a minimum percentage on a fixed turnover must be achieved to recover cost. If the turnover is significantly reduced then the percentage of overheads needs increasing to produce the required sum. If the contract period is extended, work of the same value is spread over a longer period and, as a result, turnover in a given year is reduced.

To recover his overheads the contractor may need to increase his overheads percentage, but on current work the tenders will have already been submitted and accepted, so it is too late to adjust them. This could form the basis for an under-recovery claim as illustrated in table 7.1, and this represents one way in which an overheads shortfall could possibly be substantiated, using a similar approach to that used in the Hudson formula.

*Hudson formula approach*

$$\frac{\text{head office (profit) \%}}{100} \times \frac{\text{contract sum}}{\text{contract period}} \times \text{period of delay}$$

Hence contractor's loss of contribution to overheads (profit) as a result of the delay shown in table 7.1 $= \frac{6}{100} \times \frac{800\,000}{12} \times 3 = £12\,000$. Other formulae used on occasions are Emden's and Eichleay's, as described by Thomas (1993).

Table 7.1   Under-recovery claim

| | |
|---|---|
| Budgeted turnover in 12 months | £6 500 000 |
| Estimated cost of general overheads | £390 000 (6 per cent) |

Assume that a contract valued at £800 000 is extended 3 months, where the contract was expected to be completed in 12 months. The possible value of the shortfall could be £800 000 × (3/12) = £200 000, being the value of the work that could have been done in the extended 3 month period.
Therefore the turnover in this particular year would be reduced in the manner shown:

| | |
|---|---|
| budgeted annual turnover | £6 500 000 |
| *less* calculated shortfall | 200 000 |
| actual turnover | £6 300 000 |

When tendering, general overheads are expressed as a percentage of the estimate, and once a project is obtained there is no way of increasing this percentage. Therefore the overheads recovery could be calculated as:

6 per cent of £6 300 000 = £378 000
This shows an under-recovery of £390 000 − £378 000 = £12 000

## Loss of Profit

Most contractors' claims will include some element of profit. However, a claim of this kind is very difficult to prove, since profit can be lost by inefficiency or bad tendering before any extension of time occurs.

There are various schools of thought:

(1) It can be argued that an overall shortfall of £200 000 (as calculated in table 7.1) will result in a profit shortfall in 12 months of say 3 per cent of £200 000 = £6000.
(2) Another argument is that since the total value of the work was not reduced, although it was spread over a longer period, neither was the profit reduced; thus there was no actual loss.
(3) Yet another approach could be that keeping key operatives on the site for a longer period than anticipated reduced their profit earning power elsewhere.

The definition that loss is something the contractor should have received but did not is untenable. It could possibly be argued that in case (1) the company will not be able to pay the desired return to its investors on their invested capital as a direct result of the anticipated shortfall arising from the extension of the contract. The most common result of a loss of profit claim is a compromise.

In disruption claims the biggest problem for all parties lies in the interpretation of the word 'direct' in respect of loss and expense. In modern high-speed construction, the disruption of one trade or area of work reverberates through all following trades and, if the contract is bedevilled with disruptive incidents, the tracing of individual losses is virtually impossible. Claims are then made by the comparison of tender totals of operatives and plant hours with the actual hours expended. This tends to reduce the contract to a prime cost situation and makes no allowance for the possibility of poor planning, mistakes, bad workmanship and dilatory management by the contractor. These problems highlight the difficulties inherent in the fair and reasonable assessment of claims of this type.

## INSOLVENCY OF BUILDING CONTRACTORS

### Nature and Causes of Insolvency

The author has drawn extensively on the very informative and practical guide produced by the Society of Chief Quantity Surveyors in Local Government (1981) in writing this part of the chapter and readers are referred to this publication for more detailed information. Keating (1967) has referred to insolvency as an 'omnibus term to cover various forms of individuals or companies being so short of cash as to affect their legal position.' The *Oxford English Dictionary* defines it as 'the fact of being unable to pay one's debts or discharge one's liabilities.' In law, only individuals or partnerships can become bankrupt. Insolvent companies are put into liquidation. A receiver can be appointed to a company that is not insolvent and receivership is not followed inevitably by liquidation.

In the late nineteen seventies and early nineteen eighties many insolvencies occurred among building contractors, amounting to more than 1500 in 1982 alone; this represented over 14 per cent of total liquidations in England and Wales during this period. An even worse situation occurred in the 1990–94 recession, which deteriorated still further in 1995, due to falling workloads and keen competition, accompanied by rising costs of labour and materials, forcing more firms into insolvency. These arose primarily because of wide fluctuations in the demand for building works, the existence of many small and medium sized firms often lacking sufficient financial expertise and being under-capitalised, and maybe high interest rates and restrictions on capital. Some contractual and organisational aspects can also contribute significantly to the occurrence of insolvencies, such as the inviting of an excessive number of contractors to tender for projects, the accumulation of undecided claims, excessive price-cutting by tenderers, decrease in the value of work in progress with consequently higher proportionate overheads, delay in payments to contractors arising from non-settlement of final accounts and non-release of retention, undervaluation of works, and overvaluation of works in the early stages leaving insufficient funds to complete the works.

### Duties of Receivers and Liquidators

The concept of receivership applies when the debenture holders of a company, or their trustees, consider that the security against their debentures is in jeopardy and appoint, through terms in the debenture or through the courts, a receiver. The function of a receiver is to get in the assets charged, collect rents and profits, exercise the debenture holders' powers of realisation and pay the net proceeds to the holders in respect of their charge. He has no power, as a receiver, to carry on the business of a company. The liquidation or winding up of a company is the process whereby its life is ended and its property administered for the benefit of its creditors and members.

When a receiver is appointed the company need not go into liquidation. If it does, a different person will normally be appointed liquidator. Where a receiver and manager is appointed over the whole undertaking, he will for all practical purposes replace the board of directors as on liquidation the company will probably be wound up. It is the receiver and manager that the quantity surveyor often deals with in the case of insolvencies of building contractors. The receiver and manager is not, however, liable for those contracts entered into by the company even if he oversees their performance. Following the reinstatement of a contractor the liquidator may still disclaim, if the contract proves too onerous, but a receiver and manager has no such power (SCQSLG, 1981).

### Indications of Imminent Insolvency

There are usually some indications of financial or liquidity problems on a contract before a receiver and manager or liquidator is appointed. These include the following aspects:

(1) slowing down in the progress of the works and a decrease in the labour force;
(2) reduction in the value of materials on site, increased delays in the delivery of materials and non-scheduled movement of plant off the site;
(3) non-payment of nominated subcontractors;

(4) requests from the contractor for the quicker honouring of certificates;

(5) changes in the contractor's management personnel; and

(6) unusual visitors to the site, such as head office senior management.

Hence the quantity surveyor, clerk of works and architect should all be on the alert for these warning signs. The financial and technical press should be scrutinised for any relevant reports or statements and the financial affairs of the contractor monitored, as far as practicable.

### Possible Action prior to Appointment of Receiver or Liquidator

The following actions should be taken to safeguard the employer's interests:

(1) The employer's advisers should keep all information received strictly confidential to avoid any possible escalation of the contractor's difficulties (Aqua Group, 1990b).

(2) The architect should take special care to ensure that the quality of materials, standard of workmanship and the works generally satisfy the requirements of the contract documents (2.1). He should immediately request the remedying of any defects and if the contractor fails to comply with this written notice within 7 days, the architect should employ another contractor to carry out the work (4.1.2).

(3) Similar action should be taken in respect of temporary fencing, hoardings, site protection and the like.

(4) The clerk of works should be instructed to resist the removal of any unfixed materials or goods delivered to, placed on or adjacent to the works and intended for use on them, until it is established that the employer has no legal right to possession (16.1).

(5) All insurance policies required by the contract conditions should be checked to ensure that they are paid up and operative for the whole of the contract period (20, 21 and 22).

(6) The contract requirements with regard to the payment by the contractor of subcontractors must be monitored (35.13.2 and 35.13.3).

(7) Any valuations (30.1.2) should be prepared carefully to ensure that there is no overvaluation of preliminaries or insurances, work in progress, materials on or off site or fluctuations. Materials on or off site should be rechecked as to their quality and suitability for the works.

(8) No interim certificate (30.1.1.1) shall include the value of any defective work.

(9) The clerk of works should be instructed to ensure that materials and goods on site that have been included in certificates are adequately protected (30.2.1.2).

(10) Where the financial position of the contractor is in doubt, the architect could exercise his prerogative of excluding from his interim certificate

the value of any materials or goods off site (30.3). If included in previous certificates, then the security and marking of the materials and goods should be rechecked and copies of the document vesting ownership in the contractor obtained.

(11) If the contractor stops work on the site without a receiver or liquidator having been appointed, the site should be made secure and a security watch placed upon it. Where the contractor has supplied the names of financial referees, they should be approached with a view to determining his financial position.

### Relevant Provisions of the JCT Standard Form

The following are the principal contract provisions:

(1) Clause 27.3 provides for the employment of the contractor to be automatically determined in the event of the contractor:
   (i)   becoming bankrupt or making a composition with his creditors; or
   (ii)  having a winding up order made; or
   (iii) having a resolution for voluntary winding up passed (except for the purposes of amalgamation or reconstruction); or
   (iv)  having a provisional liquidator, receiver or administrator duly appointed.

(2) Clause 27.3.3 also provides that the employment of the contractor may be reinstated by agreement between the employer and the contractor.

(3) Upon the determination of the employment of the contractor the contract is still in being but in place of the conditions of contract relating to the contractor's completion of the works, and payment by the employer, the rights and duties of employer and contractor are as provided in clause 27.5.

(4) The employer is empowered to employ and pay other persons to carry out and complete the works for that purpose, to use all temporary buildings, plant, tools, equipment and materials intended for and delivered to the works or adjacent to them. The contractor shall remove them when the architect so requires, failing which the employer may remove and sell them and hold the proceeds, less all incidental costs, to the credit of the contractor (27.6).

(5) This right cannot be exercised in respect of hired plant as the firm that has hired the plant to the contractor is entitled to remove it. If the employer wishes to retain the plant he must make arrangements with the plant hire firm.

(6) If the supplier of any of the unfixed materials on the site has reserved his title in them until he has received full payment he may repossess them when he hears of the contractor's insolvency (Romalpa (1976) 1

WLR 676 CA). A similar situation exists in respect of materials that have been delivered to the site for work by a domestic subcontractor who has not received payment from the contractor for them.

(7) The employer is also empowered to purchase all materials necessary for the completion of the works.

(8) Clause 27.6.5 provides that the employer is not bound to make any further payment to the contractor until the works have been completed and the accounts verified, and it also obliges the contractor to pay or allow to the employer any direct loss and/or damage caused by the determination.

## Procedure upon Appointment of Receiver or Liquidator

The actions to be taken can conveniently be divided into two phases.

### (1) Preliminary action

(i) No more certificates should be issued or valuations made and any payments due from the employer to the contractor should be withheld.

(ii) The name and location of the receiver and manager or liquidator should be obtained and a check made that the name and form of the company have not been changed.

(iii) The site should be adequately secured, and protected, the works and materials protected from weather, and the police informed of the situation.

(iv) Insurances should be arranged for materials and plant on site and the value of the works executed at current values.

(v) Arrangements should be made to take possession of materials off site certified under clause 30.3, probably by moving them to the site or other secure store.

(vi) The necessary authority should be obtained from the employer for action to complete the works and where there is a bondsman he should be informed.

### (2) Subsequent action

(i) Arrangements should be made for representation of the employer at any meetings of creditors.

(ii) Statutory undertakers should be informed and arrangements made for the maintenance of temporary services.

(iii) A financial assessment should be prepared.

(iv) Nominated subcontractors and suppliers should be informed of the liquidation.

(v) A schedule should be prepared detailing the state of the works, including any defective items.

(vi)   The receiver or liquidator should be given a time limit for any assignment.

(vii)   A schedule of plant should be prepared, contractor's plant not required removed and hire plant released or the hire arrangements continued.

(viii)   Outstanding letters should be acknowledged and replies sent where possible.

(ix)   Staff costs, fees and other expenses should be recorded as these may constitute a charge against the receiver and manager or liquidator.

### Completion of the Contract Works

The main aim is to secure the earliest possible continuation of the contract, since the longer the delay in arranging completion the greater the cost to the employer. There are four principal methods by which the contract works may be completed, namely: reinstatement, assignment, novation and completion by the employer using one of several contractual arrangements.

*Reinstatement*   The employment of the contractor under the contract may be reinstated in accordance with clause 27.3.3. This is normally the quickest way, but a written assurance is required from the liquidator, where appointed, that he will not subsequently disclaim.

*Assignment*   A receiver and manager or liquidator may arrange the completion of the works by assignment with the agreement of the employer (19.1). This can be an effective procedure provided that the financial standing of the assignee is satisfactory, as the assignee accepts all the rights and liabilities under the original contract.

*Novation*   In the event of assignment not proving possible, the receiver and manager or liquidator may secure a successor contractor of acceptable financial and technical standing. The quantity surveyor much check on the extent to which liabilities under the original contract are being qualified and evaluate their consequences.

*Completion by the Employer*   The quantity surveyor will obtain tenders for completion by negotiation, competition or direct labour. The following types of documentation may be used involving alternative forms of contract:

(1)   original priced bill with a lump sum or percentage adjustment;

(2)   original unpriced bill with the quantities amended;

(3)   original unpriced bill suitably amended to produce a schedule of rates;

(4)  new bill of firm or approximate quantities;
(5)  new specification with completion drawings and a schedule of rates or lump sum;
(6)  daywork schedule; and
(7)  prime cost and fixed fee.

The choice will be influenced by:

(1)  the ease of determining the final cost at the tender stage;
(2)  the nature and extent of the work remaining to be executed;
(3)  the time available to prepare documents and obtain tenders; and
(4)  any additional postcontract work that may be required.

Whichever approach is adopted, provision will have to be made for the following:

(1)  completion of work that is only partly executed;
(2)  making good the original contractor's defects;
(3)  renomination or new nominations of subcontractors, as necessary; and
(4)  a list of materials, plant and equipment on site or held in store.

### The Position of Subcontractors

In most situations the contractor employed to complete the works will make his own arrangements for domestic subcontractors. With nominated subcontractors, the determination of the employment of the main contractor results in the nomination ceasing to have effect. Their reinstatement thus depends on new nominations to the successor contractor. The quantity surveyor will need to compare the available alternatives and recommend those most favourable to the employer.

The employer may use retention money relating to nominated subcontractors' work, to meet the cost of completion of the works. Work completed by nominated subcontractors for which payment has not been made is part of the equity available to the employer but if, at final settlement, surplus monies are available then such value constitutes a debt due to the receiver and manager or liquidator.

Nominated subcontractors will be required to repay any direct payments from the employer on demand and where the employer produces reasonable proof that there was a petition or resolution concerning winding up of the company existing at the time of payment. Works of urgency or safety undertaken by nominated subcontractors after the insolvency of the main contractor will be ordered by the employer and paid by him direct.

**Financial Aspects**

Reinstatement is the quickest and cheapest method of securing completion of the works, provided the quantity surveyor receives assurance that
the receiver and manager or liquidator will discharge all the contractor's
responsibilities under the contract. If the employment of the contractor is
not reinstated, the employer should be advised to file his claim with the
liquidator as a creditor of the company in liquidation. The details will be
submitted on settlement of the final account.

Assignment and novation are the next best means of securing completion of the works, but various additional elements of cost will be incurred,
covering security and protection of the site; insurance; professional costs;
renominations or new nominations of subcontractors; and effect of delayed completion.

In all cases where the employer arranges completion of the works the
additional costs are likely to be high as, in addition to the costs listed
under assignment and novation, there are:

(1) the higher costs of engaging a different contractor to complete the
    works;
(2) the cost of making good all defects arising from the original contractor's
    work;
(3) the cost of replacing any materials removed from the site where the
    supplier had reserved his title; and
(4) the cost of new security or bond, where applicable, mainly with public
    authorities.

On the other hand the employer will have the same resources on which
to call, such as retention money, unpaid variations, unsettled claims, un-
claimed fluctuations, use of contractor's own plant impounded by the
employer, any outstanding credit for other contracts with the same contractor, and possible set-off of liquidated damages for delayed completion.

**Final Account and Settlement**

On completion of the works following the insolvency of the original contractor, two final accounts are generally required. A normal final account
determines the final cost of the contract and a notional final account sets
out what would have been the final cost of the contract had the original
contractor continued and finished. This procedure quantifies one element
of the loss sustained by the employer as a result of the insolvency, since
the difference between the two accounts constitutes a loss to the employer.

A notional account is, however, unnecessary under reinstatement, assignment and novation. The receiver and manager or liquidator will nor-

mally employ a quantity surveyor to settle the notional account with the employer's quantity surveyor.

To the direct loss, computed from a comparison of the two final accounts, are added additional staff time costs and/or fees and any indirect costs such as security. The employer will also claim against the original contractor for loss of his entitlement to deduct damages if the works have continued beyond the original completion date.

## INSOLVENCY OF EMPLOYERS

In the event of the insolvency of the employer, the contractor can determine his employment by registered post or recorded delivery (28.1), and he then becomes the creditor of the employer and the amount of the debt will be calculated in accordance with the provisions of clause 28 of the JCT Standard Form.

# 8 Cost Control of Construction Projects

This chapter is concerned with the control of the cost of projects by the design team throughout both the design and constructional processes and the methods by which the contractor monitors and regulates the cost of construction on site. The quantity surveyor has an important role in all these processes.

## OBJECTIVES OF DESIGN COST CONTROL

### Need for Cost Control

Cost control, in its widest interpretation, aims at ensuring that resources are used to best advantage. A narrower definition of precontract cost control was produced by a RICS committee (1982) as 'the total process which ensures that the contract sum is within the client's approved budget or cost limit,' but it will be shown later in the chapter that this activity extends well beyond keeping within the approved budget if it is to be really effective.

The majority of promoters of building work insist on projects being designed and executed to give maximum value for money. Value for money should ideally include securing the relevant quality and encompassing the life of the building. Hence quantity surveyors are employed to an increasing extent during the design stage to advise architects on the probable cost implications of their design decisions. As buildings become more complex and employers more exacting in their requirements, so it becomes necessary to improve and refine the cost control tools. Rising prices, restrictions on the use of capital and changing interest rates have caused employers to demand that their professional advisers should accept cost as an element in design, and that they should ensure suitably balanced costs throughout all parts of the building, as well as an accurately forecast overall cost, involving close collaboration between the architect and the quantity surveyor (Seeley, 1995a).

As long ago as 1970 the report of the Special (Future of the Profession) Committee of the Royal Institution of Chartered Surveyors described how many employers adopt cost limits for projects and are instrumental in spreading an awareness of efficiency and value for money in building; and this presages greater use of the quantity surveyor in establishing cost targets, in the appraisal of alternative solutions, and in cost control as a continuing process. This theme was reiterated in the SITE Report (1980b), which defined the role of the quantity surveyor as ensuring that the resources of the construction industry are used efficiently to serve the best interests of society.

An RICS Report: QS2000 (1991) rightly emphasised that clients need early and accurate cost advice, more often than not well in advance of site acquisition and of a commitment to build. In helping to define client's requirements in financial terms, quantity surveyors are exerting considerable influence on any resulting design. This requires not only a knowledge of construction and construction costs, but also a knowledge of the property market and an ability to anticipate and visualise client's detailed requirements. Seen in this context cost estimating is a truly professional service, requiring the exercise of a high level of expert discretion and judgement in conditions of considerable uncertainty. An important challenge is thus to improve the accuracy of quantity surveyor's cost estimates, given uncertainties over client requirements, design and future cost and price movements in the industry and in the economy generally.

Advice may be given by the quantity surveyor on the strategic planning of a project which will affect the decision whether or not to build, where to build, how quickly to build and the effect of time on costs or prices and on profitability. During the design stage, advice is needed on the relationship of the capital costs to maintenance and operating costs and on the cost implications of design variables and differing constructional techniques. The cost control process should be continued throughout the construction period to ensure that the building is kept within the agreed cost limits. Building cost can be considered as a medium relating purpose and design and it certainly forms an important aspect of design. An employer is very much concerned with quality, cost and time; he wants the building to be soundly constructed at a reasonable cost and within a prescribed period of time (Seeley, 1996).

It is, therefore, vital to operate an effective cost control procedure during the design stage of a project to keep its total cost within the employer's budget. Where the lowest tender is substantially above the initial estimate, the design may have to be modified considerably or, even worse, the project may have to be abandoned. Pressures from ten main sources have combined to stress the importance of effectively controlling building costs at the design stage:

(1) There is greater urgency for the completion of projects to reduce the amount of unproductive capital or borrowed money, and few employers have sufficient time for the redesign of schemes consequent upon the receipt of excessively high tenders.

(2) Employers' needs are becoming more complicated, more consultants are being engaged and the estimation of probable costs becomes more difficult. Furthermore, individual developments tend to be block sized rather than plot sized in scope.

(3) Employing organisations, both public and private, are themselves adopting more sophisticated techniques for the forecasting and control of expenditure, and they in their turn expect a high level of efficiency and expertise from their professional advisers for building projects and require a broad and comprehensive range of services (RICS, QS Division, 1991).

(4) The introduction of new constructional techniques, materials and components creates greater problems in assessing the capital and maintenance costs of buildings.

(5) Changing prices, restrictions on the use of capital, variable interest rates and low contractors' profit margins all make effective cost control that much more important.

(6) The move towards reduced waste and better use of resources coupled with the Latham Committee Report (1994) objective to reduce building costs by 30% by year 2000, creates the need for more cost forecasting and improved cost management.

(7) There is an increasing demand for integrated design to secure an efficient combination of building and services elements in complex developments, such as hospitals, with effective cost planning to optimise the design solution within the budget figure.

(8) Rising energy costs necessitate costing alternative heating and thermal insulation measures.

(9) More attention should be paid to life cycle costing and total cost appraisal.

(10) Demand patterns centre around the elements of value-added, finance, management and investment (RICS, QS Division, 1991).

## Main Aims of Cost Control

The main aims of cost control are probably threefold:

(1) To give the employer good value for money – a building which is soundly constructed, of satisfactory quality and appearance and well suited to perform the functions for which it is required, combined with economic construction and layout, low future maintenance and operating costs, and completed on schedule as lost time is also lost money, and in accordance with the agreed brief.

(2) To achieve a balanced and logical distribution of the available funds between the various parts of the building. Thus the sums allocated to cladding, insulation, finishes, services and other elements of the building will be properly related to the class of building and to each other.

(3) To keep the total expenditure within the employer's budget, frequently based on an approximate estimate of cost prepared by the quantity surveyor in the early stages of the design process. There is a need for strict cost discipline throughout all stages of design and execution to

Table 8.1   Examples of exceeded estimates

| Project | Estimate (£m) | Final account (£m) |
|---|---|---|
| Sydney Opera House | 2.5 | 87 |
| Thames Barrier | 23 | 400 |
| Barbican Arts Centre | 17 | 80 |
| Devonport Dockyard | 21 | 83 |
| Kingston Crown Court | 0.25 | 1.8 |
| Channel Tunnel | 4800 | 11 000 |
| DoE Trident nuclear submarine base | 1100 | 1900 |

ensure that the initial estimate, tender figure and final account sum are all closely related. This entails a satisfactory frame of cost reference (estimate and cost plan), ample cost checks and the means of applying remedial action where necessary (cost reconciliation).

The problems of keeping the costs of completed projects within the initial estimates are highlighted in the examples in table 8.1.

A HM Treasury document (1995) showed that cost and time overruns were widespread throughout public sector projects. The analysis of 803 government projects in 1993–94 showed that on average projects cost 13.1 per cent more than original budget estimates and took 6.5 per cent longer to complete. The government was looking to save £3bn per annum through the adoption of a more professional approach to buying goods and services, and the increased use of partnering, as described in chapter 19.

As a general principle, the agreed budget or cost limit should be regarded as the maximum cost, and the quantity surveyor in collaboration with the other members of the design team should endeavour to satisfy the employer at a lower cost, where possible. Where sums have been included in the approved budget for items such as abnormal site costs, which are subsequently found to be not altogether necessary, then the employer should be notified of the consequential saving.

In the application of cost planning techniques, the quantity surveyor is concerned with many issues of building economics, some involving returns as well as costs, and some examples follow:

(1) substitution between capital and running costs to secure the minimum total cost;
(2) investigating different ways of producing the same building at lower cost;
(3) finding ways of slightly altering a building so that for the marginally greater use of resources, the returns are more than proportionately increased; and

(4) investigating methods of using the same resources to produce a different building which could give greater returns (Seeley, 1996).

Design criteria apart from cost must be considered to produce a properly balanced design, and it is important that the quantity surveyor does not lose sight of this. For instance, the cost of a project could be reduced merely by using cheaper materials, finishes and fittings, despite the fact that maintenance and operating costs would probably be increased considerably in consequence. Furthermore, the lower quality materials and components may be quite out of keeping with the class of building in which they are being incorporated. An economically priced project is required, but not necessarily the cheapest as a certain standard of quality has to be maintained. A building must also be designed so that it can satisfactorily perform its required functions. It might, for example, be possible to cheapen the cost of a factory roof by introducing more columns and so reducing the roof spans. This approach would be quite ineffectual if the factory needed large unrestricted floor areas for the successful functioning of the production processes. Costs can often be reduced by disregarding the aesthetic quality of the building to be erected. Plain façades devoid of any form of embellishment could reduce costs but result in most uninspiring elevations.

## IMPLEMENTATION OF DESIGN COST CONTROL

### Principal Factors in Cost Planning

Cost planning has become more sophisticated over the last decade and it is increasingly becoming a service that the employer is expecting from the quantity surveyor, to ensure that he receives value for money and that the project cost is kept within the agreed budget. To undertake this service effectively it is necessary for the quantity surveyor to be appointed at the earliest possible stage, in order to make a positive contribution at the brief and feasibility stages.

Employers often require accurate estimates of total cost even although their brief may be ill-defined, and the information about the new project that can be made available to the quantity surveyor may be very scanty. At the feasibility stage, most of the major decisions about the type and form of development will have been made and hence it is essential that cost advice from the quantity surveyor should be available to the design team at this critical stage. The quantity surveyor can then prepare realistic estimates of cost using a variety of techniques, covering the selected alternatives. In the United Kingdom the quantity surveyor has two main sources of cost information relating to similar projects, which he can adjust for time, location, main design features, quality and market conditions. These are the details of contracts passing through his own office in the building cost library that he establishes, and on a wider front through the data circulated by the Building Cost Information Service (BCIS) of the Royal Institution of Chartered Surveyors (RICS). The latter provides detailed information on a wide range of projects, but the information supplied requires expert handling and adjustment to suit local conditions. In the future the cost planning service will become increasingly sophisticated as the quantity surveyor makes greater use of the computer and the BCIS computerised cost database and on-line facilities.

As part of the cost planning function, the quantity surveyor can forecast the cost implications of a whole range of design variables, such as size, shape, height, circulation ratio, wall to floor ratio and various other building characteristics. Most employers require a functional and economic layout, as well as a building with good aesthetic qualities, both inside and out. Complicated shapes result in greatly increased building costs often without significant advantages. High circulation ratios resulting from large amounts of floor area in entrance halls, corridors, passageways, stairways and the like are wasteful. The aim should be to endeavour to keep circulation space at not more than 15 per cent of gross floor area. Rental values are often related to usable floor space and thus high proportions of circulation space increase the amount of non-rent earning accommodation.

Costs of buildings tend to rise disproportionately with increases in the number of storeys over three, but this extra cost may be partially offset by the reduced demand for highly priced land. Hence feasibility studies need to embrace land as well as buildings, when comparing the financial advantages of alternative layouts. The disposition and siting of buildings also require careful consideration to produce an efficient, attractive and economic scheme.

With framed buildings the quantity surveyor will need to liaise with the architect and structural engineer to secure the optimum type and spacing of columns and beams. The requirements of the activities to be carried out in the building will also need to be considered. A closer spacing of columns will probably be cheaper but this will be of no avail if it interferes with the processes to be undertaken in the completed building.

The quantity surveyor must be fully conversant with the functional requirements and cost implications of a wide range of constructional methods, components and materials. As the design develops the quantity surveyor will advise the architect of the cost implications of the various alternatives, in order to assist the architect in reaching a sound decision. One of the most variable cost elements is the foundations which are influenced by the type of subsoil and loads to be carried. A variety of framed structures and loadbearing walls needs comparing and here again the quantity surveyor's knowledge and experience of costs is valuable.

Services can account for 30 per cent or more of the total cost of a building and it is desirable that the services engineer should be brought in early in the design stage. Integrated design of a building structure with the services will prove far more satisfactory than a building that has reached an advanced stage before the services are considered, resulting in a more costly and less efficient services layout with the services probably being more obtrusive. Employers are also attaching greater importance to economic and efficient lift provision in high rise buildings, which in the past have often been over-specified.

Apart from producing realistic costs of alternative proposals, it is necessary to prepare developers' budgets or feasibility studies to show the financial viability of the proposals, and particularly a reasonable rate of return to the employer who is building as an investment. In some cases the employer may wish to know the highest price that he can afford to pay for the land. The use to which land can be put is often affected considerably by planning controls, and also the permitted height of buildings, car parking requirements and the like. These calculations are particularly important when developing highly priced land in central urban areas.

An employer may have decided the amount of his budget and wishes to know the size and quality of building that can be provided for this sum of money. The quantity surveyor will compute the amount of floor area that can be provided by reference to the known costs and floor areas of similar buildings.

The quality or standard of the building can be adjusted to keep within the approved sum if the floor area subsequently proves to be insufficient. This aspect needs watching very closely as, for example, a low standard of finishes in a prestige building would not be acceptable, and it is unwise to reduce initial construction costs at the expense of future running and maintenance costs. As with many cost calculations the right balance has to be achieved.

Figure 8.1 illustrates diagrammatically how the opportunities for making cost adjustments reduce substantially as the project progresses from the feasibility stage through to the end of the construction period.

**Cost Planning Process**

The Royal Institute of British Architects (1973) formulated a suggested pattern of procedure for architects in the preparation and implementation of building schemes. This plan of work, as depicted in table 2.1 and illustrated in figure 8.2, represents a sound and practical analysis of the various operations and has been applied successfully on many contracts. It incorporates suggested activities by the quantity surveyor aimed at providing effective cost control throughout the design process, and these are now described.

(A) *Inception* This represents the employer's decision to build, the setting up of the administrative organisation, and the appointment of the architect and other members of the design team. The architect requires a site plan, details of preliminary items, erection times and any cost limits.

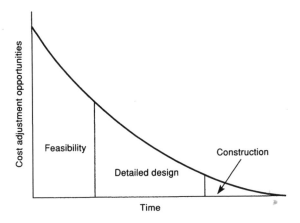

**Figure 8.1** *Cost adjustment opportunities*

(B) *Feasibility* At this stage the architect and the employer are endeavouring to establish the employer's specific requirements. The quantity surveyor supplies cost information normally based on the actual costs of previous buildings of similar type, suitably adjusted to make allowance for such factors as differences of location, site conditions, market conditions and quality of work, and prepares a provisional estimate on a comparative or interpolation basis. The design team should consider the design timetable and the tendering procedure.

(C) *Outline proposals* The employer's requirements have now been established and their viability confirmed, the site surveyed and the architect begins to consider the alternative ways in which the building can be designed and constructed. Some drawings will be produced at this stage and the quantity surveyor will be in a position to give general guidance on costs and, in particular, to evaluate the financial effect of different solutions to any specific design problem, and he often prepares an outline cost plan. It is an important stage as a considerable proportion of the final potential building cost is being formalised, thereby restricting significantly the degree of flexibility of choice in subsequent stages, as illustrated in figure 8.1.

(D) *Scheme design* During this stage the major planning problems will be resolved and the outline designs will emerge. The sketch designs will include sections and elevations, and services and finishes will be considered in addition to the form of the structural framework, with advice from the appropriate consultants. The quantity surveyor checks on his approximate estimate figure and, with the aid of extensive cost information, reappraises the initial cost plan with provisional target cost figures allocated

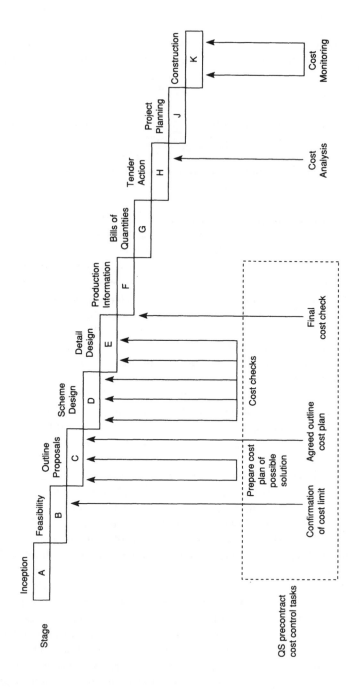

**Figure 8.2** *Sequence of design team's work*

to each element or major part of the building. The quantity surveyor may adopt an elemental approach, a comparative technique or a mixture of both.

(E) *Detail design*    Sketch plans are now finalised and some working details are prepared. Outline schemes will be prepared by consultants and designing subcontractors, with provisional estimates supplied in some cases. The quantity surveyor will provide comparative costs of different forms of construction, materials, components and service layouts and will adjust the distribution of costs in the cost plan if required. These cost studies should include probable running and maintenance costs wherever they are likely to have a significant effect on the outcome. Continuous cost checks by the quantity surveyor will ensure that the development of the design remains compatible with the cost plan; this process is sometimes described as cost reconciliation. When all the design drawings have been prepared and cost checked, a final cost review should be made by the quantity surveyor and a report submitted to the architect.

(F) *Production information*    The final working drawings (production drawings) are now prepared from which the bill of quantities can be produced. Consultants, subcontractors and suppliers supply detailed information at this stage, including realistic estimates and quotations. The quantity surveyor continues his cost checks on the data produced against the final cost plan. He will also be available to give advice to the architect on any financial or contractual matters associated with the project, including the terms and conditions of the main contract and subcontracts and on the selection of tenderers, and will be preparing for the work in stage G – the compilation of the bill of quantities (Seeley, 1996).

A useful checklist of objectives, information requirements and procedures at each stage of the cost planning process was prepared by the QS Division of the RICS (1982), together with supporting documentation.

## Cost Planning Techniques

Cost planning establishes the needs, sets out the various solutions and the cost implications of these solutions, and finally produces the probable cost of the project. At the same time a sensible relationship must be maintained between cost on the one hand, and quality, function and appearance on the other. Various methods of cost planning have been developed but there is no one universal system that can be applied satisfactorily to every type of project. Cost control, operating at various stages of the design process, will require different techniques according to the amount of information available. There are, broadly, two basic methods of cost planning currently in use, although in practice variations of these methods have been introduced.

One method is often described as the elemental cost planning system, in which sketch plans are prepared and the total cost of the work is obtained by some approximate method, such as cost per place or per square metre of gross floor area. The building is then broken down into various elements of construction or constructional parts such as walls, floors and roof, and each element is allocated a cost based on cost analyses of previously erected buildings of similar type. The sum of the cost targets set against each element must not exceed the total estimated cost. Cost checks are made throughout the design stage and lastly a final cost check is made of the whole scheme. Thus the system incorporates a progressive costing technique with the establishment of cost targets and the use of constant checks to ensure that the design is kept within them (Seeley, 1996).

Another method is generally described as the comparative system. It starts from sketch plans but does not use a fixed budget like the elemental system. Instead a cost study is made showing the various ways in which the design may be performed and the cost of each alternative approach. The cost study is usually based on approximate quantities and constitutes an analysed estimate. It provides a ready guide to design decisions and it enables the architect to select a combination of alternatives that will satisfy the financial, functional and aesthetic considerations. The selection thus made becomes the working plan and operates as a basis for the specification and working drawings. The quantity surveyor will need to carry out cost checks periodically throughout the design stage as with the elemental system, to ensure that the architect's proposals are being kept within the total cost limit agreed with the employer (Seeley, 1996).

The essential difference between these two methods of cost planning is that with the elemental system the design is evolved over a period of time within the agreed cost limit, whereas in the comparative system the design is fairly clearly established at the sketch plan stage, after the choice of alternatives has been made, and it is not generally materially altered after this stage. The elemental system has been described as 'designing to a cost' and the comparative system as 'costing to a design'. Practical examples of the application of these two methods are given in *Building Economics* (1996) by the same author. An initial cost plan with cost checks relating to a social club is illustrated in appendix L, and this lists most of the normal BCIS elements.

## Cooperation between Architect and Quantity Surveyor

The quantity surveyor frequently acts as specialist adviser to the architect on all matters concerned with building and engineering costs. He offers considerable assistance to the architect in advising on the financial effect of design proposals and so helps in ensuring that the money available is put to the best possible use and that final costs are kept within the agreed budget. It involves a team approach to design often requiring changed attitudes and relationships between the architect and quantity surveyor. The cost plan prepared by the quantity surveyor is used continuously throughout the detail design stage of the project as a means of checking that the detail design is kept within the cost framework.

Where the estimated cost of an element exceeds the cost target, then either the element must be redesigned or other cost targets reduced. Where a number of design elements would have to be adjusted to keep costs within the total cost limit and this would result in a building of undesirably low quality, the design team should request additional funds from the employer. The essence of cost planning is to enable the architect to control the cost of a project while he is still designing.

The architect must benefit from the quantity surveyor's knowledge and skill in ensuring that the design is founded on a sound economic base from the outset, by the ability to make design decisions in full knowledge of their economic consequences, and by formulating the design against an agreed cost background so that a balanced and consistent design is secured. While not wishing to inhibit the enterprise and creative powers of the individual designer, it is nevertheless incumbent upon an architect to take positive steps to secure overall economy in design, and the quantity surveyor can make a valuable contribution towards this. Each profession will in the process also obtain a better understanding of the other's problems.

The quantity surveyor's cost information will vary from major design alternatives, such as the size and height of residential blocks, to cost studies into alternative forms of construction, cladding and finishes. Fundamental questions must often be asked such as 'Is this expenditure really necessary?' or 'Is there a better way of meeting the particular need?'

## COST MODELLING

Brandon and Moore (1983) described and illustrated the use of microcomputers in building appraisal and how this necessitated the use of models, which are representations of a real life situation. Modelling can be more fully defined as an act of representing a real existing system, which is observable and measurable, in a form that can be controlled and manipulated with the aim of understanding the behaviour of the system so as to be able to predict its state for some future time. The normal approach is to identify and structure the problem and devise a method of solving it, incorporating a suitable form of measurement. These processes are ideally suited to computer execution.

Smith (1980) devised an approach whereby the design target parameters and cost targets are set out and compared with those contained in the budget model, as a quick way of assessing the chosen strategy. Once the strategy is agreed the design team develops the sketch plans within the framework of the strategic design plan, which itself was developed from the framework of the budget model.

The aim of cost models is generally to represent accurately the whole range of cost variables inherent in a building design to secure improved cost forecasts and/or design optimisation. It has been argued with some justification that the traditional approach to cost modelling gives a poor representation of costs, since they are not modelled in the way in which they occur. Furthermore, quantity surveyors normally reuse data from previously completed projects as the basis for their models. Considerable adjustment to this data is needed, frequently on the basis of supposition and presumption, as no two projects are identical (Ashworth, 1986).

## Purposes of Cost Modelling

Deciding the purpose of a cost model will affect its form and the variables which will be incorporated within it. Wilson (1984) suggested the following two categories as being of significant importance:

(1) Design optimisation models which are primarily concerned with securing value for money in building and may also be used as an important component of the cost planning process and incorporate life cycle costing facets. Their main strength lies in comparing one solution with another.
(2) Tender price prediction models, where the main aim is to forecast the likely tender sum to be obtained from a contractor and also take cognisance of the contractor's estimating variability and the factors influencing market price. Because of the variable factors involved predictive models will be less reliable than design type models and some inaccuracy in forecasting is likely.

## Risk Analysis/Monte Carlo Simulation

It is certainly not difficult to see that a significant level of uncertainty is inherent in construction cost (if less so in construction price). It may follow that a deterministic cost model used as a predictor of construction cost will only be accurate if a considerable measure of good fortune occurs. The variation in the prices of different contractors when submitting priced bills of quantities indicates that exact prediction is never likely to be achieved. It is for this reason that estimates of variance will often be given by cost models in order to limit the cost modellers' exposure to risk. This limitation of risk to the modeller has been achieved by transferring the risk to the model. Risk analyses are performed on models in a number of ways, the most common being a Monte Carlo simulation technique. In such a technique the independent variables within a model are given a degree of uncertainty by applying a level of variation to their value. This uncertainty is derived from a multitude of sources such as historical records of weather conditions. A method often used to apply the uncertainty is to define a probability profile for the variable. Examples of such profiles are rectangular, triangular and trapezoidal.

In the simulation technique a specified number of occurrences of the variable and random number generators are applied to define the precise value for each of the occurrences and these precise values are used, in turn, in the model. From these multiple runs it is possible to produce a probability profile which can give confidence for statements concerning cost.

### Network Cost Modelling System

Bowen (1987) postulated that attempts at modelling the cost of buildings have largely attempted to explain costs as relatively simple functions of different measurements of the finished building, element or component, and thereby viewing the building as a single discrete step. It is argued that such an approach ignores the fact that construction is a process consisting of separate but dependent physical activities over a period of time and subject to uncertain cost and duration. It fails to explain when and how costs are incurred (inexplicability), disregards interrelatedness (unrelatedness), and ignores uncertainty (determinism).

To overcome the above weaknesses. Bowen (1987) advocated a process-based modelling approach. It was suggested that PERT like networks can be used in a structured manner to model cost by representing expenditure as it occurs during construction activities. The proposed system can use AI (artificial intelligence) to link sub-networks representing the construction of different elements to create a complete, representative network for the entire project. For each element, different sub-networks representing different designs which are different in type, but not size or similar parameters, can be used to enable the user to compare different design/construction alternatives. Being computerised, this system can use a database containing cost information, sets of sub-networks, different default values and definitive functions.

### Expert Systems

The development of expert systems began in the mid 1980s with the development, jointly by the Royal Institution of Chartered Surveyors and Salford University, of the ELSIE system. The system, as described by Brandon (1990), encompasses the accrued knowledge of a significant number of construction industry professionals and encompasses this knowledge within the framework of an expert system.

By using the expert system, it is possible for a user, who need not be a construction professional, to respond to questions asked by the computer software and to be led, as if by a constructional professional, through the maze of interrelated data, to arrive at a cost based upon both the information provided by the user and the assumptions made by the software where the user did not have sufficient information.

Subsequently a private company was established to market and sell the software. This company, Imaginor Systems, still based at Salford, has expanded its operations from the original office development scope of ELSIE to additional fields such as industrial buildings, as described in some detail in chapter 19.

Readers requiring more detailed information on building cost modelling are referred to Raftery and Newton (1995).

## COST ANALYSES, INDICES AND TRENDS

### Cost Analyses

A comprehensive system of cost analyses and cost indices has been developed in the United Kingdom by the RICS Building Cost Information Service (BCIS) and on the maintenance side by the RICS Building Maintenance Cost Information Service (BMCIS) which has been abbreviated to Building Maintenance Information (BMI).

Cost analyses seek to:

(1) enable the design team to determine how much has been spent on each element of a building;
(2) assess whether a balanced distribution of costs has been obtained;
(3) permit comparison of costs of the same element in different buildings; and
(4) obtain cost data for use in planning of other projects.

Large sums of money are sometimes spent on the elements of structure and services which are disproportionate to the quality and efficiency that they contribute to the building. On the other hand too little may be spent on finishes to provide, for example, the appropriate acoustic quality. Adjustments in the allocation of money between different elements can often result in a better building, both functionally and architecturally.

A cost analysis of a tender records the effectiveness of the cost control exercised throughout the design stage of a project. At tender stage it will be too late to adjust the elements that are out of balance and the reductions are likely to fall on finishes and fittings, often resulting in a less satisfactory building.

Comparisons are often made between analyses of similar projects when elements such as floor and ceiling finishes can usually be compared directly, while others such as external walling must be adjusted for plan shape, and others such as foundations, upper floors and roof must be considered in relation to the number of storeys.

In the BICS standard cost analyses the element list has been divided into six groups, five of which cover the building and the sixth external works. The substructure is a single element group, while the superstructure comprises frame, upper floors, roof, stairs, external walls, windows and external doors, internal walls and partitions, and internal doors; internal finishes embrace wall, floor and ceiling finishes. Fittings appear as a collective single element. The largest single group covers services, which is subdivided into fifteen elements or components, some of which have limited application.

Two types of standard cost analysis are provided – a concise cost analysis and a detailed cost analysis. The concise cost analysis occupies only half a page and provides background information about the project and costs of the six element groupings, including the cost/m² of gross floor area.

Detailed cost analyses normally occupy four to six pages and contain considerable information about the form of the project, site and market conditions, contract particulars, contract breakdown and competitive tender list, accommodation and design features, floor areas suitably categorised into usable, circulation and ancillary uses, internal divisions, wall to floor ratio, internal cube and storey heights. There is a summary of element costs giving the costs of all elements and element groupings, expressed as both total element and cost/m² of gross floor area (gfa), with preliminaries apportioned among the elements. In addition the element unit quantity and element unit rate are inserted against each element, where appropriate, and this is a useful feature when comparing cost analyses for similar projects. Preliminaries are shown both as a separate item and spread over the element costs. Specification notes are included for each element to help with the adjustment of element prices when preparing cost plans for new buildings. Photostated reduced copies of drawings illustrate the general form of the building. The detailed analyses are a valuable aid in the preparation of initial cost plans for new projects, provided that they are used with care and skill. Unfortunately many submissions to BCIS do not include detailed analyses.

These cost analyses are available to subscribers on-line from a central computer via a telephone line. Appendix M contains a detailed elemental cost analysis of a three storey high specification office building on the 1995 BCIS format, which has been substantially revised and improved. The cost per m² of gross floor area exceeded £1000 in 1994 with 50 per cent of the cost in the superstructure and 20 per cent in services. The two columns containing total cost and cost/m² with preliminaries spread over the elemental costs have been omitted because of lack of space. Also omitted are the specification notes and drawings which were incorporated on the BCIS analysis.

The *Architects' Journal* and *Building* also publish useful cost analyses periodically. The elements differ from those contained in the BCIS analyses and the prices are rather dated as they refer to completed buildings, but they do incorporate excellent drawings, including working drawings, photographs, specification notes, unit quantities and rates, and comments on significant cost aspects.

### Cost Indices

Cost indices are of two main types:

(1) tender price indices based on a representative sample of tenders, of which there are several sources in the United Kingdom (BCIS; DoE (PUBSEC); and Davis, Belfield and Everest); and
(2) building cost indices, usually confined to certain types of buildings such as general, steel framed, concrete framed and brick construction (BCIS), and also housing (*Building*). These are related to fluctuations in the cost of labour and materials and did, until the nineteen seventies, provide a reasonably good basis for forecasting future costs.

However, in the United Kingdom, the two sets of indices diverged widely in the wake of a deep recession in 1990–95. Contractors were forced to reduce tenders far below increased costs in order to obtain contracts, mainly by reducing oncosts and profit margins and increasing productivity. Obviously this cannot continue indefinitely otherwise the already large number of insolvencies will reach unacceptable proportions and the depleted construction industry will be unable to meet the rise in demand for its services when it materialises. For example, tender price indices changed from a base of 100 in 1985 to 115 in mid 1995 (after rising, falling and rising again), whereas general building cost indices rose from 100 to 163 over the same period (continually rising throughout the period).

There are also very wide ranges in the costs of buildings of similar types in the United Kingdom. For example in late 1993 average building price ranges ($£/m^2$) published by BCIS were factories (purpose built): 123–1155; offices (concrete framed): 326–1464; shops: 127–893; primary schools: 226–958; and public houses, licensed premises: 332–1101. The respective mode figures ($£/m^2$) provide a more realistic base, being factories: 214; offices: 537; shops: 260; primary schools: 433; and public houses: 445 (BCIS, 1993).

### Cost Limits

Local authorities and other public bodies often have to operate within cost limits or guidelines issued by government departments. For instance, with public housing the *DoE* operates cost criteria which are intended to act as indicators. For educational buildings, the *Department of Education and Science* (DES) determines cash allocations nationally and controls the amount of work commenced in schools each year depending on government policy. While the *Department of Social Security* (DSS) provides design guides for hospitals and cost allowances are prescribed for different forms of accommodation and oncosts are included for differing site conditions and external works. *Housing associations* obtain significant grants from the Housing Corporation for which they have to make bids and satisfy the Corporation that they are giving value for money and meeting prescribed performance standards. The housing associations have to make up the shortfall from other sources.

## CONSIDERATION OF FUTURE COSTS

### Total Cost Concept

With many projects cost planning cannot really be effective unless the total costs are considered, encompassing both initial and future costs. The term 'costs in use' has in the past often been used to describe this technique but a more apt term is 'total cost.' More recently the term 'life cycle costing' has gained popularity to embrace the total cost of an asset over its operating life (Flanagan and Norman, 1983). Using these techniques, which both have the same objective in mind, all the costs in erecting, maintaining and using the building are converted to a single sum, which is the annual cost or present value of all costs over the life of the building. They are employed as a design tool for the comparison of different designs, materials, components and constructional techniques. They thus provide a valuable guide to the designer in securing value for money for the employer, and can also be used by property managers or developers to compare costs against the net value accruing from future rents less outgoings.

Most design decisions affect running costs as well as first costs, and what appears to be a cheaper building may in the long term be far more expensive than one with much higher initial costs. Running costs often amount to about two-thirds of the annual equivalent of first costs, and for industrial buildings can be as much as one-and-a-half times the equivalent. The proportions vary widely from one class of building to another.

In this form of calculation, future costs are discounted to convert them to present day values using valuation tables. For example, the replacement cost of a component, such as a water storage tank, in 20 years time, is found by multiplying the cost by a certain factor – the present value of one pound in 20 years time at an appropriate rate of interest, say 6 per cent. A range of relevant worked examples are given in *Building Economics* (1996) by the same author.

There are problems in assessing the lives of components, materials and buildings, discounting rates, inflationary rate, effect of taxation, changing statutory and occupancy requirements and other factors. Nevertheless, it is better for the quantity surveyor to make these comparisons rather than to ignore future costs. If small cost differences result then he will not need to pay too much regard to them, but where significant differences arise, then they warrant careful consideration at the design stage. A greater feedback of maintenance cost information to the design team is needed to improve the accuracy of the techniques.

In the United Kingdom one valuable initiative has been the establishment of Building Maintenance Information (BMI), originally BMCIS. The service distributes analyses of the annual maintenance costs of a variety of buildings converted to the cost per 100 m² of floor area as a unit of comparison, spread over selected elements and subelements. BMI property occupancy cost analyses provide information on the type, location and erection date of the building(s), management criteria, budget procedure, maintenance management and operation, building function and parameters, and form of construction. The cost is allocated to improvements and adaptations; decoration (internal and external); fabric (external walls, roofs and other structural items, fittings and fixtures, and internal finishes); services (plumbing and internal drainage, heating and ventilation, electric power and lighting, and other M&E services); cleaning (windows, external surfaces and internally); utilities (gas, electricity, fuel oil, solid fuel, water rates, and effluents and drainage charges); administrative costs (services attendants, laundry, porterage, security, rubbish disposal, and property management); overheads (property insurance and rates); external works (repairs and decoration, external services, cleaning, and gardening). Even within the same class of building there are wide variations in unit costs, but there are also wide variations in the form, construction, age and location of buildings, and the type of maintenance management, and these all have a considerable influence on the elemental maintenance and running costs.

Table 8.2 shows 1992/93 occupancy costs for seven properties with wide variations in the costs of the various elements. It will be apparent that one year's costs cannot be taken as typical, as work such as external decoration will probably only take place at four to five year intervals.

### Energy Conservation

Since the early nineteen eighties a more positive approach to energy conservation was engendered with fast rising cost of energy and a growing realisation that it made good economic sense to reduce heat loss from buildings.

For example, BRE (1993) has shown that besides short payback periods being achievable, savings from energy efficiency projects in *refurbished*

Table 8.2  Property occupancy costs

| Building type | Location | Date of erection | Floor area (m²) | Improvements and adaptations | Decoration | Fabric | Services | Cleaning | Utilities | Administrative costs | Overheads | Total |
|---|---|---|---|---|---|---|---|---|---|---|---|---|
| | | | | | | *Cost in £ per 100 m² of floor area (1992/93)* | | | | | | |
| Offices (2 storey) | Coventry | 1990 | 3563 | — | 140.34 | — | 651.14 | 628.68 | 1126.75 | 314.34 | 3266.91 | 6128.16 |
| Administrative Headquarters | N.E. England | 18th and 19th centuries | 3905 | 40.10 | 69.71 | 518.49 | 202.71 | 637.48 | 812.08 | 431.16 | 70.94 | 2742.57 |
| Multi-purpose Assembly Hall | N.E. England | 1967 | 2230 | 316.14 | 3.45 | 203.99 | 436.86 | 637.48 | 862.52 | 2144.43 | 70.94 | 4359.6 |
| University College | N.E. England | 1974 | 8874 | 182.69 | 3.43 | 307.57 | 403.99 | 902.67 | 926.54 | 2057.10 | 70.94 | 4672.24 |
| Student Residential Accommodation | N.E. England | 1971 | 1477 | — | — | 287.54 | 302.91 | 1456.29 | 933.64 | 393.41 | 66.05 | 3439.84 |
| Office buildings (3 & 5 storey) (air conditioned) | Solihull | early 1960s | 10690 | 758.78 | 140.32 | 467.73 | 701.59 | 1152.48 | 2871.84 | 2827.50 | 2797.01 | 10958.47 |
| Office building (6 storey) | London SW1 | 1899 | 6040 | — | 2649.01 | — | 741.72 | 658.94 | 1171.52 | 1648.84 | 3855.07 | 10725.10 |

Source: *BMI Property Occupancy Cost Analyses* (1995)

Table 8.3   Savings and payback periods for energy efficiency measures in refurbished industrial buildings (1993 prices)

| Activities and installations | Energy savings | Annual payback (years) |
|---|---|---|
| **Typical savings** | | |
| Draught proofing | 15–20% | 1–5 |
| Building insulation | 10–15% | 2–6 |
| Boiler replacement | 10–20% | 1–4 |
| Time and temperature controls | 5–15% | 1–5 |
| Destratification fans | 5–20% | 1–3 |
| Lighting | | |
| Replace tungsten by fluorescent lights | 40–70% | 1–3 |
| High-frequency electronic ballasts | 15–20% | |
| Use efficient luminaire reflectors | 20–50% | 2–6 |
| Install automatic lighting controls | 20–50% | 2–5 |
| Localised instead of general lighting | 60–80% | 4–8 |
| **Specific examples** | | |
| High-speed roller shutter doors | £2 700 (14%) | 2.6 |
| External roof insulation | £100 000 (66%) | 2.4 |
| Decentralising heating system | 40–60% | <2 |
| Condensing boiler installation | 10–40% | 1–4 |
| Occupancy lighting detectors | £36 000 (70%) | 1.8 |
| Time switching of lighting | £8 000 (32%) | 2.1 |
| Heat recovery for space heating | £21 000 | 1.3 |
| Convert standby generators to CHP | £22 000 | 3.2 |

Source: *BRE. IP 2/93.*

*industrial buildings,* as illustrated in table 8.3, continue for the life of the refurbishment and are largely protected against inflation through energy price increases, including VAT since 1994. The internal rate of return or net present value can as a result be favourable. Thus there is a good economic case for improving energy efficiency when refurbishing factory and warehouse buildings.

Meyer (1983) has described ways of reducing energy costs at the design stage and comparing the savings with related construction costs. He also outlines both simple and more complex methods of calculating payback periods, return on investment and net present value. These methods allow the designer to take account of a number of economic considerations, such as depreciation, taxes, financing and inflation. More detailed information on the implementation of energy conservation methods is given in chapter 10.

## POSTCONTRACT COST CONTROL

### General Procedures

The quantity surveyor's cost control function does not finish at the tender stage but continues throughout the execution of the contract. However, the quantity surveyor, under the standard forms of construction contract, has no authority to issue instructions that would affect the cost of the project. What he does, in practice, is to operate a budgetary control system which can provide the information that will enable others to control the cost, and this system should go much further than providing an assessment of final cost. The importance of budgetary control systems has increased considerably in recent years owing partly to the funding arrangements for public sector projects, partly to the reputation construction contracts have gained almost worldwide for overspending, and partly to employers' demands to start projects quickly and consequently without adequate preparation (Nisbet, 1989).

At the time the contractor starts work on the site, the quantity surveyor should have scrutinised the priced contract bills, schedules of basic rates, where applicable, contract drawings, contract conditions, contractor's master programme, insurances and any other relevant documents. He should at an early stage agree ground levels with the contractor and suitable arrangements for dealing with daywork vouchers and claims for increased costs. An accurate record of drawings should be maintained with revisions to drawings noted and costed and variation orders costed and filed, and at the same time the architect should be supplied with all relevant cost information. The opportunity should be taken on the occasion of site visits for measurements and interim valuations to note matters such as labour strength, plant in use, weather conditions and causes of delay, which subsequently have a bearing on the subject matter of claims. Throughout the contract period the quantity surveyor should maintain effective cost control arrangements by keeping a constant check on costs and by supplying cost advice to the architect in ample time for any necessary corrective action to be taken without adverse effects on the project (Seeley, 1996).

The cost plan also has its uses during the postcontract period. When a tender is accepted the priced bill can be analysed in a similar manner to that of the cost plan. A comparison of the priced bill and final cost plan is most valuable in that it shows up the differences between the cost plan and the tender, and so assists in preparing future cost plans. When work

on site is commenced the cost analysis can be used for controlling variations. The analysed tender provides a framework of costs which can help
to provide a running forecast of total costs as the project proceeds.

The main function of the quantity surveyor, once work has started on
site is one of project financial control. It is important to ensure that any
variations, claims or extras do not raise the likely final account figure
above the cost limit. It is particularly necessary to monitor the financial
effect of variations and these should, ideally, be costed by the quantity
surveyor before they are issued, as described in chapter 6. The quantity
surveyor liaises with the contractor at regular intervals for the valuation of
variations, to agree remeasured work and to discuss any claims submitted
by the contractor and/or subcontractors. It is, however, the quantity surveyor's responsibility to prepare independent valuations, although he should
consider any representations made by the contractor.

Hence cost control measures operate throughout the construction period to ensure that the authorised cost of the project is not exceeded.
Normally the contract sum will represent the authorised cost, but on occasions the sum can be varied during the contract period by the employer. Employers' requirements as to cost control can vary and hence
this aspect needs clarifying (Aqua Group, 1990b). Generally the contingency sum should be used only to cover the cost of extra work that could
not reasonably have been foreseen at the design stage as, for example,
extra work below ground level. It should not be used for design alterations, except with the prior approval of the employer. The quantity surveyor should additionally attend all site meetings and advise on contractual
matters.

It is very important that the probable cost of all variations should be
computed before the architect issues the variation orders, so that their
financial effect can be taken into account. Early consideration should be
given to expenditure against provisional and prime cost sums and the
contingency fund, and the examination of subcontractors' and suppliers'
quotations.

It is advisable to carry out continuing cost studies in constructional areas where detailed design is incomplete, and this applies particularly to
mechanical and electrical services. Additionally, the quantity surveyor will
normally produce monthly forecasts of final expenditure, and predict and
monitor cashflow. The progress is best shown in the form of a graph with
possibly a solid line depicting the anticipated rate of expenditure calculated from the contractor's master programme, usually of the characteristic S curve shape, with perhaps a broken line showing the actual rate of
progress drawn by connecting the gross values of work done at each valuation.
In this way the quantity surveyor can check on the contractor's progress
on site and the probable consequences of any delays to completion. These
form an important aspect of the financial management of capital schemes.
The receipt of periodic financial statements or reports by the employer, of

the type illustrated in table 6.8, enables him to anticipate his future financial commitments and to revise his capital budget where appropriate. They normally list authorised expenditure, details of savings and extras, actual expenditure, progress and other related matters.

Should any of the cost information obtained by the quantity surveyor prove unsatisfactory, such as the possibility of final expenditure exceeding the contract sum, urgent action must be taken to rectify the situation working in close liaison with other members of the design team.

Summing up, the employer should be informed of his financial commitment and when he will be required to make payments. The design team and, in particular, the quantity surveyor must effectively control expenditure on variations, contingency expenditure, provisional and prime cost items, set against quotations, quality control, completion to time and claims. The problem of major additional requirements not provided for in the initial contract and needing extra funding is one of paramount importance and affects all parties to the contract.

The employer can reasonably expect to be supplied with the following information:

(1) An estimate of the final account at regular intervals during the contract period, preferably monthly.
(2) A comparison of the estimate with the total allocated financial resources.
(3) If the employer's total allocated financial resources comprise components from different resource bases, an allocation of the estimate between these bases.
(4) If the comparison between the final account estimate and the resource allocation is unfavourable, he will require an explanation and an indication of remedial action.
(5) An indication of when he will be expected to pay money and approximately in what amounts.

**Review of Valuations**

The following detailed checks should be carried out:

(1) The computation has been prepared by the quantity surveyor where he is responsible for reporting valuations.
(2) The gross valuation is within the approved funds and/or financial statement figures and advise the architect/employer if there is a probable need for extra funds.
(3) The preliminaries and insurances have been calculated in accordance with established policy, such as *pro rata* to the value of work done.
(4) The work done has not been under- or over-valued by reference to the architect's progress report.

(5) The value of materials on site does not exceed the value of materials to be incorporated into the works for the forthcoming two to four months, depending on the size of contract and stage of completion.
(6) The materials valued off site have been inspected and properly identified as the employer's property.
(7) Defective work, where applicable, has been notified by the architect and the value deducted from the valuation.
(8) The value of mechanical and electrical services have been confirmed in writing by the consulting engineers, where applicable.
(9) The value of the work of nominated subcontractors and that of materials of nominated suppliers have been checked by the quantity surveyor on a random sample basis.
(10) The value of fluctuations represents a reasonable proportion of the work done in relation to the rate of inflation.
(11) The contract completion date or extended date has not been reached or notify the architect of the contractor's potential liability for liquidated and ascertained damages.
(12) The release of retention is authorised by the issue of the appropriate certificate (practical completion and making good defects).
(13) The value of previous payments has been calculated from previous certificates as issued by the architect.
(14) The arithmetic has been checked.
(15) The time spent is in accordance with manpower budget (Nisbet, 1981).

**Review of Bill of Variations**

The following detailed checks should be undertaken:

(1) Contract sum from articles of agreement (the priced bill of quantities may have a different total).
(2) Total amount certified by inspecting all certificates (not valuations) and whether the employer has made any payments direct to a nominated subcontractor and amount of payments adjusted if necessary.
(3) All prime cost sums and rates have been adjusted.
(4) All provisional sums have been adjusted.
(5) All approximate quantities have been adjusted.
(6) All architect's instructions have been dealt with.
(7) All nominated subcontractors' and suppliers' documents are complete.
(8) Dayworks are properly referenced to the final account.
(9) Fluctuations are properly referenced to the final account.
(10) All figures have been arithmetically checked.
(11) Compare the final account total with the contract sum and ensure that differences can be explained.
(12) Items without architect's instructions do not require an architect's authorisation and/or prepare a list and send to the architect requesting authorisation.

(13) Where the total of the contract bill is not the same as the contract sum check that the final account has been adjusted by the correct percentage.
(14) Additions and omissions have been correctly priced in accordance with the prices in the contract bills and contract provisions (Nisbet, 1981).

*Note*: where the local authorities edition of the Standard Form is used, supervising officer is substituted for architect.


## COST CONTROL BY THE CONTRACTOR

### Cashflow

The assessment of the profitability of a particular contract consists basically of knowing precisely the value of work executed at a specific date, compared with the actual costs incurred in achieving that value of work. The difference between the two figures will be the amount available to allocate to the off-site overheads of the company, to fund its working capital and make a profit. In an adverse situation the difference may show that off-site overheads are not being covered and that no profit is being made. In the worst situation, the actual costs of construction on site may exceed the value of the work that those costs have generated.

The usual reason given for a company's difficulties is cashflow, whereas this is more often a symptom of the problem and not the cause. Many construction companies become insolvent through bad estimating and planning, ineffective contract control or inadequate site cost control. Cashflow may be defined as the actual movement of money in and out of a business. Within a construction organisation positive cashflow is derived mainly from monies received through monthly payment certificates. Negative cashflow is related to monies expended on a contract to pay wages, purchase materials and plant, meet subcontractors' accounts and overheads expended during the progress of construction. On a construction project, the net cashflow will require funding by the contractor when there is a cash deficit; where cash is in surplus the contract is self-financing. With contracts operating under United Kingdom standard conditions with retention funds and low percentage profits on turnovers, construction firms are frequently in financial deficit for much of the contract period.

Cashflow problems can be reduced if effective procedures can be operated by the contractor in respect of the matters listed:

(1) realistic monthly assessment of preliminaries from fully documented and priced preliminaries schedules;
(2) increased costs under contracts with fluctuations kept up to date in monthly valuations;

(3)  variations to the contract accurately assessed and included in valuations;
(4)  daywork sheets completed and cleared for monthly payment;
(5)  discounts and retention monies properly claimed against the contractor's own nominated subcontractors and suppliers;
(6)  collection of all monies properly due to the contractor; and
(7)  ensuring that all claims for loss and expense are fully documented, properly presented and submitted as quickly as possible.

**Site Cost Control**

It is cost control in the context of profit or loss that is the primary concern of the contractor's quantity surveyor. In this capacity he works closely with the site manager who is basically the controller – monitoring performance, comparing it against predetermined targets and taking remedial action where necessary. The factors to be controlled include the tangible physical resources of operatives, materials, machines and subcontractors. Equally important are the non-tangible items such as progress and productivity (time) cost (money), quality, safety, information, methods and the performance of subordinate management staff (Gunning, 1983).

The main sources of data available to the contractor's quantity surveyor are now listed and described.

(1) *Contract bills*  Genuinely firm bills of quantities ease the task of the contractor. However, this poses a problem on civil engineering contracts where traditionally the bills are approximate and the whole of the work is subject to remeasurement (measure and value contracts).

(2) *Estimates of cost*  The employer's quantity surveyor will normally regard the priced bill in terms of end products of work. This is a very different concept from the cost constituents in terms of labour, materials, plant, sublet work, site establishment costs, overheads contribution and profit that may be components of the individual rates. Examination of the pricing notes that support the tender make-up should enable the cost structure of each individual billed item to be established and the determination with certainty of where monies have been distributed within the tender. The estimating papers are, therefore, not only the back-up documents to the tender but, where the tender produces a successful contract, one of the fundamental management documents.

(3) *Method statement*  The method statement prepared by construction management will form the basis of working against which the tender will have been computed. This statement will detail the assumptions made as to how the project is to be resourced with plant and labour and what is to be carried out by subcontractors and what by direct labour.

(4) *Master programme*    The master programme will set out in as much detail as possible the sequence and timing of the intended works.

During construction these four data sources will be supplemented by interim valuations; up to date accounts of labour, plant, materials and subcontracted work; salaries and all other site costs; and finally the programme of actual work executed compared with the assumptions upon which the tender was based. Supported by this back-up data, achievement of effective cost value comparison will involve the following activities.

(1) The calculation of the true value of work carried out on a cumulative basis to a prescribed cut-off date.
(2) The restatement of the true value of the work in terms that can be directly compared with costs. Some contractors refer to this restatement as the preparation of the earned allowances, whereby in relation to the work carried out, not more than a certain sum should have been spent on labour, plant and other components.
(3) Costs are collated to the same cut-off date as that adopted for the statement of true value, sometimes termed 'true selling value'. Costs will be adjusted to take account of liabilities for costs that have not yet been recorded but against which value of work has been taken, and for costs generated against which no value of work has yet been created.

Supplied with these three sets of data, it is then possible to ascertain whether profit or loss is being made on expenditure against labour, materials, plant, sublet work and site overheads, and whether the level of contribution is better or worse than that upon which the tender was based. This comparison may be made in terms of the whole project or, if data in terms of both value and cost can be accurately subdivided, into the construction elements making up the total project.

Meopham (1983) emphasised that it is equally important for the contractor to monitor the liquidity position on a contract as its profitability. If, for instance, a contractor were to take on a project of one year's duration at 3 per cent profitability, this profit would be cancelled out by retention until completion of the work, and he would have to fund the project throughout the whole of the contract period. To secure a reasonable return on capital, the contractor will ideally be seeking projects with interim payments, minimum retention and prompt payments after certification, otherwise further allowances need to be included in the tender. When, therefore, contractors are pressing for payment, it is not necessarily motivated by mercenariness or impending insolvency but could be part of sound financial management.

The key document for monitoring the value of work in progress is the interim valuation. The contractor will not, however, normally accept without question the values certified by the employer's quantity surveyor. Most contractors will themselves prepare payment applications in considerable detail. For internal accounting purposes, the contractor's application will subsequently be adjusted by any over- or under-measurement.

For example, a contractor may have been paid the full value of formwork after its has been struck, but may have priced the formwork in the bill to include final rubbing down and finishing of the concrete surface. The prudent contractor will reduce the value of the formwork to reflect the further work that is still required. There may in fact be many items where some adjustment will need to be made.

The computerised breakdown of interim payment applications into earned allowances is probably the most common application of computers to the financial management of contracts by contractors. A number of standard packages are available and, should the employer's quantity surveyor receive a payment application in the form of a computer printout, he can be reasonably certain that earned allowances have been used in preparing the application.

Another technique sometimes used to compute the earned allowances is to make the calculations in conjunction with a computerised network analysis of the project. The operations on the network are resourced in terms of money to achieve a value printout for payment purposes. The contractor can at the same time subdivide these monies against each operation into earned allowances. Provided that variations are minimal the system can work well, but if variations are extensive it quickly breaks down.

As a management tool, the contractor's quantity surveyor is looking for consistent and inconsistent trends as between the value and cost of each of the earned allowances month by month. If a consistent trend is observed, for instance by regular overspending on labour at a consistent level, then management will endeavour to identify some underlying reason. Examples include the all-in labour rate being higher than anticipated at the time of tender, or output being consistently lower than the constants upon which the tender was based. The inconsistent relationship between earned allowance and actual cost is more likely to have been caused by an isolated event, and again management will want to identify the reason and obtain a solution in each case. The key factor is to recognise that there is a problem at a time sufficiently contemporary with the event to be able to take effective action. On the cost side of the cost value comparison, it is essential that there is close liaison between the contractor's quantity surveyor and the site accountant on a very large contract (Meopham, 1983).

## Cost Value Reconciliation

The cost and value of variations must be continually monitored and assessed. A contractor must ensure that he has clearly defined procedures for identifying variations and that he conforms fully with the requirements of the contract regarding notices and the supply of supporting information and particulars. After interim valuations have been made, the contractor should list all unagreed claims, variations, daywork, remeasurement, interest on overdue sums and any other disputed items. The contractor should also assess the progress achieved by monitoring the value of work carried out against that programmed. This can best be done by constructing an S curve of the forecast valuations against time and then plotting actual value against actual time.

Accurate recording of the cost of materials, plant, labour, site staff and overheads, and subcontractors' work and claims is essential in cost value reconciliation. For instance, there is frequently a delivery charge for plant and this may not have been indicated on the initial order. This can include time of the plant travelling from its depot as well as the delivery charge itself. When operators are provided the question of overtime arises and also greasing or servicing time.

A monthly reconciliation of materials delivered to the site should be made against the quantities certified in the measurement. Allowance must be made for materials rejected, used on site but not measured as in strengthening temporary roads, used off site as in minor work for adjoining landowners, and materials that are stockpiled. If the quantity unaccounted for exceeds the estimated wastage allowance, further investigation is needed. Possible causes are errors in measurement, additional work being performed without supporting variation orders, or loss through unforeseen circumstances, such as excessive penetration of granular material into a very soft subbase.

With regard to labour costs, site staff usually have the responsibility to complete timesheets indicating the number of hours worked by each operative and the amount of bonus earned in that particular period. Management will monitor the allocation of staff on the various contracts, and will record holidays, sickness or other reasons for absence and overtime payments, and check that staff and overheads charges are kept within the intended budget.

Residual credits may occur when materials or items of plant have been purchased for a certain contract but on completion still retain some foreseeable value which can be used on other future contracts. They do, however, require careful examination and evaluation (Seeley, 1993b).

### Overheads

On a construction project a distinction is usually drawn between site overheads and company overheads. The contractor is generally reimbursed for his site overheads by means of the contract preliminaries or sometimes by sums added to the rates contained in the bill of quantities. The rate of recovery of revenue consequently may not match the rate at which expenditure is incurred on site overheads.

**Financial Reporting**

Good management practice dictates that reliable and regular financial reporting is necessary to control a project effectively and reports should be produced ideally on a monthly basis. A basic financial report of a contract should contain:

(1)  initial tender figures and expected profit;
(2)  forecast figures at completion for value and profit;
(3)  current payment application by the contractor;
(4)  current certified value;
(5)  adjustments to the certified valuation;
(6)  costs to date and the accounting period in question; and
(7)  cash received to date, retention deducted and certified sums unpaid (Barrett, 1992).

# 9 Value Management

## GENERAL PRINCIPLES

This chapter aims to show the various approaches to value management, the advantages to be obtained, a comparison with cost planning/cost management and a comprehensive case study. Readers requiring a more comprehensive study of this subject are referred to Norton and McElligott (1995).

Value management is attracting considerable attention within the UK construction industry as major clients become increasingly concerned with the achievement of value for money in their construction projects. However, as stated by Green and Moss (1993), value management often means different things to different people, and there is considerable confusion between value management and value engineering. In practice the former term is favoured in the UK and the latter term is used extensively in the United States, where it is often performed by engineers with applications to manufacturing industry. There are also a variety of different approaches, occurring at different stages in the design process, and these will be examined in some detail later in the chapter.

Value management operates within an organised schedule of procedures. This enables the functional requirements and alternative solutions, with their associated costs, to be identified and developed to a strict timetable. Value management is often undertaken in the form of an intensive workshop conducted by an independent team of experienced design team professionals acting in a consultative capacity to the client. On the completion of the workshop, they produce a comprehensive report, with recommendations, for review and assessment by the client and his project design team. This procedure does not adversely affect the project design team's responsibilities to the client. The value management team, operating in a complementary role, acts as a positive catalyst for savings and improved efficiency (Beard Dove, 1990).

In theory, value management can be undertaken at any stage of the design process. However, as a general rule, the earlier a study is undertaken the more effective it will be in providing the opportunity to rationalise design before it is so firmly established that any change will significantly increase design/planning costs. In practice the timing is often critical, although it should be stated that there can be no guarantee that overall initial capital costs will be reduced although a more efficiently designed project will almost certainly emerge.

Norton and McElligott (1995) defined 'value management' as the full range of value techniques available and 'value engineering' as value techniques applied during the design phase of a project.

A number of definitions of value management have been formulated; one meaningful one being: 'a service which maximises the functional value of a project by managing its evolution and development from concept to completion, through the comparison and audit of all decisions against a value system determined by the client or customer' (Kelly and Male, 1993).

Another useful definition is 'a systematic multi-disciplinary effort directed towards analysing the functions of projects for the purpose of achieving the best value at the lowest overall life cycle project cost' (Norton and McElligott, 1995).

### Reasons for the Client Commissioning Value Management Studies

Carter (1991/92) identified the following reasons why a client might wish to commission a value management study:

(1)  client's concern at the escalation of estimated costs;
(2)  client's concern at tenders received in excess of budget;
(3)  client losing confidence in the design team and/or project, arising from such factors as planning delays, external factors or lack of competence;
(4)  client requires an independent audit or appraisal of the project before it is submitted for sanction;
(5)  client seeks to minimise capital and/or operational costs and maximise profit;
(6)  client must achieve capital and/or operational savings to make a profit;
(7)  client wishes genuinely to seek an innovative/better solution to his project;
(8)  client wishes to experiment with a new technique that he has discovered;
(9)  a consultant recommends a new technique to the client.

Both Carter and Norton have shown that quantity surveyors are among the professionals fitted to undertake the value management role. It is evident that this technique can be of considerable value when used for large, complex projects.

### ALTERNATIVE APPROACHES TO VALUE MANAGEMENT

There are a number of different approaches that can be adopted when carrying out value management with the choice often being decided by the type and nature of the project, the timing of the operation and the make up of the design team. It is customary to prepare a job plan incorporating a recognisable strategy, which normally comprises the six phases of information: creativity; evaluation; development; presentation/recommendations; and action and feedback. The various procedures are now described.

**The Charette**

This is undertaken after the project brief has been formulated and the design team appointed but before the actual design is commenced. The client's representatives and the design team meet under the chairmanship of a value manager or facilitator for one or two days in order that the brief can be examined in detail and questions raised. The next stage is to generate ideas for rationalising the brief, when functional analysis of the space requirements can form a major component, and improving the project's cost effectiveness. These ideas are then evaluated and, if accepted, are incorporated in a revised brief.

**The 40 Hour Value Management Workshop/Study**

This is probably the most widely accepted formal approach to value management, and is used as the basis for training of value engineers as prescribed by the Society of American Value Engineers (SAVE). It is normally undertaken at about 35 per cent of the way through the design stage which is about as late a stage as is reasonably practicable. The sketch design of the project is reviewed by an independently appointed second design team, under the chairmanship of a value management or value engineering team coordinator (VMTC or VETC), the composition of this team of possibly six to eight professionals reflecting the characteristics of the project under review. For example, a project involving a substantial proportion of mechanical and electrical work could create the need for four persons with these professional backgrounds to form part of the team. The workshop normally takes place near the project site, probably in a hotel or a room in the client's office. The complete drawings are sent to the VMTC/VETC for distribution to the team during the week preceding the workshop/study. During the week of the workshop/study, the team will follow strictly the stages of the job plan (Kelly and Male, 1991 and 1993).

The 40 hour study spread over five days concludes with a number of design/construction modifications which are referred to the client for endorsement and implementation. It is claimed that savings of up to 30 per cent may be achieved in the United States, but savings of this magnitude are unlikely to be obtained in the UK with the tighter cost control procedures. However, as highlighted by Carter (1992a), it can have its drawbacks as the potential exists for confrontation and the external team's proposals can be seen to be critical of the project design team and may be resisted. The short timescale may make it difficult for the external team to fully understand all aspects of the project proposals and it leaves only a restricted period of time to prepare revised designs and for them to be fully and accurately costed.

Norton (1992b) has described a value engineering study undertaken in 1992 for a bus maintenance and storage project estimated to cost $24m in New York for the City Office of Management and Budget. It is interesting to note that a specialist value engineering consultancy firm was engaged to conduct an independent 40 hour workshop following the formal job plan guidelines, and was carried out at 10 per cent design stage to give the greatest potential for savings at the earliest stages of design. As a result of the study 32 recommendations were accepted, 23 rejected and a further five underwent further study. The implemented recommendations produced savings of about 15 per cent.

### One-Two Day Workshop/Study

Carter (1992a) has strongly advocated this approach as being more appropriate for use in the UK. He recommends that a two day study be held on a Friday and Monday, while a one day study can be held on any weekday. All members of the project design team should be represented including the client, facilities manager, letting agent and other relevant parties. At the beginning, each team member usually makes a brief verbal presentation using drawings or other suitable material, with a maximum duration of 10 to 15 minutes.

The value manager frequently records the relevant data on flip charts, and seeks to identify major constraints, which can be physical (site, ground conditions, height, light or access), operational, statutory (company or legislative), time or cost, each having an impact on the project.

The next stage involves the preparation of a FAST diagram (Functional Analysis System Technique), which will be described and illustrated later in the chapter. The quantity surveyor/cost engineer then breaks down the cost plan (where available) over the weekend, hence the choice of Friday and Monday for the study.

The FAST diagram is then examined to identify any functions which appear to have an abnormally high cost or to identify functions which can be omitted or modified. The next step is an intensive session (brainstorming) which could reasonably be expected to generate 50 or more suggestions to modify the brief, relax the constraints or modify the design/construction proposals in order to achieve a more efficient design or technical solution to eliminate unnecessary costs.

These suggestions are reviewed as being either: (a) rejected (with reasons recorded) or (b) to be developed by the project team. The latter items are then prioritised. The value manager/engineer then compiles a comprehensive report (probably of some 40 to 50 pages), encompassing all the elements of the study and concluding with recommendations as to which items are to be developed by the project team. This report is normally issued within five to seven days or at the end of the study to the client/project sponsor for implementation.

This shortened form of study is much cheaper and quicker than the 40 hour workshop, probably costing less than £10 000 (1992 prices), and is considered to be more appropriate to the UK. Carter (1992a) has undertaken studies using this approach achieving benefit ratios between 1:30 and 1:300.

## Two or Three Day Workshops

Doyle (1993) has outlined another approach to value management adopted by a joint venture of E. C. Harris and Australian Value Management and involves a planned series of highly structured think tank sessions chaired by an outside professional facilitator. The two successive workshops explore the objectives, perceptions and interpretations of the brief and address issues in a pre-emptive way.

On day one of the first workshop, arranged at the earliest possible stage, ideas which may amount to hundreds are reduced to a workable shortlist by rating their cost and functional values. On the second day, approximate cost implications are identified in groups working with the quantity surveyor and project manager. They are finally rated and prioritised for possible incorporation on the third day. After design development, a further three day workshop ensures that the project is reflecting its original aims and that cost effective solutions are being identified. It is claimed that the potential benefits using this approach are substantial and give as an example the £35m savings made on the £100m Brisbane International Airport.

## The Concurrent Study

This approach uses the existing project team under the chairmanship of a value manager or facilitator. The group meets on a regular basis during the project design phases, offering maximum continuity. However, it has the disadvantage that creativity is not so evident and it may be more expensive than the 40 hour workshop (Smith, 1993).

## The Package Review

This is often used in management forms of contract, wherein package reviews, consisting of a detailed appraisal of each package (or element or trade), are undertaken by the project team as an ongoing process, continuing throughout the design, procurement and construction phases. Discussions with specialist contractors and manufacturers form an important part of this process (Smith, 1993).

## The Contractor's Change Proposal

This is a value management change proposal initiated by the contractor after the contract is let. Under US government contracts, the contractor is encouraged to develop value engineering (VE) proposals on a voluntary basis. The contractor then shares in any resultant savings if the VE plan is implemented (Smith, 1993). The major benefit is that it permits the contractor to be proactive and to use his construction/engineering knowledge and expertise to improve a facility at the on-site stage. Whilst the disadvantage is that the contract may be delayed while the design team investigate the merits and viability of the proposed change. For this reason any changes tend to be relatively superficial (Kelly and Male, 1993).

### Design and/or Construction Audit

This process aims to define a project's objectives, by formulating a list of the client's needs and wants, and provides a clear indication of both the cost and the worth of a project. The procedure adopted often follows that of the charette or a 40 hour workshop (Smith, 1993). Kelly and Male (1993) also describe a value engineering audit, whereby a value engineer acting on behalf of a large corporate company or government department reviews expenditure proposals submitted by subsidiary companies or regional authorities, and the procedure follows that of the normal job plan.

### VALUE MANAGEMENT STRATEGY

The approach to value management (VM) can vary for each project, but it is customary to provide a job plan to establish the format to be adopted. A job plan should comprise a recognisable set of processes, as now described.

*Phase 1: the information stage* should cover the assembly of all relevant information appertaining to the project under review and the assimilation and analysis of this information. A cost benefit analysis of objectives should be undertaken, having regard to the client's or end user's method of calculating values, as for example through function analysis techniques and the construction of cost models and possibly FAST (function analysis system technique) diagrams, which are considered later in the chapter.

*Phase 2: the creativity/speculation stage* which comprises the generation of suggestions as to how the required functions can be performed or improved. Group creative techniques should be introduced such as 'synetics': the art of producing a greater end result than the sum of the individual parts.

*Phase 3: the evaluation/analysis stage* consists of the evaluation of ideas generated in the creativity phase, for example by collective or individual rating systems. It also entails the rejection of any unproductive, speculative ideas, of which there are inevitably a high number.

*Phase 4: the development stage*, where the ideas considered at the evaluation stage to have merit are examined and potential savings are costed, with consideration being given to both capital cost and the effect of operational and maintenance costs (life cycle costing). There is considerable scope for the use of cost models and computer aided calculations. Any ideas which either cost more than the original or are found to reduce quality are discarded.

*Phase 5: the presentation/proposal stage*, comprising the presentation of the refined ideas considered to be worth implementing, supported by drawings, calculations and costs.

*Phase 6: the implementation/feedback stage*, where the ideas agreed to be worthwhile are then implemented. Feedback from the sponsors of the VM exercise should ideally be passed back to the VM team to complete the learning cycle (Smith, 1993).

## COMPARISON OF VALUE MANAGEMENT AND COST MANAGEMENT

Value management can be described as a service which is provided in the earlier stages of a project where the primary goal is to determine explicitly the client's needs and wants related to both cost and worth, sometimes described as judgement values, by the use of functional analysis and other problem solving techniques, which will be described later in the chapter. Whereas in cost management, the main thrust is on cost budgeting, management and control, and embraces such activities as feasibility studies, cost planning in all its aspects, the production of bills of quantities, tender evaluation, on-site measurement and valuation and the settlement of final accounts.

Value management often precedes cost management in its timing but there can be a substantial overlap between the two activities during the inception, feasibility and design stages. In both cases an early investigation is desirable while the design is in its early stages, as the longer it is delayed, the more advanced the design, the less opportunity there is for radical changes to the project, the lower the cost savings that can be made and the higher the cost of implementing them.

Cost management has been defined by Kelly and Male (1993) as a service that synthesises traditional quantity surveying skills with structured cost reduction or substitution procedures using a multi-disciplinary team. However, many quantity surveyors would disagree with this definition, argueing that it is too restrictive in defining their cost management role and that effective cost management is not dependent on a multi-disciplinary approach, valuable though it may be. Furthermore, the quantity surveyor can make an objective client project appraisal and have regard to the client's needs and wants against a background of cost.

It cannot be denied that the value management process entails a detailed methodology aimed at achieving savings in cost and/or increased value of a construction project. By approaching the problem in a well structured and organised way, an increased number of alternative solutions are likely be found, as compared with those emerging from a typical cost management approach. Furthermore, the study is usually taken proactively, as opposed to mainly reactive cost reduction investigations which are often only carried out when the project budget is exceeded (Norton, 1992a).

Functional analysis allows the division of a problem into manageable units and promotes an alternative and more comprehensive approach resulting in the consideration of more solutions. In cost management studies the main thrust is often devoted to reducing the cost of items in which savings are readily identifiable, such as the lowering of standards of finishes or reducing the quantity of expensive components and thereby reducing quality and possibly increasing maintenance cost.

Norton (1992a) believes that value management can achieve more fruitful results than cost management techniques on their own. There is no doubt that the client would receive a more comprehensive and efficient service if both systems could be used together on the same project.

## VALUE MANAGEMENT TECHNIQUES

This section of the chapter examines some of the more fundamental and operationally important techniques used in value management, such as functional analysis and FAST diagrams, as it is considered that an understanding of these processes will be helpful to the reader when studying the application of value management to construction projects.

### Functional Analysis

*General principles*

Functional analysis is a powerful technique in the identification of the principal functional requirements of a project. In general the function of an item or system can be expressed as a concise phrase, often consisting of a verb followed by a noun, as this provides a precise and readily understandable description of the function. Useful active verbs include amplify, change, control, create, enclose, establish, improve, increase, prevent, protect, rectify, reduce, remove and support. A typical verb and noun relationship is door: v. control, n. access; and cable: v. conduct and n. current.

It should be recognised that it is not usually possible to seek alternatives to a technical solution without first identifying the functional definition. For example, light is required in a room (functional definition), and to install a component which emits light is a technical solution. A functional definition is frequently obtained by first seeking a technical solution and then defining the functional performance of that solution.

Functions can be subdivided into primary or basic and secondary. Primary functions are those without which the project would fail or the task would not be accomplished, whereas secondary functions are a characteristic of the technical solution selected for the primary function and may be non-essential, although both need identifying to fully understand the problem. Kelly and Male (1993) give the example of an electric filament lamp which satisfies the primary function of emitting light but is also accompanied by unwanted secondary functions, such as generating heat, inducing glare and looking unattractive, and these secondary functions can be resolved by further technical solutions.

Norton (1992a) has described how for function analysis purposes, most secondary functions have zero use value, but some secondary functions may be essential to the basic (primary) function, in which case they are termed required secondary functions and are allocated a value. A typical example of required secondary functions would include compliance with the Building Regulations.

Cost and worth are allocated to each function. The cost is the amount derived from the cost estimate while worth is the lowest possible cost at which the function can be performed. In practice, worth is generally derived by the value management team making an evaluation based on comparison of standards of the design component, historical cost data and/or experience. The total cost and worth of the component's functions are calculated and converted to a cost/worth ratio. Generally, when a cost/worth ratio is 2.00 or above, the component is likely to be adopted for its cost reduction effect.

Worth of secondary, non-essential functions is taken as zero. The first areas to examine for savings are those that perform secondary functions that can be reduced or deleted entirely without affecting the basic function of the component (Norton, 1992a).

### Functional analysis applied to construction projects

The function of a building is to provide an environmentally controlled space suitable for its required use, and its design constitutes a technical solution to the functional requirements of the space. All products and components used in the building perform a function. Kelly and Male (1992) have described how in a functional analysis the function of each component is examined by asking the question 'what does it do?'. In a value management study the next question is likely to be 'how else can this be achieved?'. An intensive (brainstorming) session is held and other technical solutions are then generated.

Kelly and Male (1993) have subdivided functional analysis into four phases or levels, as follows:

*(1) Task*  The client perceives a problem, which may have been identified through a study of efficiency, safety, markets or profitability. Where a client sees a building as the answer to his problem, he is likely to be faced with a building procurement decision and to subsequently enter into a building contract. An alternative is to first approach a value manager who, with representatives of the client organisation, can carry out a value audit and this will help the client to decide whether the provision of a new building offers the best solution to his problem.

*(2) Spaces*  Having determined that a building is the best solution to his problem, the next stage generally involves the architect or the whole design team preparing the brief along with the client. A full performance specification of requirements may not be available from the client and it may therefore be necessary for the design team to determine the client's space requirements through the production of sketches and cost plans.

*(3) Elements*  This is the stage at which the building assumes a structural form. As Kelly and Male (1993) postulate the purpose of an element is to enclose and make comfortable the space provided, but it does not contribute to the client's requirements.

*(4) Components*  This is where the elements become part of the built form. Contact with the client at this stage is very limited since the client value system is likely to have been incorporated at previous levels. Components are chosen to satisfy the requirements of the elements in terms of surrounding and servicing space (Kelly and Male, 1993).

Norton (1992a) has adopted a different approach and believes that function determination may not always be straightforward, particularly as the basic (primary) functions of one item may be considered at different levels. For example, a building's basic function may be for a developer to create a profit while a basic function on a lower level is to enclose space. The different hierarchial tiers at which the function may be considered may be termed levels of abstraction. It is important to know at the outset of a VM study the operative levels of the function. For instance, should alternative methods to create profit be dominant or alternative methods to enclose space. In practice, the levels of abstraction are defined by factors such as the client's requirements for the study, design stage and the like. Norton (1992a) proposes that in order to assist the identification of levels of abstraction, a hierarchy may be determined based on the 'how-why' approach.

## FAST DIAGRAMS

### General Principles

The FAST (Functional Analysis System Technique) evolved from the functional analysis approach to establish a hierarchy of functions and to identify the means by which they can achieve an end result or objective. The principal advantage of the method is that it breaks the overall problem down into individual and readily manageable components and permits a balanced analysis at different levels. It leads naturally to the identification of those items in the current brief or design which attract high cost for low functional value and those items of high importance coupled with low cost. As we shall see later the value of FAST diagrams can be much enhanced by adding the costs of the various activities.

The system revolves around 'how-why' relationships in the studies by Norton (1992a) and Kelly and Male (1991). Thus they resemble a decision tree, by answering the questions 'WHY' when reading from right to left and 'HOW' when reading from left to right, as illustrated in figure 9.1. However, Carter (1991/92) works in the opposite direction, as shown in figure 9.2, and the author does feel that this represents a more logical approach, but both methods should give the same end result, and it could be argued that the choice of method is a matter of personal preference.

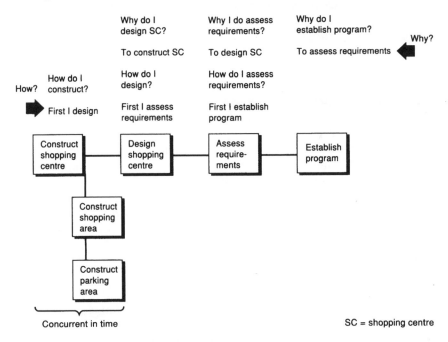

**Figure 9.1** *Extract from a typical FAST diagram (Norton, 1992a)*

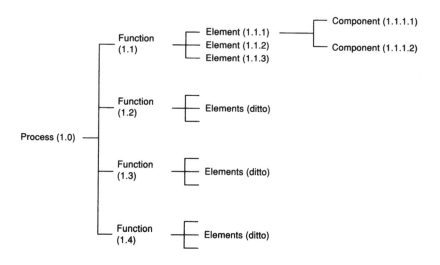

**Figure 9.2**   *Compiling a FAST diagram*

## Compiling Functional Analysis (FAST) Diagrams

The following procedural notes, kindly supplied by Tim Carter of Davis Langdon Management show very clearly the means of compiling FAST diagrams for construction projects.

*Procedure*

1. Identify *key* function(s) of project.
2. Compile fast diagram, working from left to right.
   WHY?—>HOW?
3. Divide/subdivide functions and elements into components to *appropriate* level of detail.
4. Number each item as indicated in figure 9.2.

*Typical functions (not exclusive) identified in earlier studies*

- Prepare site
- Provide temporary facilities
- Provide accommodation (A)
- Provide internal environment (B)
- Accommodate services (plant rooms, ducts, floor/ceiling voids, etc.)
- Enhance quality (prestige?)
- Reduce costs (operational and/or maintenance)
- Provide flexibility

- Provide for expansion
- Comply with regulations ('in house' or statutory)
- Ensure equipment reliability/availability
- Secure operations
- Safety requirements
- Enhance working conditions (provide acceptable working environment)
- Provide external environment (works)
- Provide welfare facilities ⎫
- Circulation facilities ⎬ can be combined
- Accelerate completion. ⎭

Not all are relevant; some may be combined or be implicit in the total building function.

*Typical subdivision of functions*

(A) *Provide accommodation*

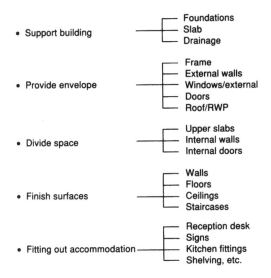

- Support building ——— ⎡ Foundations
  ├ Slab
  └ Drainage

- Provide envelope ——— ⎡ Frame
  ├ External walls
  ├ Windows/external
  ├ Doors
  └ Roof/RWP

- Divide space ——— ⎡ Upper slabs
  ├ Internal walls
  └ Internal doors

- Finish surfaces ——— ⎡ Walls
  ├ Floors
  ├ Ceilings
  └ Staircases

- Fitting out accommodation ——— ⎡ Reception desk
  ├ Signs
  ├ Kitchen fittings
  └ Shelving, etc.

## Criteria Scoring/Alternative Analysis Matrix

Norton (1992a) has described how during the analytical phase of a value management study, the ideas generated during the creative phase are sorted into a list of feasible lower cost or energy saving alternatives. The advantages and disadvantages of each idea are listed or discussed and the ideas are subsequently ranked in order of viability. Selected viable ideas are then evaluated in detail and capital and life cycle costs estimated.

Having ascertained the cost effect of the ideas, alternatives may then be

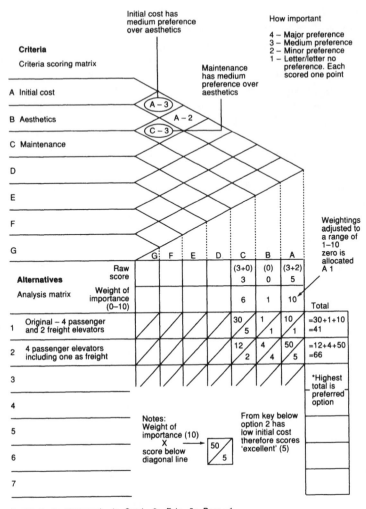

**Figure 9.3** *A criteria scoring matrix*

*Source*: Norton, 1992a.

the subject of a weighted evaluation that includes a consideration of intangible factors, such as aesthetics, flexibility, reliability and the like. Tangible and intangible criteria are listed and weighted by the value management team in accordance with the client's requirements, using tools such as the criteria scoring matrix, as illustrated in figure 9.3. These weightings may then be applied to alternatives using an alternative analysis matrix or equivalent method. The alternative achieving the optimum weighted score is considered to be the most viable option and is then presented to the client during the recommendation phase (Norton, 1992a).

## VALUE MANAGEMENT CASE STUDIES

Two excellent Value Management case studies are contained in Norton and McElligott (1995).

### Bank Processing Centre, Northern England

This case study based on papers provided by Tim Carter of Davis Langdon Management explains the main processes and operation of value management techniques as applied to a specific project, using the two day study approach.

It starts by detailing the key elements of the client's brief and, as could be expected, these call for a high quality building which is also efficient, user friendly and secures the lowest operating costs, with an 18 month timescale.

This is followed by a schedule of the VM study team members who were present on the two allotted days (a Friday and the following Monday). It will be noted that the client is represented by the project manager, facilities manager and a director. All the design team are well represented by all the senior staff concerned with the project, as is also the appointed management contractor, as this project is being undertaken as a management contract. The meetings were chaired by a partner of the value management consultancy supported by a senior project manager.

Figure 9.4 shows a FAST (functional analysis) diagram summary showing all the functions with their estimated costs, all aimed at rationalising operations and achieving lower operating costs, at an overall cost of £8.84m. The most expensive items are providing accommodation (1.2) at £1.766m and the provision of the internal environment (1.3) at £2.034m, and these are likely to be the prime areas for investigation to reduce costs without lowering standards. A cost reduction of £0.216m is scheduled.

Figure 9.5 illustrates a detailed FAST diagram covering 'providing an acceptable working environment', with extensive components, all individually priced. In practice, detailed FAST diagrams will be prepared for all the 14 functions.

The value management team examine all possible alternatives with the objective of meeting the required performance standards at lower cost or improving standards for the same cost. As is normally the case, a considerable number of options were scrutinised and rejected by the management team as failing to meet the prescribed criteria, while many more were considered and were the subject of further investigation by the project team. A schedule shows these latter items with their likely cost savings, or

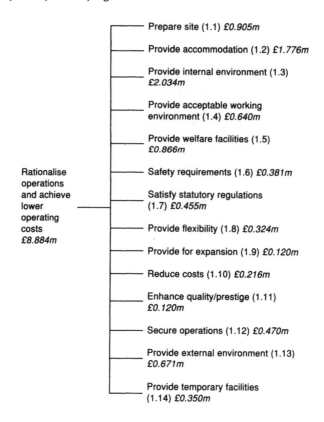

Prepare site (1.1) *£0.905m*

Provide accommodation (1.2) *£1.776m*

Provide internal environment (1.3) *£2.034m*

Provide acceptable working environment (1.4) *£0.640m*

Provide welfare facilities (1.5) *£0.866m*

Rationalise operations and achieve lower operating costs *£8.884m*

Safety requirements (1.6) *£0.381m*

Satisfy statutory regulations (1.7) *£0.455m*

Provide flexibility (1.8) *£0.324m*

Provide for expansion (1.9) *£0.120m*

Reduce costs (1.10) *£0.216m*

Enhance quality/prestige (1.11) *£0.120m*

Secure operations (1.12) *£0.470m*

Provide external environment (1.13) *£0.671m*

Provide temporary facilities (1.14) *£0.350m*

**Figure 9.4**   *Detailed FAST diagram of Bank Processing Centre*

extras where their provision is advisable to achieve the required standards and/or reduce maintenance or operating costs.

Finally, the overall cost benefit of the value management study is listed and in this particular project yielded savings of £296.1K representing an overall saving of 2.7 per cent, and taking the cost of the study at approximately £8K, a cost benefit ratio of 1:37 is obtained.

*Bank Processing Centre – Northern England*
*VM Study 16–20 March 1990*
*Key elements of client's brief*

1. Provide contingency processing operations (urgently).
2. Rationalise rented accommodation in NW England.
3. New building to be:
   • Modular
   • Single storey
   • Open plan/highly flexible

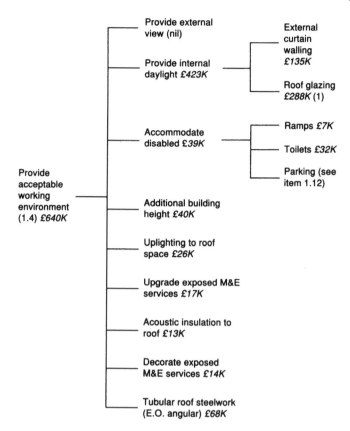

**Figure 9.5**  *Detailed FAST diagram of provision of acceptable working environment to Bank Processing Centre*

*Note*: Glazing £144K; cleaning equipment £68K; sun shading £47K.

- Utilitarian/functional
- Energy efficient
- Single standard
- Expandable (100%)
- User-friendly
- Controlled access
- Phase 1: 800–900 staff
        4:1 female/male
        Age 16–22
- Comfort cooled (not A/C) for 120 W/m²
- Without suspended ceiling
- Raised floors in operational areas.

4. New building to enable client to achieve/maintain *lowest* processing costs.

5. New building to project image of hi-tech/cost-effective client (i.e. as item 4).
6. Building *must* be operational by October 1991.

(A) *Options considered and rejected by VM team*

1. Provide site concrete batching plant
2. Reduce 600 mm concrete oversite slab
3. Bend reinforcement on site
4. Prefabricate reinforcement mesh to piled area slabs
5. Divert SW drainage into canal
6. Consider integral plant rooms
7. Reuse existing buildings for main restaurant
8. GRP/polycarbonate rooflights in lieu of glass
9. 2.00 m high block perimeter wall in lieu of metal cladding
10. Ha-ha to site perimeter in lieu of security fence
11. Client to insure all works (clause 22B)
12. Refurbish existing pumphouse; use as site offices
13. M/C to purchase temporary site offices at end of project
14. Omit tubular steelwork and use standard angles/channels.

(B) *Options considered and to be investigated by project team*

|  | Approx. (saving) extra £/K |
|---|---|
| 1. Bay sizes altered by 90 degrees | – |
| 2. PC planks to walkways in lieu of solid | 7.5 |
| 3. Renegotiate connection charges/electricity charges by NORWEB | (39) |
| 4. Reconsider allowance £108K for louvres (Item 1.3) | (20) |
| 5. Reconsider roof elements (materials, sizes, span, maintenance) | Range (50–100) |
| 6. Balance daylight/solar gain with fewer rooflights/artificial lighting | Range (8–70) |
| 7. Check external doors budget £85K (Item 1.2) | (4) |
| 8. High level clerestorey glazing to east wall in lieu of rooflights | – |
| 9. Check £32K budget for disabled toilet (Item 1.4) | – |
| 10. Catering company to provide catering equipment and rentalise and/or | – |
| 11. Reconsider scope of catering equipment £319K (Item 1.5) | (81) |

12. Reduce lightning protection and use steel
    frame (assess risks)                               (2)
13. Client to consider fire compartmentation
    policy (provision for fire wall)                   (1)
14. Check contingency of *£57K* to strengthen
    roof steelwork for sprinklers to be added
    (Item 1.8)                                          –
15. Verify 20% expansion provision in catering    See item
    equipment                                          11
16. Phase carpet installations with phased occupation   –
17. Cheaper carpet tiling to non-walkway
    'corridor' areas                                    –
18. Consider bonded carpet tiling to main areas
    and/or 'corridors'                                  –
19. Consider gravel in lieu of paving alongside canal   –
20. Low fence or rail alongside canal as
    demarcation barrier?                                –
21. Consider turnstiles (card key controlled)
    and CCTV in lieu of main reception security
    screen/oscillating doors                           (30)
22. Reassess security provision; costs v risks    See Item 25
23. Consider cheaper pavings around building perimeter  (2)
    (blocks costing *£67K* in Item 1.13)
24. Plan landscaping one season ahead to obtain savings  –
25. Question *£85K* 'electronic' security fence to
    site perimeter (Item 1.12)                         46
26. Upgrade painting specification to 'exposed'
    roof steelwork to reduce maintenance.              40

*Net saving*                            £(257.5K)

Overall cost/benefit of VM study

|  | *Budget* | VM savings |
|---|---|---|
|  | £/K | £/K |
| 1. Construction works/MC's fee add | 9515 | (257.5) |
| 2. Professional fees (15%) | 1427 | (38.6) |
| *Estimated costs* | £10 942 | (£296.1) |

VM study yielded approx. *2.7%* savings overall
Cost of study approx. *£8K*
Cost benefit ratio = 1:37
*Note*: Savings would have been much higher had director not changed
       shortly before VM study (new director reluctant to take risks!)

## CONCLUSIONS

It has been clearly established in this chapter that value management is not merely a cost cutting exercise, it also takes account of the three-way relationship between function, cost and value.

McElligot and Norton (1995) have identified three important aspects of value improvement:

- same performance at reduced cost;
- improving performance at same cost;
- improving performance at reduced cost.

In their comprehensive and practically oriented book entitled '*Value Management in Construction: A Practical Guide*', Norton and McElligott explore very fully the different job plan phases of a value management study, supported by a wide range of case studies. The different techniques and approaches are examined and compared with their potential benefits, which the author has endeavoured to outline in this chapter, and the way in which value management can complement cost planning, to provide an improved service to the client.

The timing of the value management study can be critical and figure 9.6 produced by Carter (1991/92) illustrates very clearly the optimum time for conducting such a study. Currently, such studies/workshops are conducted at between 10 to 35 per cent stage of the design process, using the design team chaired by a value management team coordinator or by an independent value management team.

Smith (1993) emphasises that, where appropriate, consideration should be given to capital cost, life cycle costs, programme, buildability and/or optimum return on money. By adopting this approach, good value for money can be achieved. It is also important that design teams should not see value management or design/cost reviews as an attack on aesthetics or quality. Equally, alternative solutions should not be viewed as criticisms of the existing design.

Brown (1992) believes that value management represents a natural progression for the quantity surveyor in leading the search for alternative technical solutions and presenting them as evaluated and costed options.

There are doubtless many benefits to be gained by adopting value management techniques and Beard Dove have identified the following aspects as being the most important:

(1) reduce project costs;
(2) improve design efficiency;
(3) optimise value for money;
(4) concentrate design effort;
(5) advance design decisions;

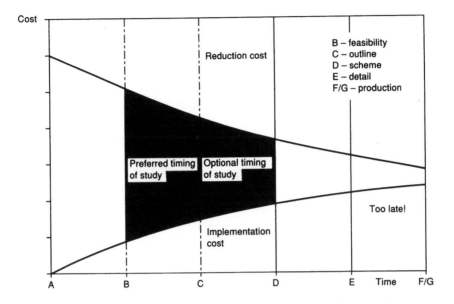

**Figure 9.6**   *Value management: optimum timing for study (Carter, 1991/92)*

(6)  highlight design options for selection;
(7)  improve ways to comply with the brief;
(8)  afford an independent functional review.

Carter (1991/92) has critically examined both the benefits and possible disadvantages of value management, and his findings follow.

**Benefits**

- Examines function and cost
- Provides opportunity for options to be considered
- Seeks better technical and more cost-effective solutions
- Identifies and reviews constraints and criteria affecting the project
- Design changes can be accommodated at minimal cost (if study conducted early)
- Opportunity for in depth project review and greater understanding for all team members
- Team building
- Identifies and can eliminate unnecessary costs
- Generates greater client confidence
- Can shorten overall programme period (longer brief/design period and shorter production/construction period)
- Assists client decision making

**Disadvantages**

- Extra work for existing project team, which is not always reimbursed, as it is at the client's discretion
- Disruption to project team
- Can incur extra fees
- Can extend design period

Carter (1991/92) also raises the very pertinent point concerning traditional projects, as in how many of these where the tender comes within budget does the design team look for any savings/improved value, even though there could be a potential saving of as much as 5 to 10 per cent on many schemes.

# 10 Development, Redevelopment and Environmental Aspects

This chapter examines the quantity surveyor's role in development, redevelopment, environmental and conservation work, as these become increasingly important issues.

## THE ESSENCE OF DEVELOPMENT

Every development whether it be for a public authority, industrialist or private investor, has a 'market value' – a potential worth or earning power. Even civic buildings, hospitals, churches and universities have an assessable value to the community – a cost above which it is not reasonable or feasible to build. Within certain limits of aesthetics, function and performance, the most economic development is that which shows the greatest return to the community for the minimum capital invested. This does not imply that the cheapest is the best; often the opposite is the case.

The art of phasing development to give an early return which can be used to pay for the less remunerative items is one of the objectives of a skilful developer. In this connection, an amalgamation of public and private agencies in development can be of great benefit to the community. It is also essential to integrate the planning of large scale redevelopment. For instance, the retention of an existing road layout even if only to provide a pedestrian precinct or parking facilities can economise in new construction, and the potential earning power of existing facilities should not be overlooked. Time is also important in the planning process; for maximum economy the time between capital expenditure and completion of a project should be kept to a minimum.

The private developer or industrialist will require a financial appraisal or feasibility study to determine the likely capital expenditure and probable revenue in order to arrive at the anticipated return on the money invested. Whether the project is to be financed by public or private funds, it is important to know the cost implications at the outset in order to be able to appraise the viability of the scheme. It is necessary for the developer to know the nature and extent of the proposed development, its cost and the time required to complete it. Indeed, the whole development process is becoming more sophisticated. Schemes need to be appraised from every aspect – aesthetic, fiscal and social. The long and frequently

293

frustrating negotiations to assemble sites, obtain planning permission, barter with local authorities and secure finance, demand a truly professional approach.

Hence a developer usually wishes to know whether the investment of capital in a project will be justified by the return which he can expect to receive. It is therefore necessary to assess as accurately as possible the value of all the expected returns and benefits and to compare them with the estimated costs. One problem is to express all the benefits in monetary terms as there may well be some indirect and intangible benefits, such as more contented employees or greater prestige value, which flow from a project. The quantity surveyor is frequently called upon to make cost comparisons of different design proposals with varying capital, maintenance and running costs. The task of the quantity surveyor is to inform the developer which is the most economical scheme after taking all these costs into account.

To this end general surveyors should have a general knowledge of building costs in addition to a detailed knowledge of values, while quantity surveyors ought to have a general knowledge of values as well as a detailed knowledge of building costs. There is an evident need for the quantity surveyor to be aware that the general practice surveyor, in making his general financial appraisal, has to consider the capital cost of the works, land purchase, compensation for extinguishment of leases, bridging finance, long-term finance, rental and capital values profitability, and maintenance and other outgoings. The key factor in any successful project is generally believed to be the triangle of valuation surveyor, quantity surveyor and architect, although other professionals may also be involved. The general practice surveyor will be primarily concerned with the broad economics of development, the architect with the design of the project and the quantity surveyor with the interaction of these two aspects of development. Some of the more important matters to be resolved by the development team include:

(1) ensuring development to maximum permissible plot ratio;
(2) planning the most economical and profitable use of available floor space;
(3) implications and suitability of different methods of procuring the building;
(4) the speed of construction balanced against financial considerations, such as the cost of bridging finance and loss of rent or interest;
(5) the effect of incurring extra capital costs including expensive finishings, balanced against additional net rental value (if any); and
(6) the effect of incurring extra capital costs balanced against a consequent reduction in future maintenance and operating costs, including depreciation allowances (Seeley, 1995a).

## DEVELOPMENT APPRAISAL

### Appraisal Techniques

In the early nineteen nineties returns on investments in the private sector were low and, in consequence, employers in the industrial and commercial fields in particular were hesitating before committing substantial reinvestment in new building stock. They were also becoming more critical and demanding in terms of both time-scale and cost/value for money and expected a truly professional and positive response from quantity surveyors. On occasions, quantity surveyor's services are required at a more strategic planning level on more specialised or complex projects.

In investment appraisals, the employer needs to be satisfied that he will receive an adequate return on the capital invested, and the quantity surveyor plays an important role where the investment is in buildings, which have their own special characteristics. For example:

(1)  it is usually significantly expensive, in relation to the employer's resources;
(2)  it is real investment, being created as a new asset;
(3)  maintenance and operating costs are an important element of total cost;
(4)  it is a prominent and long lasting asset, giving rise to social costs and gains as well as private costs and gains, and subject to various public controls; and
(5)  the resultant legal, social, financial and technical complexities are sufficiently far reaching to involve a number of different specialist professions.

The quantity surveyor cannot be an expert in all these aspects, but he should have a sound understanding of them and be able to liaise with and advise the other consultants.

Because of the cross currents of finance, politics, interest rates, taxation, inflation and employment prospects, financiers demand a more detailed and thorough appraisal of projects presented to them. Having grown accustomed to the advanced investment appraisal techniques applied to the alternatives such as equities and gilt edged securities, they expect the same professional and skilled approach to development projects. This involves the use of more sophisticated methods of site selection, demand analysis, feasibility testing, techniques of funding, cost estimating, valuing and tax forecasting, project planning and design (Darlow and Morley, 1982).

The traditional technique of residual valuation used to appraise developments, as described later in this chapter, often requires refinement in the form of projections, cashflows and probability/sensitivity analysis (Balchin *et al.*, 1995). All feasible alternatives must be detailed and evaluated, and these studies must include the employer's operational requirements and the estate management and investment aspects. Ideally, for each alternative proposal, all the costs and benefits should be quantified and evaluated, brought to a common cost basis, using a similar approach to that adopted for cost benefit analysis (cba) as described in chapter 19. For example, an investment appraisal for an existing or proposed building project should include life cycle costs such as maintenance, cleaning and rates, in addition to the statutory requirements for fire and safety. Social benefits may be difficult to evaluate in monetary terms and it may only be possible to weight the various factors.

### Developer's Finance

Quantity surveyors must be familiar with the methods of funding construction projects and be aware of the differences between short term and long term loans. Sale and leaseback arrangements are becoming more common, whereby the developer sells the completed project to an investor and takes a long lease of the development at an agreed rent. The rent is calculated at an acceptable rate of return on the purchase price. The developer then sublets the premises to occupying tenants to obtain a profit rent. For more comprehensive information on financing arrangements readers are referred to the works by Fraser (1993) and Isaac (1994).

There are also allowances and loans available in various situations, on which the quantity surveyor may be required to advise the employer. For example, there are some capital allowances for industrial buildings and structures, machinery and plant, oil wells, and agricultural buildings and works. The industrial building allowances change periodically under the provisions of Finance Acts. There is also a wide variety of special forms of financial assistance such as regional development grants, selective discretionary grants, European aid, urban development corporations and inner city aid, and a number of agencies that can offer assistance in particular circumstances, the principal one in 1995 being English Partnerships and the establishment of a single regeneration budget. Some local authorities offer loans and/or grants for businesses converting, extending, improving or modifying industrial or commercial buildings in the authorities' improvement areas. On the debit side there will be VAT on alterations, repairs and maintenance. The employer will also be faced with liability for corporation and income tax on his profits once the building becomes operative. All relevant financial factors need to be taken into account at the feasibility stage.

### Residual Valuation Techniques

Prior to the purchase of a site for development, a developer must know what forms of development will be permitted on the site and have access to a development appraisal (sometimes termed a developer's budget or feasibility study). To decide whether the scheme is feasible, he will require advice on a fair price for the land; the probable building costs; and the probable rent or selling price of completed building(s).

Readers requiring more information on development and project appraisal are referred to Seeley (1995a) and Hutchinson (1993), and to Yates and Gilbert (1989) for the appraisal of capital investment in property.

A worked example follows to illustrate the application of the concept of the development appraisal to a practical development problem. It should, however, be emphasised that calculations of this type can be very complex and in practice the assistance of a valuer will normally be required.

Planning consent has been given for the erection of an office block of 10 000 m² on a vacant building site. It is estimated that the building will produce a net income of £1 400 000 pa and will cost £680/m² to build. Assuming it will take eighteen months to build, determine the present market value of the site.

*Gross development value*

| | | |
|---|---|---|
| Net income from offices | £1 400 000 pa | |
| YP in perpetuity at | | |
| eight per cent | 12.5 | £17 500 000 |

*deduct* costs

| | | |
|---|---|---|
| Building – 10 000 m² | | |
| at £680/m² | 6 800 000 | |
| *add* architect's, surveyor's | | |
| and consultants' fees | | |
| (16 per cent) | 1 088 000 | |
| | 7 888 000 | |

One-and-a-half years
   building finance on
   half cost

$$\frac{£7\,888\,000}{2}$$

| | | |
|---|---|---|
| at 11 per cent | 433 840 | |

| | | |
|---|---|---|
| Legal and agency fees | | |
| and advertising costs | | |
| (three per cent of GDV) | 525 000 | |
| Developer's profit | | |
| (ten per cent of GDV) | 1 750 000 | |
| | 10 596 840 | |

Residue = £6 903 160

The residue represents three items: value of the land; acquisition costs (4 per cent of value); and cost of borrowing for 2 years @ 11 per cent pa on land value and costs.

Let $x$ = Land value, then £6 903 160 = $(x + 0.4x) \times 1.11^2$
£6 903 160 = $1.2814x$

$$\text{Site value} = \frac{£6\,903\,160}{1.2814} = \underline{£5\,387\,200}$$

The anticipated return of the developer is usually assessed at about seven to nine per cent of development costs, and it would be useful to check the figures in this example by the anticipated return method.

| *Costs* | £ |
|---|---|
| Building, including fees | 7 888 000 |
| Building finance | 433 840 |
| Legal, agency and advertising costs | 525 000 |
| Land cost | 5 387 200 |
| Site finance and acquisition | 1 515 960 |
| Total costs | £15 750 000 |

The net income of £1 400 000 gives a 8.89 per cent return on the development costs of £15 750 000 (Seeley, 1995a).


## ENVIRONMENTAL MANAGEMENT

### Reasons for Growing Importance of Environmental Matters

It was not until the late 1980s that the wider effect on the world's resources and the balance of the global environment became generally recognised, when considering the relationship of buildings to the environment (DoE, 1991b), while Cadman (1990) identified 'growing environmental concern' as one of the four factors, alongside market balance, the demographic trough and European integration, that will change attitudes to planning and investment in the future. The RICS (1993a) identified three main stimuli that are likely to secure a positive reaction to environmental issues: namely, legislation, public opinion and profit.

### 1. Legislation
The nature and scope of legislation are changing rapidly as evidenced by the Report on European Environmental Legislation and its Application (Metra Martech, 1991). For example, currently permitted levels of contamination may become legislatively unacceptable as standards and expectations rise.

## 2. *Public opinion*

Public interest in environmental matters is growing as illustrated by the national press. A typical example was the concern expressed over the use of timber framing after it was shown in a poor light in a *World in Action* programme, with a significant effect on residential construction practices and house prices. Many green issues have been adopted by democratic governments and it is being increasingly emphasised that there are serious risks in ignoring environmental considerations.

## 3. *Profit*

Legislation is usually effective because non-compliance often results in a monetary fine. Public opinion is frequently sought to increase the competitive edge, obtain an enhanced image and increases in demand and perceived value (RICS, 1993a).

## Environmental Definitions and Issues

The RICS (1993a) formulated the following definition of *Environmental management*: 'The management of environmental factors to enable human activity to exist productively and compatibly alongside, and enhance where practical, other global and life sustaining processes, in a beneficial, sustainable and harmonious synthesis for future generations'.

Thus, where possible, the surveyor should beneficially manage the factors that affect the environment. However, where this is not possible, he should manage the client's interest to reflect the environmental constraints. Readers requiring more information on environmental management in construction are referred to Griffith (1994).

There are many environmental issues which affect the work of the surveyor, ranging from climate to contaminated land, energy and timber.

### Environmental assessment

Environmental assessment (EA) is the process by which the environmental implications of a project are assessed. An environmental statement (ES) is the formal document which must be produced in certain cases to set out the result of the EA. Environmental information (EI) is information relating to the environmental implications of a project whether contained in the ES or in representations from statutory consultees or third parties. Proposals for sensitive developments such as quarrying, open cast mining, motorways, waste disposal sites, extensive new urban or rural developments and major out-of-town shopping schemes require detailed environmental impact assessments of their effects during and after implementation, and by 1993 approximately 1500 EAs had been prepared in the UK covering some 60 per cent of all local authorities.

Public opinion may quickly be motivated by what is seen as lack of environmental consideration, which could include such matters as loss of amenity and visual value. Commercial and environmental issues can also come into conflict in cases such as the Cardiff Bay development, where waterfront regeneration could cause the permanent loss of important estuary habitat for wildlife (RICS, 1993a).

EC Directive 85/337/EEC required member states to ensure that where projects have a significant impact on the environment, EI is obtained and considered before consent for the project is given. The main UK regulations are the Town and Country Planning (Assessment of Environmental Effects) Regulations 1988, which operate additional to the planning system by directing that, in certain cases, an ES is supplied and considered before granting planning permission. This does not cut' across a planning authority's general duty to have regard to all material considerations, including environmental impact, but is an addition to that duty.

The Town and Country Planning (Development Plan) Regulations 1991 state that: 'Most policies and proposals in all types of plan will have environmental implications which should be appraised as part of the plan preparation process. Such an environmental appraisal is the process of identifying, quantifying, weighing up and reporting on the environmental and other costs and benefits ... but ... does not require a full environmental impact statement of the sort needed for projects likely to have serious environmental effects'.

### Environmental labelling

There could be an important role for environmental labelling in property and construction, although in 1995 there seemed only limited public awareness of this process. However, environmental audits and assessments are becoming more evident in the property and construction industries and BRE have made a substantial contribution with BREEAM (Building Research Establishment Environmental Assessment Method), by assessing a building's impact on such matters as global warming, ozone depletion, acid rain, sustainable materials, recycling, local environment and indoor environment (BRE, 1990).

The EC Environmental Council of Ministers introduced an Eco Labelling system, adopting a 'cradle to grave' approach. Among some 20 product categories, the UK long term proposals encompass light bulbs, lighting, air conditioning equipment, building materials, insulating materials and water conserving devices.

### Conclusions

All surveyors have a duty to inform their clients about relevant environmental issues and to ensure that they are given adequate consideration in the decision making process. It can often be difficult to evaluate them in monetary terms and it may be necessary to use cost benefit techniques, as described in chapter 19. As more client organisations proceed along the BS 7750 environmental management route, they are likely to prefer to engage advisors who have BS 7750 certification.

## Environmental Impact Assessment of a Completed Shopping Centre

It is considered useful to the reader at this stage to examine an enlightened post contract monitoring of a recently completed building project from an environmental assessment viewpoint, including identifying the relevant criteria.

The Town and Country Planning (Assessment of Environmental Effects) Regulations 1988 stated that a shopping centre was likely to have a 'significant environmental impact' if it exceeded 10 000 m² in area, had a site larger than 5 ha, or was in close proximity to more than 500 private dwellings, although these criteria were not mandatory. Moreover, there are projects which exceed these criteria but for which no environmental impact assessments (EIAs) were carried out, and even when they were there was no guarantee that the assessments were adequate.

An EIA can be no more than a prediction of what the assessors expect to happen to the environment if the particular project is implemented, and there is no obligation for the prediction to be compared to the live situation on completion. Indeed without effective post project feedback the whole process could become quite meaningless.

In like manner the BRE Environmental Assessment Method does have some weaknesses as it is mainly concerned with the building elements, their insulation levels, use of CFCs and associated aspects. To be really effective the investigation should have a wider remit to consider, for example, whether an out-of-town shopping centre can be energy efficient if it attracts thousands of cars per day and the relative importance of energy efficiency as compared with visual acceptability in terms of environmental impact (Parker, 1993).

In a serious attempt to face up to these and many other related issues in 1993, Benoy, the designer of the recently completed £20m Waterside Shopping Centre in Lincoln, using their environmental consultancy practice, undertook a detailed evaluation of the completed project, financed by £12 000 of sponsorship including contributions from the contractor and the quantity surveyor. To put the scheme into perspective, although the total area of the shopping centre was 12 500 m², the derelict nature of the site led the local authority to decide that no EIA was required before planning consent was granted.

It appeared to the designer that developers and planners needed more relevant criteria and sophisticated analyses, and hence in the evaluation the local authority, local retailers and the general public were asked whether they believed that the Waterside project was really necessary. It was felt that an EIA should have to justify the need for a proposed development, although in many cases this does not occur, as if there is no proven need for a new building then all the other environmental criteria are irrelevant.

A check was made on whether the planners and developers considered alternative sites, which is a mandatory requirement on larger projects, and then proceeded to examine the impact of the construction work, which is often overlooked but which can be a major issue on large city centre projects.

Nearly all city centre developments have archaeological implications and as the Waterside project was only 5 m from the River Witham, the effect on local watercourses was considered in the evaluation. An important socio-economic issue to be investigated was the effect of the development on road traffic patterns and pedestrian flows. As the centre had no car parking provision, it was not expected to draw extra traffic from out of town but it could attract local residents from other retail areas and thereby contribute to their decay.

Another objective was to endeavour to evaluate the visual impact of the development, by asking users of the centre, English Heritage and the Royal Fine Art Commission to compare photographs of the undeveloped site with the artist's impressions that formed part of the planning application and finally the finished development. This would determine whether the artist's impressions misled the public and the planners. Other aspects considered included policies for the management of solid wastes, mainly packing and wrapping, as most shopping centre managers have little interest in solid waste management and there are normally few facilities for tenants to recycle their waste.

Ultimately, subject to Department of Trade and Industry (DTI) approval of Benoy's application for development funding, a much more comprehensive and sophisticated environmental assessment, including an elemental analysis of the building and its components would be available in 1995. The results of this investigation could have a valuable spin-off at a time when retail development and urban redevelopment were gaining momentum (Parker, 1993).

## CONCEPT OF ENVIRONMENTAL ECONOMICS

The need has been identified for quantity surveyors, in a role as construction economists, to be concerned with all aspects of the construction process and the underlying forces behind the various activities. These encompass the effects of public and private investment policies and aesthetic and planning factors, all of which play a part in determining the system of economic forces which underlie the construction process. A study of economics in its broadest sense entails consideration of how people behave in their activities of producing, exchanging and consuming, and the motivating forces behind these activities.

The term 'environment' embraces constructional works and their general surroundings and thus environmental economics can be considered as the study of the forces affecting the use of assets and resources in satisfying man's need for shelter and a properly managed environment. The surveyor's skill should not be confined to the measurement of physical features or entities but should be extended to measure the forces which are at work in the deployment or changed use of the resources which shape our environment. Expressed in another way, we must deal with cause as well as effect. Hence possibly the base on which the surveying profession is founded should be changed from the land or property to the environment.

The quantity surveyor as a building/construction cost consultant or building/construction economist may be called upon to advise a building client on a number of separate but interrelated issues. The main issues are now listed, together with the broad field of knowledge and expertise that is needed to give effective advice in each case.

(1) *Why* the client should build. Requires a knowledge of economics and the ability to forecast trends in the economy, coupled with an appreciation of the client's financial and production problems and a knowledge of building costs.
(2) *Where* the client should build. Involves a knowledge of national and international markets, economics of transportation and communication, population trends, taxation benefits and building and planning legislation.
(3) *When* the client should build. Requires a knowledge of the alternative ways of financing buildings, legislation affecting capital and revenue expenditure and investment allowances.
(4) *How* the client should build. Requires expertise in contract procurement planning and administration, selection of technical advisers and a construction team, and ability to maintain financial control of the contract.
(5) *What* the client should build. Involves coordination of the specialist advisers to ensure value for money and evaluation of capital and future costs of alternative design solutions, and of completed buildings for sale or lease (Seeley, 1995a).

## URBAN RENEWAL AND REGENERATION

### Basic Problems

One of the most urgent and complex problems facing many local authorities at the present time is that of urban renewal. Most towns and cities have evolved over a long period on a radial road pattern which is ill-suited for present-day traffic needs. Surrounding the pressurised inner core is often a girdle of mixed residential, commercial and industrial uses in varying stages of obsolescence, frequently termed twilight or 'blighted' areas. Comprehensive development of all these areas is essential if satisfactory layouts are to be achieved.

Complete redevelopment of town centres is necessary in some towns, particularly where extensive growth is anticipated. Redevelopment is needed to overcome present deficiencies, such as bad traffic congestion, lack of parking space, inadequate loading and unloading facilities, restrictive traffic regulations, excessive fumes and noise, dangers to pedestrians, and in some places there is a definite conflict arising from the use of particular roads by different kinds of traffic. In smaller towns, pedestrianised shopping streets and precincts with rear-loading facilities, combined with a suitable network of distribution roads and adequate car parking space will often provide the best long-term solution.

A number of important practical issues result from redevelopment. They entail extensive and costly diversion of underground services, many of which were routed under existing roads, which themselves are to be diverted or closed. The twilight areas are characterised by multiple-occupation, overcrowding, lack of essential facilities, neglect, dilapidation, mixed industrial and commercial uses and general environmental decay. There is both a social problem and one of physical decay, but it must be questioned whether the community can afford to allow the wholesale demolition of the houses, most of which could still have many years of life. Property owners may also be very dissatisfied with the method of delineating unfit properties in a slum clearance area, whose compensation is restricted to site value. However, urban renewal offers the opportunity to restore integrity and character to a depressed neighbourhood (Seeley 1995a).

The post war planning policy for inner cities resulted in large areas of derelict land and deteriorating buildings in the 1970s to 1990s, which can be very expensive to rejuvenate. Much skilful and attractive redevelopment and rehabilitation is needed to bring life, interest and attractiveness to many of these depressed inner city areas, as was highlighted by the Civic Trust (1988). There is also a pressing need to create more ecologically rich open space in inner cities, preferably connected by 'green' networks (RICS, 1992a).

The characteristics of urban decay and deprivation vary greatly from one location to another and no single form of organisation is suited to all situations. Hence it is desirable to exercise flexibility in the way the problems are approached and in the choice of urban development organisation, but preferably based on the local community provided with adequate powers to resolve the problems. The ICE (1988) wisely postulated that rational and efficient development would be secured by long term indications of future capital budgets, possibly in the order of three years, with the appraisal of grant applications conducted at one level only, preferably regionally.

## Financial Aspects

Integrated inner city policy was sought by the Government through setting up the *Urban Regeneration Agency* (URA), subsequently renamed *English Partnerships* (EPs) under the Leasehold Reform, Housing and Urban Development Act 1993, whose main purposes were:

(1) to take over policy decision making from relevant ministers or Secretaries of State, such as in DoE, Department of Employment and Department of Trade and Industry (DTI);
(2) to complement regional industrial policy of DTI and to take over the administration of the English Industrial Estates Corporation;
(3) to buy and develop inner city sites, thereby assuming the responsibility of much of DoE's Urban Programme, to award City Grants and Derelict Land Grants (DLGs) and eventually be responsible for a large slice of DoE's total budget;
(4) to administer City Challenge.

*Eurograms* were available to assist with a variety of regeneration schemes. Unfortunately, in December 1993 regeneration schemes worth millions of pounds had been cut or delayed because European Union (EU) regional infrastructure grants earmarked for run-down regions had not been taken up by local authorities before the 31 December 1993 deadline.

Local authorities blamed central government restrictions on council spending and confusion over the eligibility of private sector partners in the schemes for the low take-up of EU grants. Under EU rules, regional grants can fund up to 50 per cent of a project and the national government has to fund the balance. Local authorities claimed that government restrictions on council borrowing meant that they were unable to provide matching funds without cutting other vital services and some local authorities that had tried to raise funds through the private sector had also run into difficulties. Hence the seemingly generous Eurograms appear to be fraught with difficulties in their implementation.

Yet in 1995, despite all these government measures, the inner cities were worse off. Unemployment was higher, investment in public and private ventures lower and deprivation more severe. The main reasons stemmed from the poor state of the national economy and the reduction of the central government financial support to local authorities in large urban areas.

In practical terms local authorities are often only able to provide the infrastructure within which the private sector can operate. The availability of finance to the public sector is so small in relation to total needs, that it must be used in the most effective way and this will normally be as a catalyst. The government financial cuts in the 1990s underlined this fact. Local authorities can provide sites, services and buildings but they must, in addition, provide the right climate for investment. They should ideally give maximum encouragement to developers, streamline the planning process and, wherever possible, plan joint schemes with them.

Developers cannot in the main retain their completed schemes as rent-producing investments and must usually sell them on completion to the institutions. Most institutions are reluctant to purchase property in unattractive locations, where it is difficult to assess future growth, and with certain types of property, such as small industrial units they are reluctant to accept the management with the risk of tenants going out of business. Finance for housing is much less of a problem, since building societies are generally prepared to support new housing in inner areas, although in the early 1990s they were also suffering from a reduced inflow of funds.

Building costs in inner city areas are higher than on greenfield sites, especially when the existing roads and services have to be renewed. The cost differential in 1995 could be as much as £6000 to £8000 for identical houses. Selling prices on the other hand could be in the range of £8000 to £15 000 less than would be obtained for comparable houses in suburban locations. The differences result in £14 000 to £23 000 less per unit being available for land and profit. However, prices vary so much in different locations that these cost relationships should only be regarded as indicative (Seeley, 1995a).

### Basic Appraisals Objectives

Prior to formulating town centre redevelopment proposals it is essential to make an objective appraisal of the existing centre, having regard to its function, assets and deficiencies in terms of convenience and safety, usefulness by day and night as a shopping, commercial and social centre, and its civic character and architectural qualities. The objectives are primarily concerned with function (future size and purpose of town centre), layout (distribution and extent of main uses), circulation (pedestrian and vehicular movement) and character (retention and enhancement of town's individuality).

Plans for the future should aim at satisfying the human need on a modern scale by retaining or introducing some of the more picturesque uses which give character and vitality to the urban scene, such as pedestrian ways, arcades, street markets, small places or squares, landscaped spaces for rest and relaxation and facilities for amusement. The massing of buildings, their height and silhouette, and in detail the outline, textures and colours of materials, are all matters of vital importance in composing a street scene which is both attractive and harmonious.

It is desirable to carry out an economic survey and appraisal to assess town centre demand. This would embrace a physical survey of the existing shopping centre, analyis of existing traders, examination of recent transactions in sale or renting of premises, determination of market region, analysis of existing and potential turnover, examination of car parking facilities, and study of traffic flows and bypassable traffic. From the results of the surveys it will be possible to make a realistic assessment of shopping and parking needs and of the required scale and rate of development (Seeley, 1995a).

## Implementation by Partnership

Large scale town centre redevelopment schemes are often undertaken on the basis of a partnership between the local authority and the private sector. The local authority has powers of compulsory acquisition which are vital if the best pattern of development is to be achieved. Modern developments require large areas of land in order to make the best use of modern techniques of mixed development, enclosed shopping malls and sophisticated traffic arrangements. It is essential, for this reason, to be able to combine together numbers of small awkwardly shaped freehold sites of the nineteenth-century town. It is also desirable for land ownership to be in as large units as possible to balance the more profitable uses of land against unremunerative ones such as open space. The local authority also meets certain basic costs apart from land acquisition, such as site clearance and the provision of roads and services and execution of other public improvements.

The private developer often has a vital role to play with regard to availability of capital, knowledge of the market and ability to exploit commercial opportunities. To be successful the development must be in the right location with adequate and readily accessible parking space, satisfactory service access, correct amount of retail space for purchasing power of area and reasonable rent levels. An enclosed air conditioned shopping centre needs a minimum supporting population of 50 000 to 70 000, but trading in such a centre is likely to exceed that in a comparable open precinct by about twenty per cent. Many market towns do have extensive catchment areas for shopping purposes (Seeley, 1974).

It is necessary to prepare a developers' brief, probably prepared by a consultant, listing the essential feature of the scheme as a basis for competitive tendering or negotiation with developers. The brief would probably include such matters as the objectives of the scheme; introductory comments on the town and its region; planning proposals; plans for the centre; site plan; number and types of unit to be provided; special requirements such as pedestrian and vehicular access, parking provision, siting, height and architectural quality of buildings and landscaping; phasing; drawings required; disposal terms; and obligations with regard to displaced traders. It is essential that the local authority should retain the initiative in planning and guiding the redevelopment.

In disposing of land for development, the local authority usually has two principal aims in view. One is to secure the satisfactory development of the site in the interests of the public it represents, which in many cases will involve a major civic improvement and the solution of such problems as traffic congestion, car parking, lack of open space and other public amenities. The other is to obtain the maximum financial return from the developers, partly in fulfilment of the local authority's duty to the local taxpayers and partly in order to recover its own capital outlay in buying the land and constructing public services in connection with the development. These two aims tend to conflict and the solution will often involve a balance between them. It is essential to consider and reconcile the aesthetic and economic circumstances of each case (Seeley, 1995a).

Further details of the likely procedural and financial arrangements in implementing partnership schemes for town centre redevelopment are given in *Building Economics* (Seeley, 1995a). Quantity surveyors have a valuable role to play in urban renewal and need to be well versed in the technical, social and financial aspects and the importance of achieving high quality environmental standards and user friendly buildings of good design. Couch (1990) provides good background reading on urban renewal.

Quantity surveyors are also playing a useful role in the refurbishment and conversion of buildings and in 1982 the RICS issued a useful guide to quantity surveying documentation for refurbishment and alteration work. On the wider front there is an increasing need for the regeneration of urban drainage and water supply systems, improving road networks and infrastructure generally preparatory to commercial, industrial and residential development and redevelopment.

## BUILDING CONSERVATION

### Conservation Areas: General Background

Many parts of built up areas of the UK have been designated as Conservation Areas, and in 1993 there were no less than 7000 such areas designated in England (4 per cent of the built fabric). They are described in the Town and Country Planning Act 1990, section 277 as 'areas of special architectural or historic interest, the character or appearance of which it is desirable to preserve or enhance'.

Taylor (1991) has described how apart from the obvious central areas of historic towns and cities, conservation areas sometimes include substantial areas of mundane buildings with little architectural or civic design interest, and undeveloped areas on the outskirts of towns and villages. For example, the Hertford Conservation Area encompasses much of the town and includes numerous post-war buildings of little, if any, architectural design interest.

A RTPI Report in 1993 concluded that too many conservation areas lacked proper management and funding and that many local authorities failed to develop their aims for conservation areas, or to communicate these aims to local residents, businesses and developers.

## Conservation and Reuse of Redundant Buildings

DoE (1987) have described how in the 1980s rehabilitating old houses and the reuse of old industrial and commercial buildings, instead of demolishing them and redeveloping the sites, became widespread. There are numerous large old buildings in the UK which are no longer needed for their original purposes, especially in the older towns and cities. They frequently have more character than the modern buildings which would replace them, but if left empty they soon decay and blight the area around them, and it is often cheaper to adapt an existing building than to erect a new one although it may create more problems. In the late 1980s and early 1990s, interest grew in the conversion and reuse of old buildings, not only amongst those who wish to conserve our heritage, but also those who wish to bring back economic life into run-down areas or just to carry out their own innovative projects, as in the Lace Market area of Nottingham. This form of adaptive reuse has an important role to play in urban regeneration.

The DoE Handbook on reusing redundant buildings (1987) aptly explains what is involved in devising and implementing successful reuse schemes and provides practical information on the organisational and management aspects. It also contains 14 case studies describing how different organisations from the public, private and voluntary sectors carried out successful conversion schemes in a wide variety of circumstances, and embraces the conversion of old factories, warehouses, mills, railway stations and other buildings. There was a wide range of new uses to which 400 converted buildings in the UK had been put in the mid-1980s, encompassing workspaces (44.5 per cent), leisure/retail (32.0 per cent), housing (9.25 per cent) and mixed use (14.25 per cent). In conservation areas, particular emphasis needs to be placed on scale and the use of local materials.

The following informative and well illustrated books cover the rehabilitation and reuse of old buildings: *Saving Old Buildings* (Cantaouzino and Brandt, 1980), *The Housing Rehabilitation Handbook* (Benson *et al.*, 1980) and *The Conservation Handbook* (PSA, 1988).

## Listed Buildings

A Nottinghamshire County Council survey in 1993 revealed that one in twenty of the county's listed buildings were at risk and many were in danger of collapse.

The rate of decline of some listed properties had raised fears about the unscrupulous development of their sites. Some developers could allow buildings to fall into disrepair in the hope that they may be demolished, leaving the sites available for redevelopment.

A research report published by the RICS/English Heritage (EH) in 1993 showed that between 1980 and 1992, listed buildings had surprisingly out-performed other categories of building investments.

## A Town Conservation Case Study

The Civic Trust Regeneration Unit has been involved in the preparation of numerous urban strategies on behalf of local authorities throughout the UK in recent years, each with a strong emphasis on conservation aspects. Readers wishing to pursue this topic in more detail are referred to Just and Williams *Urban Regeneration: A Practical Guide* (1997). I have selected the small Welsh town of Llanidloes because of its great character and history and the very discerning way in which the project was approached by the Civic Trust for Wales and the Civic Trust Regeneration Unit (1992), with full regard to the conservation implications.

### Implementation of a Town Scheme

Conservation area status, which embraces much of the town and which the Civic Trust recommended should be extended, provides the local authority with additional planning powers to control new developments and to preserve the existing buildings. It was considered that Llanidloes would benefit from a more positive Town Scheme to assist in the effective conservation of the town's architectural heritage by providing grant aid to owners. Town Scheme grants (up to 50 per cent of the cost of eligible works) would be offered from a fund provided on an annual basis by Welsh Historic Monuments (Cadw) and the local authorities to help with the cost of restoring and repairing buildings of architectural or historic merit. Eligible works include structural repairs to the external fabric of the building; redecoration arising from structural repairs; and fees for professional advisors. In England, Town Scheme grants are also available for internal structural repairs and it was hoped that this policy could operate in Llanidloes.

### 'Use of care and repair' grant scheme

Damage to the quality of the built environment is not restricted to disrepair. It can and does often arise from the ill-considered replacement of traditional features such as windows and doors by modern, inappropriate fittings. The cumulative effect of such replacement is to erode local quality and to undermine the value of the careful restoration work applied to selected buildings. It therefore seems appropriate to encourage householders to follow good practice and maintain original features in addition to enforcing planning restrictions.

The Civic Trust therefore wisely proposed that the District Council establish a 'care and repair' scheme as operated by a number of local authorities, sometimes in collaboration with housing associations, as in the case of Newark and Sherwood District Council and Nottingham Community Housing Association. Grants are made available to cover up to 40 per cent of the costs of repair or replacement of windows and doors to properties within the conservation area, provided that the work is carried out to the satisfaction of the local authority's staff. The Civic Trust rightly believed it preferable to repair rather than replace wherever possible.

*Design guidelines*

Even with strong planning controls and a positive grant scheme for repairing buildings, problems can still arise. A new shopfront that is in discord with its neighbours; a quantity of new PVCU windows replacing traditional timber sashes; a traffic scheme requiring a batch of new traffic signs – all can mar local character just as much as straightforward neglect and decay. Hence there exists a vital need to establish clear design guidelines against which all new development can be judged. The Civic Trust described how good design is relatively easy to illustrate but extremely difficult to define.

Design guidelines incorporate the following objectives:

- *Retention of existing buildings and features* as part of a strategy of sustainable development to maintain and recycle old buildings where possible.
- *Simplicity and restraint* are major facets of new schemes with particular attention paid to scale, massing, roof line, materials and finish.
- *Attention to detail*, particularly when replacing shopfronts and domestic doors and windows.
- *Integrating new buildings into their surroundings* and for new industrial estates and housing developments, this requires careful selection of roofing and the use of landscaping to soften the impact of the buildings or conceal them.
- *Aiming for quality* with regard to spaces beside and between buildings (Civic Trust, 1992).

## ENERGY CONSERVATION

### Energy Audits and Surveys

Energy conservation became one of the most important factors in life cycle costing in the late 1980s and 1990s, as a means of improving long term economies and efficiency, and hence is of vital importance to the quantity surveyor. In this context, audits and surveys of energy use aim to identify opportunities for cost effective savings, and when combined with performance monitoring and targeting methods, they provide the information needed to ensure effective energy management. An Energy Efficiency Office (EEO) study of several thousand energy surveys showed an average potential saving of about 20 per cent of each property's energy bill, with an average payback period for implementing the recommended measures of 18 months (Field, 1992).

Energy audits establish the quantity, cost and end use of each form of energy input to a building or site over a specific period. A preliminary audit can assess energy use from fuel bills and meter readings, but a site survey is required to ascertain how the energy is used and how savings can be made.

Instrumentation is usually required to measure room temperatures, flow temperatures, combustion gas composition and electrical loads, but the cost of such surveys can be prohibitively expensive for smaller buildings. When carrying out preliminary audits, comparison with published figures for good, fair and poor levels of energy use, as listed in table 10.1 can be used to assess the performance of different buildings (RICS, 1993b). Accountability and responsibility for energy use at all levels needs reviewing from overall financial control to individual good housekeeping.

A survey will identify various options for investment and energy savings. Savings are usually based on the estimated percentage of annual consumption and direct saving from a reduction in fuel cost, load, operating hours or energy loss. A financial appraisal of the economic benefits of each alternative is then needed to determine the optimum investment programme. Finally, a plan to implement the selected options should be prepared and implemented (Field, 1992).

Further detailed examination of the use of energy surveys, the appraisal of the survey findings and the subsequent investment appraisal is provided in *Energy Appraisal of Existing Buildings* (RICS, 1993b). The RICS recommend that simple unadjusted payback periods (SPB) can be used for short paybacks (<2 years) and medium payback (2–5 years), while for long paybacks (>5 years) a discounted payback (DPB) method should be used, whereby instead of adding up all the net savings, one adds up their present values, as described and illustrated by Seeley (1995a) in *Building Economics*.

## ENERGY COSTS AND SAVINGS

Table 10.2 shows 1992 costs of heat losses, energy conservation work and annual savings for a typical semi-detached house with gas heating and hot water, which highlight the extent of the heat losses in monetary terms in a badly insulated house and the average costs of energy conservation work and the annual savings that can accrue.

Table 10.1  Total energy usage in typical non-domestic buildings

| Type of building | Hrs of use pa | Energy usage per year (kWh/m²) Energy performance | | |
| --- | --- | --- | --- | --- |
| | | good | fair | poor |
| Primary school (no indoor pool) | 1480 | <180 | 180–240 | >240 |
| Primary school (with pool) | 1480 | <230 | 230–310 | >310 |
| Secondary school (no pool) | 1660 | <190 | 190–240 | >240 |
| Secondary school (with pool) | 2000 | <250 | 250–310 | >310 |
| University | 4250 | <325 | 325–355 | >335 |
| Department/chain store (mechanically ventilated) | | <520 | 520–620 | >620 |
| Supermarket/hypermarket (mechanically ventilated) | | <720 | 720–830 | >830 |
| Small food shop – general | | <510 | 510–580 | >580 |
| Library | 2540 | <200 | 200–280 | >280 |
| Church | 3000 | <90 | 90–170 | >170 |
| Hotel (medium size) | | <310 | 310–420 | >420 |
| Bank or post office | 2200 | <180 | 180–240 | >240 |
| Theatre | | <600 | 600–900 | >900 |
| Crown and county court | 2400 | <220 | 220–300 | >300 |
| Factory (small) | | <230 | 230–300 | >300 |
| Factory (large with heat gains from manufacturing plant) | | <210 | 210–300 | >300 |
| Warehouse (heated) | | <150 | 150–270 | >270 |
| Clinic/health centre | 2600 | <36 | 36–46 | >46 |

*Source*: RICS. Energy Appraisal of Existing Buildings (*1993b*).

Table 10.2   Typical costs of domestic heat losses, energy conservation work and annual savings

**Heat losses**

| | Badly insulated house (£) | Well insulated house (£) |
|---|---|---|
| Roof | 75 | 10 |
| Ventilation and draughts | 65 | 35 |
| Windows | 55 | 25 |
| Walls | 110 | 15 |
| Doors | 5 | 5 |
| Ground floor | 20 | 20 |

**Typical costs and savings**

| | Capital cost (£) | Annual saving (£) |
|---|---|---|
| Hot water cylinder | | |
| 80 mm cylinder jacket and pipe lagging | 10 | 30 |
| Roof insulation | | |
| 100 mm professionally installed | 180 | 65 |
| Cavity wall insulation | | |
| Polystyrene beads | 250 | 300 |
| Mineral fibre | 400 | 95 |
| Solid wall insulation | | |
| Interior insulation | 1600 | 95 |
| Draught stripping | | |
| Windows, doors, floors, etc. (DIY) | 75 | 30 |
| Double glazing | | |
| Simple fixed secondary glazing, living room (DIY) | 80 | 10 |
| Living room only, more elaborate secondary glazing (DIY) | 200 | 10 |
| Living room only, professionally installed | 550 | 10 |
| Whole house, simple fixed secondary glazing | 320 | 30 |
| Whole house, more elaborate secondary glazing (DIY) | 750 | 30 |
| Whole house, professionally installed | 2150 | 30 |
| Reflector foil | | |
| Fixed behind radiators on outside wall | 10 | 5 |

*Source*: Energy Efficiency Office Home Energy Survey.

# 11 Engineering Work

## MECHANICAL AND ELECTRICAL SERVICES ENGINEERING

A RICS report in 1971 concluded that because of the limitations of quantity surveyors with regard to engineering services, coupled with the opinions of services engineers, the employer does not always receive adequate financial advice. Mechanical and electrical services represent a substantial proportion of the cost of most construction contracts. That such a large proportion of the total contract should be let on the basis of outline drawings and specification identified the key area of incompatibility with current requirements for preplanning and effective cost control.

However, a survey carried out by the Quantity Surveying (Engineering Services) Committee of the RICS (Wilshere, 1982) established that the use of bills of quantities for engineering services was more widespread than previously supposed. Furthermore, a growing number of employers in both the public and private sectors are commissioning their quantity surveyors to provide a comprehensive service on engineering services, as well as the architectural and structural work, on construction projects. This is primarily because of the additional advantages in cost control, contract documentation and contractual management. When engineering services form up to 50 per cent or more of the total project value, it is clearly essential for this element to be subject to the same rigorous scrutiny as the various building elements. The 1982 survey indicated that over 80 per cent of the quantity surveyors who responded undertook work in engineering services.

Mayo-Chandler (1980) examined the situation from the contractors' viewpoint and described how for many years quantity surveyors had restricted their activities to building work and how few found it necessary to gain experience of the services industries. He believed that it was inevitable that building services should be billed increasingly, in order to secure that extra degree of expenditure control that the present day bill of quantities affords. The mechanical and electrical contracting industries viewed the early invitations to tender on the basis of bills of quantities with suspicion. They resented having to submit rates for each activity, believing that this information could subsequently be used against them. However, time has shown that bills of quantities have many advantages, and there is now much less reluctance to tender by this method. Few contractors now refuse to insert individual rates, as they once did unless they were awarded the contract.

The more enlightened employers and architects are usually enthusiastic about the use of bills for engineering services, particularly those who appreciate the benefits this has on cost control. Some engineering consultants, on the other hand, do not favour the use of bills. Engineering consultants' and services subcontractors' reactions are influenced considerably by their knowledge and experience of the quantity surveyor's ability as a specialist in this work, and their resistance is usually overcome after working on the first project where a bill has been used. Indeed, the two main service contractors' associations, the Heating and Ventilating Contractors' Association (HVCA) and the Electrical Contractors' Association (ECA), have co-operated with the RICS in the preparation of model bills, one for mechanical services and the other for electrical services.

The author in *Advanced Building Measurement* (1989) described how the quantity surveyor when preparing bills of quantities for engineering services requires a sound knowledge of the underlying technology and the ability to understand the specification and interpret the schematic drawings prepared by the engineering consultant(s). The exact routing of pipework and ductwork is often left to the craft operatives doing the work, and the contractor will have in mind such factors as the location of other services, restrictions on space and the ease of maintenance of the completed work. In essence, the quantity surveyor must put himself in the position of the operative and include in the bill of quantities all items necessary for the complete installation. In like manner, the quantity surveyor will require a detailed knowledge of the IEE regulations for electrical installations, and a sound understanding of circuitry and wiring systems is essential so that trunking, tray and conduit runs can be plotted and the correct number of cables measured. Some 40 per cent of the respondents to the 1982 RICS survey indicated that they modified or simplified the requirements of SMM6 (1979), often drawing upon their overseas experience. The measurement of electrical work was further rationalised in SMM7 (1988).

The nominated subcontract is the most widely used type of contract for engineering services. Subcontract tenders are ideally obtained by quantity surveyors on the basis of bills of quantities and the successful engineering firm is then nominated within the main building contract. Where domestic subcontracts are used, the bills for engineering services are included in the main contract bills and form part of the overall bid. On process engineering schemes the engineering services may be covered by direct contracts.

The 1982 survey showed that there was plenty of scope for increased involvement at the cost planning stage for the benefit of the employer. After the first stage of engineering design, cost estimates are usually prepared on the 'treated space/points' system (services relative to treated space measured in $m^2$ or $m^3$ as appropriate, services to specific outlets by enumerating points and other services by lump sums). At the next design stage, the 'user-equipment' approach is common such as Btu/h. At the final design stage, an approximate measurement system is the most appropriate method measuring the lengths in metres, enumerating where appropriate, and making general allowances for minor items.

In the 1990s there was a move towards low energy, low maintenance systems and the quantity surveyor needs to be aware of these developments. There was also a significant increase in the number of specialist building engineering services quantity surveyors aided by appropriate educational courses, such as those offered by Nottingham Trent University. There is an evident need for greater innovation, rationalisation/simplification, and value engineering and checking with a closer integration of the design, procurement and construction processes, and full consideration of two stage pricing and the use of joint venture or other partnership approach.

## CIVIL ENGINEERING

### Scope of Civil Engineering

Civil engineering is defined as 'the art of directing the great sources of power in nature for the use and convenience of man.' Hence civil engineering works cover a large variety of different projects, some of which are of great magnitude, both in the United Kingdom and overseas. Vast cuttings and embankments; large mass and reinforced concrete structures, such as reservoirs, water and sewage treatment tanks, sea walls, bridges and power station cooling towers; large steel structures; piling for heavy foundations; jetties and wharves; pipelines and tunnels; railway trackwork; motorways and airport runways, all form the subject-matter of civil engineering contracts. These works require considerable skill, ingenuity and technical knowledge in their design and construction, and can involve a considerable element of uncertainty owing to variable ground conditions and extensive temporary works (Seeley, 1993a).

### The Role of the Quantity Surveyor

Quantity surveyors choosing to enter the absorbing and challenging civil engineering field must have a sound knowledge of civil engineering construction and procedures, and should seek to provide services similar to those that the quantity surveyor traditionally gives on building projects. Many are employed with firms of consulting civil engineers, government departments, local authorities and privatised industries in a design team advising the engineer. Many others provide a postcontract, or more correctly post tender, service to the contractor and they may be in his direct employment.

### Civil Engineering Bills of Quantities

*The Civil Engineering Standard Method of Measurement* (CESMM3) 1991 defines a bill of quantities as a list of items giving brief identifying descriptions and estimated quantities of the work comprised on a contract. An explanation of its contents and application is given by the author in *Civil Engineering Quantities* (1993a).

The valuation of variations and securing agreement to the cost of delays have generally been the main causes of dispute on civil engineering contracts, stemming largely from variations in design and unforeseen physical conditions and artificial obstructions. Furthermore, many of the costs arising from civil engineering operations are not proportional to the quantity of the resulting permanent work. *The Civil Engineering Standard Method of Measurement* (CESMM3) 1991, in accordance with which many civil engineering bills of quantities are prepared, endeavours to remove these inconsistencies by the introduction of method-related charges, which are of two basic types: time-related charges and fixed charges. For example, the cost of bringing an item of plant on to the site and its subsequent removal is a fixed charge and its running cost is a time-related charge.

The contractor can enter and price such costs as he considers he cannot recover through measured rates, such as site accommodation, site services, plant, temporary works, supervision and labour items – all are at the tenderer's discretion. Accepting that expertise in design rests with the engineer, it seems equally evident that expertise in construction methods lies with the contractor. It is logical that the contractor should be able to decide the method of carrying out the works. A blank section in the bill of quantities permits him to list, describe and price these items (Seeley, 1993a).

### Quantity Surveyors in the Design Team

Sneden (1979) aptly described how quantity surveyors acting in the design team advising the engineer are likely to be required at the precontract stage to assist in the provision of a financial budget prior to detailed design, to undertake preliminary cost studies, to monitor and evaluate the design as it evolves and to advise on contractual arrangements and assist with the preparation of contract documents, particularly the bill of quantities.

Civil engineering projects require a design solution to physical and geological problems. These problems will largely determine the cost of the solution and the engineer will not be able to provide an acceptable one within the parameters of a predetermined budget in the same way that buildings can be cost planned within agreed cost limits.

The quantity surveyor is an expert in cost appraisal, whereas this can represent only a small part of the education and training of engineers. The employer (promoter of civil engineering works) has to know the likely cost of the project in advance of construction and the quantity surveyor's experience can be usefully employed to provide the engineer with comparative costs of alternative solutions. Subsequently, the quantity surveyor may be requested to analyse and report on tenders and, in the postcontract period, to evaluate the final cost, including remeasurement, variations and assistance in claims adjudication (Sneden, 1979).

## Contractors' Quantity Surveyors

The quantity surveyor employed by the contractor will aim to secure maximum payment for the work done at the earliest possible time to avoid any possible cash flow problems. This has however to be undertaken within the provisions of the contract and the contractor cannot receive more than his contractual entitlement. The quantity surveyor will also be required to assess the cost of alternative designs of temporary works and other operations, to prepare cost and value reconciliations for internal purposes and to forecast trends. He may be responsible for negotiating and managing subcontracts and agreeing their final accounts. He can also perform a useful role in interpreting and progressing contractual issues and settlements. From the start of the contract he will be assembling data for the final account, to ensure that it will be complete in all its aspects and to give the contractor his proper entitlement under the contract. On a project of £10m or more, the project quantity surveyor is likely to have a staff of five or more quantity surveyors to assist him in his duties, as in these commercially high risk situations, civil engineering contractors have sought progressively more help from the quantity surveyor to safeguard, control and maximise the recovery of income.

## Quantity Surveyors' Involvement

Quantity surveyors are involved in many different types of civil engineering project. For example, the Mass Transit Railway Corporation in Hong Kong employed its own quantity surveying team for the Tsuen Wan extension and also engaged chartered quantity surveyors as consultants to give enhanced control of the valuation of variations and of specialist items.

As the capital investment programmes of water services companies increased in the 1990s, the need arose for independent quantity surveyors to augment in-house establishments in some cases. When making bids for quantity surveyors' services, firms are generally required to complete appraisal forms, showing that they are financially sound, have a demonstrable quality management system in place, appropriate staffing levels, have past experience of similar projects and preferably have an office located relatively close to the project. Furthermore, an EC Services Directive requires water companies to advertise this category of work in the EC Journal in order to generate an approved list.

It is considered beneficial to examine the principal types of project in the civil engineering field in which major quantity surveying practices were contributing their services in 1995. In this context, Currie & Brown and E C Harris have kindly provided information to set the scene.

*Currie & Brown:*
Water supply and treatment (£5m–£80m)
Refineries and gas treatment (10m–£15m)
Roadworks (£150m)
Power (£100m)
Nuclear power (over £100m)
Nuclear environmental (multi-million pounds)
Rail (over £100m)
Telecoms (£100m)
Ports and marine work (over £200m)
Airport terminal (over £100m)
Manufacturing plant (over £100m)

*E C Harris:*
Water, sewage treatment and sewerage schemes, including London water ring main, impounding and service reservoirs, dams, pipelines and tunnelling in both hard and soft ground
Power generation comprising fossil fuelled and nuclear power stations, hydroelectric schemes, natural gas installations, waste recycling, incineration and area heating projects.
Land reclamation and decontamination of heavily polluted industrial land and dockland sites for redevelopment, hazardous materials disposal and nuclear decommissioning
Trunk roads and motorways reconstruction, widening and improvement, railways, airports and bridge works.

The range of services which quantity surveyors can offer is very extensive and usually includes the following activities:

- Feasibility studies
- Procurement procedures
- Budget estimating
- Cost/design comparisons
- Contractual procedures
- Value advice
- Bills of quantities preparation
- Tender evaluation and reporting
- Site remeasurement and admeasurement
- Rate negotiation
- Checking of monthly payments and final accounts
- Financial reporting and cash flow projections
- Claims assessment and negotiations
- Employers' audit procedures
- Advice on and participation in arbitration

**Quantity Surveyor's Office Organisational Arrangements**

Currie & Brown have described how these arrangements depend on the size, location and complexity of the civil engineering project, ranging from a small precontract team based in one of the firm's offices under the direction of a partner to a development project group with a project partner as team leader with associates and task groups. Postcontract organisation is normally set up under a project associate or senior surveyor with a task group designed to reflect the amount of work or the work areas being controlled.

Currie & Brown normally recruit graduates for this class of work who are then given in-house training and appropriate site training. The main problem encountered in practice is the lack of contractual awareness and a lack of appreciation of cost control by project engineers.

**Contract Administration**

Civil engineering contracts are generally administered in accordance with the ICE Conditions of Contract. The employer requires value for money and the contractor needs to make a profit. To achieve these objectives, a professional approach is needed and the quantity surveyor can perform a useful role in this connection. It is desirable that the contractor receives justifiable payments promptly and, wherever practicable, justifiable and fully substantiated claims should be settled during the course of the contract to assist the contractor's cashflow.

The programme and progress chart constitutes an important document on a civil engineering contract and is usually in the form of a bar chart prepared by the resident engineer based on the contractor's programme. This chart is important as it forms an accurate visual record of progress which can contribute towards expeditious completion of the work by showing up deficiencies and can provide valuable information in the event of a request for an extension of time or a claim by the contractor. Figure 11.1 shows a typical programme and progress chart for a road contract.

A typical example of an interim valuation for a road contract is illustrated in table 11.1 to show the general format and approach, starting with the adjustment of preliminaries and general items and followed by the various work sections, nominated subcontractors, retention and materials, less the amount previously certified to give the amount due to the contractor under this valuation. The contract period is 20 months.

**Site Investigation**

The quantity surveyor must be aware of the importance of adequate site investigation to reduce risks in the earliest stages of a civil engineering project, and to appreciate that in the absence of a site investigation ground is a hazard. Furthermore, inadequate site investigation is likely to result in escalating costs and late completions.

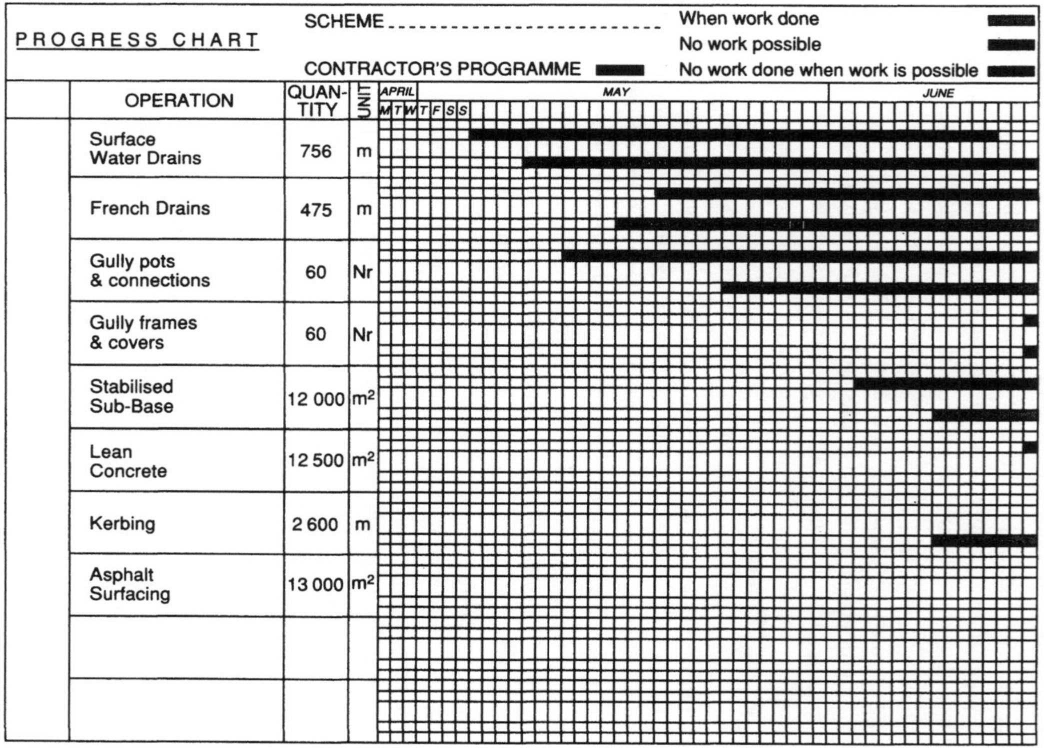

PROGRESS CHART

SCHEME

CONTRACTOR'S PROGRAMME

When work done
No work possible
No work done when work is possible

| OPERATION | QUANTITY | | APRIL M T W T F S S | MAY | JUNE |
|---|---|---|---|---|---|
| Surface Water Drains | 756 | m | | | |
| French Drains | 475 | m | | | |
| Gully pots & connections | 60 | Nr | | | |
| Gully frames & covers | 60 | Nr | | | |
| Stabilised Sub-Base | 12 000 | m² | | | |
| Lean Concrete | 12 500 | m² | | | |
| Kerbing | 2 600 | m | | | |
| Asphalt Surfacing | 13 000 | m² | | | |

**Figure 11.1** Programme and progress chart (Lee and Hayles, 1989)

Site investigations normally comprise reference to existing geological maps and other data, and soil surveys by specialists who provide borehole logs, samples, laboratory test results and a report by a person experienced in geotechnical matters, and possibly resistivity and seismographic surveys. In addition, other matters requiring investigation may include access to the site, any relevant statutory restrictions, rights of adjoining owners, availability and capacity of existing services, availability of construction materials, ease of disposal of surplus materials, groundwater level, and liability to flooding and subsidence (Seeley, 1993b).

**Land Reclamation**

Contaminated land is a subject of increasing relevance and importance to civil engineers and their quantity surveying consultants as the need to reclaim and recycle land increases, accompanied by the growing pressure on land resources and greater environmental awareness.

Contamination involves many issues, both technical and non-technical, and encompasses the use of experts in several different disciplines. In addition to civil and geotechnical engineering, expertise is required in chemistry, environmental science, geology, hydrogeology and project management, and may be needed in other areas such as chemical or process engineering, ecology, biology, valuation and other financial services and legislation.

Contamination can affect the ground, water or air and can occur through the presence of chemical substances, either in solid or liquid form, and from noxious and hazardous gases. The ICE Guide on Contaminated Land (1994) provides guidance on the investigation and assessment of such land and the appropriate remedial measures.

Abdul (1993) has described reclamation works on part of Trafford Park Development Corporation's site at Irlam, Greater Manchester, which was formerly occupied by British Steel. An environmental subconsultant was appointed to advise the consulting engineer on specific matters relating to contaminants within the site. The brief included the assessment of the potential hazards posed by contaminants within the site and the production of a decontamination report.

Of special interest to quantity surveyors was the need for accurate measurement and classification of contaminated materials excavated from the site, as there were significant differences in costs for handling the different materials. For instance, the cost of disposal of material classified as U2A (special waste under Control of Pollution (Special Waste) Regulations 1980) was about six times that of material classified as U2C (material other than U2A or U2B within a depth of 0.8 m below existing ground level)

The datum used for the final measurement was a series of existing ground levels taken after fly-tipped materials had been removed. The measurement ranged from quantifying small blocks of concrete with a tape-measure

Table 11.1   Interim valuation

---

BROADSTONE BYPASS CONTRACT
Contractor: Road Construction Ltd

*Valuation for Certificate nr 8: 15 December 1993*

| | £ | £ |
|---|---:|---:|
| Bill nr 1 Preliminaries and General Items | | |
| Performance bond | 24 000 | |
| Insurances | 122 000 | |
| Offices for engineer's staff – establishment | 6 000 | |
| Ditto – maintenance, 8/20 × £12 000 | 4 800 | |
| Attendance upon engineer's staff, | | |
| 8/20 × £15 000 | 6 000 | |
| Testing of materials | 5 000 | |
| Traffic regulation – establishment | 7 000 | |
| Ditto – operation and maintenance, | | |
| 8/20 × £22 000 | 8 800 | |
| Pumping plant – establishment | 5 400 | |
| Ditto – operation and maintenance | 6 700 | |
| Ditto – standing by | 3 300 | |
| Site accommodation – establishment | 15 000 | |
| Ditto – maintenance, 8/20 × £16 000 | 6 400 | |
| Concrete mixing plant – establishment | 16 000 | |
| Ditto – operation and maintenance, | | |
| 8/20 × £20 000 | 8 000 | |
| Hardstandings | 9 500 | 253 900 |
| | | |
| Bill nr 2 Site Clearance | | |
| Bill total | 51 000 | |
| Removal of 15 nr tree stumps, 0.5–1.0 m | | |
| diam., @£28 | 420 | 51 420 |
| | | |
| Bill nr 3 Earthworks | | |
| General excavation of material for disposal, | | |
| maximum depth 0.5–1 m, | | |
| 19 500 m$^3$ @£6 | 117 000 | |
| Ditto-maximum depth 1–2 m, 13 600 m$^3$ | | |
| @£6.50 | 88 400 | |
| General excavation of material for reuse, | | |
| maximum depth 0.5–1 m, | | |
| 15 600 m$^3$ @£4 | 62 400 | |
| General excavation of topsoil for reuse, | | |
| maximum depth not exceeding 0.25 m, | | |
| 7 200 m$^3$ @£3 | 21 600 | |
| Trimming of slopes, 6 600 m$^3$ @£0.50 | 3 300 | |
| Filling and compacting 150 mm of excavated | | |
| topsoil to slopes, 6 600 m$^2$ @£0.50 | 3 300 | |
| Imported hardcore, 15 000 m$^3$ @£8 | 120 000 | 416 000 |

Bill nr 4 Carriageway
  Granular base, 150 mm deep, 50 000 m²
    @£3.50                                             175 000
  Concrete carriageway slab, 225 mm deep,
    45 000 m² @£15                                     675 000
  Steel fabric reinforcement, 45 000 m² @£2.50         112 500
  Expansion joints, 20 000 m @£15                      300 000
  Precast concrete kerb, 14 000 m @£10                 140 000         1 402 500

Bill nr 5 Footpaths
  Granular base, 75 mm deep, 1600 m² @£2                 3 200
  Bituminous macadam basecourse, 50 mm
    deep, 1200 m² @£4.00                                 4 800
  Bituminous macadam wearing course,
    10 mm deep, 1200 m² @£1.50                           1 800           9 800

Bill nr 6 Bridges
  Bridge 1
  Bill total £230 000 – 20 per cent complete                            46 000

Bill nr 7 Culverts
  Bill total £670 000 – 25 per cent complete                           167 500

Bill nr 8 Surface Water Drainage
  Bill total £240 000 – 40 per cent complete                            96 000

Bill nr 9 Retaining Walls
  Bill total £94 000 – 20 per cent complete                             18 800

Bill nr 10 Fencing
  Chain link fencing, 900 m @£15                        13 500
  Wood post and rail fencing, 480 m @£12                 5 760
  Chestnut pale fencing, 660 m @£8                       5 280          24 540

Bill nr 11 Dayworks                                                     14 400

Variation nrs 1–13                                                      12 670

Nominated subcontractor:
Electrics Ltd – street lighting                        78 000
Add for profit, 5 per cent                              3 900
Attendance                                              6 200           88 100

                                                                     2 601 630
*Less* retention (3 per cent)                                           78 049
                                                                     2 523 581
Materials on site                                                       77 419
                                                                     2 601 000
*Less* total of certificates 1–7                                     2 196 000
Total amount due                                                    £ 405 000

---

*Source*: Seeley (1993b)

to carrying out extensive tacheometric surveys of large excavations. A total of approximately 102 100 m³ of unacceptable material was excavated from the site; of which 28 per cent was U2A and U2B materials. The final cost of the reclamation works contract, including the construction of two access roads, was £3.3m (Abdul, 1993).

## Earthworks

The majority of civil engineering projects entail a substantial proportion of earthworks and the annual value of earthworks claims is estimated to be in the order of millions of pounds per annum in the UK alone. For example, a typical major motorway construction contract could be of the value of £20–£60m over a 24-month period, of which the earthworks component could amount to as much as £10m. Although claims and disputes arise on civil engineering contracts for a variety of reasons, one of the areas most prone to disputes is that of earthworks.

Despite the major advances made in specification and contractual terms in recent times, there are still many occasions where genuine disputes can arise as a result of ground conditions which an experienced contractor could not reasonably have foreseen.

## Tunnelling

Tunnelling is notoriously prone to delays and cost overruns, as evidenced by the problems encountered on the £1.9b Jubilee Line Extension contract. Another example is Mowlem's tunnelling contract on the ill-fated Carsington reservoir scheme in Derbyshire in the early 1980s, where the contractor was required to drive through rock to form an aqueduct to abstract water from the River Derwent and feed it to Carsington reservoir. Unfortunately, the geological conditions bore little resemblance to those contained in the contract documents, resulting in the project finishing substantially over time and over budget.

## HEAVY AND PROCESS ENGINEERING

The RICS SITE report (1980) identified an extension of the quantity surveyor's traditional role by applying his skills to oil rigs and platforms, petrochemical plants, mining engineering, railway engineering and other technologies. Another RICS report in 1983 referred to the underutilisation of the quantity surveyor's role in heavy engineering. There have been a number of published papers describing the nature and scope of quantity surveyors' activities on a variety of very large and challenging projects.

## Scope of Heavy and Process Engineering

Flint (1979) listed examples of heavy engineering, such as steelworks, cement plants, sugar plants, offshore steel platforms, shipyards, aluminium and other smelter centres, energy centres and oil and petrochemical projects. He distinguishes oil and petrochemical projects, which are usually built for employers who have considerable engineering design facilities, together with their own systems of commercial controls and special conditions of contract, whereas the other types of project normally operate on a more traditional basis employing consulting engineers and quantity surveyors.

Fish (1981) analysed the main differences between industrial or process engineering and conventional construction projects:

(1) the overall scope of industrial engineering projects is generally much greater, both in financial terms and construction times;
(2) the construction process can embrace many forms of engineering, with civil engineering representing a comparatively small amount and conventional building works only a very small fraction of the overall capital value;
(3) because process engineering in any industry is always in a situation of continuing development and the construction period is very long, it is inevitable that construction work must commence long before the final design of the process plant is anywhere near complete, and so tenders must be sought on extremely limited design information; and
(4) the quantity surveying discipline is not one that is automatically recognised and accepted, but the use of quantity surveyors in this class of work at home and overseas is becoming more widespread.

## Quantity Surveyor's Organisational Arrangements and Activities

Currie & Brown kindly supplied details of a typical quantity surveying organisation for heavy and industrial engineering projects in which their practice was involved. They normally comprise a small base office team under the direction of a partner or project associate, with a project team at either the contractor or client establishment under a project associate or senior quantity surveyor. The size of the team will depend on the number of disciplines involved. The postcontract team on site operates under a project associate and the same size criteria apply.

The principal activities undertaken by the quantity surveying personnel are preparation of contract documents, tender analysis, evaluation, negotiation, postcontract administration, cost engineering valuation, variation cost control and settlement of claims.

## Types of Contract

The choice of contractual arrangement is largely dependent on the extent to which the works have been defined at the tender stage and includes the following.

*Measure and value contracts* based on:

(1) approximate bill of quantities;
(2) schedule of rates.

*Lump sum contracts* based on:

(1) performance specification;
(2) specification and drawings.

*Cost reimbursable contracts:*

(1) target cost;
(2) prime cost plus management fee.

All these arrangements can include a two-stage tendering procedure to facilitate an earlier contract start (Fish, 1981).

## The Role of the Quantity Surveyor

The quantity surveyor must be prepared to adopt a different attitude to costs, measurement and contract. He needs a flexible mind capable of assimilating a variety of conditions of contract, methods of measurement and methods of payment, and the likelihood of becoming involved in a much wider range of activities than is normally handled by the quantity surveyor. The quantity surveyor, whether acting for the employer or the managing contractor is likely to take part in initial feasibility and cost studies, followed by giving advice on preferred contractual methods, the preparation of invitations to tender, scrutiny of contractor prequalification documentation, assessment of contractors' tenders, setting up and monitoring cost control systems which are often computerised, regular comparison between achieved and budget cost, and supervision of physical progress, value of commitments, cash flow and financial requirements. All these functions are additional to the conventional quantity surveying work of measurement and certification. Contractors in the United Kingdom and overseas are coming to rely increasingly on the services of quantity surveyors (Flint, 1979).

## Quantity Surveyor's Role in the Offshore Industry

Davidson (1994) has aptly described how the typical role of the quantity surveyor in the offshore industry can be subdivided logically into the project phase and the operations phase. During the project phase an offshore structure is designed, constructed, installed, hooked up and commissioned. While during the operations phase the offshore structure, producing hydrocarbons, is operated, maintained and possibly modified.

*1. Project phase*

The quantity surveyor's role in the project phase has largely revolved around the preparation of approximate bills of quantities and schedules of rates, with several quantity surveying firms specialising in this class of work. From the mid 1990s onwards there will be relatively few new major construction projects in the North Sea and the traditional quantity surveying approach is likely to be less popular, offering new challenges to the quantity surveyor.

In a project to develop an offshore oil and gas field, the contracts for the construction of the offshore installation forms only a part of the whole project. Other contracts to be administered include contracts for design, offshore towing and installation, drilling, seismic surveys, catering, supply boats and helicopters. All these contracts require the preparation of invitations to tender, evaluation and award, and subsequent administration up to completion. Hence the operator appoints a contracts specialist who is often a quantity surveyor (Davidson, 1994).

Cost control of the whole project is a vital function and hence cost engineers or cost managers, who may be quantity surveyors, are appointed to monitor and report actual against planned cost expenditure, analyse historical cost data and predict future expenditure. There is usually a demand for claims specialists as the contracts can often be complex and diverse and may be let before the precise scope of the work has been fully formulated, and this is another area where quantity surveyors have much to offer, with their training in law of contract and constructional methods and their enquiring and analytical approach.

*2. Operations phase*

When the structure has been installed and commissioned it will be handed over to the operations group, who will be responsible for production and maintenance for the life of the field which could be between 15 and 50 years. Quantity surveyors are frequently engaged during this period providing contract strategy advice, formulating terms and conditions of contracts and invitations to tender, carrying out evaluation and contractual negotiations and administering contracts for the multiplicity of services required to maintain and operate an offshore production platform, and performing cost engineering services.

These contracts can encompass catering, helicopters, supply vessels, standby vessels, fabric maintenance, general maintenance, drilling services, engineering services and modifications (Davidson, 1994).

## Practical Applications

Brief particulars of a few projects will assist the reader in appreciating the immense scale and problems associated with this class of work. A contract at Milford Haven involved the erection of about 10 000 tonnes of pipework and other equipment, costing about £18m, excluding the value of the actual pipes and equipment. The number of bill items was limited to 27, of which two accounted for more than 60 per cent of the total cost. Measurement of the work was made by weight with the weights of the component parts obtained from computer printouts (Flint, 1979).

In a North Sea oil platform project, the work was subdivided into two main contracts – the concrete platform structure and the interconnecting pipeworks, electrics and instrumentation. With the latter contract, because of lack of complete information and the unpredictable conditions associated with working in the North Sea, payment was made on a cost plus basis. The quantity surveyor was involved with strict documentation of costs of all kinds and comparisons between budget and achieved cost. Operating costs can amount to as much as ten times those of similar work onshore. In the future more constructional work on the seabed will be carried out by remote control and quantity surveyors will be required to measure the lengths and depths of underwater work and to assess obstructions caused by different seabed conditions to arrive at an agreed cost. The quantity surveyor should endeavour to devise a fair method of measuring and describing conditions so that competitive prices can be obtained (Marshall, 1983).

Finally, Jelley and Povall (1979) have described the contractual arrangements for the North Sea Oil Terminal at Sullom Voe in the Shetlands, involving over 30 oil companies. British Petroleum was appointed to act as manager with the title of constructor. The terminal was split into two main areas – the process facilities and the offsites, and a managing contractor was appointed for each. The managing contractors were responsible for the preparation of the contract documents, sending out tender enquiries, and the assessment and subsequent recommendations with regard to tenders, employing a quantity surveyor to undertake the normal precontract work. At the postcontract stage the constructor appointed a quantity surveyor who remeasured and valued on the basis of the bills of approximate quantities and carried out a variety of functions including estimating and cost forecasting.

**Future Developments in the UK Offshore Industry**

International economics have necessitated a major review of offshore operations in the United Kingdom continental shelf (UKCS). The main developments can be summarised as follows:

- Fields yet to be developed are likely to be more marginal.
- Rates of return are likely to be substantially lower than before.
- The use of large expensive one-off platforms is unlikely to be economical.
- Many future UKCS oil and gas fields are likely to use submarine technology, tied back to mother platforms, and proprietary and standardised units.
- Maintenance costs of existing platforms will increase with age.
- Higher expenditure will be incurred to satisfy safety and environmental requirements.
- Operating costs are likely to increase faster than inflation.
- Operators and contractors will face increasing competition (Davidson, 1994).

To overcome these problems, various innovative approaches have been introduced, of which the principal ones are partnering and alliancing.

Partnering, which will be further considered in chapter 19, is a strategy whereby the contracting parties work in parallel to achieve the same goal, with each party contributing its particular strengths, and the avoidance of duplication of effort, to secure maximum efficiency. Partnering should ideally be relatively long term (three to five years) and aimed at reducing production costs.

Alliancing is a contracting strategy which is well suited to the project phase and comprises a single integrated team formed by the contributing contractors and operator.

**Cost Management in Engineering Construction Projects**

RICS Guidance Notes (1992b) assert that the poor performance of cost management in the engineering construction industry can be traced to the failure to implement thorough and systematic procedures, through formalised cost control undertaken by a cost manager. The RICS believes that the independent cost manager should be a qualified surveyor or engineer with specialist experience in cost control and management.

The Guidance Notes are intended for use by consultants employed by the promoters, the promoters themselves and engineering contractors. The basic theme throughout is accuracy and clarity. They give sound advice on the appointment of the cost manager and his functions and general duties, and the operation of cost management procedures throughout the conception and feasibility stage, design and precontract stage, procurement and postcontract stage, commissioning, use and maintenance stage, and finally the decommissioning and disposal stage. They detail the contents of project cost reports monitoring financial progress on each contract. The numerous sources of risks and uncertainties are identified and classified with provision for including their effect on programme and cost.

# 12 Project Management

## NATURE OF PROJECT MANAGEMENT

Waterhouse (1995) has aptly described how confusion often arises, particularly in the minds of clients, as to the precise meaning of project management. For many years the contractor's site agent has often been referred to as the project manager as he manages the project on site on behalf of the contractor. Clients may also appoint a professional to coordinate some of the stages of the project and this person could more properly be called a project coordinator, whereas the project manager, as understood by the Association of Project Managers (APM) and the RICS, is a single person or organisation that manages the complete process.

One very sound definition of project management is 'the overall planning, control and coordination of a project from inception to completion, aimed at meeting a client's requirements and ensuring completion on time, within the cost limit and of the required quality'. This definition clearly establishes the time frame, although even this can be extended to include the maintenance and operation of the asset throughout its working life and possibly the disposal of the asset at the end of its effective life. It also encompasses the client's essential requirements of time, cost and quality to achieve the correct balance. His requirements will also include function as the finished project must fulfil the use requirements efficiently and economically, as described in chapter 8.

Day (1994) uses the definition of 'the process of securing a facility which fully satisfies a client's requirements, within the approved time and cost', which is not dissimilar to the earlier definition and has the merit of brevity.

## ROLE OF THE PROJECT MANAGER

In large complex construction projects the lay client is faced with a bewildering array of experts, specialists, consultants, contractors and subcontractors, each with their own unique conditions of engagement and fee scale/pricing structure and, in consequence, many clients are now looking for a single person or consultant who can pull all these disparate parties and their activities together in a central coordinating role. Hence clients often identify major benefits in employing a project manager who provides a single source of reference for all matters appertaining to the project, provided the resultant savings at least equal the cost of engaging the project manager, apart from all the other advantages that can accrue as described later.

Project managers may be appointed in-house where the client has sufficient expertise in his own organisation or, particularly with complex projects which may be large or small, undertaken for an inexperienced client, then the services of an independent project manager is the most likely solution. The in-house project manager has the advantage that he is thoroughly familiar with the client organisation, while the independent consultant has to make himself fully conversant with all relevant aspects of the client's organisation and procedures, but is likely to have wider experience of project management. Table 12.1 details the wide range of possible tasks of the project manager and how the client's requirements can vary between an in-house project manager and a consultant project manager, and the lesser role that is normally performed by a project coordinator.

Carter (1992b) has listed the likely principal activities of the project manager, which now follow with some amendments:

- represent the client's interests
- define the principal objectives and rank them according to their relative importance
- develop an appropriate strategy, probably based on a project plan set to achieve the client's objectives
- define the roles and responsibilities of all concerned with the project, including the client
- appoint consultants, contractors and subcontractors
- agree the relevant terms and conditions and make payments as appropriate
- establish the necessary rules and procedures to ensure smooth and effective operation
- coordinate and integrate ex-contractual parties
- monitor and control the project team
- obtain all required information and make decisions
- provide team leadership.

The main priority of the project manager is to provide the client with the best possible service. To achieve this goal the project manager (PM) must gain the respect of the project team, be willing to listen, but have the strength of purpose ultimately to decide for himself. With regard to the project team, the lead discipline must be selected first, preferably by the PM, and thereafter the PM and the design team leader will together select the other members of the design team. The PM must make a conscious effort throughout to become a good manager of people as this is his principal task and one in which he must be proficient.

The project manager, as shown earlier, has a major coordinating role with regard to the numerous participating parties involved in implementing a project. The wide scope of this task has been clearly illustrated by Carter (1992b) in respect of the Croydon Computer Centre project where the principal participating parties and their primary functions were as follows:

- land vendor (site)
- solicitors (contracts)
- planners (consents)
- adjoining owners (interference)
- estate agents (disposal)
- British Rail (consents)
- SEEboard (electric power)
- British Telecom (telecoms)
- architect (design)
- landscape architect (design)
- structural engineer (design)
- services engineer (design)
- fire consultant (advice)
- fire officer (consents)
- lighting consultant (design)
- contractor (management)
- subcontractors (construction)
- subcontractors (design/construct)
- quantity surveyor (cost control/documents)
- security consultant (design/procurement)
- security contractor (installation)
- computer supplier (hardware)
- services contractor (maintenance)
- janitorial contractor (cleaning).

Figure 12.1 shows the contractual responsibilities of the main participating parties for the £15m computer centre at Croydon for a US bank
client, where a project manager was employed and the main construction
contract was let to a management contractor, and the 52 work package
contractors worked under the direction of the management contractor, as
described in chapter 3. While figure 12.2 illustrates the communications
network for the same project and the important coordinating role of the
project manager.

The question is often posed by members of the construction team as to
which discipline is best fitted to undertake the role of project manager.
Even within the RICS a number of divisions from quantity surveying through
to building surveying, general practice and planning and development all
believe that they have a major contribution to make. It must, however, be
accepted that no project manager can justifiably claim to be an expert in
all fields, although he/she should possess the skill to determine priorities,
control the pace and direction of activity, and ensure that the use of all
resources is optimised.

It is considered useful at this stage to examine the credentials of the
three main construction contenders for a PM appointment.

Table 12.1  Project manager's tasks and client's requirements (Waterhouse, 1992)

| Project manager's tasks | Client's requirements | | | |
| --- | --- | --- | --- | --- |
| | In-house PM | | Consultant PM | |
| | Project management | Project coordination | Project management | Project coordination |
| These vary between projects as does his responsibilities and authority. However, the main ones are as below: | | | | |
| • Be legal party to contract | • | | • | |
| • Assist in client's brief | • | | • | |
| • Develop project manager's brief | • | | + | |
| • Advise: budget/funding arrangements | • | | • | |
| • Advise: site acquisitioning, grants, planning | • | | + | |
| • Arrange feasibility study/report | • | + | • | + |
| • Develop project strategy | • | + | • | + |
| • Prepare handbook | • | + | • | + |
| • Develop consultants' briefs | • | + | + | + |
| • Devise project programme | • | + | • | + |
| • Select project team members | • | + | + | + |
| • Coordinate design processes | • | + | • | + |
| • Establish management structure | • | + | • | + |
| • Appointment of consultants | • | + | • | + |
| • Arrange insurance and warranties | • | • | • | + |
| • Select procurement system | • | • | • | + |
| • Arrange tender documentation | • | • | • | + |
| • Contractor prequalification | • | • | • | + |

| Duty | | | |
| --- | :-: | :-: | :-: | :-: |
| Evaluate tenders | • | • | • | + |
| Contractor selection | • | • | • | + |
| Contractor appointment | • | • | • | |
| Organise control systems | • | • | • | • |
| Monitor progress | • | • | • | • |
| Arrange meetings | • | • | • | • |
| Authorise payments | • | • | • | + |
| Organise communication/reporting systems | • | • | • | • |
| Provide total coordination | • | • | • | • |
| Monitor safety procedures | • | • | • | • |
| Coordinate statutory authorities | • | • | • | • |
| Monitor budget and variation orders | • | • | • | • |
| Develop final account | • | • | • | • |
| Precommissioning/commissioning | • | • | • | • |
| Organise handover/migration | • | • | • | • |
| Organise maintenance manuals | • | • | • | + |
| Plan for maintenance period | • | • | • | + |
| Develop maintenance programme | • | • | • | + |
| Plan facilities management/staff training | • | • | • | + |
| Arrange for feedback monitoring | • | • | • | + |

Key  • = typical duties
     + = possible additional duties

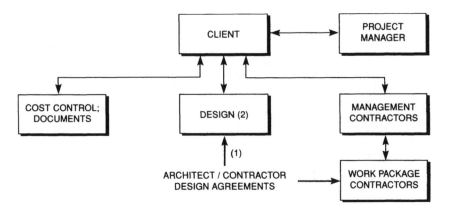

**Figure 12.1** *Croydon Computer Centre: contractual responsibilities*

*Notes:*
1. As appropriate
2. Single contract for design with Architect who 'subcontracted' structural, building services and landscaping design.
*Source*: Carter, 1992b.

*Architect* the traditional team leader who does not always satisfy the client's basic requirements of designing buildings which are functional, flexible, easy and economical to construct and operate, and who lacks the required management capability and financial/contractual skills. Many clients place a high priority on cost control and value for money and this is an area in which most architects tend to be rather weak.

*Contractor* has undoubted management skills but tends to be weak on precontract procedures, being mainly interested in the construction process with little regard for the procurement procedures, with the exception of design/build contractors.

*Quantity surveyor* can give good advice on all forms of building procurement and cost control but would benefit from concentrating more on value engineering, cost control information retrieval, and cost data processing (Carter, 1992b).

## BASIC REQUIREMENTS OF PROJECT MANAGEMENT

Carter (1990) has given extensive and very sound advice on the principal basic requirements based on many years of personal practical experience, and the following requirements are mainly derived from his contribution.

(1) Getting to know the client intimately and identifying his needs; considering fully the building users and occupants, how the building

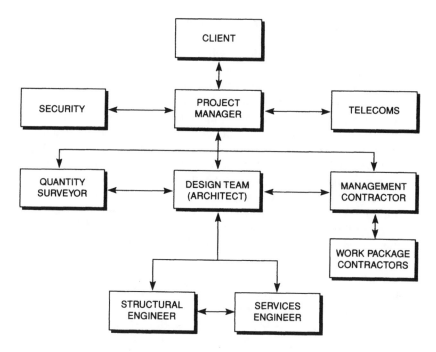

**Figure 12.2**  *Croydon Computer Centre: communications network*

*Note*:
Formal lines of communication shown. Informal links sometimes occurred to expedite communications (exception, *not* rule).
*Source*: Carter, 1992b.

will be operated, maintained and cleaned, and future flexibility of use and form; defining the client's priorities with regard to time, quality, cost and operational capability/function; and being client oriented, always remembering that the PM is undertaking his job for the client and to his satisfaction, involving a close working relationship,

(2) Obtain authority and use it wisely; earn the respect of the client and the project team by establishing credibility as a sound, discerning and knowledgeable PM.

(3) Manage and coordinate the efforts of others in the most effective way, involving project team members as far as practicable to achieve mutually acceptable decisions. In the event of conflicting views, refer the various options to the client with the PM's reasoned preference.

(4) Take positive action at all times, set achievable target dates and monitor them closely, and ensure that decision-making by the client's top management conforms to programme. Be prepared for lack of preconstruction programme input from the contractor and to learn a great deal about programming.

(5) Always be prepared to learn from others, such as by visiting earlier similar schemes, talking to participants and particularly building users and seeking views and advice where this would be helpful.

(6) Avoid adopting a 'not my problem' reaction; define the roles of all project team members and ensure that all know what is expected of them and monitor their actions closely and regularly. Help others to solve their problems and encourage them to help with yours.

(7) Plan well ahead, keep all aspects of the project under regular review, considering the various stages, and identify the areas of risk and uncertainty and how to resolve them. In particular, anticipate future problems and consider how they should be overcome.

(8) Make good positive decisions, obtain the views of all parties involved, and check the source and authority of all information received. Consider earlier alternatives that have been considered and identify why they were rejected, re-examine problems that have previously been resolved if not satisfied with an earlier decision, and be prepared to defend decisions to the client.

(9) Be prepared to admit ignorance and seek advice from the project team or others where appropriate.

(10) Be innovative such as by stimulating original thinking by the project team, but accepting that innovation can be risky and that there is a possibility of failure.

(11) Never try to defend the indefensible; where a mistake has been made, admit it and take immediate steps to rectify it, as attempting to justify a mistake wastes time and destroys credibility.

(12) Be impartial and use your integrity fully and wisely, as the only member of the project team who has freedom of action. All decisions and actions must be project and client oriented. Encourage feedback from the project team, find out what others are thinking and respect confidential discussions.

(13) Team building is important and should be done as quickly as possible. The PM should develop close working arrangements with team members, and these can be accelerated by sharing some meals and/or social events. Support the team when it is subject to criticism and accept criticism if it is justified and take appropriate action, acting throughout as team leader.

(14) Develop efficient and effective administrative and communications procedures, so that all get to know everything of which they should be aware. Distribution and mailing lists should be prepared as soon as possible and modified as the project develops. Be meticulous in your paperwork as the client relies on the PM to know everything about the project and this includes a first class filing system which provides easy access to information.

(15) Understand the depth of your involvement which is so very different from the other professional and traditional disciplines.

(16) Assist the project team members to achieve required objectives by personal involvement and displaying ample tact and diplomacy.

(17) Close monitoring and achievement of targets is vital for the success of the project and these include construction cost, completion date and possibly quality of specialist accommodation, such as a computer suite.

(18) Safeguard the client's interests at all times, such as monitoring local press coverage and making regular inspections beyond the site boundaries to ensure that there are no unreasonable effects on adjoining owners and the general public. Well maintained hoardings, safety lighting and courtesy notices help to project a good image of the client and the whole development team.

(19) The PM requires a sense of humour amidst many tense and testing situations, when he/she may have doubts and make mistakes. The duties entail substantial sacrifices of time and effort and result in high stress but are rewarded by the satisfaction of leading and coordinating a talented team of fellow professionals to the successful completion of the project.

## PROJECT MANAGEMENT QUALIFICATIONS AND ASSESSMENT

The principal body in this field is the Association of Project Managers which represents the major industry sectors, ranging from construction and engineering to education and finance, and in 1995 had over 3000 UK members. Carter (1990) stated that it is unusual for a full member of the APM to be under thirty and that a good project manager must be able to motivate, communicate and lead by example and that this only comes with experience.

The APM operates a certification scheme which is intended to be a test of competence in managing projects and not a measure of academic ability. To be eligible a member must be managing or have managed a multidisciplinary project. The certification process comprises a self-assessment submission and 200 word precis of an appropriate project, a 5000 word project report and an interview with two assessors. The APM maintains a list of certificated project managers and also a detailed body of knowledge as an aid to the self assessment task.

In addition, the RICS together with the College of Estate Management (CEM) established the first construction project management qualification – the RICS Diploma in Project Management, which is a two year distance learning course with the first graduates in 1984. There are also a number of Masters' courses in Project Management.

In 1994, the RICS General Council considered a proposal for a certification scheme for RICS project managers based on the APM certification programme but administered by the RICS for its own members. Despite the desire of RICS members to promote their project management skills in a positive way, the proposal was not considered the best method of achieving the objective and was not approved. It was also felt that the certification approach needed examining on a wider front to encompass other specialist areas.

An enterprising group of people under the auspices of APM and UMIST established a very active Women in Project Management section which has developed some excellent courses and conferences, which is most encouraging as women play an increasing and effective role in construction management.

### Project Management Agreement and Conditions of Engagement

This document issued by the RICS and the Project Management Diplomates Association was most opportune to meet the growing demand for a clearly understood and agreed appointment document for use by project managers and their clients.

The Project Management Agreement and Conditions of Engagement (1992) provide a model and are widely used by clients and project managers. The project manager has only relatively recently emerged as an independent professional and there are widely differing views as to his responsibilities and the services he provides.

Under the Agreement, the client agrees to engage the project manager in accordance with the Conditions of Engagement and the project manager agrees to provide the services inserted in Appendix A of the document and in accordance with the Conditions of Engagement. The project manager acts as agent of the client and is responsible for the administration, management and coordination of the project.

Furthermore, the project manager:

(1) communicates the requirements of the client's brief to the consultants;
(2) monitors the progress of design work, and the achievement of function by reference to the client's brief;
(3) monitors and regulates programme and progress;
(4) monitors and uses his reasonable endeavours to coordinate the efforts of all consultants, advisers, contractors and suppliers directly connected with the project;
(5) monitors the cost and financial rewards of the project by reference to the client's brief.

Appendix A in the document has been left blank to enable the client and project manager to agree the services to be provided and incorporate them into each Agreement. It is important that the services required are clearly defined to remove uncertainty and to identify the responsibilities of both parties. Appendices B–E, dealing with the matters connected with the project requiring the written consent of the client, payments, adjustment of fee, site office accommodation and other equipment and services, and disbursements and expenses, are also left blank for completion as appropriate.

It is anticipated that the project manager will frequently be appointed from the earliest phases of a project when the extent, nature, description and timescale have not been determined. The Agreement cannot be fully completed until sufficient development work on the project has occurred to identify the brief, the extent of the involvement of the project manager and the probable timescale. It may be appropriate for the parties to rely on a partially completed Agreement supplemented by other documentation for final agreement when all necessary details are available.

The project manager is required to perform the services with reasonable skill, care and diligence, as is expected of all professionals.

The RICS Project Management Agreement and Conditions of Contract: Guidance Notes (1992) contains a schedule of services, which are listed later, which may be provided or arranged by the project manager. These services may be specified in Appendix A of the Conditions of Engagement, but not all services will apply to every appointment, and other services may be required which are not listed. In addition, the extent of services provided may be varied to suit the requirements of the project. The client and the project manager should, as far as possible at the time of the appointment, determine which services shall be provided and any subsequent change shall be formally recorded between the parties.

The schedule of services in the Guidance Notes contains the following activities, which are each described in some detail in the Guidance Notes.

1. Site selection
2. Analysis
3. Agency, valuation, funding and relocation
4. Legal services
5. Consultant appointments
6. Brief, design and quality control
7. Reporting and meetings
8. Programming
9. Capital budgeting
10. Construction economics
11. Cashflow
12. Local authority and planning approvals
13. Contract procedures
14. Contract management
15. Building management, commissioning and maintenance
16. Tenancies and fitting out.

## SELECTION AND APPOINTMENT OF THE PROJECT MANAGER

In 1995 the RICS Project Management Skills/Practice Panel issued *Selection & Appointment of the Project Manager: Guidance Notes* to assist those who wish to commission project managers in the public and private sectors and is intended to be used with the *Project Management Agreement and Conditions of Engagement*, described earlier.

It is recommended that selection should be made on the basis of location, size, experience, scope of service and financial status, preferably by issuing a questionnaire or by interview or both. A letter of invitation to attend an interview should include a brief description of the project and a preliminary list of services required. At the interview, consultants should be asked to address the following issues:

- company details
- previous experience, particularly of comparable projects
- project management proposals, giving team structure and outline of strategy
- full details of the proposed project manager
- professional indemnity insurances
- references.

Interviews should ideally not exceed one hour, with 40 minutes for presentation and 20 minutes for questions. No more than three participants from either party should attend, and these should include the proposed individual project manager. Appendices to the Guidance Notes contain an example of a letter of invitation to attend an interview and an example of an interview marking scheme.

Consultants are selected by competition or by negotiation, in the manner outlined in the Guidance Notes, and fee quotations may be obtained on a lump sum, percentage of building cost or time charge basis. In the public sector, tenders are sometimes invited on the 'two envelope system', with the outer envelope containing the tenderer's method statement, resource statement, management strategy and details of key staff, while the inner envelope contains the fee bid, which is only opened if the consultant adequately meets the stated requirements. This enables the most suitable consultant to be selected on a qualitative basis only.

As from 1 July 1993, EU Directives relating to the public procurement of architectural, engineering and related technical consultancy services required public bodies wishing to procure project management services with a fee greater than ECU 200 000 (about £133 000) to advertise in the *Official Journal of the European Community*, on the basis of open, restricted or negotiated tenders.

## PRACTICAL APPLICATIONS OF PROJECT MANAGEMENT

It is felt that some practical illustrations showing how project management is operated by a number of organisations could be of value to the reader.

For example E C Harris Project Management believe that effective project management maximises efficiency and optimises performance, and thereby adds value to the end product. E C Harris provides a portfolio of services which embrace every aspect of the development and construction process, as listed in the schedule in table 12.2.

Turner and Townsend have been appointed project manager on a wide range of different project types, from the £240m South Yorkshire supertram, £95m South Leeds supertram and £68m Nottingham light rapid transport system to the £50m Inland Revenue office complex in Nottingham. In the latter project, Turner and Townsend worked alongside architect Michael Hopkins & Partners and services and structural engineer Ove Arup & Partners on the radical traditional six rectangular, 4-storey office blocks with their advanced natural ventilation and lighting provision.

David Day in his comprehensive book 'Project Management and Control' describes in detail the application of project management techniques, with reference to the Queen Elizabeth II Conference Centre, Westminster, for which Day was project manager. He covers all the principal functions in a most informative way, starting from the project manager at work, through to relationship with the client, composition of the PM support team, selection of consultants, the lead discipline, client involvement, design stage planning, feasibility studies, design process, use of specialist advisers, PM's role in construction stage, establishing cost control, time management, and completion and handover.

### Croydon Computer Centre

Carter (1990) described in detail his involvement as PM in a £15m computer centre project at Croydon for a US bank client. The PM spent several weeks with the client preparing the outline brief, followed by regular visits to the client's London office as the tender documents were prepared, bids invited from architects and contractors (management contract) and the project team appointed.

Extensive efforts were made to become thoroughly familiar with the client and his organisation and to formulate the brief in all its many aspects. The design/construction team was selected with great thoroughness and care and included the submission of prequalification bids, visiting recent projects managed by those bidding and the firms' offices, and a similar procedure was adopted for contractors. In the design development stage, options were considered from the constructability aspects and costed by the quantity surveyor, and the PM maintained close contact with the

Table 12.2   Summary of project management services provided by E. C. Harris
Project Management

---

- Definition of the Brief with the client (and with funding and letting agents)
- Liaison with client departments to assemble a consolidated statement of User Requirements
- Feasibility studies and risk analysis
- Practical problems arising from site acquisition
- Negotiation of dilapidation and early vacation settlements
- Setting up lines of communication and motivating a productive team spirit
- Selection and appointment of the professional team, formalising terms of engagement and checking professional indemnity
- Resolving the special demands associated with the provision of medical and healthcare facilities
- Programme review and integration of constraints from the Brief (possession, decanting, early access, partial completion)
- Setting up and controlling serial contracts for housing and refurbishment
- Implementation of practical obligations under funding, financial and joint development agreements
- Consideration of likely construction sequences and early ordering requirements
- Establishing appropriate design, tendering and contract implementation routes
- Provision of interface between client and design, construction and marketing teams
- Interview and assessment of contractors and specialist subcontractors
- Production of control programmes to monitor progress
- Assessment of resources: anticipating and troubleshooting potential problem areas
- Coordination of final occupation process
- Preparation of whole project 'audits' or 'snapshots' to provide for other clients an independent overview of their problem projects. Recommendation and implementation of remedial measures

---

design/construction team throughout, and participated in obtaining planning permission and minimising the cost of planning gain requirements. Changes to the brief can prove costly and delay consequences with a 'knock-on' effect. Fifty two subcontractors were engaged using in each case the procurement system most appropriate to the work package of the management contract. Extensive vetting of specialist subcontractors for capability and performance of products was undertaken.

The client introduced a third party audit which took the form of a value management workshop, as described in chapter 9, and this introduced some changes. The PM, as the client's representative, dealt with possible third party delays in a very positive and assertive way with very satisfactory results. The PM also devoted much time and effort monitoring quality control, assisted by numerous mock-ups and samples, which helped to resolve complex design/installation coordination issues and interface problems. The PM worked long hours on site and was readily accessible to all members of the workforce, which on occasions enabled him to identify and resolve many problems almost before they happened.

As shown earlier in the chapter, many participating parties were involved thereby increasing the coordination task. Commissioning the project proved to be a complex and protracted process, partly because of a late major variation by the client. As the project neared completion, tenders were invited and contracts placed for planned preventive maintenance and janitorial/cleaning services.

### Royal Armouries Museum, Leeds

This £42.5m project has been included as it contains features which are very different from those normally encountered in construction contracts and shows the wide and diverse range of tasks that can confront the PM. For example, buying the displays from numerous model makers, artists and film makers was included within Heery International Project Manager's management contract for the fit out, which necessitated reconciling construction's culture of paying subcontractors a month in arrears, depending on progress, with artists and artisans who are unable to finance their work unless paid in advance. In consequence, contrary to normal construction practice, Heery negotiated a compromise under which they received stage payments in advance to cover defined phases of work, with Heery retaining 10 per cent of the contract sum until completion of the whole contract.

Another important issue was quality control, as specifications were much less precise than for construction work and Heery had to trust the skilled craft operatives to use their artistic abilities to produce the best results. Much of what went into the model animals or films was hard to assess subjectively.

Probably the most complicated of all was managing procurement of the films and audio visual material for the museum for which Yorkshire Television (YTV) won the £1.2m contract. YTV lawyers were unfamiliar with a JCT management contract, and its standard contract for film production was incomprehensible to Heery. Furthermore, YTV was accustomed to including a non-negotiable 10 per cent in its contract price to cover contingencies like bad weather, which could affect filming schedules. Heery refused to accept this because it was committed to working within a fixed price lump sum contract. Reconciling media contract law with the requirements of a construction contract was a major problem, and this also included copyright on films.

## COMPUTER AIDS TO PROJECT MANAGEMENT

Day (1994) has described how there are a number of companies with project management packages on the market, all of whom are happy to give details of the systems and arrange demonstrations. One potential problem area is the possible incompatibility of the new system with the one in use, and the details of the latter must be included in any specification going to the supplier.

Prominent project management software suppliers include Asta Development Corporation, Project Management Software Centre, CS Project Professional and Forecast Ltd. Asta Power Project graphical software for windows was used by the Birse Construction project team on Leeds Prison extension, as they considered it would be the most adaptable for the complexities of the contract.

Asta give the following as the main advantages of Power Project software: unique segmented linked bar charts; powerful annotations communicating plans effectively; comprehensive resource management features with dynamic on-screen histograms and resources modelling; integrated costings for budget control and cash flow forecasts; and baselining, providing immediate analysis on past performance.

Day (1994) considers spreadsheets to be the most useful tool for project managers. The uses vary from preliminary cost exercises prior to the involvement of a design team to the detailed budgeting of a project which is likely to embrace hundreds of contracts, together with all the other resource costs. Costs can be checked through the various stages of development and the total effect on fees and contingencies identified immediately. Time calculations are also possible with spreadsheets, enabling expenditure forecasting and monitoring to be readily undertaken.

# 13 Other Quantity Surveying Activities

In the 1980s and 1990s, many quantity surveyors were providing services outside their normal functions. Some of these are considered sufficiently important to devote whole chapters to the specialisms, such as value management in chapter 9 and project management in chapter 12. A wide and diverse range of other activities in which the quantity surveyor may be engaged are detailed in this chapter, ranging from construction management to the settlement of construction disputes.

## CONTRACTING AND CONSTRUCTION MANAGEMENT

### Contracting Arrangements

A RICS report in 1983 postulated that 'with tighter margins prevalent in the contracting side of the construction industry, the quantity surveyor's role is changing. His skills and abilities in relation to overall management techniques are becoming increasingly recognised with the result that many are employed in resource and manpower planning, construction and project management and allied areas. Quantity surveyors occupy an ever-increasing number of senior management positions within major contracting organisations,' as described in chapter 15.

Millwood (1983) aptly described how employers in both the public and private sectors are concerned to know the end cost and the completion date of a construction project with a reasonable degree of certainty. They would certainly prefer to know the end cost with certainty, rather than that the initial price is the lowest that can be obtained in competitive tendering, leaving the final price to be calculated in accordance with the contract. Hence employers, particularly in the private sector, are looking with increasing interest at the newer forms of contractual arrangement, such as design and build and management contracts, as described in chapter 3. These can each be successful in particular situations and quantity surveyors should be ready to respond to these developments and to offer services to the employer appropriate to the circumstances, recognising that there can be considerable benefit in introducing a contractor at the formative stage of a contract, and that in management contracts the main contractor will be working with the employer's professionals to ensure that the cost and time targets are met.

## Construction Management

Some important aspects of construction management from the contractor's viewpoint were examined in chapter 5. Quantity surveyors engaged in contracting organisations will be well versed in latest management techniques and their applications for programming and progressing, network analysis, risk analysis, forecasting costs, budgeting, cost value reconciliation and other control mechanisms.

A comparison of design and contract procedures in the United Kingdom and the United States (1979) revealed a number of significant differences in approach to construction contracts. In the United States buildings are designed and constructed more quickly than in the United Kingdom, although the construction costs in comparable terms did not vary significantly. Specialist contractors have greater influence over detail design in the United States, resulting in simpler details and fewer construction processes. Scaffolding is rarely used as components are fixed from inside the building, services are installed very early in the construction process and the structure erected around them, site welfare facilities are almost non-existent and safety provisions are minimal.

The United States' equivalent of the United Kingdom quantity surveyor is the construction manager. Construction management practices place greater emphasis on the range of services that they can provide and often employ many specialists, including architects, engineers, builders, public relations officers, graphic artists and site agents. They devote more effort to marketing themselves and regard aggressive advertising as an essential part of their business development strategy, using specialists for this purpose and producing very professional advertising material. Quantity surveying practices in the United Kingdom increasingly use public relations firms to produce publicity material for them and some of the brochures, as described in chapter 17, are particularly impressive. The United States practices use computers as a matter of course and have developed effective cost and time control systems.

The strength of United Kingdom quantity surveying practices lies mainly in their ability to provide tightly integrated and effective contract and cost control systems in close liaison with the employer, with a strong involvement at the design stage. The United States objective of a total design and construction service, with responsibility for time and costs, is a commendable one. It brings the reader back to Millwood's starting point and is a matter of such fundamental importance that it must receive greater consideration by the quantity surveying profession, by examining ways in which it can offer a complete construction management service to employers.

The present dilemma is well summed up in the 1983 RICS report – 'There is an urgent need for quantity surveyors to become more conscious of the different approaches involved in the management of construction projects. Management technology is advancing and clients are becoming increasingly aware of the need for effective management. Construction management with the quantity surveyor accepting responsibility for cost, time and contractual matters is therefore an area where present involvement needs to be encouraged and expanded.'

The RICS report, *QS2000* (1991), believed the challenge of construction management is for quantity surveyors to become involved either as cost consultants or as construction managers.

## EXPENDITURE STATEMENTS, TECHNICAL AUDITING AND COST ACCOUNTING

Quantity surveyors can be called upon to assist with the compilation of building costs for a variety of purposes. Surveyors in private practice are generally reluctant to act for contractors, particularly those operating in areas where they practice, because of the possible conflict of interest at some future time. It is, however, likely that quantity surveyors will in the future forge closer links with other classes of surveyors, particularly building surveyors, general practice surveyors and planning and development surveyors. Together they can offer a wider and better service to the employer.

### Preparation of Statements of Expenditure

Accountants and tax inspectors often require evidence of the cost of new building work or of alterations or adaptations to existing buildings. The quantity surveyor involved with a project is able to supply all the required cost information suitably categorised from his own records. In the case of developments where no quantity surveyor has been employed, he can compile the necessary cost data from information held by the employer and contractor.

### Technical Auditing

Employers sometimes require a technical audit of building costs to satisfy themselves that they are being asked a fair and proper price for the work undertaken. It may be that the tender was submitted on the basis of drawings and specification and that variations have occurred during the course of the work. Alternatively, the contract may be on a prime cost plus percentage basis and the employer wishes to be assured that the costs have been correctly calculated. The quantity surveyor can investigate all the costs incurred and determine a fair and reasonable price, having regard to all the relevant circumstances. This can involve examining all available records, interviewing the contractor's personnel, visiting the site and investigating the procedures which have been used, including an assessment of the effectiveness of the contractor's internal control system.

In construction management, which was being used considerably in the mid 1990s, the construction manager deals with payments and the quantity surveyor audits these payments. It is advisable to forewarn the construction manager that the auditor will make regular reports to the client on performance. The auditing quantity surveyor will decide what checks to make and will vigorously pursue the audit so as to be able to report to the client whether work is being carried out satisfactorily.

Hemsley (1995) has referred to the advice prepared by the Institute of Chartered Accountants (ICA) for its members on managing professional liability, which can be summarised as follows:

- auditors should define in an engagement letter the extent of the responsibility undertaken
- reference should be made to the information to be supplied
- the specific tasks to be undertaken should be set out in detail
- responsibility for the accuracy of the information supplied should remain with the client
- tasks not to be undertaken should be defined
- where limited work is undertaken, a warning should be included stating the extent to which the client can rely on the information.

By adhering to these criteria, quantity surveyors acting as auditors are able to narrow the expectation gap between what clients hope for and what an audit can realistically provide.

An audit can be defined as 'a strategic enquiry that aims to provide a reasonable expectation of detecting fraud and other irregularities or errors'.

## Current Cost Accounting

Current cost accounting requires that fixed assets shall be stated in a company's accounts at their value to the business by reference to current costs, which can be assessed as net current replacement cost valued at open market value or depreciated replacement cost. The principal objectives are to ensure that the balance sheet reflects the current value to the business of the fixed assets divided as to 'land and buildings' and 'plant and machinery', and that the operating results are calculated after charging depreciation at current costs for buildings (and land in certain cases) and plant and machinery so as to maintain the operating capability of the business. The land and buildings may be held as fixed assets by a company for occupation by the business, as investment properties or as surplus to the requirements of the business.

With regard to buildings, the valuer will decide whether to assess the gross current replacement cost by applying a suitable index to the original cost or a subsequent assessment, or to make a fresh current assessment. The accountants' guidance notes show there is now a greater realisation of the applicability of the depreciated replacement cost approach as a means of arriving at the value of a property to the business. This is an area where quantity surveyors could provide a service to valuers.

## VALUATIONS FOR FIRE INSURANCE

The risks of fires and the cost of the damage resulting from them in the last decade have increased significantly. For example, changes in the working environment in offices with more elaborate equipment, greater use of electricity, and increase in bulk and height have produced their own special problems affecting not only the cost of replacement, but also means of escape and smoke control. An examination of fire statistics for all types of buildings shows continually increasing risks and replacement costs.

Both insurance companies and property owners require professional advice on valuations for fire insurance. This service is provided in two main areas:

(1) valuation of replacement cost to ascertain the sum to be insured; and
(2) the preparation and negotiation of submissions to insurance companies for reimbursement following a loss.

A property owner almost invariably takes out a fire insurance policy to reimburse him in the unfortunate event of the destruction of the premises by fire. The insurance company will only reimburse the full cost of replacing the building, or even some part of it, provided that it is adequately insured; otherwise only a proportion of the cost will be paid. Hence it is essential that the property owner makes certain that the value for which the property is insured each year covers the cost of replacement and the necessary demolition work. Building costs generally rise each year but not necessarily in line with the general rate of inflation and vary significantly in different parts of the country. The property may consist of an amalgamation of buildings of different types, sizes and conditions. In these circumstances it would be wise to employ a quantity surveyor to assess the current rebuilding cost for insurance purposes. The purchase price of a building is rarely a very good guide as this price will include the value of the land, which will not be significantly devalued by a fire, and since property prices tend to fluctuate considerably, in the case of older properties, these may bear little relation to the cost of replacement. When assessing fire damage the surveyor will need to visit the site as quickly as possible to quantify and evaluate the damage and check on the details in the policy.

At least one of the larger quantity surveying practices is now providing a range of construction industry related insurance services through an associated company, including advice and investigations regarding bond and surety, loss adjusting and risk appraisal. With the constantly increasing risks associated with major developments these services are invaluable. The associated company advises the client on the most cost-effective method of obtaining adequate insurance cover. Following an initial examination, a report and tender documentation are prepared on the specific insurance requirements and the company then seeks competitive terms and prices on the insurance market.

## ADVICE ON FUNDING, GRANTS, CAPITAL ALLOWANCES AND TAXATION

The main sources of *funding* for construction projects are banks, building societies, insurance companies, mortgage brokers and pension funds, and they can be conveniently broken down into short, medium and long term loans as described by Seeley (1995a), Isaac (1994) and Fraser (1993). The London Interbank Offered Rate (LIBOR) is the rate which prime banks offer to make Eurocurrency deposits with other banks for a given maturity. LIBOR is generally used as the benchmark for bank loans and debt issues, with the interest rate being quoted at, for example, 1 per cent to 2 per cent over LIBOR. However, LIBOR changes on a daily basis.

The quantity surveyor may be involved along with the client's financial advisers in determining the most effective funding arrangement for a particular project, and hence needs to be familiar with the main funding sources, their characteristics, merits and demerits. It is necessary to have regard to the degree of risk involved and to study the funding documents very carefully.

If a proposed development is in an area which requires urban regeneration and an increase in local employment, it is sometimes possible to obtain a *government grant* which may be sufficient to make an unviable scheme viable and enable the scheme to proceed. These grants may however be conditional upon local authority support and financial contribution. Hence the possibility of grants must be fully explored and, where obtainable, included in the development appraisal using the format illustrated in chapter 10. Subsequently, the quantity surveyor may make application for the grant on behalf of the client and satisfy himself as to its accuracy. Eurogrants are also available for some schemes of urban regeneration subject to some UK public funding.

In certain situations, such as in the enterprise zones, created to foster development and economic regeneration, provision was made for the use of a simplified planning procedure, exemption from local rates for a ten year period and 100 per cent capital allowances for certain types of building. The distribution of regeneration grants has now been transferred to English Partnerships, highlighting the need for the quantity surveyor to keep abreast of changing legislation and regulations.

The quantity surveyor must also be familiar with the *taxation* aspects of a proposed building development, in order to ensure that the most tax effective option is implemented, and the client's tax liability kept as low as possible. This could also affect the future profitability of the scheme. For example, with industrial buildings used for manufacturing purposes, capital expenditure on the building development qualifies for a *capital allowance* in the form of an industrial building allowance. This comprises an initial allowance, possibly in the order of 75 per cent and subject to annual review, plus an annual writing down allowance of around 4 per cent, until the residue of the unrelieved expenditure has been eliminated, at the appropriate rate of tax.

On most industrial and commercial buildings, a significant part of the expenditure on plant and machinery can qualify for capital allowances. The allowance is a writing down allowance of 25 per cent per annum on the reducing balance. There is no statutory definition of plant or machinery, but considerable guidance can be obtained from decided cases, most of which relate to plant rather than machinery, probably because machinery like escalators, lifts, heating and air conditioning installations is more readily identifiable. For example, in the John Lewis case (Cole Brothers Ltd v. Phillips, 1982), the following electrical items were not disputed, presumably because they were ancillary to machinery: trunking for the telephone system, wiring to heating and ventilating equipment, fire alarms, clocks, TV workshop, cash registers, lifts, escalators, burglar alarms, smoke detectors, and the electrical appliances department.

Other decided cases won by the taxpayer give further guidance and include the following items: moveable partitions in a shipping agent's office; swimming pool in a caravan park; dockside grain silos of a grain importer; transformers and switchgear to bring power from the National Grid to a department store; and decor and murals in hotels and licensed premises. Some of these components do not appear to be items of plant or machinery despite the decisions made. An overriding factor is that the items in question must be required to enable the building to perform its prescribed use.

A quantity surveyor may also be asked to advise on whether VAT is payable on building work or possibly specific items. In general, new building work is zero rated, whereas maintenance and repair work is charged at the full rate.

## BUILDING MAINTENANCE MANAGEMENT

It is not inconceivable that quantity surveyors could become involved in building maintenance management encompassing building use, assessing the life of buildings and the effect of maintenance upon rental and capital values, maintenance surveys, and planning, programming, controlling and costing maintenance and repair work. This work can include preparing and advising on building maintenance estimates, annually assessing the building maintenance requirements and investigating and possibly implementing preventive maintenance schemes. The preparation of drawings, specifications and contracts for minor capital works may also be involved, although this function more often falls within the sphere of the building surveyor.

Maintenance work is easy to delay and produces immediate revenue savings, but these are false savings, for the building will inevitably deteriorate still further and the real cost of the repairs will rise exponentially. It is considered advisable to follow with some explanatory information on types of maintenance, planning and management of maintenance work, and maintenance feedback.

### Types of maintenance

Figure 13.1 shows a method of classifying the main arrangements for maintenance work, with a primary classification of planned and unplanned maintenance.

The predominant characteristic of maintenance is the variety of factors that affect its incidence. These range from the initial design and cost involving the quality of materials and workmanship, the intensity of exposure, to the efficiency of the maintenance organisation. Their interaction directly affects the durability of the buildings and their components and the resultant maintenance work. The control of maintenance, if it is to be effective, should therefore commence at the time the building is designed and continue throughout its life.

### Planning and management of maintenance work

A maintenance plan should be formulated within the context of a maintenance policy, which itself has to be comprehensive, covering all types of work and all properties under maintenance. The policy should set out standards for the provision of the maintenance service in terms of amenity level (redecoration cycles), quality of work (inspections and lists of selected contractors), and day-to-day service (discretionary repairs and response times). The respective responsibilities of landlord and tenants should also be detailed.

The planning process has been subdivided into the following four stages by NFHA (1989): identifying needs; establishing priorities; developing the plan, possibly encompassing a five year period on a rolling programme; and monitoring results with feedback into the on-going plan.

Seeley (1987) has described suitable arrangements for the planning and management of maintenance work. Full records should be kept of each property stating the geographical location, age, condition, construction details

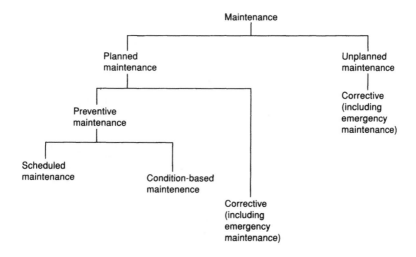

**Figure 13.1**  *Types of maintenance*

*Source*: BS 3811: 1984.

by elements, details of services, floor area and cubic content, accommodation provided, current use, and any proposals for the area by the local authority which might affect the property, all based primarily on a stock condition survey. Small organisations may use card records, but most organisations will use computers for ease of recording, updating and accessing information.

It is good policy to require contractors on new projects to supply maintenance manuals giving a physical record of each building as built, inspection and maintenance cycles for each element, list of specialist subcontractors and suppliers and information and instructions on maintenance for occupants.

Other useful sources of reference are CIOB (1990) Maintenance Management and BMI (1995) Condition Assessment Surveys.

*Maintenance feedback*

Maintenance feedback should be an essential part of any maintenance administration. Feedback may be mainly injected into the system in the following two ways:

(1) directly to the design team; particularly information on design faults, faulty workmanship and materials failures;
(2) by general discussion within the maintenance team, when solutions to problems should be documented and passed on to all appropriate personnel.

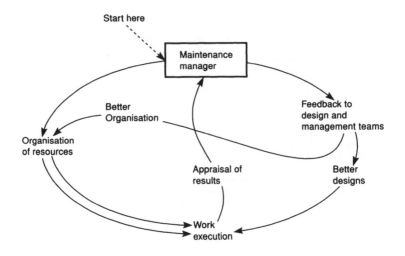

**Figure 13.2**  *Maintenance feedback*

*Source*: Seeley, 1987.

A visual representation of feedback is illustrated in figure 13.2, and this shows some of the major stages in the operation of a maintenance scheme:

(1)  management organisation of resources;
(2)  work execution;
(3)  appraisal of results; and
(4)  corrective action through feedback to design and management teams.

To assist in the feedback of information, site defects are suitably recorded showing the symptoms, diagnosis, prognosis (projection of defect performance in time), and the agreed remedy (Seeley, 1987).

## ADVICE ON HEALTH AND SAFETY, QUALITY ASSURANCE AND GREENER BUILDINGS

All these aspects are becoming increasingly important in the construction industry and clients' expectations are rising, hence the quantity surveyor must be thoroughly familiar with legal requirements and methods of implementing acceptable standards. Some quantity surveyors may give specialist consultancy advice in these areas.

### Health and Safety

The background, main provisions and method of implementation of the CDM Regulations were detailed in chapter 1. A leading international quantity surveying practice, E C Harris, has shown how quantity surveyors can perform an effective role in promoting health and safety on construction projects by setting up a specialist Health and Safety Division. This can act as planning supervisor and advise clients on their CDM responsibilities, develop and monitor the implementation of a comprehensive project health and safety plan and all the other CDM safeguards throughout the project.

### Quality Assurance

The principles and procedures relating to quality assurance and management in the construction industry were described in chapter 1. Many quantity surveying practices are certificated by BSI or other registration body and have developed quality management systems to improve the quality of their service to clients, and some offer a quality assurance consultancy service to clients and other professionals within the construction industry.

### Greener Buildings

The desirability and means of securing greener or environmental friendly buildings should be promoted by the quantity surveyor in his work as a member of the design team or a contracting organisation, or as an independent adviser to a client.

Johnson (1993) has described how a building's environmental impact extends from global factors such as ozone depletion to the quality of the environment within the building. Furthermore, these impacts arise from the decisions made at all stages of the building's life, including materials manufacture, site selection, design, construction, occupation and ultimately demolition.

Enlightened design can drastically reduce the harmful impact of buildings on the environment. Careful, creative planning will also produce a more comfortable, healthy workplace for employees. The selection of the correct materials will have a major influence on how environmentally sympathetic the building can be. In general, the closer the material or product is to nature, the more environmentally friendly it is likely to be. Another environmental issue is whether the material has been or can be recycled at the end of the building's life (Miller, 1992).

Built form has a key role to play in the eventual energy efficiency of a building. The building should ideally optimise the balance between maximising daylight, solar heat gain and ventilation, and at the same time minimising the surface area of the walls and roof. For example, a shallow plan will permit full use to be made of natural daylighting, ventilation and solar heat gain. A deep plan may require mechanical ventilation and more artificial lighting, which will increase both the capital expenditure and the running costs. The main rooms should preferably have a southern aspect. The roof should protect the walls from excessive wetting as this causes unnecessary chilling of the building fabric. In addition, care should be taken to control the amount of heat loss from the north of the building and heat gain from the south (Miller, 1992).

Finally the effect of the building on the surrounding environment should be assessed. Criteria would include such factors as fire hazard, waste generation and disposal, vehicle and pedestrian traffic generation, intrusiveness of external and internal lighting, noise generation, and external landscaping around the building (Wordsworth, 1992). A precise assessment of these factors could however entail carrying out cost–benefit studies of the form described in chapter 19.

## FACILITIES MANAGEMENT

### Nature and Scope

The quantity surveyor, along with other surveying disciplines, may be involved in facilities management, since the use of this technique is becoming more widespread as clients realise the importance of the proper management of their property assets. In QS 2000 (1991) it is recommended that quantity surveyors seeking to offer this service need to assemble information and skills in premises management and, in particular, maintenance needs and costs.

Readers requiring more detailed information on this important activity are referred to *Facilities Management: An explanation* by Park (1994), Property Helpline (1994) and CIOB (1994), while BMI (1993) produced a useful information guide for facilities management.

Facilities are generally defined as any property where people are accommodated and work, or where an organisation conducts its business, while management concerns all aspects of providing, operating, maintaining, developing and improving those facilities.

The Association of Facilities Managers has described how facilities managers typically carry responsibility for strategic planning: briefing, selecting and managing in-house and contracted resources; oversight of work, environment design, construction and fitting out of buildings, and their day-to-day operation and maintenance. In addition, they are regularly involved in property and estate management; space planning, fitting out and furnishing. They can be responsible for reprographic, communication, information technology and other services such as building maintenance, security and fire prevention, catering and transport. Hence their brief can be very wide indeed.

The RICS Facilities Management Skill Panel (1993) considered that FM embraces all the principal functions of management: planning, organising, staffing, directing, controlling and monitoring, and the bringing of all these components together in a coordinated way. It also postulated that FM consists of the following three distinct but interrelated areas, which can account for 30–40 per cent of the overall cost budget:

- the management of support services
- the management of property
- the management of information technology.

Watts (1992) in a keynote address at the RICS described the technique in an interesting and discerning way. 'Effective facilities management has to provide a complete service, encompassing the whole lifestyle of buildings. Everything needs to be taken into account, from security to catering, maintenance, energy management, space planning and project management. A theme running through all these issues is sustainability and the aim of making buildings as environmentally friendly as possible. In aiming to give buildings a long, productive and efficient life, facilities management must be in tune with the spirit of the age. We are seeing a shift away from short term exploitation of natural and built resources, towards a greater emphasis on stewardship and sustainability'.

While Park (1992) believes that FM embraces the control and the most appropriate and effective use of property resources, dealing with many interrelated aspects such as space planning, space costing, asset tracking, life cycle costing, maintenance and component specifications.

## Implementation of Facilities Management

Park (1992) has emphasised that it is imperative to secure the right usage of the building; wasted space, inefficient departmental interfaces together with an unattractive working environment can far outweigh the effects of a periodic rental appraisal and maintenance costs. Effective FM prolongs usefulness and therefore the effective life of property and slows down the rate and timing of decline. Buildings in this context are therefore resources of finite supply that need to be used in a controlled manner.

The operational flow chart in figure 13.3 illustrates the four phases in FM strategy, namely brief, research, implement and liaise, with the various related activities. Many of the client's requirements or conventions are cost oriented such as:

- space allocation with costs of rent, rates, energy consumption, cleaning and other related aspects
- component costs for tax accounting purposes, depreciation and replacement funding

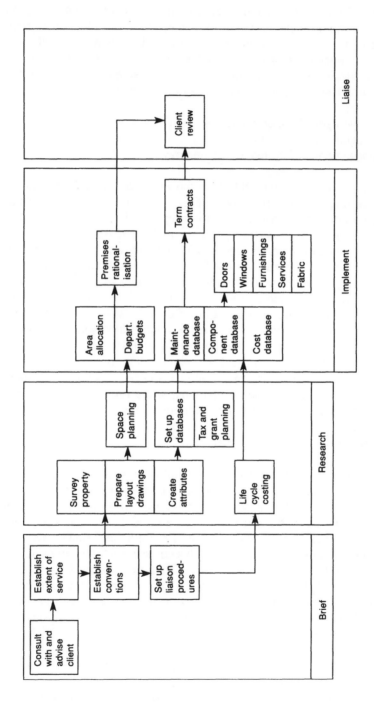

**Figure 13.3** *Facilities management: operation flowchart*

*Source*: Park, 1992.

- maintenance costs often obtained from the analysis of term contract rates to establish quarterly and annual budgets, monitor variances against cashflow and provide reports to the client's financial controller.

Park (1994) has described how the introduction of FM to an organisation can readily effect a ten per cent reduction in workspace costs, realistically rising to twice that amount in the medium term. FM is not confined to controlling cost, as there are several significant activities that can be managed through FM systems to assist in the smooth operation of the organisation. These activities include health and safety monitoring, maintenance and life cycle costing, specifications of components, selection of systems and software to collect, store, analyse and retrieve data, and the management of a variety of services encompassing heating, lighting, power, ventilation, telephones and communications, security systems, waste disposal, catering and staff welfare, creche provision and contracting out key support services.

Property is a finite resource that needs to repay its capital cost on an investment basis. Those buildings that serve a public need but cannot satisfy the investment return principle can only be funded with government grants or other non-commercial financial assistance. Buildings generally have a value cycle based on demand and hence changes in demand over time can result in a diminution of the building's value. The operation of FM that monitors the use of the building and adjusts it and the occupation to match current demands will slow the decline in value. Furthermore refurbishment, alteration and even change of use can prolong the life of the building. Where, however, the building has a defined life span, the facilities manager's objective will be to extract the maximum benefit from the property over its planned life without incurring excessive running expenses towards the end of the period, in accordance with good life cycle costing practice (Park, 1994).

The economic benefits of FM will show through improved productivity, better production quality and overheads control. For example, the monitoring of both space related costs like rent and rates in relation to workspace output, and the more difficult one of controlling location related costs such as energy consumption to major items of manufacturing plant and the extent of demand for office related items, such as coffee and tea vending points, photocopying, stationery and computer floppy disks, through adequate feedback and analysis. Realistic budgeting is all important to successful FM, but this must also encompass the feedback and monitoring of expenditure against the budget, be it quarterly or annually, and covering all occupancy and production costs, including maintenance, energy consumption and cleaning, with a view to achieving more efficient and economical processes.

## STRUCTURAL/BUILDING SURVEYS AND IDENTIFICATION AND RECTIFICATION OF BUILDING DEFECTS

### Structural/Building Surveys

Quantity surveyors may become involved in the carrying out of structural/ building surveys provided they have adequate experience in this field. It is essential that structural/building surveys are undertaken by persons with the necessary knowledge, skill and experience, as highlighted by the high incidence of negligence claims in this class of work, which is indicative of a high risk factor. The terms building survey and structural survey are often interchangeable in practice; since 1996 the RICS has recommended 'building service'.

The term 'structural survey' has been defined by the RICS (1983) as 'the inspection of visible, exposed and accessible parts of the fabric of the building under consideration'. The client frequently requires a broad appraisal of the condition of the building and general advice as to its suitability for the intended purpose. Nevertheless, clients' expectations will vary widely and must be clarified at the briefing stage. Where a client requires advice on the suitability of a building for a specific use, it may be better to undertake a feasibilty study as a separate assignment (Seeley, 1985).

A RICS Guidance Note (1991) on structural/building surveys of residential property contains the following list of the surveyor's responsibilities.

(1) assessing the client's needs;
(2) determining the extent of the investigations to be made and obtaining instructions from the client for any additional services required;
(3) undertaking the survey of the property in the form required by the client to establish the condition of the property and reporting to him in the detail and format necessary to provide him with a balanced professional opinion;
(4) complying at all times with the agreed instructions which form the contract between the client and the surveyor.

Readers requiring information on structural/building survey procedures and the format of the ensuing reports are referred to Seeley (1985), Noy (1995) and Hollis (1991).

### Identification and Rectification of Building Defects

Clients of all kinds often require an independent professional report on building defects which are causing them concern. This is an area where the quantity surveyor's knowledge and experience may fit him/her to undertake such assignments, but if he/she has any doubt as to his/her professional competence in this class of work he/she must decline the commission.

The majority of building defects centre around settlement and subsidence, dampness penetration, condensation, lack of structural stability, deterioration of materials, fungus and/or insect attacks on timber, paint failure, leaking services, defective services installations and blocked drains. On occasions, the source of the problem may be well away from the visible defect, as with dampness, and a thorough investigation is needed to identify the cause before a suitable remedy can be applied. In recent years, two other major problems of landfill gases and sick building syndrome have exercised the minds of many clients and surveyors.

### Landfill Gases

The increased use of landfill sites for building purposes has been accelerated by the shortage of well sited greenfield sites. Landfill can comprise deep deposits as with industrial and mining waste, shallow fills in urban areas and disused docks, claypits and quarries. These can give rise to chemical attack on foundations and the release of potentially harmful amounts of landfill gases such as methane and carbon dioxide and radon.

When landfill gas is encountered, the most common method of protection is to provide a gas-proof membrane in conjunction with a reinforced concrete floor slab above a granular venting layer. This is vented either by means of a trench around the perimeter of the building or by pipes and vertical risers to release gas at roof level which is a costly process.

In 1993 radon had been traced in Cornwall, Devon, Somerset, Derbyshire, Northamptonshire and North Yorkshire. Polythene sheeting is generally accepted as a popular and cost effective barrier but great care is needed in its installation to prevent perforation. Another solution is to apply a liquid asphaltic compound.

### Sick Building Syndrome

The main symptoms of sick buildings are occupants suffering from such ailments as headaches, lethargy, loss of concentration, watering or dry eyes, throat irritation, dry skin and chest complaints.

Known cases have the following common features:

- air conditioning: although some naturally ventilated buildings have sick building syndrome (SBS), it is more common in certain types of air conditioned building
- loss of environmental control: clerical staff who have less control over their working environment tend to be affected more than managerial staff
- worsening symptoms: as time passes tolerance levels decrease.

*Causes and cures*

Air quality: may be the prime cause of SBS and it is necessary to ensure suitable airflow, properly sited air intakes, and cleaning and filtration of air supplies to air conditioning plant. Air temperature should be a minimum of 16°C and 19°C is a reasonable comfort level, with a humidity of 40–70 per cent.

*Chemicals* Building products and office furnishings contain volatile organic compounds, such as solvents and components of glues and paints which emit gases and these can lead to respiratory problems and eye and skin irritations.

*Micro-organisms* Airborne micro-organisms tend to be a problem in recirculation systems in the event of poor filtration standards or inadequate maintenance.

*Ions* Low levels of negatively charged particles in air (negative ions) or high levels of positive ions appear to cause stuffiness.

*Planting* Air quality is of prime concern and hence plants which remove stale air products and replenish oxygen levels could help to overcome SBS.

*Lighting* Most office buildings are artificially lit, usually by fluorescent lights, whose inherent flickers can cause headaches. Ideally, individual control of lighting and maximisation of daylight should be provided, with the installation of blinds where sunlight could cause glare in the workspace.

*User controls* SBS studies show that occupants would prefer control over their local environment.

*Conclusions* SBS results from a combination of causes. Problems can be avoided by the exercise of greater care and understanding by designers, installers and building operators in the services they provide, and quantity surveyors can benefit from a greater understanding of the problems and their resolution, and should endeavour to influence the design team's approach and to advise clients accordingly.

There is a wealth of published work available on this subject and readers requiring more information on building defects and their rectification are referred to Seeley (1987), Robson (1991), DoE (1989), Cook and Hinks (1992) and Lee and Yuen (1993).

## MAINTENANCE AUDITS

This important subject, although peripheral to the quantity surveyor's main activities, has not been given the attention in practice that it deserves. Robertson (1983) described very effectively the nature and purpose of a maintenance audit. It forms an important part of management control which ensures that resources are obtained and used effectively and efficiently and that the organisation's objectives are accomplished. Measuring performance is now widely recognised as a fundamental part of the management control process and is a continuing process.

A maintenance audit should desirably comprise the following activities:

- the technical audit which assesses the level of maintenance work that is achieved
- the condition audit aimed at obtaining an overall view of the condition of the organisation's estate
- the energy audit detailing the quantity and cost of energy consumed and determining whether it is being used efficiently, as described in chapter 10
- the management audit which examines the management function, its policies, planning and procedures as an aid to securing best value for money
- the design audit whereby the maintenance management organisation, having implemented its own rigorous examination, can make a valuable contribution to the design process of new buildings.


## DILAPIDATIONS

Quantity surveyors are sometimes involved in the preparation of schedules of condition and schedules of dilapidations, the first at the commencement of a tenancy and the latter at the expiration of a tenancy giving details of the outgoing tenant's liability. When determining the nature of the required repairs, the surveyor must have regard to the class of property involved. He must be able to readily interpret the contents of legal documents, such as leases and tenancy agreements, and be able to fully identify their requirements and implications, with particular reference to covenants to repair and liability for repairs. He should also be well versed in appropriate standards of repair, be able to identify fair wear and tear, distinguish between fixtures and fittings and be familiar with the remedies for breach of contract to repair as described by Seeley (1985). In his recommendations, the surveyor must adopt a fair and impartial attitude and thus present reasonable proposals, having regard to the class of property, its condition and the terms of the lease or tenancy agreement.

The term 'dilapidations' refers to the disrepair or dilapidated condition of land and buildings, in situations where a legal liability is imposed upon the person(s) responsible. The person whose acts of omission or commission have caused the dilapidations is normally one with a limited interest in the property, such as a tenant for life or a lessee under a lease, whose neglect to keep the property in a good state of repair will have detrimental consequences for those who take possession of the property when his interest terminates.

The RICS (1991) has published some very helpful guidance notes on dilapidations which provide an excellent guide to good practice. It contains sound, practical advice on the preparation of schedules of condition, interim schedules of dilapidations and terminal schedules of dilapidations, together with useful examples. Seeley (1985) also provides a specimen schedule of dilapidations and illustrates the use of Scott schedules in settling dilapidations claims.

A lease may contain a provision that a tenant need not keep or leave the property in better condition than at the commencement of the term. In these circumstances a schedule of condition should be agreed between the parties and be attached to or form part of the lease to record the condition of the property. Plans and photographs may be used to supplement the schedule (RICS, 1991).

Hence the surveyor's dilapidations work can include the following basic elements:

- lease interpretation
- site inspection
- preparation of the schedule of dilapidations, or checking the schedule for a tenant
- costing the schedule
- negotiations.

## ARBITRATION

### General Background

Arbitration is a formal process for the settlement of disputes, provided for in all the standard forms of construction contract and subcontract, as an alternative to court proceedings. The basis of arbitration is that the parties to a dispute select a person on whose judgement they are prepared to rely, and agree to abide by the decision that is reached. A number of professional bodies, including the RIBA and RICS, maintain panels of arbitrators, and will at the request of the parties appoint a suitable person to act as arbitrator. Every arbitration, unless the parties express a contrary intention, will be controlled by the provisions of the Arbitration Acts of 1950 and 1979.

The advantages claimed for reference to arbitration are:

(1)  the process is generally a voluntary one;
(2)  the proceedings are conducted in private;
(3)  the decision is final; and
(4)  it is generally believed that the process is quicker and less costly than a court hearing, although some arbitrations have lasted for years with mounting costs (furthermore, the arbitrator has to be paid and a room hired for the hearing, whereas in the High Court, the judge and court are free).

The pace at which a case proceeds is, however, mainly dictated by the parties. A decision of the High Court can be taken to the Court of Appeal and to the House of Lords, and although costs may be awarded to the successful party, he will not recover the whole of his expenditure, or anything approaching it. With the recession in 1990–95, contractors were tending to settle their disputes by agreement to avoid the cost of arbitration, except where legal aspects were involved.

**Arbitration Procedure**

Procedure in arbitration follows very closely the procedure in the courts, for the same rules apply. After the arbitrator is appointed he will first hold a preliminary meeting, attended either by the parties (claimant and respondent) or by their solicitors, at which he will become familiar with the nature of the dispute. At this meeting he will give directions as to the conduct of the case and its timing, including:

(i)  preparation of the claimant's case (points of claim);
(ii)  respondent's answer to the claimant (points of defence) and the counterclaim if the respondent feels that he has a claim against the claimant; and
(iii)  claimant's reply to the defence, and a defence against the counterclaim where there is one.

These three sets of documents are referred to collectively as the pleadings. When they are complete a procedure termed discovery is initiated. It is a requirement of English law that any documents, such as correspondence and drawings, held by either party which touch on the matters in dispute and which may be used in evidence shall be disclosed to the other party. The hearing is conducted in the same manner as a case in the courts. The claimant opens his case and calls his witnesses, who may be cross examined. The respondent then follows in the same manner, and at the conclusion of his case addresses the arbitrator. Finally the claimant addresses the arbitrator. In all but the simplest cases evidence is given on oath, and the rules of admissibility of evidence are the same as in a case in the courts. At the conclusion of the hearing the proceedings are adjourned for the arbitrator to make his award. When the award is made and published to the parties the case is at an end. There is no appeal against the award of an arbitrator and the courts will enforce a valid award (Waters, 1980).

## Arbitration in Building Disputes

The types of building dispute that come to arbitration include the following:

(1) refusal to accept an architect's instruction;
(2) failure to agree the amount of a final account;
(3) failure to issue certificates;
(4) claim for extension of time;
(5) claim for additional loss/expense; and
(6) determination of employment of the contractor.

Quantity surveyors are often appointed as arbitrators in connection with building disputes as their qualifications, training and experience make them very suitable persons to act in this capacity.

Readers requiring more information on arbitration practice in construction contracts are referred to Stephenson (1993).

## The Expert Witness

Quantity surveyors are often called upon to act as expert witnesses at arbitration proceedings and public inquiries. An expert witness is a person who has special knowledge acquired, for example, in the course of professional training and experience. The knowledge enables the witness to assist the arbitrator or tribunal in coming to a decision by giving his opinion on the facts under consideration. The expert witness is frequently called upon to produce a proof of evidence, which takes the form of a written statement of what he will say, covering all his intended evidence. It will deal with the facts, which the witness may have elicited himself or which may have been made available in other ways, and the conclusions that he has deduced from those facts. If the expert witness has had the opportunity of examining the reports of opposing witnesses, his proof should also state where he differs from them and his grounds for doing so.

Quantity surveyors may be required to act as expert witnesses in respect of a wide variety of building disputes, which could embrace the adequacy of work, contractual and cost aspects. As a typical example an employer may have engaged a contractor to erect a house in accordance with 1:100 scale drawings and to the employer's satisfaction. The quantity surveyor may be called upon to give evidence of work that has been omitted and other work that is unsatisfactory and to give his opinion on the cost aspects of these matters for the benefit of the arbitrator or other adjudicator.

Readers requiring more information on the role of the expert witness are referred to Reynolds and King (1992).

## ALTERNATIVE DISPUTE RESOLUTION

Alternative dispute resolution (ADR) techniques form an important alternative to arbitration and litigation in the settlement of construction disputes. They are considered particularly appropriate for resolving disputes where there is an ongoing working/professional relationship and both parties desire and will benefit from swift resolution of their difficulties. The principal techniques are conciliation, mediation and mini tribunals, and Latham (1994) also includes adjudication.

ADR has proved to be a quick and economical way to settle disputes without recourse to formal proceedings. However, it does rely on there being sufficient trust between conflicting parties who must be genuinely willing to negotiate and to arrive at a mutually acceptable settlement. The main characteristics of ADR are: predisposition to settle, non-binding proceedings, senior management to play an active role, and guided by commercial interest and not the rule of law (Kwayke, 1993).

Where the parties are willing to settle their differences, but there are genuine obstacles to settlement, it may be possible to close the gap between the parties and obtain a settlement by the process of *conciliation*. It involves the appointment of an independent third party, mutually agreed by the parties, to hear both parties' points of view. The conciliator will usually be a recognised expert on the matters in dispute and will examine the evidence and listen to the arguments put forward by each party, in addition to contributing his own ideas on the merits of the case. All parties have to be present at his meetings and they can include the parties' legal advisers where these have been appointed, but the parties can meet each other without the conciliator being present. The conciliator's aim will be to bring the parties together to discuss fully the matters in dispute and lead them to an acceptable settlement. The conciliator does not make decisions but may make recommendations, and it is then up to the parties as to whether or not they accept them as a basis for agreement. If a settlement cannot be reached, the parties are free to pursue the matter in arbitration or litigation (Thomas, 1993).

The most popular ADR process is *mediation*, which is similar to conciliation, except that the mediator normally meets the parties separately and he may be empowered, if the parties fail to agree, to make a recommendation on the matters in dispute. Any confidential information supplied to the mediator at private meetings with one party cannot be divulged to other parties. The mediator will endeavour to find common ground at the separate meetings, at which each party can bring lawyers or other experts to help in presenting its case to the mediator, with a view to reaching a settlement. Subsequently, a meeting with all parties present will be arranged and it is vital that each party's representative(s) have authority to agree and settle the dispute. Once a genuine mutually acceptable solution is found and terms agreed, the mediator will bring the parties together and set out the agreement in writing in the form of a 'consent award'.

If the parties fail to reach agreement at the mediation, then nothing divulged can be used in evidence at subsequent arbitration or litigation, nor can the mediator be called as a witness at such proceedings. By the mid 1990s mediation was claimed to have been successful in over 90 per cent of cases and has proved to be a relatively low cost, completely confidential solution (Bullock, 1994).

*Mini trials* represent another form of dispute resolution procedure. A neutral person sits on the panel along with a senior executive from each party, and a hearing is arranged. At the hearing the parties are represented and each party presents its case referring to the 'core bundle' of documents. The panel may ask questions, but otherwise there is no cross examination as such. After the hearing the parties will endeavour to settle the dispute, using the neutral person as a mediator or conciliator as appropriate. After agreement has been reached, a joint written statement is prepared and is signed by the parties. The neutral witnesses the signatures and the agreement is then legally binding (Whitfield, 1994).

Latham (1994) advocated the use of *adjudication* as a means of settling construction disputes. If a dispute cannot be resolved first by the parties themselves in good faith, it can be referred to the adjudicator for decision. Separate adjudication was not provided for in JCT 80 in the mid 1990s, but Latham saw no reason why this procedure should not be introduced into the standard forms. Ideally, the adjudicator should be named in the contract before work starts but called in when necessary. In addition to dealing with disputes between clients and main contractors, the contract documents should specify that the adjudicator shall also be able to determine disputes between contractors and subcontractors, and between subcontractors and subcontractors. The authority of the adjudicator must be upheld and his decisions implemented at once. Latham did however accept that it would be difficult to deny a party which feels totally aggrieved by the adjudicator's decision the opportunity to appeal either to the courts or to arbitration.

## Benefits of ADR

Kwayke (1993) has usefully summarised the benefits of ADR as follows:

- disputes can be settled quickly
- saving in expense resulting from reduction in settlement time and possible elimination of legal costs
- flexible in terms of formalities: place, date and time can be arranged to suit convenience of disputants
- object and outcome of dispute decided by disputing parties themselves and not by or under the influence of lawyers
- direct participation by the disputants and the understanding of the strengths and weaknesses of the other party's case lends itself to a creative and amicable outcome
- joint effort made by the disputants to find a mutually beneficial commercial, rather than legal solution which may be too abstract, restrictive or inappropriate to their respective needs
- disputants free to find a solution which promotes ongoing business relationships without loss of face
- disputants able to avoid unwarranted publicity and can protect their trade secrets.

## Limitations of ADR

- only effective when disputants genuinely wish to negotiate a settlement and prepared to compromise
- complex construction dispute may not be resolved in a few days unless one party is prepared to compromise and this may lead to financial loss for the sake of continued business relationships
- neutral adviser may be biased, lack understanding of the technical context of the case and may be without the skill and experience required to shape the possible settlement
- the approach is non-binding and so an unscrupulous party can enter into negotiations knowing that there can be no agreement but that the final payment be delayed
- cannot be used for complex disputes which require legal opinion or in a case where public hearing or legal precedent is required
- neutral adviser may experience difficulties in extracting the truth of an event of the dispute which may protract the settlement to a point where litigation may be the only valid option
- disputants may be influenced by self interest, defensiveness and legal advice on the strength of their case and hence be unwilling to compromise (Kwayke, 1993).

These summaries highlight the many advantages of ADR, particularly with regard to the probable large savings in time and money, yet the limitations should also be considered when making a choice as to the best dispute solution route in a specific situation.

### ADR and the Quantity Surveyor

In 1995 the RICS Arbitrations Dispute Resolution Skills/Practice Panel established a dialogue with the Centre for Dispute Resolution (CEDR). CEDR is a non-profit making organisation supported by industry and professional advisers to promote and encourage the use of ADR. It was launched by CBI in 1990 and is generally recognised as one of the leading commercial mediation services in Europe.

Having regard to the importance of establishing a core of chartered surveyors with the appropriate training and expertise who can offer ADR services, it was proposed that the RICS should become an affiliate member of CEDR, and in return for recognition of CEDR's position as the leading training organisation in this field, that CEDR offer RICS members a streamlined training programme on advantageous terms. The criteria for the appointment or nomination of chartered surveyors as mediators would include accreditation by CEDR or any other recognised body in this field approved by the RICS.

# 14 Quantity Surveying Services in Europe and Overseas

## INTRODUCTION

As long ago as 1979 the RICS published the *Principles of Measurement (International) for Works of Construction* to provide guidance on the preparation of bills of quantities, with adaptations as necessary, for use overseas where existing rules are inappropriate or where no rules exist. This constituted an important service provided by quantity surveyors who recognised the need to provide a sound basis for the financial control and management of building contracts to keep pace with the substantial expansion of construction throughout the world, much of it in hitherto undeveloped areas.

UK surveyors also played a significant role in overseas developments through their representation on the International Federation of Surveyors (FIG) and the Commonwealth Association of Surveying and Land Economy (CASLE), both of which are examined later in the chapter.

The world today is an increasingly international one, as so many activities be they political, financial, trade or property, are performed regardless of national boundaries, and hence the profession needs to recognise the changing world in its future planning. There is an ever increasing demand for universal standards, including those of professional practice.

The RICS had a non-European membership of approximately 10 000 in 1995, and the countries with the greatest concentration of chartered surveyors and where the professional designation has greatest recognition are the commonwealth and ex-commonwealth countries. The RICS has accepted that it has to play a leading role in ensuring that the skills of chartered surveyors and the chartered designation are recognised throughout the world.

The RICS has recognised the increasing global nature of many working practices and accordingly undertook a review through its International Committee in 1994 and prepared an international strategy. This was followed in 1995 by the formulation of a detailed implementation plan analysing the resource implications.

The three main objectives of the RICS international strategy were:

*Organisation and support*: establish an organisational structure which will enable the profession to respond to its needs within a particular country or region and provide good quality support services to members in all countries;

*Education and training*: ensure sufficient high calibre recruits enter the profession through the extension of education and training opportunities worldwide;

*Promotion*: promote the skills of chartered surveyors worldwide to client markets, governments and other decision makers.

Langford and Rowland (1995), when examining overseas construction contracting, described how the decision to seek work overseas may have been made primarily in reaction to conditions in the home market, or in response to learning of an opportunity in a particular area, or stemming from a particular potential client or project. The initial action is likely to be a clearly defined and limited marketing initiative and to decide whether to proceed further in the light of the degree of success of this initiative. Key decisions at this stage are: the choice of area and potential client, often involving a desk study and site visit; the choice between invited tender and negotiation; the extent of the establishment, which can involve opening a local office, appointing a local agent or relying on communications from the UK supported by visits; and the tender sum made up of estimated cost, assessed risk and commercially judged margin (positive or negative), followed by sustained and well directed follow-up action. The majority of these criteria apply equally well to quantity surveyors contemplating providing their services in overseas locations.

A number of quantity surveying practices in the UK have shown considerable initiative in developing work overseas. The work of these practices encompasses many major overseas projects, including hotels, housing, industrial complexes, international and domestic airports, and arbitrations on major internationally funded capital projects. In the mid 1990s, some quantity surveying firms were making inroads into Eastern Europe and China. Overseas services are provided on a variety of bases. For example, the practice may receive a direct appointment by the client or, alternatively, may form part of a consortium with major contractors and other professional consultants.

In this context it is considered useful to analyse the overseas operations of two large UK based quantity surveying practices. For example, E. C. Harris has offices throughout Europe, South Africa, the Far East and Australia, and has been involved with international projects in Europe, Africa, the Middle East, Far East and Australia. The services offered are many and diverse and include a number of recently developed techniques, as listed in table 14.1.

Currie and Brown also provide quantity surveying services in many parts of the world and operate on a truly international scale. They have extended the core quantity surveying activities to providing consultancy to principal oil companies and design and management contractors, and have

Table 14.1  Summary of services offered by E. C. Harris

| | |
|---|---|
| ■ Construction consultancy | ■ Expert witness advice |
| ■ Management consultancy | ■ Construction management |
| ■ Employers' representative | ■ Resource programming |
| ■ Project assembly | ■ Value engineering |
| ■ Project management and coordination | ■ Financial engineering |
| ■ Feasibility studies | ■ Hazardous material and environmental audits |
| ■ Cost planning and control | ■ Strategic IT consultancy |
| ■ Financial analysis | ■ Computer systems analysis and software development |
| ■ Investment risk analysis | ■ Computer consultancy |
| ■ Mechanical and electrical installation condition surveys | ■ Property condition appraisal |
| ■ Insolvency and corporate restructuring | ■ Maintenance planning and implementation |
| ■ Claims consultancy | |

also been involved in joint ventures, encompassing projects ranging from US$ 60k to US$ 10m. Staff can be recruited from both expatriots and third country nationals. The expatriots may be recruited from the company's own existing staff resources, freelancers having a good track record for the company overseas, or locally recruited expatriots with a known track record. Third country nationals can be obtained on recommendation, by local advertising or through the local labour agency.

## EUROPE

### Overview

On 1 January 1993 the Single European Market came into being. This resulted in the end of trade barriers and the relaxation of customs regulations, enabling free movement within the European Union. The twelve separate domestic markets were replaced by a European domestic market serving over 350 million people. The construction market alone is worth more than £300 billion, constituting 30 per cent of the world construction market.

This major change poses a threat to those who are not willing to adapt but provides tremendous opportunities for those are who prepared to meet the challenges. Increased opportunities will be accompanied by greater competition resulting from the mobility of labour and abolition of barriers. Enterprising surveyors will become more European in outlook and larger firms who are not already established on the mainland of Europe are likely to consider setting up bases there, involving the investment of time and money in other European countries to obtain the greatest benefit. Fluency in a second language will become increasingly important. The directive on the mutual recognition of professional qualifications aims to make it easier for professionals to move around the European member states. Cooke and Walker (1994) give some valuable background information on the European Union and its impact on the UK construction industry.

In general, the skills of quantity surveyors are not separately recognised in other member countries. Their role is undertaken by engineers, architects or contractors, subject to the growing exceptions which are considered later in the chapter. It is possible that such services as cost planning, construction economics and project management could profitably be marketed as separate disciplines. Cooke and Walker (1994) examine in some detail the role of the quantity surveyor in France, Germany, The Netherlands, Denmark and Portugal. In addition these authors provide much information on the construction industry and construction procedures and techniques in each of the five countries.

### Needs of Those Preparing to Set Up Office in Continental Europe

Speakers at the RICS Annual Conference in Paris in 1989 gave their views on the necessary requirements for those contemplating setting up offices in continental Europe. Research, flexibility, good contacts and the ability to think as a European were considered vital for a successful business operation. Myers outlined the following four point plan for success:

- careful research into the new market
- encourage joint ventures with local experts and establish good professional contacts
- only proceed with opportunities which are as good or better than those in UK
- understand the local market.

There is an evident need for hands-on experience and fluency in the relevant European language. The chosen market should be researched thoroughly and it is desirable to appoint a national of the country of operation as second in command, as local knowledge and expertise is vital. Support from the firm's base is of paramount importance, including regular visits from the senior partner. The legal and tax factors must be considered before a particular type of company or partnership structure is decided upon, and salary and reward arrangements should be considered carefully as variations in costs of living can create disparities between offices in different countries.

## Construction Economics European Committee (CEEC)

CEEC constitutes the Construction Economics European Committee/Comité Européen des Economistes de la Construction, and was founded in 1979 by the Royal Institution of Chartered Surveyors, the Society of Chartered Surveyors in the Republic of Ireland and Union Nationale des Techniciens Economistes de la Construction. With the establishment of the single market and the need for close working arrangements between construction economists in all the EU member states, the Committee performs a most valuable role. Each member body has a maximum of three representatives on the Committee with an elected president, two vice presidents and an honorary secretary. In 1992/93 Brian Drake of the UK and a former chairman of BCIS was president.

The aim of the Committee is:

To facilitate the exchange of experience and information between professionally qualified persons who are responsible for construction economics in the EU member states and to initiate studies with a view to:

(1) promoting the training and qualification of persons who are responsible for construction economics and drawing up proposals for the harmonisation and acceptance of standards of training and qualification;
(2) establishing guidelines for the definition, content, control and practice of construction economics;
(3) ensuring adequate representation of qualified persons who are responsible for construction economics in the EEC Commission and in other European institutions;
(4) studying existing and proposed legislation and regulations relating to construction economics, with a view to their harmonisation;
(5) coordinating working methods;
(6) establishing European statistics relating to costs and types of construction, procedures and materials.

## Other Sources of Information

Readers requiring more information on European construction costs are referred to *Spon's European Construction Costs Handbook*, while those seeking more general information on the construction industries in other European countries could find the guides issued by the Construction Industry Research and Information Association (CIRIA) helpful. The RICS has set up an office in Brussels to maintain closer contact with other EC member states, with particular regard to building and surveying matters and reports of its activities appear in *Chartered Surveyor Monthly (CSM)* (Seeley, 1995a).

## FRANCE

Traditionally, the detailed design of a project in France is not finalised before tender but is carried out by the successful contractor as part of his obligations under the contract. Tender submissions tend to be lump sum based on each contractor's solution to the outline design and specification, which have proved difficult to assess, analyse and compare. In recent years, particularly on larger projects, there has been a move towards the UK approach (Gleeds, 1994a).

Historically, as described by Meikle and Hillebrandt (1989), the nearest French equivalent to a quantity surveyor has been the *metreur-verificateur* who has technical status and whose main function is to check variations and valuations of work on site. Precontract estimates have often been provided by a friendly contractor, bills of quantities were rarely prepared and projects were often well in excess of budget.

Wheatley (1989), who has practised as a quantity surveyor in Paris since 1980, works mainly for British clients with developments in France and his projects are subdivided between property development and investment, owner occupiers and tenants, and the public sector which requires effective cost control. His practice has also established a project management company with limited liability, working almost exclusively for non-French clients, who need someone to steer their projects successfully through the French system.

Gleeds (1994a) has described how in recent years, coinciding with the entry of more British practices, the designation *economistes de la construction* is being increasingly applied and is represented by the Union Nationale des Economistes de la Construction (UNTEC). There are however only a few higher education courses in France providing training for this role.

Despite the proximity to the UK, French law, culture, construction methods and even the construction industry are very different to those encountered in the UK. Wheatley (1989) stressed the need for adaptability in order to succeed.

Most quantity surveying firms that have successfully set up in business in France do not offer the same services as their 'technical' counterparts, but have concentrated on enhancing the overall value of the cost consultant to the construction process. Unfortunately, this has proved difficult and expensive, as it is not possible to secure the required fee level in the first instance. Hopefully, the deficiencies can be recouped from satisfied clients on future projects. Furthermore, professional fees in France are on average some 30 to 40 per cent lower than those in the UK (Gleeds, 1994a).

## GERMANY

In 1994 the German construction industry appeared set for significant growth, despite the sharply rising national debt, partly resulting from the large subsidies being fed into the former East Germany. Having invested DM38bn in modernisation, rationalisation and acquisition of new machinery between 1987 and 1993, the German construction industry was leading Europe in the value of production per employee, but suffered from a workforce of ageing craft operatives (Gleeds, 1994b).

Gleeds (1994b) have described how the British quantity surveyor has no equivalent in Germany, as his traditional work is performed by architects. The UK cost planning technique is not used in Germany and any pre-estimates are based on either the square metre or cubic metre basis. A total estimate of costs is needed to obtain building permission and this is submitted in a format similar to a BCIS analysis.

Depending on the chosen method of procurement, a bill of quantities may be prepared, although the trend is towards *General Unternehmer* or General Contractor arrangements, whereby the main contractor carries out the carcassing, and sublets and manages all the other trade packages. GU contracts are often based on a performance specification which usually leads to considerable horse trading during the contract negotiations and subsequently with the final account, resulting in the client often paying more than he originally anticipated (Gleeds, 1994b).

In 1994, several British firms of quantity surveyors were in the process of establishing themselves in the German market. Some were forming joint ventures with German engineering practices, bringing with them in most cases UK or foreign clients that appreciated the independent service of the quantity surveyor. However, Gleeds (1994b) believed that the independent quantity surveyor in Germany faced a number of obstacles, including resistance from architects who have traditionally worked without cost consultants. It is recommended that in order to succeed the quantity surveyor should adapt fully to the German system for cost planning, tender and contract documentation and postcontract cost control.

## SPAIN AND PORTUGAL

Reynolds and Sheppard (1989) found that the construction industry in *Spain* was very healthy at about half the UK output in 1989, and growing at about ten per cent per annum. However, they considered that Spain was a difficult country for professionals to establish practices. The country is divided into 17 autonomous units, each with powers to operate its own building regulations and planning systems. There are nearly 30 000 construction companies, but only 200 build large scale public and privately financed projects. The latter are tending to specialise in project design and quality control, placing the specialised work sections with sub-contractors.

In Spain the architect is paramount and it is obligatory to engage one on most building projects. He is responsible for design and supervision and employs the engineers and other professionals. Technical architects are responsible for site organisation, quality control and safety, and they also undertake most of the traditional functions of the UK quantity surveyor. These duties include the preparation of costed bills of quantity, measurement, cost control and general building economics advice.

There are three common types of contract in operation: the fixed price lump sum with no allowance for inflation adjustments or variations and is most used for single family housing; unitary quantities, the most common form of contract in which the unit price of construction elements (such as number of doors and cubic metres of concrete) are agreed in the contract with the final price based on measurement of work done (an inflation adjustment may be included); and the management contract where the fee may be a fixed sum or percentage, usually between 10 and 20 per cent.

*Portugal* is much smaller than the countries described previously with a population of around nine million and much less construction activity. Unlike Spain, the general arrangements are more open and adaptable and British professionals can operate without much difficulty. There is no standard form of contract and hence the contract conditions can vary from project to project. Engineers have the dominant role, being involved in design and quantifying as well as administration and coordination of a project. A team of *fiscals* are usually appointed to oversee the financial and quality control of the project. This is similar to the role of the British quantity surveyor but is carried out by engineers or technical engineers.

### EASTERN EUROPE

In eastern Europe, former eastern block countries are making the transition to a market economy, led by Poland's ambitious and fast track privatisation programme. There is enormous demand for the technical, managerial and financial skills of western contractors to help rebuild rapidly the outdated and deteriorating infrastructure of the eastern European countries. There is also the advantage of a large, well educated and potentially highly skilled workforce available. Making inroads requires considerable time and patience and should be regarded as long term investments, with an emphasis on joint ventures.

The main areas of British quantity surveying activity are Czechoslovakia, Hungary, Poland, Russia, Eastern Germany, Armenia and Bulgaria. For example, Gardiner and Theobald established a permanent office in Prague, and they have emphasised the need to form a good working relationship with local professionals, who are keen to learn western management methods. Cost management is not a separate discipline in eastern Europe, but is taught as part of architectural education albeit in a rudimentary form. E. C. Harris International has offices in Budapest and projects in various eastern European countries. Other quantity surveying firms undertaking work in eastern Europe include Franklin & Andrews, MDA and Dearle and Henderson.

Gleeds established a quantity surveying practice in Warsaw in 1989, which serves a number of clients on various types of project, including offices and hotels. Their advice to those contemplating such a move, was to set up a small office, learn the system and engage Polish staff. This probably entails several years with no fees. Gleeds' Polish office in 1995 employed five Polish nationals and four expatriots. Some British firms have been deterred by the lack of Polish funding. However, most projects passing through British offices are funded by western investors, mainly for their own occupation. The European Bank for Reconstruction and Development has provided venture capital for several projects and Polish banks started providing funds in 1995.

Gleeds were among the first British quantity surveying firms to establish an office in Prague in 1991. It is committed to a long term presence and has built up a substantial number of high profile clients, including local authorities, major landowners, the European Bank of Reconstruction and Development and private companies from Japan, USA and Europe. The traditional quantity surveying role of the firm has changed, as the offices contain many disciplines, providing business plans, property and plant evaluation, rental assessments as well as project and property management. This enterprising practice is also involved in development competitions, where developers compete for state-related work, and insolvency advice. The Warsaw and Prague offices of Gleeds augment those previously established in Paris, Madrid, Brussels and Braunschwig in Germany.

D. G. Jones and Partners opened offices in Sofia (Bulgaria), Bucharest (Romania) and Kiev (Ukraine) in the early 1990s. It is interesting to note that the three Bulgarian staff in the Sofia office speak fluent English and Russian in addition to Bulgarian, and all the staff in the Bucharest and Kiev offices are equally fluent in a variety of languages. The staffs undertake quantity surveying and project management duties and the projects include ambassadors' and embassy residences and commercial, industrial and residential development.

## RICS INVOLVEMENT IN EUROPE

The RICS Brussels office monitors closely the work of the Commission and the European Parliament. It sends relevant documents to the Institution and arranges high level meetings, retains contact with the media and deals with many queries.

By 1994, branches of the Institution had been established in Germany, France, the Netherlands and Portugal. While in the previous year the General Council of the Institution formally approved the creation of the European Society of Chartered Surveyors (ESCS) with the general aim of facilitating the promotion of the chartered surveying profession in Europe and the exchange of experience and information between professionally qualified chartered surveyors.

In 1994, the Institution issued a policy document covering central and eastern European countries (CEECs) and the new independent states of the former Soviet Union (NIS). The Institution's objectives are as follows:

- to assist members of the profession to take full advantage of the business opportunities available
- to promote the use of UK methods of practice suitably adapted and developed to suit local conditions, laws and culture
- to assist in the development of an indigenous profession of the land
- to promote or assist in the development of courses leading to a professional qualification or licence to practise and which may subsequently lead to membership of the RICS.

## THE MIDDLE EAST

In the early 1990s activities in the Middle East were dominated by the Gulf War and its legacy of destruction in Iraq and Kuwait. This resulted in an extensive programme of reconstruction work although a large proportion was given to the United States because of their bigger contribution to the relief of Kuwait. Prior to this the British construction industry's involvement in the region had reduced considerably as a result of the ending of the oil-induced boom. Other countries in the region, such as Egypt and Turkey, may offer good constructional opportunities (RICS, 1991).

D. G. Jones & Partners have extensive experience in the Middle East and outlined the main problems in establishing and operating offices as high cost of initial set up, constant management support is essential in the early years, and expensive travel and communication costs. Vacant appointments are advertised in 'Building' and successful applicants must have good qualifications and experience.

D. G. Jones started the practice in the Middle East out of Lebanon in the 1960s and in 1994 the firm had offices in Lebanon, Bahrain, Qatar, Abu Dhabi, Dubai and Muscat, and had previously had offices in Saudi Arabia, Kuwait, Iraq, Jordan and Egypt. The bulk of the work was quantity surveying but the practice was increasingly being asked to undertake project management.

Currie and Brown were kind enough to identify what they considered to be possible problems in establishing and operating in the Middle East and similar locations, and these are now listed.

## Establishing

- prequalification to work for some oil companies will not be given without having first obtained a licence to trade in the country concerned
- lengthy procurement period (3 to 12 months) to obtain a trade licence
- policy or procedural changes at the overseas government's whim
- unhelpful sponsors
- undefined requirements or procedures issued by the overseas government when applying for licence forms
- office premises, telephone, fax, bank account and signboards cannot be obtained without having a licence to trade
- difficulty of communicating with and understanding the 'average' non European
- in most cases a year's rent in advance is required to secure premises.

## Operating

- the obtaining and renewing periods for visas and labour permits for expatriot staff
- changes in the overseas government law relating to visas
- communicating with the UK due to time difference and working week
- understanding local labour laws and payments for sponsoring, hiring and firing
- making temporary financial arrangements while the licence application is being processed.

## FAR EAST

In 1995 UK quantity surveyors were stepping up efforts to win work in the Far East as prospects for traditional quantity surveying work in western Europe did not look too promising. For example, E. C. Harris believed that income from the Far East could account for half its turnover by 1998, while Northcroft also identified the region as its key growth area. E. C. Harris further added that Asian projects were often very large in size and that their primary selling lines of project management and construction management had proved very popular. Apart from Malaysia and Singapore, the firm identified a substantial growth market for quantity surveying services in Thailand, Indonesia and Vietnam. It was however stated that clients were becoming increasingly shrewd and the 30 to 40 per cent margins enjoyed by contractors working on office tower blocks in the early 1980s had been eroded significantly.

## HONG KONG

This country possesses a very different culture to western Europe, with a vibrant, high rise outlook, and a flourishing construction industry. In 1995 there were 877 qualified quantity surveyors, most of whom were members of the RICS and the Hong Kong Institute of Surveyors (HKIS), while the SST had about 400 technician quantity surveyors as the HKIS did not at that time have provision for a technical grade. The quantity surveyors were employed in private practice, with contractors and in the government service, with about 40 per cent of the total HKIS/RICS Hong Kong branch membership employed in government and quasi-government organisations, such as the Housing Authority, Mass Transit Railway Corporation and Kowloon–Canton Railway Corporation.

In 1995 quantity surveying in Hong Kong was very traditional, with the greater part of fee income coming from the production of bills of quantities. Procurement methodologies were also very traditional, although design and build was gaining ground. The 'traditional' quantity surveying market was largely dominated, at least in terms of size, by Davis Langdon & Seah and Levett & Bailey, each of whom employed around 200 to 250 staff. There were also a number of smaller practices and most of the large UK and Australian firms were represented. The Chinese market formed a large and growing part of the workload, and some quantity surveying practices had established branch offices in Chinese cities such as Shanghai and Beijing.

A substantial number of quantity surveying practices specialised in litigation and dispute resolution. While formal value management, including traditional value engineering studies had been used on a few projects, but many doubted the value of their use. The traditional rivalries existed between quantity surveyors and civil engineers, although some quantity surveying firms acted as subconsultants to civil engineering practices on some of the new Airport core projects. Bills of quantities for engineering services were increasingly being used, following a decision by the government architectural services department to demand them on all new projects. There was, however, a considerable shortage of M&E quantity surveyors and some of the larger practices employed their own in-house building services engineers to do the work.

The Hong Kong Institute of Surveyors (HKIS) was likely to become the primary body for surveying in Hong Kong, and was the only professional surveying institution safeguarded in the HK Basic Law. HKIS is organised on the RICS model with land surveying, quantity surveying, building surveying and general practice divisions. The growing importance and need for indigenous surveying bodies in developing countries has long been recognised by the RICS. It is interesting to note that in 1995 there was an active joint RICS/HKIS continuing professional development (CPD) programme organised by each of the divisions. There were excellent surveying education facilities at Hong Kong University, City University of Hong Kong and the Hong Kong Polytechnic University.

It was very refreshing to see how active HKIS was in promoting surveying in South East Asia, and it had reciprocity agreements with several other institutions in the region. It was a founder member of the Pacific Association of Quantity Surveyors (PAQS) (founded in 1993), together with the New Zealand Institute of Quantity Surveyors, the Australian Institute of Quantity Surveyors, the Singapore Institute of Surveyors and Valuers and the Japanese Institute of Building Surveyors. Furthermore, in 1995, HKIS was very active in China and had regular contact with surveyors and other construction professionals working there, and also with appropriate Chinese government ministries and other organisations involved in construction. Finally, Hong Kong has a Surveyors Registration Ordinance operated by a statutory Surveyors Registration Board, providing for suitably qualified surveyors to apply for registered professional surveyor status.

In 1994 the HKIS celebrated its tenth anniversary, although the history of the surveying profession in Hong Kong goes back to 1843 with the arrival of the first Surveyor General. A Hong Kong branch of the RICS was established in 1929 and HKIS maintained close links with the RICS throughout. The HKIS publishes a scale of professional charges for quantity surveying services for building works in Hong Kong and was, in 1994, updating the Standard Method of Measurement for Building Works and the Joint Form of Contract. A total of 64 CPD events were organised in 1994–95, which is indicative of the enthusiasm and enterprising character of the Institute's members in a heavily work committed environment (HKIS, 1995). HKIS is noted for its very high number of surveying students, showing the popularity of the surveying profession in Hong Kong with a wide range of career prospects and the opportunity to work on large and exciting projects.

## MALAYSIA

Economic growth in Malaysia in 1994 had outstripped predictions in the country's sixth five year plan, resulting in the infrastructure being inadequate to support the Malaysian economy's 8 per cent annual growth. Impressive planned projects in 1994 included the new Kuala Lumpur International Airport, a second road bridge to Singapore, the Kuala Lumpur light rail transit system, redevelopment of Kuala Lumpur city centre and facilities for the 1998 Commonwealth Games. In the early 1990s, the construction sector was growing at more than 10 per cent per year. Malaysia was attracting much international investment because of its stability, favourable climate, relatively low cost and labour availability, and the surveying profession has much to contribute.

Despite the vast amount of construction work in progress, contractors' margins are cut to the bone. It is considered vital to develop personal relationships with clients as most of them are open to offers of cheaper or more efficient ways to build their properties. Even the government is open to any ideas that promise to deliver its infrastructure programme under budget. Although Malaysia has a well educated population, there was in 1994 a shortage of trained staff and an opportunity for UK firms to participate in joint venture schemes and supplying skilled personnel. For example, quantity surveyor, Currie and Brown together with its local partner, offered administration services as a package to overseas and local contractors (Macneil, 1994).

The Institution of Surveyors Malaysia (ISM) was founded with RICS support in 1961 and brings together valuation, quantity surveying and land surveying disciplines. The surveying profession in Malaysia is tightly regulated and each discipline has a registration board. ISM was established to promote the general interests of the profession and to maintain and extend its usefulness for the public advantage. Among its objectives are the following which relate to the work of the quantity surveyor:

- provide financial appraisal of real property development, determining cost and economics of design, technical auditing and advising on construction contracts
- contract and financial management of construction projects including measurement and valuation of construction works.

The main activities undertaken by ISM on behalf of members are as follows:

- organising of CPD talks, seminars and technical/social visits
- organising consultative meetings with other institutions, such as Institution of Engineers Malaysia, Master Builders Association Malaysia, Board of Quantity Surveyors Malaysia and government agencies
- conducting the test of professional competence (TPC) jointly with the Board of Quantity Surveyors Malaysia
- conducting the ISM professional examinations.

In 1995 the approximate numbers of qualified/registered quantity surveyors in Malaysia were:

| Sector | Number of quantity surveyors |
|---|---|
| Government | 160 |
| Consultants | 340 |
| Contractors | 50 |
| Government Statutory Board | 15 |
| Institutions of Higher Learning | 40 |
| Others | 76 |
| Total | 681 |

## SINGAPORE

Singapore has received substantial regional investment from various countries, including Japan, Hong Kong, Taiwan and Indonesia. In consequence,
the surveying profession has obtained new opportunities. The market in
Singapore is highly competitive but quantity surveying skills are much in
demand. Quantity surveyors are regulated by statute.

The Singapore Institute of Surveyors and Valuers (SISV) looks after the
interests of surveyors in a very competent and effective way. It performs a
model role operating as a family of surveyors with a good understanding
of members needs through education and throughout their professional
careers. SISV arranges many professional meetings and seminars which
are all very well attended by enthusiastic groups of hard working members. In 1995 there were 450 quantity surveyor members of the Institute.

SISV defines the main roles of quantity surveyors as follows.

- advising clients, as well as professionals in the project team, on the
  most effective contractual arrangements to attain the project's objectives
- advising on tendering arrangements to select the contractor for a project
- compiling the contract documents prior to execution of the project
- valuing work on the construction site and analysing and evaluating claims
  for additional costs submitted by the contractor
- providing estimates of the construction cost of the building and exercising effective cost control at all stages
- providing valuations for fire insurance
- undertaking feasibility studies, environmental assessments and cost benefit studies
- providing expenditure statements for tax and accounting.

In Singapore, quantity surveyors have been involved with many large
and exciting developments, including the Singapore mass rapid transit system
(MRT) with the sections in Singapore city being constructed underground,
the continuing development of Changi International Airport, the construction of new towns and extensive high rise housing developments, substantial leisure complexes and a large programme of commercial and industrial
development.

## JAPAN

Yamamoto (1994) has described how many Japanese private companies
prefer design and build lump sum contracts. By contrast, public sector
clients operate competitive tendering procedures generally separating design and construction. Sometimes the nominated design and build contractors compete on the basis of an initial design proposal with cost and

programme plans. Once the design is under way, the successful tenderer agrees the detailed costs and programme. Clients nominate a specific contractor to proceed with the work based on a letter of intent and appoint the firm officially when contract conditions and price are agreed. To spread risk and train local contractors, a joint venture approach with other contractors is often used on major contracts.

Japanese building contracts are quite simple with only the scope of work, cost and completion date agreed prior to commencement. As a result of construction scandals involving bribery and corruption of local government officers, many local authorities are adopting open competitive tendering and calling for performance bonds. With regard to consultants, most architectural practices confine their activities to domestic buildings and there are few structural or M&E engineers and even fewer quantity surveyors.

At the precontract stage, Japanese clients evaluate the contractor's commitment to the project and require assurances based on the company's past performance. Based on past experience, contractors tend to include provisional sums and/or contingencies in the original lump sum contract. In general, Japanese contractors spend more time on design, preplanning and site management, resulting in higher design and build costs than occur in the UK. In consequence, Japanese clients expect the contractor to complete the project on time with no claims for extras, unless additional works have been requested. Once the building is completed, Japanese contractors often provide continuous aftercare. Maintenance requirements are fully detailed in the building contract and it is quite common for the contractor to undertake term maintenance as part of the overall aftercare commitment (Yamamoto, 1994).

Japanese project managers are highly qualified; often holding an architect's licence and having a good understanding of structural engineering, building services, cost control and time management. Site teams develop design details into working drawings and identify design discrepancies. Alternative design details can be proposed by the construction team to improve production.

D. G. Jones & Partners, a UK firm of international quantity surveyors, described in 1994 how they initially combined with a Japanese overseas project management company in Tokyo, undertaking mainly contract documentation for overseas projects. Subsequently the practice linked with architectural companies with overseas projects and carried out work for Japanese contractors who were very active overseas in the mid 1980s, coupled with project management and cost control services for private investors on overseas contracts. In the period 1989–94, the firm's main client base in Tokyo was mainly banks and financial institutions.

## UNITED STATES

In 1993 there were approximately 160 quantity surveyors registered with the RICS as working in the United States. A very informative questionnaire survey was undertaken by the East and Midwest USA QS subgroup to obtain information about the activities of these quantity surveyors, and over 50 responses were received. The predominant influences on quantity surveyors coming to the US were financial/economic and increased experience. Some were transferred by their home company, some directly approached companies in the US and others responded to advertisements. They entered into longer contracts of employment than the common one or two years contracts in the Middle or Far East.

Few quantity surveyors have acquired American professional qualifications and those that have usually opted for membership of the American Association of Cost Engineers (AACE). They worked for a variety of organisations mainly ranging from cost consultancy/quantity surveying type practices to general contractors and construction managers. Approximately one half held high level management positions, indicating that career prospects for quantity surveyors are good. Norton (1993) has, however, identified significant obstacles to be overcome by both individuals and companies prior to successful operation, as the US market has little or no experience of the quantity surveying role. Various components of the traditional quantity surveyor's work is performed by other well established disciplines in the US construction industry. For example, estimating, project strategy advice, tender and contract documentation, postcontract cost and contract control and administration are usually provided by the architect, construction manager or a combination of both.

Respondents often complained that they were only performing limited roles, being mainly recruited for their estimating skills, and were not able to make full use of their expertise. Norton (1993) suggested that a possible remedy was to enter the US market with a recognisable title, such as construction management or project management consultant and then set out to improve on the shortcomings of the services normally performed under these titles. Cost management is becoming increasingly important to US clients and this should provide scope for an expansion of quantity surveying activities.

## AUSTRALIA

The main professional body for Australian quantity surveyors is the Australian Institute of Quantity Surveyors (AIQS), which in 1994 had 1118 corporate members and 770 non-corporate members, giving a total membership of 1888, of which 428 were working overseas. There was a good complement of Australian universities offering Bachelor of Applied Science degrees in Construction Economics or Quantity Surveying.

In 1994 the national president of AIQS believed that the Institute was redefining the quantity surveyor's role for the next decade, in a similar way to that being pursued by the RICS Quantity Surveying Division in the UK. As Australia recovers from the downturn, the quantity surveyor should be a reliable practitioner whose advice is sought by a much wider range of clients than ever before. Hence the quantity surveyor has to grasp the opportunity that exists and ensure that his place in the marketplace is secured not only in the area of essential traditional services but also to meet the new demands. To meet the future challenges the following steps were being instigated:

- establishment of a strategic plan as a guide towards a planned development of the Institute's aims
- examination of the Institute's governance system to enable faster and more effective responses
- review of strategic relations with other professional organisations and or industry groups to obtain the best possible representation on these bodies to enhance the image of quantity surveying
- introduction of systems to ensure that quantity surveying maintains its high standard of recognition among fellow professionals
- introduction of a promotional and training programme to raise the profile of the profession, in the eyes of both clients and the public
- bringing together groups within the Pacific region to identify areas of common concern and to speak as a unified group (AIQS, 1994).

The well formulated mission of the AIQS was to:

LEAD: the development and promotion of the discipline of quantity surveying/construction economics
INFORM: the community of the benefits of total cost management
DEVELOP: and maintain standards of excellence and best practice in the profession
PROMOTE: quantity surveying as indispensible to value added services in the construction process
ENCOURAGE: the efficient, effective and sustainable use of all construction resources.

The Institute has developed an effective organisation through its Council and Executive, supported by four boards comprising strategic alliance, communication and policy, professional standards and publications and marketing material. There are also six chapters covering the geographical areas of Queensland, New South Wales, Australian Capital territory, Victoria, South Australia and Western Australia; and six divisions covering North Queensland, Tasmania, Northern Territory, Gold Coast, Singapore and

Thailand. These help to indicate the vast areas administered by the Institute on behalf of its members.

The Institute has representatives on a wide range of national construction and related bodies including the following:

Australian Council of Professions Ltd (ACP)
Australian Universities Building Education Association (AUBEA)
Construction Industry Development Council (CIDC)
International Cost Engineering Council (ICEC)
National Building & Construction Council (NBCC)
National Surveyors House (to share costs of Canberra HQ and Museum of Australian Surveying)
Association of Consulting Quantity Surveyors, Australia (ACQSA)
Pacific Association of Quantity Surveyors (PAQS)
Construction Forecasting Committee (CFC)
International Construction Information Society (ICIS)
NATSPEC Pty Ltd (develops and markets national standard construction specification)
National Committee for Rationalised Building (NCRB).

The wide ranging operations of AIQS in 1994 included examining and assessment of candidates for entry; admission and elevation of members; conducting education and training courses; assessing tertiary courses; sponsoring conferences and seminars; providing career advice to students and employment advice to members; writing and publishing Australian standards; publishing the *Building Economist*; compiling and publishing statistics; participating in international, national and state industry organisations; cooperating with overseas and national kindred professional associations; administering its code of professional conduct; providing information to members of the public; facilitating good relations among members; and investing members' funds to preserve their value and contribute to operating expenses (AIQS, 1994).

The AIQS rightly attached great importance to the fostering and operation of continuing professional development (CPD), aimed at providing a stimulus for members to develop new or enhanced skills, leading to broader and better services for clients and employment opportunities for members. It was recognised by the Institute that its success required the commitment of individual members, as well as the support of employers to create opportunities for staff training and the development of skills. CPD in the UK is dealt with in considerable detail in chapter 18. In 1994 the downturn in the construction industry in Australia seriously depleted the opportunities for students to obtain appropriate work experience and the Institute responded with a scheme to provide limited work experience in quantity surveying practices at a modest cost to employers.

The AIQS promoted the value of the quantity surveyor, aided by technological advancements in information technology (IT), to offer a complete management service, which could be offered piecemeal or as a package, and could include facilities management, value management, cost management, cost modelling, economic forecasting, life cycle management, documentation management, critical path networking and programming. Looking to the future the Institute recognised that technological developments will result in some of the technical tasks being carried out by electronic means. It is anticipated that the quantity surveyor will devote more time and energy to professional tasks such as research, analysis, policy advice, economic forecasting, and various management services. This will result in an overall reduction in quantity surveying practitioners, but those who do carve a market niche will adopt a much higher profile and will be highly regarded in whichever industry they are advising. Complementing this should be an increase in the number of quantity surveyors directly employed by contractors and other commercial organisations (AIQS, 1994).

*New Zealand* also has a well established and organised quantity surveying profession which has been registered since 1968 and the Quantity Surveyors Institute of New Zealand was formed in 1943.


## SOUTH AFRICA

In 1995 the approximate numbers of qualified quantity surveyors and non-registered qualified surveying technicians and their distribution was as follows.

| Sector | Nr of QSs | Nr of technicians |
|---|---|---|
| Contracting | 172 | 3 |
| Private practice | 1860 | 117 |
| Public sector | 70 | 2 |
| Retired | 96 | – |
| Total | 2198 | 122 |

Quantity surveyors in private practice in South Africa are all involved in general consultancy work, which includes project procurement and, at every opportunity and in increasing measure, appointment as principal agent within the project design team. In 1995, the profession at large did not play a major role in project management, although a small number of prominent firms were experienced in this field and interest was growing. Each housing project subsidised by the State required the appointment of a 'project manager' and those firms who were prepared to take on the role of mentor/manager/skills trainer/project coordinator enjoyed the title of 'project manager' with specific reference to housing delivery. Civil engineers were, however, the leaders in project management when associated with government construction schemes in their entirety. Value management was not perceived to be a specialised activity and tended to be regarded as 'something all quantity surveyors do as a matter of course' in the interests of the client, although probably not viewing the process in the manner outlined in chapter 9.

In the case of civil engineering and building services, certain engineering firms retained the in-house services of quantity surveyors to their evident advantage. Most of the large contracting organisations employed qualified quantity surveyors and technicians in their estimating departments. In addition, a number of the senior managers/directors in major South African contracting organisations, with international affiliation, were qualified quantity surveyors.

The Association of South African Quantity Surveyors (ASAQS) is the leading professional body for quantity surveyors in South Africa. The Association is very ably governed by a Board of eminent quantity surveyors, supported by committees covering construction economics, continuing professional development, education loans, engineering matters, fees advice, finance, general practice, housing, membership, model preambles, model preliminaries, prizes and awards, public relations and benevolent fund (ASAQS, 1995).

The Association is also represented on the Alliance of Development Professions (ADP), Building Industry Advisory Council, Construction Council of South Africa, International Cost Engineering Council, Joint Building Contracts Committee (JBCC) and the Standard System Joint Committee.

It is also worthy of note that the Quantity Surveyors' International Congress in 1996 took place at Bloemfontein to coincide with the 150th anniversary of that city. The Association maintains an education support fund which makes grants to the long established quantity surveying departments of the universities of Pretoria, Witwatersrand, Natal, Port Elizabeth, Bloemfontein and Cape Town. These grants are used to supplement computer hardware and software, to subsidise staff attending overseas conferences, etc., in recognition of the valuable role that these distinguished universities play in the education of quantity surveying undergraduates and graduates. The Engineering Matters Committee has issued model bills for electrical and air conditioning installations and a booklet on tendering for quantity surveying services on engineering work. The Fees Advisory Committee conducts an annual salary survey in order to provide a feedback to participating firms (ASAQS, 1995).

## KENYA

The Board of Registration of Architects and Quantity Surveyors in Kenya has 310 registered quantity surveyors; registration being necessary to practise as a principal. It is likely that about 20 per cent of those registered are resident outside Kenya. The principal professional body for quantity surveyors in Kenya is The Architectural Association of Kenya, Chapter of Quantity Surveyors, which had a corporate membership of 191 quantity surveyors in 1996. The Department of Building Economics at Nairobi University produces approximately 20 quantity surveying graduates per year and has been doing so since 1969.

The approximate breakdown of quantity surveyors by employment is 65 per cent in the public sector, 25 per cent in private practice and 10 per cent in contracting. The activities of quantity surveyors are similar to those in the UK and project management is emerging strongly with the quantity surveyor being a major player. However, value management/engineering is little understood or adopted and building services tend to be handled by services engineers. An area where the quantity surveyor predominates in Kenya is in the Kenya Branch of the Chartered Institute of Arbitrators, which is very active in promoting both arbitration practice and in conducting courses and examinations.

## NIGERIA

The quantity surveying profession in Nigeria has made rapid progress since the first Nigerian qualified in 1963. In 1996 there were about 1200 qualified quantity surveyors and some 140 firms. There were also six universities offering quantity surveying degree courses with two offering postgraduate courses, supported by 16 polytechnics providing HND courses in quantity surveying, making it a substantial growth area and indicating the importance attached to the profession.

The Nigerian Institute of Quantity Surveyors (NIQS) was formed in 1969 and formally registered by the government in 1970, and the Quantity Surveyors Registration Board of Nigeria was established by decree in 1986. The Institute is a leading body in organising seminars within the construction industry, is a member of CASLE and spearheaded the formation of the African Association of Quantity Surveyors in 1996.

In Nigeria, quantity surveyors are employed in every sector of the economy, including the government service. While they are denied participation in major civil engineering projects because of opposition from engineers in the civil service, many quantity surveyors have proved an indispensable part of engineering contractors' organisations. By the mid 1990s, quantity surveyors had made significant inroads into the measurement of mechanical and electrical engineering work.

In matters of arbitration within the construction industry, the NIQS is approached to appoint arbitrators more frequently than members of the kindred professions. Quantity surveyors also hold their own in the area of project management and replacement cost valuation.

## COMMONWEALTH ASSOCIATION OF SURVEYING AND LAND ECONOMY

This association (CASLE) was established in 1969 at the instigation of the Commonwealth Foundation and largely on the initiative of the RICS, to foster the development of all surveying specialisms in what were then the newly independent countries of the Commonwealth. It undertook this task with considerable success, partly due to the strength of its administration with the RICS providing the secretariat and partly due to the developed Commonwealth's support for CASLE.

The association prepared some extensive guidance notes on relevant subjects such as project management and outline syllabuses for possible degree courses in surveying disciplines, including quantity surveying. CASLE also carried out accreditation of courses in the surveying disciplines in a very thorough and painstaking way and organised extensive conferences on current surveying topics with papers presented by speakers from a wide range of member countries in selected locations hosted by the member country.

In 1988/89 CASLE member societies decided to increase subscriptions to finance an independent secretariat headed by a part time salaried executive director to provide increased levels of activity. Unfortunately, by 1993 the association's ongoing activities had been seriously eroded with provision for only a few lectures and workshops in developing countries and the issue of occasional newsletters. This resulted in the only societies from developed countries that remained in membership being the RICS, who contributed 80 per cent of CASLE's subscription income, and the New Zealand Institute of Surveyors.

The RICS regretfully felt obliged to withdraw financial support as from 1994/95 and recommended that CASLE/FIG collaborative and other mechanisms for taking forward worthwhile CASLE objectives should accordingly be actively and quickly pursued. Readers will see from the next section in this chapter that CASLE and FIG shared some common objectives.

## INTERNATIONAL FEDERATION OF SURVEYORS

This federation (FIG) was founded in 1878 and is the only international body that represents all surveying disciplines and that is concerned with professional development and with the promotion of both the practice of the profession and professional standards. FIG has no permanent secretariat as this rotates at four year intervals amongst its members and the administering society can have a substantial influence on the federation's direction and development. At the end of 1995 the RICS took over the administration of FIG for a four year period. FIG's administrative structure is headed by its Bureau with its officers nominated by the administering society.

In furtherance of FIG's aim and objectives, the RICS FIG Management Group agreed a mission statement for 1996–99 focussing on the surveyor's response to political, economic, technological and environmental change. The emphasis is on promoting professional development and assisting surveyors to develop the necessary skills and techniques to cope with changing markets for their services, and to be properly equipped to meet the needs of society and the environment.

The RICS led Bureau is to implement the following important measures:

- promote FIG and the full range of surveying services throughout the world
- help associations of surveyors to gain recognition by their governments, not as trade associations or solely learned societies, but rather as regulated or self-regulating professional bodies
- improve communications between the Bureau and member associations and, through them, to their individual members, using the opportunities created by information technology
- provide opportunities for continuing professional development, especially by organising seminars and workshops in developing countries
- progress the work that has already been initiated to establish a permanent secretariat, with the great advantage of securing continuity and improved control
- seek the financial resources needed to support FIG's professional and technical structures and a permanent secretariat.

It was recognised that the membership base needed expanding to make FIG more representative of all surveying disciplines, and particularly construction economics and valuation/property management. The RICS established a FIG Forum which meets quarterly and circulates the RICS FIG newsletter also quarterly. The 1998 FIG congress will be held in Brighton with considerable input from the RICS.

# 15 Quantity Surveying Organisation and Practice

## OFFICE AND STAFFING ORGANISATION

### General Procedure in a Quantity Surveying Office

It is essential that an office is well organised so that work is dealt with in a satisfactory and logical manner. At the start of a new project, all architect's drawings should be stamped with the office stamp and date of receipt, listed and carefully examined by all staff concerned with the measurement and cost planning work. Figured dimensions on the drawings should be checked and any omitted dimensions calculated and inserted on the drawings. It will assist the subsequent measurement if walls and partitions are coloured in different colours according to type and thickness for ease of identification. It is also good policy to insert on the general location drawings, normally drawn to a scale of 1:100, a list of component details. Any queries on the drawings or supporting documentation should be entered on query sheets for subsequent clarification by the architect. Where reference is made to materials, components or proprietary systems with which the quantity surveyor is unfamiliar, he should obtain full particulars from the manufacturer.

It is sometimes the practice on large projects to subdivide the taking-off work between different quantity surveyors or even separate groups. The subdivision could, for instance, take the form of (1) structure of the building(s); (2) joinery and finishes; and (3) services and external works. Another and probably better alternative is for the whole of the taking off work to be undertaken by a single group of staff under the supervision of a senior surveyor or team leader, the group consisting of possibly three to six staff according to the size of the project. Furthermore, there are distinct advantages in arranging for this group to undertake all the work from inception to completion of the project, encompassing cost planning, contract document preparation, tender assessment and post contract work. This procedure enables the staff to obtain a wider and more interesting experience and is likely to result in improved efficiency through greater familiarity with all the details of the scheme. With very large schemes it may be necessary for separate groups to work together to rationalise resources and use them more effectively.

After the draft bill is prepared, the important task of examining and editing it by a partner or senior surveyor follows. The proof bill from the printers also requires thorough checking, particularly with regard to quantities

399

and descriptions, preferably involving two members of staff. The final documents will be despatched to tenderers with a covering letter stating the number and nature of the documents; the date, time and place where the tenders are to be delivered, often in an envelope enclosed for the purpose; where and when the contract drawings can be inspected; how the contractor can visit the site; and usually a request for acknowledgement of receipt of the documents. Great care should be taken to ensure that each contractor receives a complete and correct set of documents.

With office procedures, it is advisable to be forward looking, retain an open mind and not to think that the customary approach is always the best. On the other hand new methods may not be foolproof and require careful examination and possibly a trial period of implementation on a small scale in the first instance. Nevertheless, we are passing through challenging times and there is a need to periodically reappraise many existing procedures relating to a wide variety of activities, including such matters as cost advice and research, feasibility studies, the pooling of cost and other data for the common good, rationalising of technical information, measurement and cost control of mechanical and electrical services, new methods of construction procurement, new methods of bill preparation, new office equipment, increased use of computers, and automation of measurement. Many of these activities could be operated more efficiently and economically given the time and motivation for their analysis and improvement.

Construction cost management now forms a much more important part of the quantity surveyor's work as described in chapter 8, and this includes the keeping of extensive cost records of all projects passing through the office, to assist with cost forecasting of future projects, and often the establishment of a comprehensive cost database with computerised data manipulation. The preparation of accurate cost estimates will always form a vital part of the work of a quantity surveying office.

### Management in Practice

Management is concerned with the effective use of resources so that work is done efficiently and objectives are met within the prescribed time scale. Of prime importance among these resources is the management of people. It is not just a question of using a person's skills and aptitudes for the benefit of the firm or other employing organisation, but also of providing encouragement and motivation for the people concerned. One very effective method of achieving this aim is by delegation of duties and responsibilities, but having full regard to the abilities, strengths and weaknesses of the staff concerned. The manager can then concentrate on management activities while leaving the subordinate free to undertake his new assignments within the limits of his authority and subject to a monitoring system (Muir, 1983).

The style a manager adopts often reflects his attitude to others. Negative attitudes will lead to a more autocratic style – the manager believing that people are basically lazy and need firm control. The democratic manager has a more positive attitude to the team. He sees them as responsible, keen and capable of initiative and self control. He listens to their ideas and encourages them to become involved in the decision making process (Fryer, 1983). In practice many organisations tend to adopt a middle course, combining encouragement of initiative and a corporate spirit with some measure of overall control.

Garnett (1979) believed that the success of professional organisations depended on the ability and enthusiasm of the professional staff. Their motivation is currently made more difficult because of problems of size, lack of real financial reward, changing nature of trade unions and effect of government legislation. Partners and their counterparts in the public sector and contracting should take positive steps to ensure the commitment and enthusiasm of their staff, and these include:

(1)  a clear structure of accountability where no one person is responsible for the motivation of more than 15 people;
(2)  clear targets for staff who can then see what is expected of them and accordingly obtain a sense of satisfaction;
(3)  some simple instructions to section or team leaders to enable them to draw out the aptitudes, talents and ability of the staff for whom they are responsible; and
(4)  to have a systematic and regular communication mechanism through partners and team leaders so that all staff know what is happening and why.

In the wake of government action with regard to the professions, Allsopp (1979) believes that the challenge is to evolve as a more competitive and efficient profession, while retaining the integrity and skill that every true professional relies on for his sense of vocation and self esteem. This can be done only by improved standards of management in selecting, planning, coordinating, controlling and reviewing every resource and activity in which it engages.

Senior management need to constantly keep under review such aspects as:

(1)  organisational structure, including definition of roles, division of work among staff, workflow and interrelationship of tasks, and degree of specialisation;
(2)  managerial style, including suitability to tasks undertaken, staff experience and needs, and constraints:
(3)  planning procedures, including communication and information flow and feedback mechanisms;
(4)  human factors, including manpower policies, reward systems, motivation and satisfaction; and

(5) staff development, including the recruitment, selection, training and career prospects.

Staff promotion is often one of the more difficult tasks facing an organisation, as apart from technical ability and willingness to work hard, other qualities are required such as:

(1) ability to carry out several tasks at the same time without panicking or letting the work suffer;
(2) facility for getting on well with other staff at all levels;
(3) skill in identifying problems that should be referred to senior management for decision and those that can be settled by the individual;
(4) ability to train others and keep senior management informed of their progress; and
(5) enthusiasm.

Ideally heads of organisations should know all members of staff as individuals – their personalities, temperament, leisure pursuits and domestic circumstances. Staff should feel free to discuss their personal problems with a partner or equivalent. Reprimanding staff should normally be done in private while praise in public has much greater impact. Congratulatory letters should always be circulated to all the staff involved.

### Statutory Requirements

All offices are subject to a number of statutory requirements which are now listed:

(1) *The Offices Shops and Railway Premises Act 1963*
This Act applies to most offices and imposes duties to ensure health, safety and welfare.
(2) *Fire Precautions Act 1971*
Reasonable fire precautions are required in all offices but certain offices require fire certificates covering fire escape routes, fire-fighting equipment and fire alarms.
(3) *Health and Safety at Work Act 1974*
Employers and employees are required to take reasonable care for the health and safety of themselves, those under their control, and others who may be affected by their acts or omissions at work. All accidents to employees should be recorded.
(4) *Equal Pay Act 1970* and *Equal Pay (Amendment) Act 1984*
Men and women must receive the same pay for like work, and the entitlement to equality extends to other terms of employment such as sickness benefit.
(5) *Employment Protection (Consolidation) Act 1978, Contract of Employ-*

*ment Act 1972* and *Trade Union Reform and Employment Rights Act 1993*

Under these Acts employees have various rights with regard to holidays and payments. For instance employees leaving to have a baby are normally entitled to maternity pay. There are also industrial tribunals which deal with a wide variety of cases ranging from complaints of individual employees of unfair dismissal or incorrect redundancy payments to accusations of discrimination on grounds of race or sex.

(6) *Sex Discrimination Act 1975* and *Race Relations Act 1976*

These form a code of good practice to provide equal treatment for both sexes and all races.

(7) *Disability Discrimination Act 1995*

This provides guidance on the employment of disabled persons.

There are also certain obligations implied in all contracts of employment. The employer owes the employee a duty to provide work, pay wages, take reasonable care of his safety and indemnify him against liability in the proper performance of his duties. The employee's duties are to act reasonably, honestly and faithfully and to use reasonable care and skill in performing his work, and to indemnify the employer against liability incurred as a result of a breach of this duty (Greenstreet, 1994).

Employment legislation requirements are also influenced by regulations which are issued from time to time and decisions of the European Court of Justice.

## General Office Management

Secretarial and clerical staff should be given clearly defined duties and responsibilities yet, at the same time, be encouraged to work as an effective team with a sense of loyalty and pride in their work. For instance, it will be one member of staff's responsibility to ensure that adequate stocks of all types of paper, stationery and other required articles are held at all times. Telephone messages should be properly recorded and the messages relayed to the persons concerned as quickly as possible.

Letters should always be drafted with care and the following guidelines observed:

(1) there should be no possible doubt as to the true intent and meaning of the letter;

(2) the wording should be as simple and concise as possible;

(3) where extensive information is entailed, it is probably advisable to send a brief letter accompanied by a detailed report or schedule, and most of the comments on letters also apply to reports, which are normally subdivided into three separate sections – introduction, body of the report, and conclusions and recommendations;

(4) technical terms should be avoided as far as practicable when writing to lay persons;

(5) letters must be free from grammatical and spelling mistakes and have the correct punctuation; and

(6) impersonal language should be avoided as it lacks emphasis – hence 'I/We consider' is preferable to 'It is considered'.

When despatching correspondence, a check should be made to ensure that all relevant enclosures are sent with the letter, and a suitable note on the letter or a small enclosure tab will help to avoid their omission, which can be frustrating to the recipient. Careful checks are also required to ensure that the correct letter is inserted in the correct envelope and that it bears the correct postage.

All incoming correspondence and copies of outgoing letters must be carefully filed in a system that permits easy retrieval. The main file heads in a quantity surveying organisation relate to projects, but on very large projects some degree of subdivision is normally required, and this may encompass the employer, architect, subcontractors, suppliers and others involved in the project. All correspondence should be filed in date order with the most recent at the front of the file. Letter references often include the project reference and the initials of the writer and typist.

Other essential office documents include staff diaries recording appointments and notes of meetings, and records of staff time spent on each project. Minutes of meetings, site records, project details, cost information and other supporting data must all be carefully preserved. Another important section of any quantity surveying organisation is the information centre or office library containing trade literature, reference books, appropriate Acts of Parliament and statutory instruments, British Standards and the like, all suitably classified and indexed.

### Office Equipment

There is now an extensive range of sophisticated office equipment on the market including dictating machines, word processors, microprocessors and electronic typewriters. Great care is needed in the selection of equipment having full regard to the managerial and operational aspects. A number of precautions should be taken in evaluating new equipment, such as do not be misled by slick sales talk, watch carefully how your sample workload is dealt with, allow potential operators to try out the equipment, ensure that special requirements are thoroughly demonstrated, and ask existing users if they would buy the same equipment again. It is unlikely that the preferred system perfectly matches all requirements and so it is probably better to opt for a compromise based on a well integrated design with a few weak areas, rather than a mediocre one offering all facilities. Similarly with photocopying, there is an extensive range from small desk top copiers to very large copiers, with facilities for A3 size copying, reduction and enlargement, collated sets, document handling and automatic stapling. Hence it is necessary first to identify and analyse the needs of the office and then to investigate and compare the equipment that appears to best meet these needs.

Furnishing systems are often selected with modular components to meet all needs and provide maximum flexibility. Automated office routines based on data and word processing can result in changes in workflows and some reorganisation of the office layout. Individual work stations result in lower worktop heights as compared with a standard clerical desk. Security aspects need considering when deciding on storage facilities for filing papers and computer media, but fire and vandal proof storage is much more expensive than conventional filing facilities. Coupled with office furnishing is the whole question of office layout and the advisability or otherwise of open planning, often combined with some measure of landscaping. Cassette magnetic tapes or microfilm are usually filed in shallow drawers or on shelves which are fitted with dividers. Microfiche, on the other hand, are more frequently mounted on pocketed pages in binders, usually on shelves within easy reach of the reader. Floppy disks are stored on edge, while computer disks are stored flat in their cases (Chambers, 1983). Some of the principal forms of computer software used in quantity surveying offices are described later in the chapter.

## PRIVATE PRACTICE

### General Framework

The majority of private practices aim at providing efficient financial management of projects and the provision of a cost consultancy service to employer and designer during the whole construction process (RICS, 1970). An RICS QS Committee report in 1971 extended this description of the quantity surveyor's work to emphasise his distinctive competence in measurement and valuation in the field of construction in order that such work can be described and the cost and price be forecast, analysed, planned, controlled and accounted for. However a later RICS report, *QS2000* (1991) commented on the relative decline in the use of building contracts requiring formal bills of quantities and the increased need for management expertise, information technology, automated tender documentation possibly linked to computer aided design (CAD) systems, and risk analysis.

The significance of quantity surveying in the private sector is shown by the fact that in 1984 (after the RICS/IQS merger), 53 per cent of corporate members in the Quantity Surveying Division of the enlarged RICS was employed in private practice and thus formed the largest single group. A RICS report on quantity surveying practice in 1974 showed a preponderance of small to medium sized practices, which was confirmed in the RICS QS Division report in 1984, which indicated that over 50 per cent of quantity surveyors in private practice were employed in offices with five quantity surveyors or fewer and over 70 per cent in offices with ten quantity surveyors or fewer. It was recognised that the medium sized firms (11 to 50 technical staff) were often healthy and dynamic, and that large practices (employing from 51 to 300 staff) usually employ a number of specialists, but are generally organised in small working groups under the supervision of a partner or associate, and thus operate very much as a federation of smaller firms. The author's investigations showed that even very small practices can offer a very effective service to the client stemming from the enthusiasm and initiative of the senior partner. At the other extreme the very large practice has substantial resources in specialised knowledge and expertise, computerised data and other facilities, backed up by continuing research and an excellent service to employers, such as the circulation of a quarterly economic survey.

It is very satisfying to see the substantial progress that has been made in the scope and nature of the service that is now being offered to employers by the profession compared with that in the immediate post war years. More attention is now paid to public relations, the marketing of quantity surveying services and the identification of the corporate image, by the issue of attractive brochures and in various other ways, to meet the needs of a changing society. This aspect is covered in more detail in chapter 17. Valuable working arrangements are also being forged with overseas practices, which must be to the benefit of the profession at large.

## Operational Aspects

Partners perform a vital role in any private practice as they are responsible for the control and operation of a section of the work of the practice. The ratio of partners to technical staff varies considerably from 1:2 or less in small practices up to 1:10 in very large ones. There is no optimum number as so much will depend on the size and type of practice and the nature of the work that it undertakes. With larger practices a ratio of 1:8 appeared to be the preferred arrangement. It is vital that the partner is fully cognisant of all that is happening in his section, that all staff under his control feel able to discuss any matters with him, and that he liaises closely with other partners, normally through partners' meetings. General overheads and profits vary considerably from one practice to another but are likely to be in the order of about 90 per cent of basic staff salaries.

Quantity surveying practice overseas tends to follow the British pattern but normally uses simpler codes of measurement than in the United Kingdom. For example in Kenya in the mid 1980s there were about 12 quantity surveying practices in Nairobi usually employing 6 to 8 staff, including probably two partners. Practices are compelled to rely to a considerable extent on the public sector for work, as the government generates about 30 per cent of the total construction programme, but they usually have a varied range of projects in hand from housing to hospitals, offices and factories.

Few practices now favour the separation of pre- and post-contract work and cost planning. The organisation of teams or groups, referred to in some South African practices as the cluster system, which undertake all activities from the inception to completion of projects, offers many advantages, as outlined earlier in this chapter, On more specialised work such as civil engineering, heavy engineering, mechanical and electrical services, cost benefit studies, refurbishment and urban renewal, there are distinct advantages in using staff who specialise in these classes of work, as each requires its own specific approach and the building up of a body of knowledge and expertise. However, all systems need periodic review to take account of changing circumstances, employers' requirements, the training of junior staff and many other related matters. For example, with cost planning some practices have found the need to establish a small central cost planning unit to augment the cost planning work undertaken in the various project groups. Also the groups may tend to specialise in certain types of project such as health projects, educational buildings, commercial schemes, and residential projects. All the technical groups are normally supported by a central administrative unit. Similar systems operate in many overseas practices.

Quantity surveying practices also undertake a considerable amount of work apart from the generally recognised areas of feasibility and initial cost advice, cost planning, advice on contract procedures, preparation of tender documentation and postcontract cost control, as described in earlier chapters. Some of the more commonly encountered additional services provided by practices included project management, facilities management, valuations for fire insurance, insurance consultancy, schedules of dilapidations, mining, process engineering, and off-shore work. Some of the South African quantity surveying practices have made an important contribution to coal, gold and diamond mining and process engineering. The effect of the computer on quantity surveying work will be considered later in this chapter.

Most of the larger practices operate branch offices often at a considerable distance from the main office, under the control of a resident partner. They are generally established to satisfy a clearly identifiable local demand and have been of considerable value in attracting work from the regional offices of the Health Authority and the former Property Services Agency. The recession of 1990–95 highlighted the desirability of practices having a diversified workload.

The subject of efficient management was dealt with earlier in the chapter and most practices clearly recognise the need for effective planning, organisation, coordination and control, and particularly the comparison of planned and actual performance. Many practices arrange periodic staff meetings which provide staff with the opportunity to put forward ideas and for them to be kept informed of the firm's policy and developments. Another useful aid to communication is the production of an office hand-book detailing procedures for handling projects at all stages and keeping it updated. An investigation of different practices highlighted the need for flexibility of organisation and also the value of a range of standard forms and letters to cover repetitive situations, as well as standard preliminaries and preambles.

Employers expect quantity surveyors to be sound, accurate and posi-tive, yet sufficiently flexible and adaptable, displaying trust and fairness in all dealings and showing a measure of commercial awareness, enterprise and resolution in pursuing the employer's interests. The employer should ideally have ready access to the same senior person who gives regular, accurate and impartial advice in advance of events. A satisfied employer should feel that he has obtained good value for the fees paid.

In 1984, about 7 per cent of UK private quantity surveying work was overseas, public bodies accounted for 36 per cent of the workload, pri-vate clients for 48 per cent and contractors for 12 per cent. By the mid 1990s the proportion of public sector work had reduced substantially. The ratio of technicians (technical surveyors) to professional staff varied con-siderably from practice to practice in 1995 with a range of none, where trainee surveyors and graduates were employed, to 1:1 or 2:1 depending on the class of work involved and the office policy.

## Administration

Administrative arrangements are becoming more complex and can en-compass a very large number of activities in the larger practices, as listed in appendix N. It is important that a member of staff should be responsi-ble for each activity. With the larger practices the question arises as to whether an office manager or partnership secretary should be appointed to oversee all the administrative arrangements and so free partners for their professional work.

Williams (1979) has argued strongly for the appointment of a competent and experienced administrator as partnership secretary in the larger sur-veying practices, but was particularly referring to general practice survey-ing. The partnership secretary must be present at all partners' meetings as he/she must be a party to the policy making discussions and will subse-quently be responsible for their implementation, monitoring performance, providing feedback and being aware of problems as they arise. It is essen-tially a coordinating role with overall control of general administration, finance, personnel, office management, advertising and other promotion. These activities will encompass the important functions of information dissemination, cashflow projections, production of annual financial accounts, monitoring of employment legislation, dealing with professional indemnity insurance and related matters.

Fiber (1986) emphasised the need for improved practice management encompassing organisation of the practice, monitoring outputs against inputs for each project, marketing and personnel, and the need for flexibility and periodic review of activities. Generally, the principal quantity surveying practices operated quality management in accordance with BS 5750 in 1995. Office accommodation in $m^2$/member of staff varied from 10 to $16m^2$ in the practices approached by the author in 1995.

## Professional Consortia

Professional consortia or consortium partnerships comprise different professionals, for example, architect, quantity surveyor, structural engineer and services engineers, who collaborate to offer their combined services to the employer in connection with a specific project for a consolidated fee. For example, an all-in fee might be made up of architect: 3.6 per cent; quantity surveyor: 2.5 per cent; structural engineer: 2.4 per cent; mechanical and electrical engineer: 2.0 per cent; total fee: 10.5 per cent. It requires all members of the professional team to have the utmost trust and confidence in each other and to agree to one of their number being the spokesman to the employer and coordinator of the project. This type of arrangement is particularly well suited for a fairly specialised or unusual form of project, where each of the professionals has considerable experience in the class of work involved. The consortium is able to offer a variety of professional skills, while the individual professionals still retain their identity and practice.

Some consortia provide for the formation of a company between the various members of the design team operating with limited liability. This is of significance as, should any member of a consortium be negligent, it is the group that are responsible and not the particular member. A limited company is a continuing organisation subject to the requirements of the Companies Acts and open to public scrutiny. Firms can join and leave the consortium with relative ease through the transfer of shares. The control of the company is, however, clearly defined in the Memorandum and Articles of Association, and there is a statutory requirement for an annual return. Some quantity surveyors have expressed reservations about a member of the consortium, other than the quantity surveyor, reporting on financial matters to the employer, as the information might be distorted and consequently incorrectly interpreted by the employer. This could possibly happen in any event through the architect under the orthodox arrangement. The most common type of consortium is that of a mixed professional team which submits a competitive tender for the design and construction of a large project, such as a power station or harbour.

Professional consortia should not be confused with group or integrated practices which have the support of the RIBA. In this arrangement the architect and quantity surveyor normally operate from one combined office, often leading to shared overheads and some combined services, and a loss of independence could result. Another interesting development has been the increasing number of direct appointments of quantity surveyors by employers.

The operation of multi-disciplinary practices and mergers of quantity surveying practices with other organisations is considered in some depth in chapter 19.

## PUBLIC SECTOR

The public sector comprises primarily central government, local government, the National Health Service and other statutory bodies. In 1984, 22 per cent of all chartered quantity surveyors were employed in this sector, but by 1990 this had fallen to 16 per cent and continued to decline throughout the 1990s.

### Central Government

Most quantity surveyors in the central government service were in the 1980s employed in the Property Services Agency (PSA) of the Department of the Environment (DOE), with a total professional and technical staff of many hundreds, with a headquarters directorate, situated principally in Croydon, and regions throughout the United Kingdom. However, in the early 1990s the PSA was privatised in accordance with government policy at the time, based on the premise that this would result in a more efficient service with greater value for money.

### Local Government

County councils normally operate quantity surveying sections within county architects', technical services, construction or other departments, as the organisation varied considerably between different counties, with an average of about 12 professional and technical staff in 1995. It is unusual for the chief quantity surveyor to progress beyond a third tier appointment. The section is often subdivided into groups dealing with specific services such as education. They offer a full range of quantity surveying services and also let out a proportion of their work as necessary, although it can be relatively low. County councils generally work to rolling programmes which are subject to annual reappraisal when the council's budget is being determined around February each year. It is important to keep costs within the budget and in most county councils the project quantity surveyors carry out the cost planning. The contractual arrangements will normally be selected to suit the particular project, but there is a general preference for selected competition using computerised bills of quantities. Cost monitoring and cost data resources are generally of a high order. Full use is made of computer aids, standard forms and documents.

District councils generally employ much smaller quantity surveying sections than county councils and they generally form part of the architect's department but may, in some instances, be part of the department of technical or environmental services. They normally offer all the traditional quantity surveying services and often have a good information and cost data resource. Their major projects are normally in the housing field and in the early nineteen nineties they were undertaking a considerable amount of rehabilitation work. The number of private quantity surveyors engaged is generally small.

## COMPULSORY COMPETITIVE TENDERING (CCT)

From 1 April 1996 onwards at prescribed dates, all local authorities delivering defined activities in-house have had to demonstrate that these are cost effective when measured against competitive bids from the private sector. The rules, made under the Local Government Act 1988, permit local authorities, to retain in-house services up to 35 per cent of the total service value or £450 000 pa, whichever is the greater. The defined services of 'construction and property services' include property management, architecture, quantity surveying, engineering and landscape architecture.

The government's objectives were to create a level playing field, to generate private sector interest in these tenders and, through competition, to promote increased value for money with significant volumes of work likely to be transferred to the private sector. The DoE produced CCT guidance notes covering the tender process, timetable, evaluation criteria and anti-competitive behaviour.

Grierson (1995) listed the tasks to be performed by local authority surveyors in preparation for CCT as follows:

(1) Define sensible client/contractor splits and decide on activities to be exposed to competition.
(2) Define specifications, contract documentation, shortlisting and tender evaluation procedures.
(3) Roles have to be ascribed for both the client and in-house teams.
(4) The tender procedure must be managed to conform to the CCT rules on anticompetitiveness and the EC services procurement directives.
(5) Bid teams have to become 'match fit' by analysing the competition, comparing themselves against private sector benchmarks and reviewing management systems resources and working practices.

Table 15.1  Issues for the CCT tender procedure

| | Applicants | Tenderers |
|---|---|---|
| Information made available by the authority: | Advertisement<br>Briefing notes | Invitation to tender<br>Specification & Appendices<br>Briefing meeting(s)<br>Documents for inspection<br>Queries & responses<br>Tenderers' initiatives |
| Information provided to the authority: | Financial information<br>References<br>Evidence of ability to perform:<br>• track record<br>• resources utilised<br>• qualified management. | Evidence of ability to perform<br>Proposals to comply with contract:<br>• method statements<br>• staffing proposals<br>• IT proposals<br>• quality mgt. arrangements<br>• subcontracting proposals.<br>Bid & pricing documents<br>Interviews & presentations |
| Method of assessment by the authority: | Applicants' submissions<br>Pre-shortlist interviews | Tenderers' submissions<br>Tenderers' response to tender process<br>Tenderers' interviews & presentations. |

Source: Chapman Hendy Associates

Table 15.1 illustrates the main issues and activities involved in the CCT tender procedure. It will be appreciated that CCT represented a major shift towards the concept of local authorities as enablers rather than direct service providers. The new approach involved the professional staff in extensive activities of which they generally had no previous experience.

The RICS (1992c) described how the complex nature of local government can be divided into two principal functions: (1) the democratic process which includes corporate policy development and review, and (2) the delivery of services, including planning, monitoring and ancillary functions. Surveyors in local government often give advice to elected members on the development of corporate policies and their implementation. This latter work cannot be defined precisely and is not therefore a suitable area for compulsory competition.

The Construction Industry Council (CIC) (1992) rightly urged the government to adopt a system of 'competition on quality tempered by price when putting professional services out to tender.' 'Cheapness is seldom synonymous with value for money; it may appear to show an early saving but it usually leads to higher costs in construction, operation and maintenance. A reduced quality of service is often a consequence of competition based on fees alone.'

The CIC proposed a very realistic framework for the acceptable use of CCT in design work as now listed.

- A shortlist of fully acceptable firms should be prequalified before going out to tender
- Tenderers should be provided with the same information on which to base their bids
- Advance notice should be given of any weighting criteria in the assessment
- The results of the tendering process should be made public.

The two envelope tendering system has been developed for the competitive appointment of consultancy services and takes full account of the quality of the people providing the service. For example, in the World Bank system envelope 1 contains the quality bid and envelope 2 the fee bid. The quality bid details needs, skills, previous experience, project leader CV, time parameters and similar matters in the form of a schedule which is then scored against a predetermined marking policy and a threshold requirement. Only those bids passing the quality threshold are considered further. Only the highest scoring quality bid has its accompanying fee bid envelope opened. Negotiations then take place and if successful an appointment is made. If unsuccessful the second highest quality bid's associated fee bid is opened.

The European Union (EU) system uses a similar approach but all the fee bids associated with quality bids that pass the threshold are opened. A weighting is then applied to both bids, the quality bid receiving a weighting in the order of 85 per cent and the fee bid 15 per cent (Pigg, 1993).

In general, public service employees share the common value of dedicated public service of which they are justly proud, and it is important that these virtues should not disappear. Swanston (1994) has aptly described how one of the greatest risks facing the public sector is the loss of specialist skills. Yet this is what some public bodies are doing in privatising design and estate management services, despite the fact that these departments also provide expert advice for the other client departments in the management and use of their property holdings. At the extreme end of CCT with 65 per cent of the work contracted out, the remaining 35 per cent may not be adequate to maintain an efficient in-house team, particularly with the smaller authorities faced with reduced budgets in the mid 1990s. Furthermore, sufficient expertise is needed to formulate and oversee the CCT tendering arrangements.

## CONTRACTING

### General Background

Quantity surveyors are undertaking an increasingly important role with contracting organisations and, in 1990, 19 per cent of all chartered quantity surveyors was employed by contractors. As long ago as 1974, a RICS survey found that many of the chartered quantity surveyors in the construction industry were either directors or managers of large departments undertaking a range of services greater than traditional estimating, measurement and settlement of final accounts. In 1978 almost 40 per cent held director appointments.

Cottrell (1979) described how the duties and even the title of the contractor's quantity surveyor varied considerably from company to company, according to the particular company's policy and requirements. While there is usually a common base involving measurement, valuation and financial reporting, other duties that may be carried out include involvement in estimating, buying, bonus incentive schemes, scheduling and monitoring materials, enquiries and placing orders with subcontractors, debtor control, site costing and plant control. The quantity surveying management structure may operate independently but in parallel with contracts management or may be integrated with the contracts management team, and there are a variety of other arrangements as discussed later in the chapter.

### Main Functions of Contractors' Surveyors

Probably the most authoritative account of the quantity surveyor's role in a contracting organisation was contained in a paper by Johnston to a quantity surveyors' triennial conference in 1978. He pointed out how prudent contractors have for many years employed quantity surveyors to protect their interests, particularly in the more controversial contractual areas. Apart from agreeing interim valuations and final accounts with the employer's quantity surveyor, the contractor's quantity surveyor has many other important duties to perform:

(1) He can, for instance, be involved in the placing of subcontract orders for both nominated and domestic subcontractors. It is essential that the contractor's quantity surveyor ensures that the correct contractual conditions and financial details are embodied in the orders and subcontract arrangements. Sometimes instructions are issued to place nominated subcontract orders based on conditions at variance with the main contract, subcontract periods may be incompatible with the agreed main contract programme, proper cash discounts may be omitted and quotations may be months out of date. Once the order is placed, the responsibility lies with the contractor, creating the need for his surveyor to ensure that the arrangements are correct.

(2) The contractor's surveyor is required to report to management, usually monthly, on the overall financial state of the contract, giving comparisons of actual cost with the corresponding internal value for the various cost codes and subtrades. This report will highlight major loss situations at an early stage, thus providing contracts personnel with the opportunity to consider remedial action or, if it is too late, management can at least budget for future losses. There is the added benefit of feedback to estimating and contracts staff, hopefully preventing a recurrence of similar problems on future contracts. The term 'internal value' refers to the differences that may exist between the nett prices inserted in the bills of quantities and the target values established for the use of contracts and surveying staff immediately prior to the commencement of works on site. Where contractors are required to submit their priced bills with their tenders, the employer will receive a bill on which pricing was probably started some 7 to 10 days before the date of tender. Almost invariably the quotations that arise just before the tender is submitted are the most competitive and, as it is too late to alter individual rates in the bill, an overall financial adjustment can be and often is incorporated in the preliminaries bill.

Adjustments to labour targets, as a result of changes in market conditions after the estimate was compiled, or economies resulting from reduced target construction programmes, can also substantially alter the rates inserted in the bill. There can be a delay of as much as 3 to 6 months between the submission of a tender and the commencement of work on the site. The supply of more accurate target values to contracts and surveying staff enables management to monitor more accurately the efficiency of the progress of works on site, involving the contractor's surveyor in both internal and external valuations.

(3) The contractor's quantity surveyor can advise contracts staff on significant aspects of the contract conditions, in their dealings with the architect.

(4) Many of the medium to large construction companies are becoming more involved in the precontract development stage, including the use of design and build, develop and construct, and package deal contracts. The contractor's surveyor has a major role to fulfil in these contractual arrangements, advising design staff on the most economical form of construction and cost planning and monitoring the development as a whole.

Readers requiring more information on cost control of construction work by contractors are referred to chapter 8 and Pilcher (1992).

### Study of Quantity Surveying Function in a Large National Contracting Organisation

It was considered useful to examine the organisation and activities of the quantity surveying function in a large national contracting organisation. The commercial director has absolute function responsibility to the main board of the company for all matters relating to estimating, quantity surveying and procurement of subcontractors' work and materials suppliers. The business of the company is operated regionally to cover the north, midlands, west and south of the country and a fifth division provides design and management services.

Each of the main operating businesses has a regional surveying director with line responsibility to the regional managing director and functional responsibility to the commercial director. They have regional responsibility for quantity surveying and procurement with an appropriate structure reporting to them on the turnover, type of work and related matters. Each business also has a regional estimator also with line responsibility to the regional managing director and functional responsibility to the commercial director.

Each region operates on a divisional structure based upon circa £25m units. Each division has a divisional surveyor supported by a surveying team of varying seniority and experience based upon the type of workload. An average division will have circa 8 to 12 surveyors, but the special contracts division needs a greater number than average because of the smaller size of contracts.

The main activities and staffing requirements are as follows:

*Estimating*: a £50m pa regional business will have approximately 5 or 6 estimators plus 2 or 3 clerical/administrative staff.

*Buying*: a £50m pa regional business will have 2 buyers with one support staff.

*Quantity surveying*: each contract is a self-contained unit and has responsibilities for subcontractor procurement, measurement, valuation of variations, agreement of the final account and subcontractor accounts, claims, payment of subcontractors, preparation of management accounts, forecasts, cashflows and related matters. The number of quantity surveyors on each contract will depend entirely upon specific need and will usually be site based.

It is the policy of the company to promote professional qualifications in surveying and building and approximately 50 per cent of the commercial staff have a recognised professional qualification. A limited number of experienced measuring surveyors complement the more senior surveyors, particularly on very large contracts. Alternatively, trainees within the organisation undertake this role as part of a broader experience gaining programme.

In the 1990s the company employed less people per unit of turnover because of the more difficult and exacting conditions. The company also placed much greater emphasis on the need for accurate projections of cashflow and end profit in order to maximise return in an extremely difficult marketplace. Far more commercial awareness was sought so that notices in respect of delay and loss and/or expense were issued promptly and there was a trend towards greater communication and negotiation skills. Action was taken to restrict clerical/administrative activity and thus increase the time available for strategic management of contracts. All the major commercial operations have been computerised in whole or in part and this reduces non-productive time, ensures greater efficiency and provides a link between the varying activities that allows the instant retrieval of data. In 1995 each of the companies was obtaining BS 5750 accreditation and the company was moving towards the implementation of a total quality management initiative.

## Civil Engineering and Mechanical and Electrical Engineering

There has been a significant increase in the number of quantity surveyors employed in civil engineering and mechanical and electrical engineering contractors' organisations. In the early nineteen sixties these companies had few, if any, quantity surveyors, but the changes in the last three decades have been extensive, with some civil engineering companies employing as many as eight quantity surveyors on a major motorway contract. A company with an annual turnover of £60m in 1995 employed around 40 quantity surveyors. The production of numerous claims in the civil engineering industry has provided another role for the quantity surveyor, whose basic training makes him invaluable in analysing contract conditions, site instructions and related documents. The mechanical and electrical engineering industry has many firms with quantity surveyors employed in senior management posts, advising engineers on contract conditions, documentation, application of the price adjustment formula and other related matters.

## Subcontractors

Another section of the construction industry to benefit from the use of quantity surveyors is that of specialist subcontractors. Relatively simple trades like roof tiling, floor tiling, plastering and painting have found a need for quantity surveyors, particularly in the implementation of contract conditions and the application of the price adjustment formula for price fluctuations.

## Directorships

Among the many attributes required for directors Johnston (1978) identified the basic need for a good blend of technical skills coupled with a suitable attitude and temperament to produce a harmonious working relationship. The financial management and control of a company is an extremely important aspect, and this is a function that many quantity surveyors are able to fulfil. The quantity surveyor's familiarity with costs and prices enables him to assess reports on the progress of contracts on a financial basis and, when added to a good contracts management eye for an efficiently run project, provides the ideal combination for monitoring and ensuring the satisfactory progress of a contract.

Johnston (1978) believed that there was still considerable potential for quantity surveyors who wish to aspire to directorships in construction companies. Although the large and most of the medium sized companies already have at least one surveying director with adequate senior staff for progression, there are a large number of small to medium sized firms who could benefit from the services of a quantity surveyor on the board of directors. They can help to eliminate some of the troughs from successive years' balance sheets, particularly at a time when the 1990–95 recession has underlined the need for greater accuracy in an industry that often exists on very small profit margins. Many of the smaller contractors rely to a large extent on speculative building, and this is another area of contracting that is in urgent need of the services of quantity surveyors.

## TIME AND COST MANAGEMENT

During the 1990–95 recession when the volume of construction work was drastically reduced, coupled with the increased use of competitive fees, many quantity surveyors were offering significant discounts on the scale fees which were no longer mandatory. This situation produced an urgent need to carefully monitor and manage office costs, to eliminate the misuse or underusage of resources of staff, equipment and materials. This in its turn created the need for the proper recording of time spent and other resources used on each project in order that they can be effectively costed and the costs continuously monitored.

Each member of staff has a responsibility to use his/her time as efficiently as possible and to be continually looking for ways to undertake tasks more effectively and in less time. One way of accelerating this process is by introducing profit sharing or end of year bonuses as an incentive to greater efficiency. The allocation of staff between projects needs to be done in a realistic and carefully considered way to avoid an inbalance between allocations to different projects of varying size and complexity.

The planning and coordination of staff time may be portrayed on bar charts which show how each member of staff's time has been used and which can then be costed and charged to the particular project. Staff costs will include such related costs as national insurance and pension contributions, holidays and sick pay and other oncosts. The monitoring of the staff costs for each project should be performed regularly to provide a comparison between estimated and actual costs, and to show whether the costs of the work on the project are within target. They can subsequently be used as a guide when preparing future fee bids. Each member of staff will also keep a diary showing how time has been spent each day and will encompass non-productive time as well as productive, the latter will be split between the different projects and activities, such as measurement, cost planning and preparation of valuations. This will help to identify methods of securing greater efficiency in the future.

Progress charts provide another useful aid to effective management by indicating current progress on projects in hand and also future estimated commitments. Hence the overall situation is identified and appropriate action taken as necessary.

## CASHFLOW PROJECTIONS

The client will need to know the likely sums which he will be required to pay the contractor and when they will occur. To meet this need the quantity surveyor will prepare a schedule of the likely amounts due and their timing, working from the contractor's approved programme of work which he submitted with his tender or shortly after its acceptance. The results are normally displayed on a computer spreadsheet for ease of assimilation by the client.

Design variations by the architect can result in changes both to the amounts due and their timing. Where the changes are significant they should be brought to the attention of the client. Delays in the execution of the contract work will also affect the amount and timing of payments.

The quantity surveyor will need to prepare similar cashflow projections for his own organisation, showing the project expenditure on staff and related costs set against the fees received. In this way he will readily see the financial position throughout the design and execution of the contract and can take steps to have adequate funds available where there is an excess of expenditure over income. This latter situation will usually be offset by income accruing in respect of other projects which are operating on different timescales. Hence it would be unwise to consider a specific project in isolation as it is the overall picture that is required, and this can be obtained from a summation of the financial consequences of all projects in hand at any given time.

## FINANCIAL REPORTING

This aspect was considered in chapter 8 which dealt with the cost control of construction projects. It is included here because of its utmost importance in a quantity surveying organisation. It is vital that the quantity surveyor keeps the client fully informed on all cost aspects of a project, and particularly where changes are made to the design or sequence of the works during their construction.

Fortunately, the quantity surveyor is notified of such changes before they are implemented on the site. Hence he has adequate time to prepare detailed cost comparisons and can inform the client of the likely cost consequences of the changes, including the probable impact on other sections of the works. As more detailed information becomes available on variations, mainly from measurements, quotations or dayworks, the quantity surveyor can refine his original estimates and update his financial reports to the client. The quantity surveyor must also have regard to the client's detailed requirements on financial matters relating to the project, as they can vary significantly between different clients.

For example, a periodic financial report to a client may need to identify the financial consequences of architect's instructions (AIs), and to include such matters as the adjustment of prime cost (PC) items and provisional sums; remeasurement and valuation of remeasured work; dayworks; price fluctuations where applicable; and the provision for claims and any likely future changes.

Financial reports may also be required by clients on a variety of other matters such as feasibility studies on proposed large scale developments; environmental impact assessments (EIAs); the effect of adopting a 'greener building' approach; the choice between refurbishing existing buildings and demolishing them and redeveloping the site; the choice between using in-house personnel and an outside contractor; the choice between traditional and system building; the provision of a bridge or an underpass; or whether to construct large sewers in tunnel or open cut.

## PARTNERSHIPS

Many newly qualified chartered surveyors see the achievement of a partnership as their ultimate goal, in the same way that an aspiring academic hopes that he will one day become a professor. It is a wide ranging and important subject which can be dealt with only in outline in this book owing to limitations of space. It should be noted that practices can only describe themselves as 'chartered surveyors' or 'chartered quantity surveyors' when the majority of partners are corporate members of the RICS.

### Nature of a Partnership

The Partnership Act 1890, which is the main Act concerned with partnership law, defines a partnership as 'the relation which subsists between persons carrying on a business in common with a view to profit.' A business is defined as 'including every trade, occupation or profession.' The prime requirement for a successful partnership is mutual trust between the partners (RICS, 1980) The duty of good faith which must exist between the partners overrides all other considerations (Harmer and Camp, 1982). It is frequently said that a partnership is as important a relationship as marriage and much the same criteria apply when choosing a partner. Furthermore, each partner is jointly and severally liable for the partnership debts, limited only by the extent of his private estate, and hence a partnership should never be entered into hastily.

There are basically two ways in which chartered surveyors may practice together:

(1) formation of a limited or unlimited liability company in circumstances defined by the RICS General Council; and
(2) partnership.

The only true partner is an equity partner, who is responsible for and derives benefit from the assets and business of the practice. Aspiring partners are often offered positions as associates or salaried partners, who receive some share of the profit of the practice in addition to a salary. In this way improved status and income are secured although some believe that it is mainly a device to retain the services of very competent members of staff. Their inclusion on the notepaper of the practice could render them liable for the firm's debts in the event of the assets of the equity partners being insufficient to meet them, hence they need to be adequately indemnified by the equity partners against this liability. Consultants are senior partners of a practice who have passed the normal retiring age and who, in return for continued service to the practice, draw an income as opposed to a direct pension or annuity.

### Reasons for Establishing a Partnership

The principal reason for establishing a partnership is to enable two or more surveyors to pool their knowledge, experience and resources for their mutual benefit. The individual retains his professional identity and there are often tax advantages. The expansion of a practice provides increased capital, the possibility of more work and greater specialisation with the provision of an improved service to clients. The principal disadvantages relate to the totality of a partner's obligations in respect of practice debts and the possibility of conflicts of interests or actions between partners.

### Partnership Capital

Normally a prospective partner will be expected to pay a certain sum for his share of the partnership, which is usually related to his share of future profits and the existing capital structure of the practice. In some cases there may be an additional monetary requirement for goodwill, based on the reputation built up by the existing partnership, but this practice is now on the decline and mainly arises in connection with retired senior partners. Davey (1983) considered that the level of fixed capital provided by partners should be such as to enable the profits earned in any accounting year to be withdrawn in full (after proper provision has been made for taxation) within a period of, say, six months from the end of the accounting period. He also described how capital is required to finance the firm's debtors (to the extent that these are not financed by the firm's normal creditors) and fixed assets such as office furniture and equipment, and cars. Any ways in which the holding of these assets can be reduced will have a direct impact on the amount of partnership capital that is required. Capital is returnable if the partnership is dissolved in a solvent state, or in the event of a partner retiring from or otherwise leaving the partnership (Harmer and Camp, 1982).

Finance for the acquisition of a partnership share is normally obtained in one of three ways:

(1) direct loan from the existing partnership to be repaid over a fixed term by the new partner receiving a reduced income and profit share during that period;
(2) personal loan from the applicant's own bank or the practice's bank; or
(3) raising of finance in times of restricted credit through life assurance annuities and other financial institutions (RICS, 1980).

## Preliminary Action by a Prospective Partner

A prospective partner should consider carefully what positive contributions he can make to the practice in the form of more projects, the right personality, being able to work harder and make difficult decisions, having sufficient enthusiasm, technical skill and organising ability beyond that possessed by his senior colleagues, being able to express himself concisely and clearly at client's meetings and having a practical plan for raising the capital needed for the partnership share.

Before any agreement is reached, full professional advice should be sought from a solicitor and an accountant as to the legal and financial implications. Clearly a junior partner cannot alter significantly the terms of the agreement presented to him, but at least he should be fully aware of the implications and can then make a reasoned decision based on this information (RICS, 1980).

## Partnership Agreement

There is in law no obligation upon the partners to set out the terms of the partnership agreement in writing, but it would be imprudent for any partner to rely on oral evidence of the existence of the agreement. Disputes and difficulties may easily arise in the absence of a written agreement (Harmer and Camp, 1982). In the absence of a written document the provisions of the Partnership Act will apply and these will not always reflect the true intentions of a professional partnership. On the death, retirement or expulsion of a partner, all concerned will wish to know where they stand (RICS, 1980).

Every partnership is different and there is no standard form of agreement. The main contents of a partnership agreement generally comprise details of the practice; the arrangements for present and future contributions of capital; profit/loss sharing; the basis of each partner's drawings; any restrictions of partners' activities outside the practice; management procedures from accounting and responsibility viewpoints; arrangements for future partners; provision for retirement; and provision for distribution of a partner's share in the event of retirement or death. In addition the agreement often contains details of voting arrangements within the practice, and provision for reviewing the partnership arrangements at regular intervals is becoming increasingly common. Other matters such as annuities, goodwill, insurance and provision for the children of existing partners may also be included. Readers requiring further information on the drafting of partnership agreements and on partners' professional responsibilities are referred to Chalkley (1994).

**Operation of the Partnership**

Partners need to be kept informed of the level of profitability of the practice and, to this end, full financial accounts should be prepared on a quarterly basis, including a work in progress valuation. To assist the management, budgets and cashflow statements should ideally be prepared for at least two years in advance.

Clearly defined provisions in the partnership agreement should set out the terms and conditions for the retirement of a partner, including the withdrawal of his capital and financial provision for his future needs.

## INCORPORATION WITH LIMITED OR UNLIMITED LIABILITY

As the move towards company status for quantity surveyors has grown the RICS withdrew the earlier restrictive regulations and introduced more flexible arrangements. The main advantages of company status are the greater freedom in the ways in which finance can be raised to improve efficiency and the ability to compete with other professional organisations which are companies.

A member may carry on practice as a surveyor through the medium of a company provided that he complies with the RICS Regulations, whereby a surveyor who is a director of the company must ensure that a provision is included in the Memorandum of Association or equivalent constitutional document of that company stating that 'Any business of surveying for the time being carried on by the company shall at all times be conducted in accordance with the Rules of Conduct for the time being of the Royal Institution of Chartered Surveyors'. It is now possible for a firm of chartered surveyors to be partly owned by, for example, a commercial organisation, although chartered surveyors must have full responsibility for the whole of the surveying work.

## STAFF SELECTION

The normal procedure for selecting staff for employment is by interview following the submission of an application form or curriculum vitae (CV). Vacant posts are usually advertised in the appropriate technical and professional journals and possibly local newspapers, or national newspapers in the case of senior staff. Those requesting application forms should also be supplied with a comprehensive job description and details of the employing organisation.

The employer needs to be satisfied that the applicant is able and willing to do the job and that he/she can work satisfactorily with the rest of the team. A useful approach is to break down the job requirements into major responsibilities, skills and knowledge, Langford *et al.* (1995) have listed what they see as the likely important characteristics or a selection of them depending on the nature of the vacant post. The listed items are as follows:

*Personality*: drive and ambition; motivation; communication skills; leadership/cooperation skills; energy; determination; confidence.
*Professional*: reliability; integrity; dedication; pride in work well done; analytical skills; listening skills.
*Business*: efficient; economical; understanding and acceptance of procedures; understanding of profit.

The interview panel normally consists of two or three people interviewing one candidate. Alternative but less popular methods are selection boards and group selection. The interview should be essentially a conversation with a purpose, which aims to 'draw out' the candidate to talk freely about him/herself and his/her career. The interviewers must ideally plan, direct and control the conversation to make an accurate prediction of the candidate's likely future performance in the vacant post and the candidate must be able to satisfy the interviewers that this is the case, without appearing arrogant, self-conceited or presumptuous. Typical questions cover the reasons why the candidate took his/her current job, the tasks undertaken and knowledge and skills acquired, and the reasons for applying for the new post.

An interview can rarely be completed in less than 20 to 30 minutes but can prove partially unproductive if it exceeds 40 minutes, except perhaps in the case of senior appointments. Allowance has to be made for the giving of information to the candidate and putting him/her at ease and for candidate's questions. Interviewing techniques vary considerably but the use of a checklist of pre-prepared questions with a vacant column for comments can be useful. A decision may be made at the end of the interviews in order to finalise the arrangements or, alternatively, candidates may be told that they will be informed of the decision within a short time. It is good policy to ask candidates at the close of their interviews whether they will accept the post if it is offered to them.

## OPPORTUNITIES FOR WOMEN IN QUANTITY SURVEYING

The Equal Opportunities Commission (1986, 1987a and b) believed that inequality between sexes resulted in women being concentrated in a few industries, receiving less opportunities to learn skills, and receiving less pay. The Commission also argued that occupational segregation should be eliminated because it results in a waste of human resources, lower morale in the workforce, unlawful discrimination and breaches of the Equal Pay Act 1970. It is important that employers have an equal opportunities policy that is published, understood and operated and that women employees should be treated in exactly the same way as men.

It can be stated categorically that the RICS and CIOB have for many years past encouraged the entry of women into their professions and there are no impediments to entry on the grounds of sex or race. Both professional bodies have periodically expressed their disappointment at the proportionately low numbers of women being recruited into surveying and building, as they both believe that they are losing a great deal of talent for which their professions are the poorer.

In 1994 women made up only 2 per cent of the total quantity surveying membership of the RICS, of which 4 per cent (22 in number) were partners. The profile of women chartered quantity surveyors in 1994 showed 58 per cent in private practice, 16 per cent in contracting, 16 per cent in the public sector, 6 per cent with commercial/industrial companies and 2 per cent in education. Women quantity surveyors have held academic headships at Kingston upon Thames and Port Elizabeth, South Africa. In private practice, Kay Williams was appointed the youngest of 26 partners with a 300-strong national quantity surveying practice in 1993, based on a good track record of working in senior positions under pressure and her ability to develop good relations with clients. She has worked on a variety of exciting and complex projects including on-shore and off-shore heavy engineering contracts in the UK and overseas, and by virtue of her enthusiasm and hard work must surely be a great inspiration to all women who aim to secure top positions in the profession.

Greed (1991) felt that women were often held back by the role they are given and that men really believe that they are trying to do their best for their female employees but appear to have little idea of what women really want. She gives the example of the woman minerals surveyor who wanted to be directly involved in mineral extraction surveying whereas her employer thought she should work on spoil heap reclamation; she commented, 'doing landscaping with pretty trees and hedging, he thought that was more ladylike'. Greed also made constructive suggestions to enable women surveyors to have life long careers, rather than being seen as assistants or short term helpers, and thus to achieve more senior posts and help shape the structure of the profession and the built environment for the future.

The Latham report (1994) recommended that equal opportunities must be vigorously pursued by the construction industry, with encouragement from the Government. It was further recommended that the Construction Industry Council (CIC), Construction Industry Employers Council (CIEC) and the Constructors Liaison Group (CLG) should produce action plans to promote equal opportunities within the industry and to widen the recruitment base. A Latham report working party on equal opportunities in 1995 recommended a construction industry code of practice on equal opportunities; guidelines to improve workplace conditions; and guidelines on recruitment for the industry.

Also in 1995 the CIOB and the Institute for Employment Studies (IES) published a DoE sponsored report, based on responses from 466 women members and former members of the CIOB, that cited a variety of barriers that deter women from entering the building industry and prevent promotion within it, including: its poor image; perception among employers that women are not up to the job; women lacking confidence; and the difficulty of combining a career with family life. The CIOB put forward a very constructive 14-point action plan to encourage more women to go into construction and to promote career progression, coupled with extensive publicity and to more ably support and equip women in the industry to play a more effective role.

Directors of some of the leading national contractors made some very pertinent comments on the valuable role of women in the industry, and these included the following: 'they get on better with clients and, on site, what happens is the men's attitude changes – the site becomes more civilised'; 'women's core skills – presentation and communications are better, I find that women combine intellectual ability with an attitude that is more conducive to good relationships'; 'more women would mean more human-friendly employment; better client relationships, an improvement in conflict and more civilised sites'; 'attitudes are also being forced to change as clients are increasingly represented by women'.

## SALARIES AND BENEFITS

### Salaries

In 1994 the RICS conducted a salary survey based upon self-completion questionnaires sent to all RICS members which were subsequently analysed overall and by divisions. The accuracy of the survey will be affected significantly by the level of response with over 30 per cent coming from quantity surveyors. Table 15.2 shows typical salaries of quantity surveyors affiliated to the RICS in 1994. It should however be noted that there are

Table 15.2   Typical salaries of quantity surveyors in 1994

| Category of surveyor | Main salary range |
| --- | --- |
| Trainee surveyors | £10k – £15k |
| Assistant surveyors | £10k – £20k |
| Senior surveyors | £15k – £25k |
| Partners/directors | £20k – £50k* |

* Note a few are paid in excess of £100k.

*Source: RICS salary survey 1994.*

wide variations in salaries paid; which are influenced significantly by geographical locations and the ages of respondents. In general the salaries compare favourably with those received by general practice surveyors and architects, although there was a general dissatisfaction with the renumeration packages.

## Benefits

About two thirds of respondents had a company car, with the most common value being £10k–£13k. This was sometimes supplemented by business mileage allowances (16 per cent) or private mileage allowances (47 per cent). Nearly 38 per cent of respondents qualified for an annual bonus, although 60 per cent received under £1000. Between 10 and 20 per cent of respondents received individual private medical insurance, family private medical insurance, permanent health insurance and RICS subscriptions, while over 50 per cent had company pensions available to them, which were mainly contributory. A small percentage of respondents received a subsidised mortgage and/or part of a share option scheme.

With regard to company cars, the report pointed out that with changes in tax legislation, there was an increasing tendency for car allowances to be offered and this accounted for approximately 30 per cent of respondents with a usual allowance of up to £3000 pa.

## REDUNDANCIES

In the exceptional and very severe recession of the early 1990s many surveyors felt that they had not only lost their jobs but also their self respect, with immense fears for their future in a time of uncertainty. Widespread redundancies also have a demoralising effect on remaining staff who feel a sense of loss coupled with some guilt and worries about their own positions. A considerable number of employers turned to outplacement agencies for help through counselling and advice, to assist redundant staff in making a new niche for themselves and sometimes to change career direction. The RICS has also given what assistance it could and encouraged firms to introduce job sharing where this would ease a difficult situation, and with financial advice or assistance through the RICS Benevolent Fund.

There were complaints that some employers tended to take on young and inexperienced staff at lower cost to fill gaps in their slimmed down organisations to the detriment of experienced redundant staff. It is important that where employers are faced with making staff redundant that they perform the unpleasant task professionally with adequate care and understanding. At a time when the construction industry is rightly trying to promote a more positive and professional image, it is bad practice and false economy to behave like the firm that took just five minutes to dismiss the construction director with 24 years' experience. Employers have a duty to at least give those made redundant a feeling of hope for a fresh start.

## APPOINTMENT OF A QUANTITY SURVEYOR

As long ago as 1983 the RICS (Quantity Surveying Division) encouraged the appointment of independent quantity surveyors to provide clients with truly independent and authoritative advice. While the same body in 1991 advocated the early involvement and appointment of consultant quantity surveyors, as quantity surveyors were increasingly being approached direct by clients in the first instance.

In all cases it is essential that there are comprehensive yet precise forms of appointment in order that neither the client nor the quantity surveyor is in any doubt as to the services to be provided by the quantity surveyor and how and when he is to be reimbursed. The RICS (1992) has issued three related documents for this purpose which cover all aspects admirably. It is desirable to use these documents unless the quantity surveyor or client has a strong preference for using his own documents.

The three documents for use in England and Wales are as follows:

- Client guide
- Form of enquiry and fee quotation
- Form of agreement, terms and conditions.

These three constitute a complete set and ideally should be used together, but each can be used individually as, for example, where a client prefers to use some or all of his own documentation. The form of enquiry and fee quotation provides for alternative procedures, the 'simplified' and the 'detailed' depending on what the user wants. It is not expected that all the contents of the form of enquiry will apply to each and every scheme.

The detailed procedure encompasses in considerable detail all relevant matters as now outlined:

- Project information (employer/client, project location, general project description, project programme, project budget and project tender documentation/contract forms).
- Quantity surveying services requested/offered (the wide ranging list covers inception and feasibility; preliminary and/or other estimates; cost planning; advice on procurememt method/contract; pre-tender estimate; tender documentation; negotiating tender; tender report; tender revisions; interim valuations; cost monitoring; final account – preparation and agreement; attendance at meetings; provision of printing/reproduction; and other services).

- Fees (including adjustment as appropriate; additional services; expenses with the method of charging; basis of charging fees; instalments; interest; time charges showing the hourly rates of the different categories of surveyor; and percentage lump sum fees).

The form of agreement sets out details of the parties, the services to be performed and the fee to be paid. The terms and conditions cover all relevant matters relating to the contract between the employer/client and the quantity surveyor. These matters embrace payment for the quantity surveyor's services, within 28 days of sending the invoice and payment of the stated rate of interest thereafter; communications between the employer/client and the quantity surveyor; insurance; assignment; suspension or termination by the employer/client; suspension or termination by the quantity surveyor; consequences of suspension or termination; copyright; and liability of the quantity surveyor.

The client guide gives recommendations from the RICS as to how a quantity surveyor is to be selected and appointed and indicates in particular that a quantity surveyor should not be appointed purely on the basis of his fee level. Anticipated performance, quality of service and the fee are all equally important. It is hoped that clients selecting a quantity surveyor will follow the stages set out in the guide and are warned against unregulated fee competition activity. It emphasises that a full and precise description of the quantity surveying services is to be provided and gives guidance on the basis of the fee and method of payment.

The client guide illustrates the advantages of appointing the quantity surveyor at the inception stage of a scheme in order that he can give advice on costs and contractor procurement. It describes three methods of selecting the quantity surveyor: ie based on knowledge of a particular quantity surveyor, from a panel or from a short list.

As quotations should be sought only from quantity surveyors considered suitable and capable of providing the service required, normally the lowest quotation should be accepted provided that it is clear that the service required is possible at the level of fee. Where quotations are sought on a competitive basis they shall be analysed and compared on a basis which incorporates all the component parts of the quotation into a single index in order to facilitate comparison. The resultant indices should then be tabulated so as to indicate the lowest and its relationship with the others. A useful schedule in the guide contains a practical example of how this is done.

## QUANTITY SURVEYING FEES

When submitting fee bids for professional services a quantity surveyor must comply with the following RICS Bye-law:

RICS Bye-law 19; Regulation 15 (1995) relating to inducements for the introduction of clients states that 'No member shall:

(a) directly or indirectly exert undue pressure or influence on any person, whether by the offer or provision of any payment, gift or favour or otherwise, for the purpose of securing instructions for work, or accept instructions from any person on whom he has reason to believe that undue pressure or influence may have been exerted by a third party in expectation of receiving a reward for the introduction;

(b) quote a fee for professional services without having sufficient information to enable the member to assess the nature and scope of the services required;

(c) having once quoted a fee for professional services revise that quotation to take account of the fee quoted by another member of the surveying profession for the same services;

(d) quote a fee for professional services which is to be calculated by reference to the fee quoted or charged by another member of the surveying profession reduced by some proportion or amount.'

A new Regulation 15(3) in 1995 permits exceptions to the above requirements in respect of third party favours under certain specified conditions.

For many years the RICS issued detailed sets of scale fees to encompass the various services and classes of work applicable to quantity surveyors, based upon and embodying the collective experience and wisdom of members and, where appropriate, agreed with client representatives as being fair and reasonable. They proved useful to clients with little experience of construction projects who needed some guidance on professional fees. Although the scales of fees still remain they cannot be mandatory because of the ruling of the Monopolies and Mergers Commission Report in 1977, whereby fee competition has become commonplace, particularly in the public sector. RICS Bye-law 19, as quoted earlier, was revised to take account of the M&MC recommendation in 1983.

A RICS report (1991) called into question the relevance of many RICS recommended fee scales. Given that the benefits of professional services are increasingly assessed in terms of the value added to the client's business, it was suggested that the remuneration basis should reflect that trend. Hence the practice of subsidising say, relatively valuable cost advice and cost planning from the fee earned for the relatively mechanical preparation of bills of quantities appears particularly irrelevant. In 1995 the RICS Quantity Surveying Divisional Council was considering whether further guidance should be provided for members/clients on methods of charging, inviting and responding to tenders, assessing and presenting the extent and range of services provided and any other related matters.

In the 1990–95 recession many quantity surveyors were charging fees at 50 per cent or less of the scale fee in cut-throat competition in order to secure work. This unfortunate practice, caused by an horrific trading climate, could cause irretrievable harm through staff cuts, pay freezes, closed offices, reduced investment in new technology and techniques, less staff training and corporate marketing and a reduced service to clients both in quality and in the range of services offered. Quantity surveyors were also being asked to prepare feasibility studies on work which might not materialise for many years. The only real and lasting solution was a change of priorities by the government by investing more money in much needed buildings and the infrastructure.

## QUALITY ASSURANCE IN QUANTITY SURVEYING

Quality assurance related to the construction industry was examined in chapter 1. In this chapter the aim is to examine it in relation to quantity surveying practices.

We have seen that quality assurance (QA) is a systematic way of ensuring that organised activities are implemented in the way in which they are planned. It is therefore a management discipline concerned with creating attitudes and controls which are designed to prevent problems arising, as described in chapter 1. Its main purposes are:

- to meet client requirements
- to meet industry requirements
- to improve management control with regard to the quality of service
- to increase efficiency
- to gain marketing advantage.

It costs money to set up and run, but against this must be set the savings which result from more effective management, improved methods of working, and an enhanced level of performance.

A large quantity surveying practice commenting on the adoption of QA stated 'There is no doubt that our initial motivation was a combination of client pressure, and seeing our competitors moving into QA. We felt it essential to keep our edge in a keenly competitive market. Now, however, at final assessment stage, we are in no doubt that QA has improved systems and management in the practice. QA is not easy, especially in a recession. It takes time and resources, but we are undoubtedly better organised'.

By 1990 formal quality management systems and quality assurance procedures continued to be considered and implemented by the partners of a considerable number of quantity surveying practices. Having developed formal quality management systems, practices may wish to consider certification by a third party to demonstrate compliance with BS 5750 (ISO 9001/EN 29001). The practice's quality system must address all activities, services or resources which could adversely affect the services to be provided.

Practices who wish to pursue assessment and certification are faced with the task of demonstrating that their own unique procedures comply with a standard which was originally drafted for the manufacturing industry, and thus usually requiring some adaptation. A number of certification bodies have been approached by practices seeking third party certification. To assist them the RICS (1990) published appropriate guidelines giving an overview of the standard with some relevant examples.

It was considered helpful at this stage to include the quality management policy statement of Currie and Brown, an international quantity surveying practice. It is the policy of the practice to:

- provide the highest possible standard of service to clients
- operate whenever possible a quality management system which has been developed in accordance with BS 5750 Part 1 for professional quantity surveying services
- ensure that personnel using reasonable skill and care, work in accordance with the procedure detailed in the quality manual
- provide documentation to demonstrate the foregoing has been achieved and evidence of training and instruction as necessary to staff
- periodically prepare quality reports and reviews of procedures in order that the quality management system may be improved
- provide facilities for clients' representatives to verify compliance with the quality management system.

## INFORMATION TECHNOLOGY

The RICS report *QS 2000* (1991) made some pertinent forecasts concerning the future use of information technology (IT) in quantity surveying organisations. It was considered that information flows in construction will increasingly be made electronically, and that the development of new IT tools for quantity surveyors, and databases in particular, will need to progress alongside that of other systems used in design and construction, particularly computer aided design (CAD) systems. To improve speed and efficiency, the fundamental nature of much of modern quantity surveying activities needs to be examined. Much of this involves the management of information: sifting, analysis and synthesis of large amounts of project data together with communications with other project participants. It was believed, with justification, that information management and communications systems are ideally suited to facilitate this.

The report also asserted that the development of new services will need to exploit information technology. In particular, emphasis needs to be given to those applications with potential in areas of high uncertainty, such as forecasting. The increasing automation of measurement may eventually be absorbed into CAD and lead to the computerised production of tender documentation. Quantity surveyors must play a part in these developments and learn to use CAD systems. Developments in expert systems, particularly those related to early cost advice, could also accelerate the use of CAD by quantity surveyors.

Hawkings of E. C. Harris sounded a note of caution when he stated 'The 1980s explosion in information technology had the laudible aim of making organisations in both public and private sectors more efficient. Many have taken full advantage of the newly available tools to transform their productivity, but too many others have made expensive investments in hardware and software with disappointing results'.

Stoy Hayward (1991) found when looking beyond the declared IT plans of medium sized organisations, there was a wide diversity in the type of IT applications that they were planning, ranging from sophisticated CAD/CAM systems and telecommunications to relational databases and the use of predictive software tools. Despite this diversity in planned applications two distinct, yet interlinked, common strands of thought within IT plans could be identified:

(1) Organisations are increasingly recognising the importance of information within an organisation and the concept of using IT to better manage strategic information.
(2) An increased emphasis on 'the better use of what we have' thereby increasing the use and scope of IT.

Two RICS members' focus groups considered that information technology had become paramount to the chartered surveyor in achieving business efficiency. Internal economies are a strong incentive for firms that have a strong branch organisation. Larger firms are enthusiastic users of networks which are especially useful in helping branches to keep in touch and allowing partners with specialist knowledge to work on a single document. Firms were finding that they tended to maintain the same level of service with fewer people or lower grade personnel needed to process data. However, closer supervision is needed by a senior person, but where this is done the middle management can be reduced. The greater availability of information results in a greater call on professional judgement.

Practitioners from small practices have quite a different decision-making process and costs are very carefully considered. As one sole practitioner explained: 'My investment has been based on need. I need a project which is going to pay for the investment in terms of time and money'. The problem of keeping down costs may be solved by linking up with other small practitioners to buy hardware and software.

In 1995 many members felt that the RICS should provide on-line services to members in the same way as BCIS. This included RICS policy documents and library services. One member commented 'If you are working on, for example, a hospital project, it should be possible to dial and ask what experience others have had'. It was suggested that the RICS could provide basic services and charge for specialist ones.

### E-mail (Electronic mail)

E-mail was being introduced in 1995 and was making substantial changes in the pattern of communication, both within firms and with clients. It has meant large savings in time; providing immediate response and cutting down on secretarial and partners' time. In one large practice, e-mail operates throughout the whole of the regional network. Files can be sent and two people can be looking at the same file and discussing and modifying its contents. In 1995 the RICS was connecting to the Internet (the Information Super-highway) and thereby providing electronic mail in/out services and the ability to send and receive documents directly to and from members (with Internet access) for the cost of a local telephone call. Four formats of information are available: e-mail, world wide web (published information), forums (talking to other people interested in the same subject by typing out points), and file transfer protocol.

### Cordless Technology and Teleworking

Advances in communications technology are radically changing the ways of working across time and space – and consequently the demand for and design of buildings. One of the major advances is the development of cordless technology and data transfer, whereby office workers are no longer bound to their offices where space is often becoming significantly underused and can vary their work times in a more convenient location. This can result in homeworking or teleworking from satellite offices or 'telecentres'. Tapping (1995) believes that the office of the future is more likely to be a 'virtual office' which capitalises on the increasing use of cordless laptop and notebook computers and mobile telephones to provide a flexible working environment. The physical business office will still exist but in a much reduced form. A further examination of this topic is included in chapter 19.

## USE OF COMPUTERS

All quantity surveying organisations use computers to assist in carrying out their diverse range of activities, normally having analysed their requirements in detail and assessed and costed the various options for satisfying them, without being susceptible to the promotional approaches of software companies. The activities covered include accounts, salaries, job costings, personnel resources, general typing and word processing, general spreadsheets, general graphics, bills of quantities, cost and financial reporting, statements, cashflow, drawing registers, programmes, estimating, and contract administration. The use of computers offers the principal advantages of speed of output of both technical and administrative material, reduced staffing requirements, and increased availability and exchange-ability of information.

A number of the larger quantity surveying practices have developed their own computer systems. For example E. C. Harris developed CATO which was the first fully digitised system, with plan dimensions read auto-matically and entered directly into the computer. CATO has since evolved into a family of software tools with applications in estimating, cost control and procurement. In like manner, Currie and Brown wrote bespoke soft-ware to meet the practice's needs and those of clients and is especially proud of its 'Costfax' and 'Live Bills' programmes.

'Costfax' is a versatile cost control system and is capable of quickly calculating costs relating to the budget, control estimate, expenditure and forecast. Its main purpose is to provide access to costs even on a day to day basis. While 'Live Bills' provides a fast and accurate tender analysis, assisting the early appointment of contractors. In its postcontract role it facilitates the assessment of progress and the financial management of contracts.

The hardware requirements of a large quantity surveying practice can be very substantial indeed. For example, Currie and Brown in 1995 had around 120 IBM PC AT 386/486 compatible, comprising desktops, laptops and notebooks all with appropriate laserjet and dot matrix printers and plotters.

Readers requiring more information on information technology and com-puterisation are referred to Hartigay and Dixon (1991), Feenan and Dixon (1992) and Dixon (1988).

## ADVERTISING

### General Background

Wheeler (1983) aptly described how chartered quantity surveyors acting as true professionals had refined and improved the services they provided to clients, but had acted with the utmost decorum and modesty in not publicly announcing their achievements and abilities. In this respect the profession had been subject to considerable outside criticism for continu-ing with what were regarded as outmoded practices. Wheeler believed that chartered quantity surveyors should go out into the market place and, albeit with due and proper regard to the need for decent, correct and inoffensive behaviour, make known to the public at large what they can do, what they have done and how their services benefit the user. He saw the spearhead coming from forceful marketing and promotion by indi-vidual concerns, both public and private, to bring about a wider knowl-edge of the existence of quantity surveying practice and techniques and a realisation of their full value and potential. There is admittedly, a widespread ignorance of the services that the quantity surveyor can offer, and the previous tight restrictions on advertising inhibited the development of a better public awareness and understanding. It is of benefit to the prospective client, and ultimately to the construction industry, to become better informed about the management, administrative, coordination and cost control functions that the quantity surveyor can provide.

The initial breakthrough came in 1979, when chartered quantity surveying practices were permitted for the first time to send brochures to prospective clients. Since the introduction of fees competition, the tide of opinion has changed among construction professionals in general.

**Advertising Regulations**

Since 19 February 1988 there is no separate set of regulations for advertising and publicity. The relevant rules are set out in Bye-law 19, Conduct Regulations 15 to 20 (1995). They represent a substantial relaxation compared with the former rules, especially in the area of inviting instructions, and are more in keeping with the increasingly competitive climate.

The key requirements for general advertising and publicity are accuracy, professionalism, clarity and a proper respect for the confidentiality of client's affairs. Members of the RICS are still limited in the ways in which they may invite instructions, for example door knocking at private addresses and, in general, telephoning private numbers is prohibited. But they now have greater freedom to invite instructions by letter. Such letters must, however, contain provisos designed to warn the recipients of the danger of liability to two fees if one surveyor is supplanted by another (Regulation 20).

The RICS (1994b) believes that the need for close supervision of all published or circulated material has been evidenced by the number of instances reported to the RICS Professional Conduct Committee of breaches of the regulations relating to advertising and inviting instructions. In many cases, the principals, although familiar with the rules themselves, were not aware of the contravention because of the absence of machinery in the firm for monitoring their application.

To overcome this problem, it is advisable that all advertisements should be approved by a principal or other nominated senior member of staff before publication, who should also sign any circulatory letters issued by the firm. If it is not possible for a partner or director to see all advertisements before publication, it has been suggested that a chain of responsibility should be established and made known throughout the firm.

However, in 1995 a RICS Working Party recommended that the current Rules on inviting instructions and advertising should be abolished and that members should comply only with the law of the land applicable to all by not being inaccurate, or misleading or likely to cause public offence.

# 16 Professional Ethics, Standards and Conduct

## PROFESSIONAL ETHICS AND STANDARDS

Apart from the requirements of the RICS Rules of Conduct, all members have a moral duty of care when dealing with clients and their affairs and to exercise the utmost honesty and integrity in all their dealings. Clients can rightly expect that professional men and women will possess a reasonable measure of competence and skill in their particular calling and will use these qualities to the benefit of the client.

This is reinforced by some of the RICS Bye-laws and Regulations, such as Bye-law 19(3) and Regulation 3 which prescribe that 'No member shall be connected with any occupation or business in any way which would, in the opinion of the General Council, prejudice his professional status or the reputation of the Institution'. Chalkley (1994) described how in times past earlier bye-laws sought to prevent chartered surveyors from engaging in commercial activities which were thought to be incompatible with the concept of professionalism. Although times have changed and the profession has now to operate in a fierce commercial and competitive climate, professional ethics must still be preserved.

The chartered surveyor when offering commercial services alongside his professional activities, must ensure that they do not inadvertently harm the client's interests or the reputation of their profession. It should be noted that under Regulation 3 (1)(a) a chartered surveyor practising in quantity surveying is not permitted additionally to manage or control any organisation which is wholly or partly engaged in building or civil engineering contracting.

The upholding of professional ethics can also be read into the primary objectives of the RICS Rules of Conduct as now listed:

(1) That members of the Institution shall discharge their professional duties to their client and to the wider public interest in accordance with the objectives of the Institution as prescribed by its Royal Charter.
(2) To this end members shall:
    (a) ensure that where there is a conflict between the interest of the member and that of the client or the greater public good, the interest of the client shall prevail unless it is at odds with the wider public interest;
    (b) ensure that they perform their duties
        (i) with integrity and honesty;

(ii) competently and diligently; and

(iii) by adopting personal and professional standards which enhance the reputation of the profession and the Institution.

It is important that members when practising should confirm to their clients in respect of an instruction that they are competent and have experience in the type of business and the geographical area concerned. The member must not hide any relevant facts from the client of which he should be aware and which could prevent the member securing the commission, as this would be patently dishonest and unethical. In like manner, a surveyor must not claim to achieve the highest standards when he knows that it is not possible, as he would be misleading the client.

The RICS (1994c) postulated that all chartered surveyors view themselves as professionals, as they consider themselves to be highly qualified and skilled. They would certainly wish the outside world to view them as professionals. There is, however, a danger for any profession that its members may seek to claim that compliance with specific rules excuses them from having to have regard to their more fundamental professional and ethical obligations.

It is worthy of note that the RICS (1994c) believed that ethics should be taught as part of university surveying degree courses and this must be of benefit to the students, their future clients and the profession at large. It could lead to higher professional standards and enhanced attitudes to the need for greater honesty, dedication, care and trust in all professional relationships with clients, other surveyors and the general public, and the avoidance of making decisions which are morally wrong.

Jonas (1992) in his RICS Presidential address described how the British professional has built a reputation for integrity and objectivity which is the envy of much of the world. Thus chartered surveyors have to fulfil their clients' high expectations of their standards which in the real world have to be balanced against the surveyors' price levels and the value for money which they provide. He also asserted that pressure on surveyors' prices cannot justify a lowering of standards. Jonas believed that the best way towards increasing influence was for the chartered surveyor to regain the high ground of public respect and trust in all areas of practice.

The Reverend David Jenkins, former Bishop of Durham, recognised the wider remit of a professional man which often goes far beyond the client's brief, when he wrote 'A professional man is one who, in the judgement of his peers has proved himself competent in the exercise of the work he has undertaken. He is one who is not limited in the performance of his duties by a timetable or, when he understands his work aright, by the ability of those he serves to pay him. He does not practise his skill as a mere technician, but as a human being, conscious of the fact that he is dealing with human beings in the complexity of human situations.'

Clark (1987) in his RICS Presidential address made some very relevant points from his long and wide experience which the author believes to be well worth quoting. 'The fundamental qualities which have always been expected of a professional are integrity and detachment. The truth is that professional men and women, with very few exceptions, represent and practise these unthinkingly. Professionalism is unquantifiable, hard to define, but instinctive.'

Clark believed that the protection of the client's interests remained paramount but the author believes that there can be occasions when it is necessary to consider the wider aspects, as with environmental impact analysis, as described in chapter 10. Clark further considered that it is the duty of a profession continually to review the standards of competence to be expected, to help members to achieve those standards with the provision of further education and information, and then to monitor the level of success. He saw the professions in this country as a powerful stabilising influence. They hold the middle ground and with their instinct for fairness and balance, they are a counterpoise to the forces of extremism.

Finally to round off this section I think I can do no better than quote two short sections from a foreword by Sir Winston Churchill to an old but undated hardback publication entitled *The Chartered Surveyor: His training and his work* published by the former Chartered Surveyors' Institution.

'The importance of high standards of technical efficiency is obvious: but even more important is that sense of professional honour and fidelity which springs from prolonged vocational training and a lively sense of corporate responsibility. To be a Chartered Surveyor is not to be merely an individual who has acquired certain knowledge and aptitudes: it is to be a member of a body sustained by the comradeship of equals and with a collective dignity and authority'..... 'A love of justice and equity and a firm resolve to deal with every issue on its true merits without fear, favour or affection, are inseparable from the rightful discharge of a surveyor's duty'.

The situation in which a quantity surveyor is employed by a contracting organisation entails a rather different approach, as here the quantity surveyor is dealing with fellow professionals as distinct from laymen, and can have a reasonable expectation that these professionals are entirely familiar with the techniques and procedures being used and their implications, which accordingly do not have to be explained in detail. He must, however, perform his duties with integrity, honesty, diligence and competence, and never knowingly attempt to deceive other parties.

## RESPONSES FROM QUANTITY SURVEYING UNDERGRADUATES

Morledge and Hogg (1995) carried out a valuable survey at The Nottingham Trent University to obtain the views of students in the final years of part time and thick sandwich courses for the BSc (Hons) Degree in Quantity Surveying on ethical standards, attitudes and practice, and 133 responses were received. All students had experience of working with quantity surveyors, approximately one half with contractors, one third in private practice and smaller numbers with local authorities and subcontractors.

Attitudes to loyalty varied with the majority (84 per cent) believing that they owed loyalty to their employer, while 61 per cent of those employed in private practice or with local authorities believed they owed loyalty to the client as opposed to 12 per cent of those employed by contractors and subcontractors. Irrespective of type of employment, 53 per cent considered that they owed their first priority to their employer.

The majority (91 per cent) of those in private practice or with local authorities stated that they would accept a gift from a contractor and most of them thought it would not affect their impartiality, although it was affected significantly by the nature and value of the gift (87 per cent acceptance of bottle of whisky to 15 per cent taking cash).

The majority of quantity surveying students employed by contractors considered that the primary function of their employment was to make profit, and many of them would not be concerned if their objectives disadvantaged the client. Practically all respondents (96 per cent) would enhance remeasured quantities of dayworks if asked to do so by their superior, and a large proportion (85 per cent) would attempt to beat down the price of a subcontractor by using technical or intellectual advantage. They are more likely to believe that commercialism has precedence over an ethical stance and a minority (16 per cent) expressed concern at the level of ethical standards they put into practice at work.

The majority of undergraduates believed that membership of a professional body would not affect their ethical standards at work. Approximately one half expressed awareness of the rules of professional conduct, although their knowledge was generally vague and confined mainly to professional indemnity insurance requirements, advertising restrictions, prohibition of bribes and fraud and need for fidelity to clients.

The findings of this survey must be of concern to the profession and point to the need for the inclusion of professional ethics and standards on surveying courses. It would also be advisable to include it as a component of the Assessment of Professional Competence (APC). The importance of upholding high professional standards in all activities cannot be over emphasised.

## AIMS OF CODE OF PROFESSIONAL CONDUCT

The RICS is governed by a Royal Charter which imposes obligations on the Institution, including the maintenance and promotion of the usefulness of the profession to the public advantage. The self-regulatory role of the Institution over the activities of its members is largely implemented through the Rules of Conduct.

The primary aim of these Rules of Conduct is to establish a set of ethical principles which promote the duty of members to discharge their professional obligations to their clients and employers while, at the same time, having due regard to the wider public interest, as described earlier in the chapter. To achieve these aims, certain standards of behaviour are required of chartered surveyors in their business activities. It has also been suggested that these standards should also be exercised in their private lives although this is more questionable, as possibly being very desirable but not mandatory.

These standards require members to:

- discharge their duties to their clients, employers, employees, colleagues and others with due care in accordance with the provisions of the Institution's rules and guidance
- perform their duties with competence, diligence, honesty and integrity
- fearlessly and impartially exercise their independence of professional judgement to the best of their skill and understanding
- not discriminate on the grounds of race, sex, creed, religion, disability or age or on any other basis and promote equality of opportunity within the profession
- adopt personal and professional standards which enhance the reputation of the profession and the Institution.

## PROFESSIONAL CONDUCT REQUIREMENTS

### Introduction

Jonas (1992) rightly pointed out how modern economic strains put huge pressures on standards. This applies equally to standards of professional conduct for chartered surveyors, yet this is one of the main criteria, alongside professional expertise, which distinguish them from unqualified surveyors. Hence it is essential to be able to demonstrate that the code of conduct is supported by all RICS members and is enforced rigidly, promptly and openly when it is transgressed.

It is vital that the Institution retains the right to self-regulation in a manner which is relevant to the needs of clients, employers and members, ensures that the conduct of members is measured by achievable standards, and ensures that members' compliance with the rules/codes of conduct can be monitored in a fair and cost effective way.

### Rules of Conduct for Chartered Surveyors

Chartered quantity surveyors are subject to a set of stringent rules of professional conduct to which they must adhere at all times. The operation of these rules ensures that no chartered quantity surveyor can be a party to unprofessional conduct which could bring himself and the profession into disrepute, without making himself liable to disciplinary action involving heavy penalties including, in the case of a serious breach, the risk of expulsion from the Institution. Dann (1983) in his RICS presidential address, emphasised how chartered surveyors 'as members of a profession perform a skilled activity and subscribe voluntarily to a code of principles of conduct. These two characteristics are inseparable; they have always been so, and they must ever remain so. The public needs professional bodies to provide machinery for maintaining minimum standards.'

RICS Bye-law 19 and the associated Regulations together comprise the Rules of Conduct which are contained in a booklet. The rules and particularly the regulations are subject to periodic review, involving amendments, and these amendments are published in *Chartered Surveyor Monthly*, and copies of the latest rules can be obtained on application to the Institution. It is imperative that all members are thoroughly familiar with the Rules of Conduct and that they abide by them.

The most fundamental rules are that no member shall conduct himself in a manner unbefitting a chartered surveyor or carry on practice as a surveyor under any such name, style or title as to prejudice his professional status or the reputation of the Institution. Another vitally important rule is that no member shall be connected with any occupation or business which would in the opinion of the RICS General Council prejudice his professional status or the reputation of the Institution.

Other paragraphs of Bye-law 19 and associated Regulations in 1995 were concerned with conflict of interest; keeping separate bank accounts for clients' money and accounting for all moneys held (examined later in the chapter); keeping required accounting records (examined later in the chapter); complying with Institution practice statements; maintaining insurance against claims for breach of professional duty (examined later in the chapter); connected businesses (examined earlier in the chapter); trade discounts; financial interest; incorporation with limited or unlimited liability (described in chapter 15); site boards; status and designations; inducements for the introduction of clients (described in chapter 15); professionalism, accuracy and clarity (described in chapter 15 under the heading of advertising); avoidance of claims of superiority; advertising in association with non-surveying undertakings; inviting instructions (described in chapter 15); contraventions procedure; continuing professional development (CPD) (examined in chapter 18); references to the Institution; confidentiality of clients' affairs; and notification of practice details. Bye-law 20 deals with the responsibility of members for their firms, and particularly relates to the publicising of partnerships and designatory letters on notepaper or in advertisements.

The main features of the Regulations associated with Bye-law 19, which are not covered elsewhere, are now outlined.

## Conflict of interest

Where a conflict of interest arises or may arise between a member's interests and those of a client, the member shall:

(a) disclose promptly to the client the relevant facts;
(b) inform the client that neither he personally nor his firm/company can act or continue to act for him unless requested to do so, having advised the client to obtain independent professional advice; and
(c) confirm the position in writing to the client.

A similar provision operates where two or more clients are involved.

## Trade discounts

A member in his professional capacity cannot accept, other than for the benefit of the client, any trade or other commercial discount or commission from any trader providing goods or services used in the building construction or maintenance industries, unless the discount is on goods or services ordered by the member on behalf of the client, who has been notified.

## Financial interest

A member shall not take a financial interest in any matter in which he acts in his practice as a surveyor nor shall he accept instructions on terms which could be construed as taking a financial interest, except in the circumstances listed in the regulation (Reg. 9).

Furthermore, a member performing the function of quantity surveying shall not, without disclosure to all parties, permit himself to be named in the contract or elsewhere if he is associated with one of the parties, whereby he may gain or lose financially, apart from his normal professional fees, according to the amount at which the contract is settled.

## Site boards

A member shall not exhibit his name, or the name of his firm or company, or the name and position of a member employed by a public authority, on a building site unless the Institution's standard name board is used, or a similarly unostentatious board with suitably sized lettering.

## Status and designations

In any list of partners and/or staff in a firm, or directors and/or staff, carrying on practice as surveyors, every member named shall ensure that his status within the firm or company is clearly stated and that chartered designations are correctly used.

### Avoidance of claims of superiority

A member shall not make a public statement claiming an advantage or superiority over other firms, unless this statement can be substantiated to the satisfaction of the RICS General Council.

### Advertising in association with non-surveying undertakings

A member shall not publicise his services or allow them to be publicised, in association with any goods or services available from any other source in such a way as to raise doubts as to the independence of his professional advice or give rise to a conflict of interest.

### Contraventions procedure

Where it appears that a contravention of the Regulations relating to accuracy and clarity, claims of superiority and inviting instructions may have been committed by a member, following a complaint made by another member, the Institution shall not investigate it unless the complainant has notified the member concerned direct or through named officers of the appropriate branch or divisional branch and invited the member's comments, and the member has failed to satisfy the complainant that there are not sufficient grounds for taking any further action, or unless the complainant satisfies the General Council that there are sufficient grounds for the Institution to act.

### References to the Institution

A member shall not:

(a) purport to represent the views of the Institution unless expressly authorised so to do; or
(b) publicise the Institution or its members generally in terminology which has not appeared in an Institution advertisement or been approved by the Institution.

### Confidentiality of client's affairs

A member shall not, without his client's consent, disclose personal information about the client. This information includes financial or business activities or other information of a sensitive nature which is not generally known.

*Notification of practice details*

Every member shall, within 28 days of a request, supply the Institution with particulars of his practice or employment in the required format.

## PROFESSIONAL NEGLIGENCE

Oddy (1995) has emphasised how the expectation of clients is very high in respect of surveyors' competence to undertake professional work. Applications for inclusion in the list of potential tenderers for professional services now frequently require the completion of a detailed application form and questions can include details of the operative professional indemnity insurance policy and quality assurance procedures. Some clients are also imposing their own terms of engagement on firms of surveyors, setting out in precise terms the services to be performed, the fees which will be paid and the way in which the contract for services can be terminated. The firm of surveyors may also be asked to give details of the persons who will undertake the work and the client will then reserve the right to approve any changes of personnel.

Furthermore, clients expect to be listened to and to have their complaints dealt with in an appropriate manner by responsible professional bodies. The RICS receives on average some 2000 complaints a year of which approximately one third relate to negligence/incompetence, which gives an indication of the scale of the problem.

Clark (1987) believed professional incompetence to be a social danger, and that the responsibility to protect the public from it lies with the profession. It is important to distinguish between negligence and incompetence. Negligence is an instance of a lack of proper care and attention, while incompetence is a display of inadequate ability or fitness. A competent person can be negligent and as a result is culpable to a greater or lesser degree, but Clark posed the question 'Which of us have never made a mistake through inattention?' Professional negligence has been well defined as such a neglect of a professional duty of care as to render the professional person committing the act, error or omission of neglect, liable in law to a client or some other third party who sustains loss by reason of that neglect (Jess, 1982).

A major difficulty is deciding the appropriate standard of care to be applied in a negligence action and the following definition provides a useful guide: 'The test is the standard of the ordinary skilled man exercising and professing to have that special skill. A man need not possess the highest expert skill; it is well established law that it is sufficient if he exercises the ordinary skill of an ordinary, competent man, exercising that particular art'. (*Bolan v. Friern Hospital Management Committee* (1957) 1 WLR 582).

This approach was also taken in *Lanphier v. Phipos* 1838 AER 421 with the statement 'Every person who enters into a learned profession undertakes to bring to the exercise of it a reasonable degree of care and skill. . . . There may be persons who have a higher education and greater advantages than he has, but he undertakes to bring a fair, reasonable and competent degree of skill'.

The case of *Glasgow Corporation v. Muir* (1948 AC 448) also provides a good guide to the limitations of a defendant's liability. 'Legal liability is limited to those consequences of our acts which a reasonable man of ordinary intelligence and experience so acting would have in contemplation'. It is thus acknowledged that experts must expect to be judged by the standards of experts, but the principle remains the same: what would a man of average professional knowledge and experience have done? (Lavers, 1983).

The case of *Hedley Byrne & Co Ltd v. Heller and Partners Ltd* (1964) AC 465, (1963) All ER 575 HL, established the possibility of a duty of care in relation to a negligent mis-statement resulting in economic loss and opened the way for the law of tort to provide remedies for professional negligence concurrent with, and at all times extending far beyond, contractual remedies (Dugdale and Stanton, 1982).

This principle was subsequently highlighted in *Henderson and Others v. Merret Syndicates Ltd* (1994) 3 All ER 506. It was decided that where a person assumes responsibility to perform professional or quasi professional services for another who relies on those services, the relationship between the parties is itself sufficient to give rise to a duty on the part of the person providing the services to exercise reasonable care and skill.

It follows from the *Henderson* case and also from *Murphy v. Brentwood District Council* (1991) 1 AC 398, that architects, engineers, quantity surveyors and other consultants contractually employed by clients will owe those clients not only contractual duties but tortious duties as well. In practice, the significance is that generally the client has a more favourable limitation period for a claim in tort. The limitation period for a breach of contract claim usually runs for six years from the date of the breach; for the tort claim it runs from the date the building is physically damaged as a consequence of the negligent act or omission. Under the Latent Damage Act 1986, the tortious liability can, in some circumstances, run for 15 years from the date of completion (Akenhead, 1994).

A quantity surveyor is someone possessed of a special skill who undertakes, irrespective of the contract, to apply that skill for the assistance, *inter alia*, of the architect. The quantity surveyor cannot only reasonably foresee but probably expects that the architect will rely upon his judgement and skill (and it is reasonable for the architect so to rely). A duty of care therefore arises to ensure that the valuations prepared and submitted to the architect for certification have not been made negligently. If, for instance, the quantity surveyor deliberately undervalues the works for the client's benefit, he renders himself liable to the architect, and to the contractor in tort, for misrepresentation, reasonableness in this case including mathematical errors (Davies, 1979).

Further guidance on professional negligence in the construction industry can be obtained from Lavers (1996).

Some useful accounts of negligence claims made against quantity surveyors and the final outcomes have been provided by RICS Insurance Services Ltd (1989). The nature of these claims follows but the reader is referred to the RICS publication for more detailed information.

(1) A housing association lost part of a grant owing to the failure of a quantity surveyor to include the appropriate item in the bill of quantities. A settlement was agreed before costs were incurred in legal proceedings.
(2) A claim was made against quantity surveyors for the additional cost incurred in completing outdoor sports facilities as a result of their miscalculation of the volume of material to be excavated. The assured acknowledged that they were at fault and the final additional cost was borne by the assured to the extent of their policy excess and the balance by their underwriters.
(3) In preparing a tender for a builder client a firm of quantity surveyors included an item priced at well below the proper figure owing to a clerical error and received a claim for the amount consequently irrecoverable from the principal. Calculation of the loss was straightforward and the claim was settled under the professional indemnity policy subject to the assured's excess.
(4) An attempt was made to hold quantity surveyors responsible for loss on the basis of their rough estimate of probable cost of an industrial development based on scanty information, but the proceedings were ultimately dropped.
(5) Quantity surveyors advised the liquidator of a building company to carry on uncompleted contracts to completion, and faced a claim when they proved to be unprofitable. Although the assured maintained they had not been negligent in giving their advice, both parties agreed to a settlement at a reasonable figure to avoid the cost and worry of a court case.

## PROFESSIONAL INDEMNITY INSURANCE (PII)

### Introduction

Oddy (1994) aptly described how the Institution on behalf of its members accepts a responsibility towards the public; professional indemnity insurance (PII) cover provides a likelihood of recompense being available in the event of a surveyor being found negligent in the performance of professional services. It also enhances the image of chartered surveyors, particularly with the requirement for run-off cover as in the case of a retired partner, which operates in effect as part of an after sales service connected with professional work.

Oddy (1994) argued that a compulsory code of insurance was necessary for the following reasons:

(1) it gives the Institution some control over the quality of cover;
(2) a market based cost system geared to a mandatory quality of cover impinging on all the professionals in the market, tends towards fairness as between private practitioners, so that those who have taken steps to maintain high quality services are likely to be regarded as low risk and, in consequence, pay lower premiums;
(3) it enables the Institution to claim the public relations advantage for its members that all are insured;

The consequences of no insurance cover can be serious and far reaching and encompass the following possibilities:

(1) there is a risk that a surveyor's business and/or his personal assets could be obliterated in trying to meet a claim;
(2) adverse publicity would be generated if claims against members were not met;
(3) absence of insurance could result in a prudent professional being unable to protect himself against the consequences of a sudden market change (Oddy, 1994).

Chalkley (1994) stated 'The Compulsory Professional Indemnity Insurance (CPII) Regulations represent one of the main 'pillars' of the Institution's Rules providing protection for the public. We are all human and make mistakes. The Regulations seek to ensure, in so far as is reasonably possible, that member's clients do not suffer as a result. But the Regulations provide considerable benefit to members also. The cost of litigation today can be enormous. The mere bringing of an action by a client can impose considerable financial strain on members, even if they are not found to be negligent. The insurers will pay the member's legal costs in defending actions, provided the member complies with his obligations under the policy and notifies the insurers at the outset of the possibility of a claim.'

PII therefore has the function of protecting both a surveyor and the surveyor's client and it is likely that through recourse to the policy, the client will be fairly treated.

The combination of bringing a profession into disrepute through a serious act of negligence and the impact of a poor claims record, is likely to result in the inability to obtain insurance, thereby leaving the public unprotected. This must incur penalties and, if necessary, exclusion from the professional body altogether. It follows that the inability of a member to obtain professional indemnity insurance must, at the very least, be a matter for investigation. In consequence, a professional body places an obligation on its members to conform to a set of rules defining the minimum standard of PII required. The ability to obtain the required PII operates essentially as a guarantee of good performance, whereas a failure to obtain the cover leads to sanctions on the careless and unprofessional (Oddy, 1994).

### RICS Compulsory Professional Indemnity Insurance Bye-Law and Regulations

RICS Bye-law 19(8) requires that 'Every member shall, in accordance with the Regulations, be insured against claims for breach of professional duty as a surveyor'. This is supported by a number of Regulations and the most important are now summarised for the benefit of the reader, who is reminded that these Regulations are subject to periodic review; these summaries show the position in 1995.

Insurance cover is compulsory for members who are, or are held out to be, principals in private practice, including consultants, and employees who undertake work for persons or bodies other than their employers, in respect of that work. Run-off cover is also compulsory for members who have undertaken work in any of these capacities in the previous six years, and have since retired or left the practice, or the practice has ceased to trade. A number of retired members, particularly former sole practitioners, have complained bitterly, with justification, about the high cost of insurance premiums required to comply with the run-off provision.

Members are required to insure by means of a policy no less comprehensive than the RICS Professional Indemnity Collective Policy, as issued by RICS Insurance Services Ltd. RICS Insurance Services is a company that was set up at the instigation of the Institution and in which the Institution has a small shareholding. It is a specialist company that deals exclusively with liability insurances for the surveying profession and has been in operation since 1976. The company has developed the collective policy referred to in the Regulations and places the insurance with a range of insurers who appreciate the importance of the profession and its insurance needs.

Minimum cover for policies renewed on or after 1 April 1995 is £500 000 where gross income exceeds £100 000, £250 000 where gross income is up to £100 000, and £100 000 where gross income is up to £50 000. The premium payable will depend, *inter alia*, on the range and type of service to be undertaken by the practice, its fee income, the location of the practice, and other related matters. The Regulations also prescribe the maximum amounts of uninsured excess a member may be required to pay before indemnity is granted.

Under the Institution's monitoring arrangements, every member is required to provide the Institution with a signed certificate confirming details of his current indemnity insurance policy within 28 days of effecting the policy. In addition, members shall provide the Institution, within 28 days of being required to do so, with evidence that the member is not subject to these Regulations or that he has complied with them.

Members who are employees are advised to ensure that they are indemnified by their employers against claims arising from alleged professional negligence in the course of their employment.

Members are recommended to give serious consideration to obtaining public liability insurance. Such insurance is concerned with liabilities to third parties for injury to their persons or property which fall outside the scope of PII.

Members are reminded that if a claim is made against them the insurance must be sufficient to meet not only the damages awarded, but also interest on such damages and the costs of the person making the claim. Interest and costs can increase significantly the amount payable in respect of claims. The amounts specified in the Regulations are minimum amounts, and members may need much higher levels of cover.

The RICS has published helpful memorandums of guidance on professional liability for chartered surveyors in employment, in the public sector, and in the corporate sector.

### Quantity Surveyors' Insurance

In 1994, a London based broker stated that a quantity surveyor will pay on average about one per cent of fee income for PII compared with ten per cent or more for those in higher risk categories, such as commercial valuers. In 1990 RICS Insurance Services reported 91 claims notified by quantity surveyors, dropping to 67 in 1991. These compare unfavourably with prerecession notifications which averaged between 20 and 40 per year.

The feeling among insurers in 1994 was that a highly litigious climate, together with a tendency to cut corners when times are hard, was largely responsible for the increased number of claims. Diversification into areas of work carrying higher liability also had a part to play. Claims relating to over-certification, where interim certificates are overvalued and clients sue for negligence, are the most common, followed by construction management, project management and project coordination. Where quantity surveyors are moving into non-traditional areas, it is advisable for these elements to be separately rated for insurance purposes. In an extreme case a quantity surveying firm could find itself paying 50 per cent of its PII premium for what amounts to only 5 per cent of its work, calling into question the financial viability of offering certain high risk services to clients. Policy holders can take heart from the fact that up to 50 per cent of claims do not stand up to investigation (Davis, 1994).

### Managing Professional Indemnity Risk

There are a number of important steps that surveyors can take to reduce their exposure to claims. Good practice procedures are held in a firm's favour when renegotiating insurance, and brokers/insurers should be informed of any risk management procedures that are in place. Although it may be difficult to quantify significant savings in premium, risk management will reduce claims and therefore time spent in dealing with them. Points of particular importance are:

- a thorough understanding of client's instructions, ensuring that they are absolutely clear
- ensuring that the client gets what he wants
- confirming instructions in writing before undertaking the work – if instructions are continuing, then confirm all variations
- allocation of work to the right people, bearing in mind their abilities and experience – if work is passed to a trainee or someone with limited experience, make certain that the work is supervised
- keep proper notes, telephone messages, etc., as well as proper records of meetings and decisions taken
- make sure reports and certificates are completed in accordance with any RICS in-house guidelines
- review the firm's systems regularly and make changes if they are found to be inadequate – this shows underwriters that the firm is aware of problems and is willing to act when necessary.

A firm does not have to be quality assured under BS 5750 to lay down procedures and to follow them – let the insurer have a copy of the firm's procedures manual and the quality control standards adopted as evidence of quality control. It is also worth remembering that if instructions were confirmed in writing and proper notes kept during the period when work was being undertaken, approximately one half of the claims notified would not arise.

## MEMBERS' ACCOUNTS REGULATIONS

### Introduction

Many chartered surveyors, including some chartered quantity surveyors, hold money belonging to other people. It is therefore appropriate that there should be regulations dealing with members' accounts and the handling of clients' money.

The purpose of Members' Accounts Regulations is to ensure that monies entrusted to chartered surveyors are:

- paid into a separate designated account
- properly recorded in the accounts maintained by the member
- properly monitored.

If the Regulations are complied with, it should be impossible for a member to confuse his client's money with his own or inadvertently to make improper payments. Under the Regulations the books of account are required to be kept written up to date; in order to determine whether or not the books are properly written up it is necessary to reconcile cash books with bank statements to ensure that the funds available agree with the liabilities to clients.

In a partnership or company the responsibility of maintaining a proper book-keeping system is shared by all partners or directors. Any misappropriation or error by one partner or director, or member of staff is the responsibility of every partner or director.

### RICS Members' Accounts Regulations 1993 and Bye-law 19(4)

Bye-law 19(4) reads as follows:
'Subject to the Regulations every member shall:

(a) keep in one or more bank accounts separate from his own, his firm's or his company's bank account and client's money held by or entrusted to him, his firm or his company in any capacity other than that of beneficial owner;

(b) account at the due time for all monies held, paid or received on behalf of or from any person (whether a client or not) entitled to such account and whether or not after the taking of such account any payment is due to such person; and

(c) keep such accounting records as are specified in the Regulations and maintain them in accordance with the Regulations.'

Under Regulation 1(2) client's money is defined as follows:
'Client's money means any money received or held by a member or his practice over which he has exclusive control and which does not belong solely to him or his practice or a connected person.' Thus, for example, if a client can separately withdraw money from an account to which a member is also a signatory it is not client's money for the purposes of the Regulations, as the member does not have exclusive control.

At the request of the Quantity Surveyors Division an amended note to the Regulation, giving examples of clients' money, was approved, stating that 'Monies held or received by a member from a client which is due to other consultants or other third parties if they have been appointed by the member on the instructions of the client.' This note excludes the situation where a member is instructed by a client to undertake a job but then decides to subcontract part of it, as were the client to pay the member who then failed to pay the subcontractor, a debtor/creditor relationship might arise. The Regulation does not apply to contracting work.

Regulation 3 requires members to make certain disclosures to clients if they are signatories to accounts which contain monies which do not belong to them or their practice but which is not clients' money as defined by the Regulations. Members must notify their clients in writing that:

- the account is not a client account as defined in the Regulations and the money held is not clients' money
- the Regulations do not apply to the account and the money held in the account is not covered under the RICS clients' money protection scheme.

Regulation 4(3) refers to disclosure of identity of client account. Under this regulation members must, at the request of clients, provide them with written details of the client account including the full name and address of the bank or building society, the account title and whether or not it is an interest bearing account.

It was not considered necessary to give a detailed account of the many regulations in the Chartered Surveyors' Rule Book, to which the reader may make reference if he requires more information. The more important regulations cover client's instructions to overdraw a client account; requirements to show current balance; requirement to keep register of client accounts; and preservation of accounts for at least 6 years from date of last entry.

Members who hold client's money are also required to deliver to the Institution once a year, a certificate and accountant's report in the form set out in the schedule to the Accountants Report Regulations 1993.

## RICS CLIENTS' MONEY PROTECTION SCHEME

The Institution's Scheme of Protection for Clients' Money is an insured arrangement which provides a scheme of last resort in the event of losses of money up to certain limits entrusted to chartered surveyors. The scheme has two main purposes:

(a) to uphold the good name of the Institution and its members and the profession's right to be self-regulating; and
(b) to provide a remedy to members of the public who lose out as a result of having entrusted their monies to a chartered surveyor. It assists in fulfilling the Institution's object as set out in the Royal Charter 'to maintain and promote the usefulness of the profession for the public advantage'.

The main elements of the scheme are that it reimburses:

(a)  direct pecuniary loss by any member of the public;
(b)  monies lost as a result of fraud or dishonesty of any member or principal or employee of any qualified firm (i.e. a firm with at least one RICS member partner or director);
(c)  losses of up to £30 000 each and every claim up to an overall limit of £2m in any policy year.

The scheme is designed to afford some protection to members of the public by making funds available for their reimbursement when all other avenues are exhausted. Chartered surveyors and qualified firms can themselves receive no financial benefit or protection under the scheme. In the event of a loss the primary obligation is on the defaulting member or qualified firm to make full restitution. It is only in cases where the member or firm concerned is unable to make full restitution that the scheme can take effect. It is a precondition of the admission of any claim that the name of the alleged defaulter is reported to the police authorities on behalf of the claimant.

The scheme is funded by the payment of an annual levy of £55 (1996 rate) per principal (i.e. partner or director) where the firm holds clients' money or where the Institution has not received a certificate to say that it does not.

## INADEQUATE PROFESSIONAL SERVICES

### Introduction

Oddy (1995) defined inadequate professional service (IPS) as one where there is a lack of knowledge on the part of the person who carried it out, a lack of performance by some one who possessed the appropriate knowledge, or some deficiency in the delivery of the service. This lack of professional competency may at one extreme leave the client with a smarting sense of dissatisfaction and at the other of being liable in negligence.

In 1994 there was a general feeling in the profession that it was no longer acceptable for the Institution to say that it will investigate only those complaints which involve alleged misconduct on the part of chartered surveyors. Hence the Institution should also intervene where it appears that a chartered surveyor has failed to provide a service which it is reasonable to expect from that person as a chartered surveyor.

The concept of IPS is derived from the statutory requirements imposed on solicitors. Examples might include a chartered surveyor in the following situations:

- taking on work knowing that he is insufficiently experienced or skilled in that field to provide services of a reasonable standard
- taking on work knowing that he has insufficient time or resources to deal with it competently
- not confirming instructions properly
- not confirming fees properly
- not completing work in a reasonably prompt manner
- not keeping his client informed.

The underlying message is that a chartered surveyor must carry out his professional work with due care and skill, proper diligence and promptness. If a complainant considers that this duty has not been honoured, then the Institution should have the power to investigate. Such a power would build on RICS Bye-law 19(1) and enlarge the scope of the disciplinary procedures, so as to encompass the investigation of performance as well as conduct (RICS, 1994d).

### RICS Proposals and Consultations

The concept of IPS and how it should be handled was discussed by the RICS General Council and extensive consultations conducted with divisional councils, branches and the membership at large during the period 1993 to 1996. A green paper was issued in April 1994, covering methods of investigating inadequate professional services (IPS), and this proposed that the Institution should investigate complaints of IPS, which may fall short of negligence and which may or may not cause financial loss.

Following receipt of numerous comments on the green paper, a working party of the Standards and Practice Committee was set up in early 1995, which reported its findings to the General Council in April 1995. The report was fully debated by General Council which then arranged for widespread consultation prior to further consideration of the matter by General Council in 1996. The wide ranging and lengthy process of consideration and consultation shows the complex nature and sensitivity of the issues involved.

The author believes the issue of IPS to be of such paramount importance to the future well being of the profession and its relationship with clients, that a summary of the main findings of the 1995 working party is now listed, although it will be appreciated that at the time of writing the book these recommendations had not been ratified.

- A minority of chartered surveyors fail to provide a good service which satisfies the needs of their clients, either consistently or occasionally. The Institution can no longer afford NOT to take appropriate action in these cases and to provide a remedy for the public in the event of a proven inadequate service.

- The costs of providing a redress system must not place too great a burden on the profession as a whole; hence any redress must be payable by those at fault.
- The Institution should do more to encourage the provision of a competent service by members to clients and others.
- The Assessment of Professional Competence (APC) should include a short compulsory course on client and job management and there should be greater encouragement by the Institution for undergraduate courses to embrace basic principles of the Rules of Conduct and client management.
- The Institution must have power to investigate allegations of IPS on the receipt of a complaint from a client.
- The definition of an IPS must be tightly drawn to cover the delivery of the service and to exclude negligence.
- The Institution would not intervene in allegations of negligence where there is loss that is capable of being actionable through the Courts.
- Members should be required to notify clients in writing of their terms and conditions of engagement, including fees unless otherwise known.
- There should be a requirement on members to have their own complaints handling procedure to which clients would be referred in the first instance.
- There should be considerable streamlining of the disciplinary process.
- Where a member is found to have provided an IPS there should be a proper means of redress to the client by the introduction of a new penalty of a power to order the repayment or remission of a fee up to a limit of £2500.
- A small administration fee of not more than £50 should be charged on clients wishing to make an allegation of an IPS against a member, which would be returnable if the allegation was found proved.
- If the client had not paid the surveyor's fee due to be paid for the work undertaken then, prior to consideration of the complaint, the fee would be paid into the Institution.

In 1996 the RICS General Council was considering a new complaints handling procedure which centred around an in-house system in surveyors' offices, based on mediation and/or compulsory arbitration.

## ARBITRATION SCHEME

### Complaints against Chartered Surveyors

The Institution advises that complaints against firms of chartered surveyors (where at least one partner is a chartered surveyor) should be sent to the RICS Professional Conduct Department, and should include permission for the RICS to send a copy of the complaint and supporting documents to the chartered surveyor involved. Complainants are also advised to consider whether to use the Surveyors and Valuers Arbitration Scheme or whether the complaint should be handled through a court of law.

The RICS can investigate complaints alleging professional misconduct including unjustifiable delay in dealing with the client's affairs; failure to reply to letters; disclosure of confidential information; and failure to disclose conflicts of interest. If chartered surveyors are found to have breached RICS bye-laws or code of conduct they can be reprimanded, suspended or expelled from membership, resulting from disciplinary proceedings in which the member has a right to be represented and to appeal against a decision. The RICS is not, however, entitled to comment on, or investigate, cases where the law provides a remedy, nor to assess and award compensation. In a case such as alleged professional negligence or breach of contract, or where there is a claim for compensation, the client is advised to consult a solicitor or the Citizen's Advice Bureau.

### Surveyors and Valuers Arbitration Scheme

The Chartered Surveyors Arbitration Scheme was launched in 1989 and replaced by the joint RICS/ISVA 'Surveyors and Valuers Arbitration Scheme' in 1993. It offers a quick, inexpensive and informal method of settling disputes between chartered surveyors and their clients, involving claims relating to legal liability, including the alleged negligence or breach of contract of members or their employees. It may be used with the consent of both parties. Where members wish to agree to use the scheme they will usually need to obtain the consent of their professional indemnity insurers.

The arbitration scheme is independently administered by the Chartered Institute of Arbitrators (CIArb) and is funded partly by the registration fee, paid by both parties, and partly by the Institution. Most cases are dealt with on a documents only basis; very occasionally a site visit has been necessary. Since the initial scheme was introduced, the awards have been evenly split between those in favour of the claimant and those in favour of the respondent. The awards made ranged from £550 to £13 500. Most of the applications concerned mortgage-related work and a few involved full structural surveys.

In 1996 the RICS General Council considered replacing this voluntary arbitration scheme, which had a poor rate of take up of about 10 per annum since 1989, with a compulsory arbitration scheme which could deal with disputes over fees and had the support of insurers.

# 17 Marketing of Professional Services

## CLIENTS' NEEDS

Before considering the nature of marketing and how it can be used to best advantage, it was felt desirable to examine clients' needs, as it is now generally believed that provision of chartered surveyors' services are best market demand led. Hence there is an increasing need to identify what clients want and how the quantity surveyor can best satisfy those needs effectively and efficiently. In most cases this data will form one of the major criteria to be examined before a marketing strategy is formulated.

Lay (1991) identified the core role of chartered surveyors as adding value for owners and occupiers in the creation, management, development, maintenance and improvement of, and investment in their stock of assets, and its augmentation. This definition aims to cover all types of chartered surveyor as it was believed that the demand is for a broad and comprehensive range of services. The demand is becoming increasingly sophisticated and changing demand patterns stem from the growing realisation of the significance of property as a critical corporate asset and as a key element in business strategy. Demand patterns tend to centre around the elements of value-added, finance, management and investment.

Lay (1991) advocated promoting our professional activity on a wider front than any one surveying specialism, under the collective heading of 'the property profession', which was subsequently adopted as the RICS 'strapline'. The author did however press for its enlargement to 'land, property and construction' which he considered would more readily identify the inputs from quantity and building surveyors. Lay also foresaw the increasing need for a one-stop service and the likelihood of an increasing number of mergers in the private sector to provide greater resources and the ability to extend the range of services on offer.

The RICS Report *QS 2000* (1991) asserted that many clients want readily purchasable design, procurement and management of construction, and that professional services were increasingly assessed in terms of the 'value-added' to the client's business, an aspect shared with Lay. It was found that quantity surveyors' pragmatism and realism – 'common sense' – is the basis of what many clients value highly about quantity surveying services. The quantity surveyor's intimate knowledge of projects is at the root of the contribution which he makes to the value of the client's business.

The same report asserted that quantity surveying is very much a client led profession to which it must respond, but also to develop more on its own initiative. The changing requirements will demand a range of new services and improved efficiency in such vital services as early cost and procurement advice, construction cost control and project management. A key challenge facing the profession is how to demonstrate the value and indispensability of its services and how it is able to produce added-value to the client's built assets.

A RICS Quantity Surveying Study in 1984 also emphasised that clients expect the quantity surveyor to identify with their objectives, the context in which they operate, and the constraints the client must expect. The report also identified the following minimum expectations regarding the quantity surveyor's advice, which are still relevant in the 1990s.

- impartial/independent
- timely and regular forecast, in advance of events
- primarily on cost, but giving time implications and interaction of time and cost
- appreciative of cost of production and final value to the client of construction work
- inclusive of both building and services engineering work
- inclusive of options, which are available for the client to choose
- inclusive of recommendations for action.

Most clients and surveyors interviewed identified the following factors as significant in the successful provision of a quantity surveying service.

- early and continuing involvement by a person who was sufficiently close to the work; clients generally wanted a partner or senior level contact
- service performed on time and clients generally wanted a quicker service than they often received
- fee for the service provided seen as good value in relation to the benefits achieved
- initial advice at feasibility and conceptual stage had been accurate
- all advice derived from skilled consideration of all aspects
- accurate, regular, timely advice, with recommendations for action throughout the entire project
- amount of detailed work carried out by the quantity surveyor had been in balance with needs of project.

In 1984 a RICS Quantity Surveying Division report listed many items of unsatisfied demand for quantity surveying services and these included the following duties, many of which were being increasingly performed in the mid 1990s. The services included building services, civil and industrial engineering; maintenance work; energy conservation; building insurance evaluation; taxation and investment grant advice; technical auditing; contractors' estimating; contractors' procurement; project management and coordination; construction management; general management services; lead consultancy; staff secondment; multi-skill advice on procurement; multi-skill advice on all aspects of land/buildings; and overseas work. Other potential services included space-use planning and analysis (part of facilities management); work study; evaluation of bonus/production schemes; production cost planning; risk analysis/management; and programming/networking.

It is interesting to note that activities such as value management; development appraisal; refurbishment work; environmental management; urban renewal and regeneration; dilapidations; arbitrations; and alternative dispute resolution (ADR) found no place within the potential services in 1984.

## MARKETING OF SURVEYORS' SERVICES BY THE INSTITUTION

In the 1990s the RICS increased its activities in promoting the chartered surveyors as the professional to make land, property and construction work better for people. Watts (1992) described how larger firms required vision and needed to be recognised as being the consultants to go to when government or other major clients are preparing long term property or land use strategies. The RICS President takes every opportunity to speak to the press and give TV and radio interviews to raise the profile of the profession; and to meet members in their own workplace to listen to their views on business issues and the performance of the RICS.

The Institution formulated an enhanced external communications strategy in 1993 aimed at promoting chartered surveyors as the property profession, with the main objectives of increasing the total demand for chartered surveyors by raising public awareness and respect; increasing their influence in national policy making; and increasing understanding of property as an essential factor of production. The target audience was the public, business and government. The methods used included media and public relations, published information, government relations, and corporate identity. The evaluation encompassed media coverage and market research.

Property, which the RICS believes incorporates construction, is regarded as one of society's vital assets and must be managed effectively at every stage in its lifecycle. The property lifecycle is depicted in figure 17.1, wherein the main interests of quantity surveyors are contained within planning, funding and construction, although it will be appreciated that they can also be, and often are, involved in refurbishment and redevelopment work.

The Institution also prepares an annual marketing plan which shows the wide range and intensity of the public relations work undertaken by the Institution on behalf of members. In 1992, the Institution recognising the profession's need to meet the demands of the market, to identify new opportunities and to sell its services more vigorously, established two new

**Figure 17.1**    *The Property Lifecycle*

market-based networks – market panels and skills panels. The latter were subsequently renamed practice panels. This development has enabled the RICS to more effectively focus its work on the profession's markets, streamline decision taking, improve services to the membership and increase member participation. The market panels comprise six small groups of experts drawn from the Institution's membership and include one panel on construction. These panels aim to ensure that the Institution's policies match the market's demands in each sector and that it reacts quickly to change. They will identify opportunities and competitive threats. The skills panels aim to develop the skills of the profession in relation to market needs and opportunities.

The RICS markets and communications department aims to increase the demand for chartered surveyors' services. It achieves this in a number of ways, but mainly by influencing the markets in which surveyors operate. The market sector campaigns, of which there were 13 underway in 1995, refine this work through more direct targeting. There is also an international committee, which is one of the standing committees of the general council, with responsibility for initiating and implementing the Institution's overseas policies and activities. As from 1996, the former European committee was combined with the international committee, as surveyors' activities increased in eastern Europe.

The Junior Organisation of the RICS (1991) believed that it is important to market the profession of land, property and construction in a multidisciplinary approach, coupled with a single designation of 'chartered surveyor', in order to encourage a marketable corporate identity and also an effective decision making structure. Not all divisions of the RICS are entirely happy with this integrated approach, but it is a logical step forward. For example, the Institution's lion symbol is an important part of the professional identity of all chartered surveyor firms, as well as of the Institution itself.

On a narrower front, an RICS Quantity Surveying Division report in 1983, recommended that the Institution should ensure that the major financial institutions, such as the World Bank, the Asian and other development banks and aid agencies, are continually appraised of the quantity surveyor's capabilities. These bodies should be encouraged to appoint consultant quantity surveyors to provide value for money and effective cost control on all projects from the outset, with particular emphasis on developing countries. Equally, more effort should be made to ensure that quantity surveying services are better understood and appreciated by those with whom first contacts are often made, such as embassy staff. In the mid 1990s, consultant quantity surveyors extending their activities into eastern Europe and other parts of the world, felt that they lacked any noticeable British government support, which could be beneficial.

## NATURE AND PURPOSE OF MARKETING

Many professional people have been brought up to talk little about their qualities and achievements, and have tended to think that the best recommendation is by word of mouth from satisfied clients. This was also influenced by the earlier RICS restrictions on advertising. We are now in a fiercely competitive climate and surveyors can no longer stand aside from the harsh realities of the business world if they want their organisations to be successful.

Probably the best definition of marketing is that produced by the Institute of Marketing which reads as follows: 'The management process responsible for identifying, anticipating and satisfying customer requirement profitably'. *Identification* of appropriate requirements and markets involves the carrying out of market research, which will be explored later in the chapter. Briefly, it entails identifying potential client organisations and, within them, knowledge of which individuals have the most influence in the decision making or taking process. *Anticipation* implies forward planning so that the appropriate services are available at the required time and at an acceptable and yet profitable price. A combination of clear identification and good anticipation will lead to *customer satisfaction* (Yates, 1987).

A modern quantity surveying practice will therefore accept the need for marketing and have a good understanding of what it involves and the input necessary to achieve satisfactory results. Greenhalgh (1987a) advised that marketing should not be seen solely in terms of promotion and advertising, but within a wider context of blending the firm's structure and skills with the client's requirements. The central objective of marketing quantity surveying firms is to create a positioning strategy within the total market and to aim the firm's publicity and promotional activities towards a chosen market sector, as described later in the chapter.

## MARKET RESEARCH

Bevan (1991) has rightly distinguished between market research and marketing research. Market research examines the physical aspects of the market place, whereas marketing research looks at its behavioural aspects. Thus market research investigates the size and nature of the market place and the extent of competition. Marketing research is concerned with geographical and economic influences and the likelihood of changes in demand. Quantity surveyors will usually seek to identify the target market on which they will then proceed to concentrate their promotional activities. In some cases the market will be well known to them but in most cases some exploratory work is required to identify potential new sources of work. Pearce (1992), when considering construction marketing, suggested a number of sources of information on possible clients including reference books, on-line data systems, official statistics and publications, the press, networks, company information, commercial leads services and the firm's own sources.

Bowman (1986) advocated the use of a professional researcher to help identify the location and distribution of target audiences, the nature and size of their activities, their present use of the firm's type of service and their likely future requirements, their attitude to the firm's profession and perhaps even its own practice. He can then give a complete picture providing a framework for decisions on the appropriate promotional and marketing policies, helping to pinpoint relevant objectives, identifying misconceived attitudes that require correction and values that are inadequately perceived, and reading the trends of the market place.

Yates (1987) has approached this activity from the viewpoint of a practising quantity surveyor and suggests that the following questions need answering in order to identify the target market.

- Is the target market readily identifiable?
- How large is it?
- What are its components?
- At what level or levels do they operate?
- What will persuade them to come to you rather than go to a competitor?
- What message would you wish them to receive?
- How will you measure the results?

Watts (1992) wisely advised chartered surveyors to find out what their markets wanted and then offer them a very limited range of services in which the firm is genuinely expert and they actually need. It is impossible for a company to offer all services to all clients, as some degree of specialisation or segmentation is essential. This may be geographical, technical in terms of concentrating on certain specialised sectors of the market, or client oriented when a practice may concentrate on certain key clients because of their in-depth organisational knowledge. Greenhalgh (1987a) believes that an essential requirement before a market segment is decided upon is an internal management audit to establish the firm's own technical, managerial, geographical and resource based skills. This will form the basis of choice since it is essential to marry these core skills with the most suitable market segment.

## MARKETING STRATEGY

It has been suggested that all marketing strategies are based on a mix of the three variables of cost, differentiation and focus. *Cost* must be distinguished from price, for example prices may be cut in the short term to buy a market share by providing a price advantage over competitors, but this can only be maintained if it is based on a cost advantage. *Differentiation* is where the service on offer is different from that offered by competitors, with some distinctive advantages based on professional expertise. *Focus* is closely allied to differentiation, as it would be pointless differentiating a service if there was no clearly discernable market for it. Focus comprises the identification of a definable, specific market for a service, which is then tailored to meet the wants and needs of that market (Bevan, 1991).

A *SWOT analysis* is used to analyse the firm's operations and that of competitors in terms of strengths, weaknesses, opportunities and threats, as shown in table 17.1. A SWOT analysis is usually the first step in the preparation of a marketing strategy, from which a firm can identify and plan to use its strengths to exploit opportunities. It can also recognise and endeavour to redress or avoid its weaknesses, and take defensive action against any known threats. Pearce (1992) points out that it is easy to confuse strengths with opportunities and weaknesses with threats. Strengths and weaknesses are characteristics of the firm's own organisation, being internal matters which are to some extent under its control, whereas opportunities and threats are external factors, but ones which the firm may be able to use to its advantage.

The next step may be to prepare a mission statement, which encapsulates the aims of the organisation in a few succinct sentences. Its principal aims are usually to motivate staff and to inform clients of the firm's aims and commitment to providing a high quality service.

The firm is then ready to prepare its market strategy which comprises a carefully reasoned plan identifying the best clients, what the firm has to

Table 17.1   SWOT analysis

| *Strengths* | *Opportunities* |
|---|---|
| local connections | larger contracts |
| local expertise | enhanced fees |
| intelligence | more projects |
| local track record | specialised services |
| lower costs | |
| | |
| *Weaknesses* | *Threats* |
| lack of national connections | competition from national firms |
| lack of breadth of experience | breadth of national firms' experience |
| little economy of scale | fee pressure from local competition |
| limited access to advanced technology | new firms setting up in practice |
| lacking in management expertise | |

*Source*: adapted from Bevan (1991)

Table 17.2   Possible list of headings for preparing a marketing strategy

| Plan | Plan for preparing the strategy |
|---|---|
| Data collection and assembly | SWOT analysis |
| | Services on offer |
| | Clients |
| | Competitors |
| | Resources |
| | Pricing and profit |
| | External factors |
| Analysis | Key factors influencing strategy |
| Strategy for clients and services | Positioning |
| | Segmentation |
| | Services |
| | Clients |
| | Main advantages |
| | Bidding |
| | Volume and market share |
| Action, organisation, resources and budget | Bidding strategy |
| | Organisation |
| | Budget |
| | Marketing action plan |
| | Arrangements for review |

*Source*: adapted from Pearce (1992)

offer them, how they should be approached and what they will be told to convince them to accept the firm's services rather than those of a competitor. A possible list of headings for preparing a marketing strategy is listed in table 17.2, but it will be appreciated that these will vary from one organisation to another.

The headings can also form an activity list on which a plan for preparing the strategy can be based. A common procedure is to examine the activities which the firm does well and those at which it is not so good, and the external factors which may help or hinder the firm's activities. It is then possible to take a broad view of the business and to decide where its main thrust should be and its overall policy. Then by examining, classifying and recording the firm's services, clients, competitors and resources, and the factors influencing the volume, cost and profitability of services, a complete picture of the business is obtained and key factors that influence strategy can be identified. The strategy can then be formulated having regard to the key factors (Pearce, 1992).

## IN-HOUSE MARKETING TEAM V. EXTERNAL CONSULTANTS

Of those quantity surveying practices which operate a clearly defined marketing policy, some maintain professional in-house staff, the majority employ outside public relations consultants and a few combine both. The instant response and intimate knowledge of the in-house team has to be balanced against the wider range of expertise provided by external consultants. Whether a practice decides to use an outside agency or to build up its own internal marketing team could depend very much on individual preference, budget and workload, and it is not so much the method by which marketing is carried out, but the results which are important.

The in-house team has a good working knowledge of the firm and how it is structured, its aims and how the individual members operate. A team can be assembled to cover advertising, design, and public and press relations. It has the advantage of a quick response and its scale of operations will only be restricted by its budget. Thus the use of an internal team ensures identification with the practice and its ambitions and promotes a direct understanding of its aims. It is also argued that the team is in a better position to obtain the information it requires than an outside agency.

However, an in-house team will have problems to overcome as there is likely to be a lack of understanding by staff of what the team are trying to do. Some people see marketing as just creating literature and it is vital to the success of the team that marketing has to be embraced as a philosophy by the whole of the practice. Yates (1987) believed there could be a danger once in-house skills have been established of them becoming stereotyped and it could prove expensive to obtain new ideas from outside sources.

When engaging external consultants, it is customary to have a nominated senior member of the quantity surveying practice acting in a coordinating role. Eldridge (1987) recommends the preparation of a simple outline brief, to include the aims, target audience and the services to be offered, before calling in two or three consultants to listen to the firm's problem and prepare outline proposals, including the method of handling the assignment and costs involved in fulfilling the firm's criteria. After examining the proposals and selecting a consultant, a programme of activity will be agreed.

The cost of engaging outside agencies is bound to be high, being assessed on the time and effort required for each individual task. It does offer the opportunity for new ideas to be constantly sought and brought forward from a highly experienced organisation. Some quantity surveyors see the consultants' most valuable contribution in contacts with print and media representatives. Quantity surveyors in an in-house team require training in handling basic relations with the press, as they tend to be over cautious and fail to appreciate what makes a good news story. They are, however, good at briefing consultants and outlining practice performance. External consultants may, by contrast, face a possible conflict of interest when handling the marketing affairs of several clients in the same profession.

Both quantity surveying practices and their outside consultants have identified an inevitable learning curve after the initial appointment of a marketing team, whether in-house or external. While quantity surveyors need time to come to terms with the essential activities of defining aims and demands and satisfying them, marketing professionals may not understand immediately the issues facing the quantity surveying profession or readily appreciate that it has its own special needs and problems.

## CORPORATE IDENTITY

Having identified the target market, established the message to be put over and assessed the required resources, the next step is to create a single corporate identity, which delivers the firm's message, created by a skilful visual design in a readily recognisable form, and which is then used consistently throughout all marketing and promotional material (Yates, 1987).

The identity is expressed in a small number of elements, namely names, symbols, colours and lettering. A very critical component is the logotype or 'logo', being the symbol chosen to sum up the firm's identity. Typical examples in the construction industry are Taylor Woodrow's tug of war team, Bovis' hummingbird and Tarmac's Ts. Leading quantity surveyors seem to favour red squares or rectangles before or after the firm's name. Most organisations have a house colour or system of colours which need to be appropriate and applied consistently, accompanied by a suitable style of lettering (Pearce, 1992).

The graphic design of corporate identities and their components is a specialist, skilled activity, and the designer is thoroughly conversant with printers and the ever-changing processes of printing and graphic reproduction. The best approach is probably to ask the designer for an appraisal of the firm's existing emblem, where it exists, after he has seen all existing material and had a guided tour of the firm's premises. Samples of the designer's work will also help to assess the quality and suitability of his work. The next step is for the designer to supply recommendations and an estimate within an agreed timescale. This process is usually conducted in parallel with two or three more designers. Following approval, the selected designer will proceed to produce the final 'logo'. This may seem to be a rather long winded process but it does constitute a very important marketing concept (Bowman, 1986).

## PROMOTIONS AND PUBLIC RELATIONS

### Promotions

The main aim in preparing all promotional material is to achieve a high standard of quality of presentation which, coupled with quality of personal service from the firm, will help to develop further business by favourable client referral.

Greenhalgh (1987a) has listed the main objectives of promoting the firm as:

- creating an awareness by potential clients
- persuading prospective clients to invest in the service offered
- correcting false impressions about the service or firm.

Depending on which of these objectives the quantity surveyor wishes to achieve at a particular time and the profile of the prospective client, a certain mix of communication methods can be chosen. The various promotional methods are examined in the later sections of this chapter.

### Public Relations

Well managed public relations can help to create a favourable impression of a firm. This can be substantiated or otherwise by conducting properly arranged internal and external polls. It has been suggested that public relations is little more than organised and systematic common sense. The Institute of Public Relations has defined public relations as 'the determined, planned and sustained effort to establish and maintain mutual understanding between an organisation and its public'. This is a very discerning and comprehensive definition which sums up the activity extremely well. The emphasis on sustained effort and the maintenance of mutual understanding, highlight the need for continuing attention as opposed to a one-off activity.

Public relations has also been described as an activity designed to create a climate of opinion in which it is easier to carry on one's business. Admittedly, being better known and understood may help in the following ways:

- increasing the number of clients, especially those who are seeking the services that the firm is best equipped to provide
- ensuring that when and if things go wrong – as one day they are likely to do – people will know enough about the practice to be able to offset the temporary difficulty against the firm's background of continued success and trustworthiness
- adding weight to the views and contribution of the firm in professional circles
- easing the path of negotiations with local authorities and other bodies
- making the firm a more attractive and discerning employer (Cole-Morgan, 1986).

It is important that all employees are aware that the impression they make on clients rebounds on the firm as a whole. As clients may subconsciously be asking themselves whether the person with whom they came into contact was polite, knew his/her job, and was interested in the client. If the answer to any one of these questions was 'no', then that employee is doing the firm a disservice. Hence it is important that everyone in the firm feels that they are part of a team, are aware of its achievements and standards and can talk about the firm with sincerity, confidence and pride. Visits by partners to the offices of staff at all levels will help to promote these attributes. Publicity should be given to important events such as the winning of major contracts or commissions and the launching of authoritative reports.

## PRESS, RADIO AND TELEVISION

Yates (1987) has rightly described how close and constant contacts with members of the technical and news press can probably be one of the most significant aids to promotion. With large practices this may be achieved either through an external press relations consultant or an internal press office, which issue press releases giving news and information on all newsworthy activities of the firm. In smaller practices it is more usual to designate a senior member of staff as the firm's spokesman. The press is almost invariably interested in names and prices, which may be classified as confidential information.

It is important that all members of staff are fully aware of the news potential of the schemes in which they are involved and of the benefits of securing press coverage. It is good policy to encourage staff to establish close relationships with individual journalists, which could develop into a two way transfer of information to the benefit of both parties. When contributing press articles they should be prepared with the intention of showing the firm to be leaders in the market place, with members who are both professional and authoritative (Yates, 1987).

Cole-Morgan (1986) has emphasised that the first step in press relations is to convince the press that the spokesman has something interesting and worthwhile to report. What is news will vary with the intended audience and sometimes the same story will have to be presented in several different ways to suit the different types of media, such as national newspapers, local newspapers and trade outlets. For untrained persons, news identification often proves to be the most difficult part of press relations. This can be helped by a study of the coverage of the various media, not just for news but magazine articles and documentaries as well.

Pearce (1992) describes how in dealings with the press, it is important to realise that most journalists want to get their facts right and that they usually have a lot of space to fill with worthwhile news. Recognition of these factors will assist in communications with the media to obtain coverage for positive, newsworthy information about the firm's activities in national, regional and local newspapers, weekly technical journals and daily news programmes on national, regional and local radio and possibly broadcast and cable television as well.

It is important to be aware of what is happening both within the local community and within the wider professional world and to be available to provide quotes, even on difficult subjects. Regular press exposure of individuals increases their credibility to the public and the standing of the firm. Another useful promotional activity is the supply of well written articles on matters of professional, general or local interest for publication in appropriate journals (Bevan, 1991).

## SPONSORSHIPS AND SCHOLARSHIPS

*Sponsorship* aims to keep the firm's name firmly in front of the appropriate target audience and to do so in a way that further business may be developed. Yates (1987) describes how sponsorship while portraying an image of public involvement and support, may be little more than philanthropy; a form of activity which only buys goodwill. Hence before entering into the sponsorship of a specific activity, a realistic assessment should be made of the likely benefits to be achieved. Thereafter, every effort should be made to derive the perceived benefits from the activity, including the entertainment of actual or potential clients.

A typical approach is to sponsor sports events or art exhibitions, for which the Central Council of Physical Recreation and the Association for Business Sponsorship of the Arts can provide information on how these activities can be done successfully as a mutually beneficial business arrangement (Cole-Morgan, 1986). Another approach is to act as a sponsor for new academic courses or research projects of value to the profession and the firm.

Funding of *scholarships* and even professorships, particularly in the new universities, has become popular with some of the large surveying and other professional practices. It is important that the firm maintains close contact with recipients and shows an interest in their progress.

## CLIENT ENTERTAINMENT

Entertaining is an important part of the process of fostering business relationships. There will often be opportunities where it is possible to entertain prospective clients in a situation which is more congenial than just a normal business lunch. Race meetings, seats at the theatre or opera, cricket matches and other forms of entertainment can provide a suitable environment for getting to know a captive audience.

Pearce (1992) suggests, with good reason, that a clear policy should be established for personal entertaining, with regard to who undertakes it, approval procedures, quality and venue, and nature of entertainment, to avoid blatant misuse. It should always be borne in mind that the purpose of entertaining is to improve personal relationships and so enhance the prospects of conducting further business.

Cole-Morgan (1986) described how clients prefer to do business with people they know and like. Regular lunches and dinners of up to twelve people can be used systematically to extend the range of the firm's friends amongst such groups as leading members of related professions, leading industrialists and developers, constituency MPs, civic dignitaries and representatives of the media. The luncheons should desirably be structured to encourage constructive debate on matters of common concern and interest. The host, who is normally the senior partner present, can towards the end of the meal say a few words about the firm, its hopes and ambitions, and perhaps ask the guests to comment on any changes they would like to see. This approach could result in the guests feeling that they are being taken into the firm's confidence as respected friends of the practice. Some professionals believe that the best promotional approach is a social one by meeting potential clients and other professionals as fellow members of golf clubs and the like.

## EXHIBITIONS

It is possible to promote the image of the firm and the expert and specialist services it offers in an imaginative, creative and impressive way at an exhibition, where space is taken in the form of a trade stand. This does, however, require detailed consideration as to the nature of the exhibition, the exhibitors and likely visitors to it. It must be carefully costed to include all expenses and staff time in attendance. It would be unproductive and a waste of resources to put on a first class display and then fail to back it up with staff with the relevant expertise. Decisions will have to be made as to whether senior highly skilled staff are better employed at the exhibition or on their normal responsible assignments.

## PRESENTATIONS

Where a potential client is contemplating proceeding with a building project and requires advice on which procurement method to use, he may approach a quantity surveyor for such advice. It is possible that the best approach for the quantity surveyor is to give a presentation of high quality showing the various options available with their merits and demerits. He will conclude the presentation by giving his recommendation for the method to be used for the client's project with the reasons for his decision.

Pearce (1992) has detailed the wide range of visual and audiovisual techniques available to support presentations. They include overhead projectors and 35 mm slides, tape/slide presentations, video, film, and flip charts. With modern developments, almost all presentation material can be prepared in-house using a desktop computer with a graphics package, but taking great care with the quality of design and production. Linking the slides with an audio tape is useful where a standard message is being given or where there are two or three projectors in use. Videos are now more widely used as prices are reduced and they become more accessible, but Lovell (1990) found that they could be counter-productive with some quantity surveyors' clients and suggested that their main use could be with uninformed clients and overseas clients who were not familiar with the quantity surveying service.

## SEMINARS AND LECTURES

Seminars and lectures can also provide another useful promotional technique, which enables the firm to convey a specific message in an area of its own expertise and to assemble a carefully chosen audience, where personal relationships can be developed and contacts made more familiar with the services on offer. The events can be made more attractive by inviting a highly respected and authoritative guest speaker. It is important to choose a topic for the seminar/lecture which is well researched and presented and which will create extensive interest.

## REPORTS, NEWSLETTERS AND AUTHORSHIPS

### Reports

Some surveying firms produce authoritative and valuable reports at regular intervals covering their activities and also current market trends and other useful information. Some reports are confined to topical professional matters where detailed, first hand information is very sparse. The purpose and content of reports requires careful and skilful consideration. Yates (1987) believes that the best procedure for preparing a report is as follows:

(1) identify a gap in existing coverage;
(2) be satisfied that the gap is worth covering and identify the appropriate target audience;
(3) issue the report in such a way that it is timely, informative, well presented and, where possible, backed up by editorial comment in the press.

### Newsletters

Newsletters are best circulated to selected interest groups of clients, architects, engineers and throughout the quantity surveying practice itself, to convey the desired image to clients, other professionals, the public and employees. They enable the firm to talk directly to both clients and potential clients in words of their own choosing (Greenhalgh, 1987b).

Lovell (1990) conducted a survey of quantity surveying practices who produced newsletters and found that they were generally sent to clients quarterly and contained information on recent projects, regional offices, cost trends and the construction climate. Quantity surveyors found that they were well received by clients and they found them to be a useful means of keeping in touch with former clients. The clients in the survey found them of greater interest than brochures or videos and commented that 'it keeps a practice's name in the forefront'. They were particularly interested in the technical content, such as regional tender price variations, graphs showing projected building costs, recent relevant case law, taxation, and consequences of contract types on tender prices. In addition, they would be interested in some general information about the practice as a whole, although this was generally thought to be of minor importance compared with the technical content.

## Authorships

Another promotional concept is the authorship by the practice or members of the senior staff of good, practical technical books on topics of professional interest, as these keep the practice's name in the minds of the readers and must be good publicity. In some cases surveying practices have produced their own books, but more usually they have engaged the facilities of an established and reputable publisher who produces good quality technical books at modest prices.

## PERSONAL CONTACTS

It is important to keep an accurate and up to date list of all the personal contacts of members of the firm. This ensures that any information issued by the firm is sent promptly to appropriate contacts in a personal capacity. It must always be remembered that a single incorrect mailing, resulting from an inaccurate or misspelt list, can do disproportionate harm to the firm's relationship with a key client. Continual updating of the contacts list should accordingly be afforded a high priority (Yates, 1987).

## ADVERTISING

Advertising is probably one of the most obvious forms of promotion and can be general/non targeted advertising or specific target advertising. The latter category is influenced strongly by the selected market and the firm's objectives. General media advertising reaches a wide market at relatively low cost. More specific targeting necessitates a more selective choice of publication in which to advertise. The best form of corporate advertising is to constantly promote a range of high quality services and this requires a high standard of design in its presentation. Site boards and office name plates are a form of advertising but of limited value.

The quantity surveying practices surveyed by Lovell (1990) all conducted some advertising but generally on a limited scale and often restricted to sharing advertising space with a client to publicise a project. In a few instances they had advertised on their own initiative in property, financial press and business magazines, although they were not convinced that it was worthwhile. Clients displayed an indifference to advertising and stated that they would be more receptive to articles about a practice in a journal or the editing of a price book. The general feeling emerging from the survey was that advertising was unlikely to have an immediate response but might help in the long term, if it increased public awareness.

A firm that is considering its advertising strategy, needs to make decisions on the timescale, services to be offered and amount to be spent.

The essence of advertising is repetition, involving expenditure over a period of time. The public is not waiting to receive the firm's message and constant advertising means that many cannot help but notice it. It would be unwise to try and cut corners on cost or to make the advertisements unduly glamorous. Probably a modest advertisement every week in a local paper read by part of the target audience would have more impact than a single large advertisement in a popular national newspaper (Bowman, 1986).

Pearce (1992) has rightly emphasised that advertising is a sophisticated activity and can involve the expenditure of considerable sums of money. Hence a firm must know what it wants to say and to whom. The next step could be to identify a competent advertising agency with experience in the quantity surveying field, who will need a simple marketing brief about the firm's objectives and the target market, and the part that advertising is expected to play.

## BROCHURES

A skilfully produced high quality brochure implies a first class service. However, unless the distribution is carefully arranged, there is a danger that it may not reach the full target audience.

All the quantity surveying practices surveyed by Lovell (1990) produced brochures which they distributed to existing and potential clients. The less enterprising practices produced brochures containing a limited input from graphic designers and were updated at approximately five year intervals. By contrast, the more enthusiastic practices updated their brochures every two to three years with the help of specialists. All the practices regarded the brochures as important, but in terms of securing a commission they believed the brochure had less impact than personal contacts.

The clients interviewed by Lovell considered brochures to be of passing interest and suggested that they would be more interested if the brochures were specific to their type of projects such as, for instance, housing developments or marine works. Some clients were sceptical of brochures which contained photographs of completed buildings with no commentary; one stating, with some justification, 'pictures don't tell me whether it was built to the right time or cost'. These comments suggest that it could be worthwhile producing brochures for market sectors and providing a brief case study alongside photographs, containing a description of the quantity surveyor's role. This concept has also been taken up by Pearce (1992) wherein he describes the differing requirements of clients in different construction market 'segments' and he gives examples of an industrial client, a property developer and a local authority.

The author was given the opportunity to examine the brochures produced by three national quantity surveying practices, all of which were of extremely high quality and illustrated with excellent coloured photographs of impressive construction projects.

For example, E. C. Harris described how the practice offered services in quantity surveying, project management, risk management, property management, engineering services, environmental services, contract management, computer management, and value management. Excellent individual brochures were produced covering specific market sectors as diverse as health care and civil and heavy engineering. Further publications of interest to clients included 'Outlook', the firm's annual report and a very informative newsletter headed 'Economics Survey'.

By contrast, Currie and Brown described their main project areas ranging from airports and defence installations, to hospitals and offices, and the same brochure described their principal services, ranging from estimates and cost planning through to value engineering and corporate recovery and insolvency. Some well prepared and illustrated brochures are produced in different languages as the firm has offices in some two dozen overseas countries with an extensive corporate structure.

Turner and Townsend's attractive brochure emphasised the practice's commitment to securing flexibility and quality. A substantial list of services offered by the firm is provided, with emphasis on cost advice, development monitoring, cost management, risk management, grants and tax allowances, innovative procurement systems, energy audit surveys and value engineering. A separate brochure covers the firm's project management service.

In recent years many local authorities, particularly county councils' property services or architectural services departments, have produced impressive brochures showing potential clients a wide range of professional services that they are able to offer the private sector. These services include all the normal quantity surveying functions.

## MARKETING AUDIT

Pearce (1992) has advocated that a full marketing audit should be carried out as required to take stock of every aspect of marketing activity and of other factors that affect it, in the nature of a post mortem and monitoring operation. He suggests that the audit should address the following four questions:

- What were the firm's original marketing intentions?
- What did the marketing strategy actually do?
- What has it achieved?
- What can the firm learn from the answers to the first three questions that will help in deciding what should be done to achieve more?

The timing of such audits can be influenced by a number of factors, including changes in external market conditions, new opportunities, an internal reorganisation, emergence of new competition, and some services reaching the point of decline. The audit can be undertaken by internal management staff where they have adequate expertise, otherwise it will probably be necessary to employ an external consultant.

## CONCLUSIONS

Lovell (1990) posed the question 'Is the marketing of quantity surveying practices really worthwhile?' and 'Is a market oriented practice more likely to secure a commission?' The clients in Lovell's survey responded 'those who just sit back haven't much hope; on balance it can only increase their prospects.' However, being good at marketing does not guarantee a commission; the practice may be granted an interview but get no further if, for instance, it makes extravagant claims. It is necessary for a practice to make itself familiar with the client's specific needs and to identify its services with these needs. In general, Lovell found that many of the larger quantity surveying practices were taking marketing seriously, but less evident was the impact this will have in the future.

# 18 Education, Training, Professional Development and Research

## CAREERS PROMOTION

A RICS Report in 1980 recognised the importance of recruitment in realistic numbers and of the right quality, despite the competing claims of other professions; while a RICS QS Report in 1983 highlighted the effect of market forces and the limited control of numbers which can be exercised by the RICS, being subject to the variations in national and local educational policies.

Over the years the RICS has produced a series of careers pamphlets aimed at informing and hopefully encouraging school leavers to choose a surveying career. They describe the tasks undertaken, the high level of job satisfaction that can be obtained and future prospects in the different employment sectors. They also describe how Britain pioneered the development of the profession and how chartered quantity surveyors now work in over 120 countries worldwide on projects as varied as the Singapore Metro System, an urban development in Turkey, a zoo park in the Sudan, an airfield in the Falklands, and a leisure complex in the Carribean. The Institution also issues careers stands and a film to be used at careers conferences and visits to schools. As a further initiative the RICS produced a teacher's resource pack relating to the property profession as a teaching aid in connection with stage 4 of the National Curriculum. These packs are available to teachers on request and have proved popular. The schools initiative network had matched 1500 chartered surveyors to the 2600 sixth form schools by 1993.

The qualified surveyor achieves high professional standing by years of study, training and experience. The road to success is built on integrity, common sense, sound judgement and a genuine interest in people and their needs. Young surveyors with these qualities will find endless scope to develop managerial talent and technical ability.

Entrants to the profession ought ideally to take decisions with their brains rather than their feelings; be happy at working on their own where necessary; have an aptitude for figures; be enthusiastic with intellectual ability and mental agility; have an enquiring and perceptive mind; be dedicated, hard working and adaptable; have plenty of common sense; be able to communicate effectively; and be keen to serve the public, meet people and work outdoors on occasions. The skills of communication are particularly important and encompass oral, written and graphic aspects.

## EARLY DEVELOPMENTS IN QUANTITY SURVEYING EDUCATION

The principal methods of study for the examinations of the professional body in the 1940s, 1950s and early 1960s were by correspondence course, day release at a local technical college, or possibly a combination of the two, coupled with training in a quantity surveying office. As time passed some members of the education committees of the appropriate professional bodies recognised the desirability of establishing full time courses of study in suitable centres. Study by the part time route was often a lengthy, hazardous and exacting process with many imperfections as many older quantity surveyors can recall. A student carrying out a responsible job in a quantity surveying office could be fully stretched and working under considerable pressure. He needed some spare time for recreation and relaxation and hence the time available for study was restricted. The combined pressure of full time employment and part time study was very great indeed and could, on occasions, adversely affect the student's health. Furthermore, the subject areas covered in employment could not possibly embrace all those contained within the professional examination syllabus and satisfactory integration of theoretical study and practical work was difficult to achieve. In many cases the part time student received little personal tuition or guidance with his studies and needed a large measure of self discipline (Seeley, 1976).

Later developments were the introduction of full time courses for professional examinations and external degrees and subsequently diploma courses approved by the professional bodies. The next development occurred in the late 1960s and early 1970s with the introduction of quantity surveying degree courses in universities and the former polytechnics. Ever since the raising of the education entry requirements to two GCE 'A' levels and the issue by the RICS of the Wells Report in 1960 and the Eve Report in 1967, there had been a growing awareness of the desirability of establishing a network of degree courses throughout the country which would in time form the principal method of entry into the profession. Sixth form pupils were encouraged to believe that they were all potential graduates and that degrees formed the bridge to most professions and careers. In the absence of ample good quality degree courses in quantity surveying, the profession was likely to lose many good recruits to other professions where there were adequate degree courses available, although some employers felt that quantity surveying was a very practical profession which necessitated entrants being trained in quantity surveyors' offices.

A further RICS Report prepared by the Brett-Jones Committee in 1978 contained a number of vitally important recommendations which included the maintenance of both full time and part time routes to qualification, improvement in communicative skills of potential surveyors, more positive guidance to be given to employers on the supervision of candidates with a suggested model training agreement, support of graduate entry as the normal route into the profession, concessions towards membership of those whose experience was solely in research or teaching, a strengthened test of professional competence, mandatory continuing professional education for corporate members, and support for the Society of Surveying Technicians (SST). Further important recommendations were contained in another RICS report (1989b), which formed the background for the subsequent 1994 report.

## RICS EDUCATIONAL POLICY

Far reaching changes took place in higher education between the mid 1980s and the mid 1990s with the advent of mass higher education and enormous growth in student numbers, independence of colleges, upgrading of polytechnics to universities commonly known as the new universities, and changes in the content and structure of courses and in the funding arrangements for higher education. These developments have transformed the character of higher education which now includes a wide range of universities and colleges differing in size, scope and development. Surveying education provision increased substantially as a result of the increased allocation of resources to higher education up to the mid 1990s, and not because of any increased demand by employers (Kennie, 1995).

The RICS Education and Membership Committee in 1994 produced a most comprehensive and forward looking education policy discussion document titled 'A Strategy for Action', which was approved by the General Council. It described how in the early to mid 1990s employers had become much more selective and demanding when recruiting graduates as they became more cost conscious resulting from the recession. Smaller employers placed greater emphasis on the need for graduates who were equipped with immediately usable vocational skills, whereas the training schemes of larger employers enabled them to concentrate on academic high achievers. Recruitment and investment in training, estimated at £30 000 to £40 000 per graduate in 1994, aimed at long term retention of the recruit and integration into the ethos, structure and future of the organisation.

An examination of the needs of employers identified the following general trends:

(a) employers' needs vary considerably depending upon their professional activity and market niche;
(b) different groupings of employers have varying ideas of what specific knowledge, skills and competencies should be exhibited by graduates.

To clarify the situation, the RICS commissioned and distributed the findings of research into what prospective employers believed to be the essential components of surveying courses. Research undertaken by Nottingham Trent University (RICS, 1993c) indicated that vocational surveying skills were placed in a lower order of priority than personal skills, such as self management, written and oral communication and the ability to solve problems. Other research undertaken independently by the Quantity Surveyors Division (RICS, 1992) supported this but concluded that it was not possible to rank the need for, or separate, personal skills and technical knowledge.

Hence the Committee recognised that there was a need for a wide spectrum of academic courses ranging, for example, from the highly academic to those which are broadly based with a significant input of business awareness and relatively low surveying content, to those with a more traditional, technical, vocational base. Such a spectrum of courses will enable potential employers with widely varying needs to recruit what they perceive to be the most relevant types of graduates. Clearly, the output from such courses will have a range of skills. The marketplace for surveying services is extremely wide, from sole practitioners providing a range of services, niche practices offering specialist support to the corporate and public sectors and international multi-disciplinary practices, all with varying needs. As the spectrum of courses continues to widen there will be a need for educational institutions to ensure that employers are aware of the main contents and aims of each accredited course. The author also believes that there is a need for the RICS to publish a summarised version of surveying courses, updated periodically, outlining their principal objectives and characteristics for the guidance of school careers teachers, potential students and employers alike.

The Department of Surveying at Nottingham Trent University responded to the Quantity Surveyors Division request for debate and discussion when they held the Ivor Seeley inaugural debate in 1994, when the motion 'the emphasis on traditional quantity surveying skills is no longer consistent with the needs of the profession' was defeated with a predominance of final year students in the audience. Noel McDonagh proposed the motion and David Hoar opposed it, each supported by students.

The RICS 1994 Education Policy Report also dealt with the Assessment of Professional Competence (APC), structured training, continuing professional development (CPD), Society of Surveying Technicians (SST), research and related matters. These matters will all be examined in some detail later in the chapter. The Report also contained figure 18.1 which shows the main routes to membership and transfer to fellowship of the RICS, although it omits the SST route which had not been finalised at that stage. Transfer to fellowship will be further considered later in the chapter.

In the late 1980s and early 1990s the RICS made significant advances

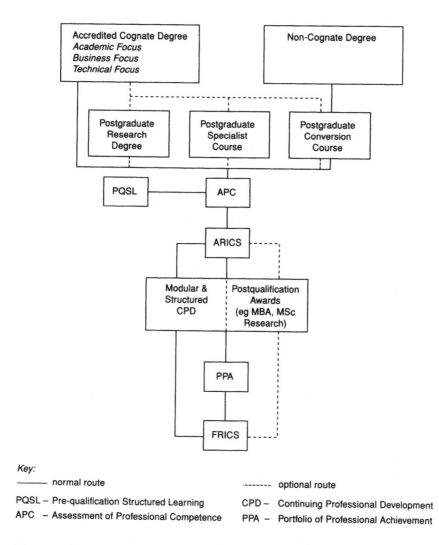

Key:

——— normal route                          ------- optional route

PQSL – Pre-qualification Structured Learning     CPD – Continuing Professional Development
APC  – Assessment of Professional Competence     PPA – Portfolio of Professional Achievement

**Figure 18.1**  *Routes to membership and transfer to fellowship of the RICS*

*Source*: RICS Education Policy: A Strategy for Action (1994)

on a number of educational fronts, ranging from approval of the first common degree course allowing entry to three divisions at Hong Kong University and of European courses, links were forged with educational
establishments through to RICS teaching fellowships, the introduction of a
RICS education seminar, the production of RICS Education News, and
entering into reciprocity agreements with overseas surveying bodies.

## Accreditation of Courses

The RICS maintains a significant measure of control over higher education establishments through its accreditation procedures whereby it ensures that satisfactory academic and professional standards are upheld. Each course is assessed on its merits as satisfying minimum threshold requirements for the education of prospective surveyors.

In 1993 the RICS moved away from the accreditation of individual courses towards centre accreditation. This recognised that course providers should have the freedom to develop the diversity of courses which the profession needs, and accommodates the move by educational establishments to modularisation of courses, to give greater flexibility and student choice. Accreditation is undertaken under the auspices of the RICS Surveying Courses Board with review visits normally being undertaken at seven year intervals. If an institution is well resourced, has sound management practices and established systems for internal quality control, then the courses it offers should be intrinsically sound. Visiting panels focus on the strategic plans of a centre, the deployment of financial and physical resources, staffing provision and quality, staff development and research, entry standards, links with the profession, and quality of graduate output. With individual course programmes, visiting panels will test the programmes on offer against the Board's criteria, and obtain an overview of the philosophy and general direction of courses.

## Integration of Educational Courses

A Latham Report working party proposed that common subjects should be taught to mixed groups of architects, engineers and surveyors, as it saw great benefit in combined modules such as basic business and team building skills. However, the RICS questioned whether they could realistically be accommodated in surveying undergraduate curricula and considered them to be inappropriate for post graduate courses. The proposal reiterates a recommendation in *Building Britain 2001* for more shared knowledge of the design and construction processes with cross fertilisation between disciplines adopting unified courses in the early stages of higher education. In years past quantity surveying students at Nottingham Trent combined with architectural students at Nottingham University to undertake a live school project annually for Nottinghamshire County Council, with beneficial results for all the students. The main advantage was the opportunity of observing first hand the constraints facing the designer and the impact of cost control on the building designs, coupled with the direct contact of the students of two interrelated disciplines.

## Core Skills and Knowledge Base of the Quantity Surveyor

The RICS Quantity Surveyors Divisional Council in its 1992 report on this subject made the following recommendations:

- quantity surveyors when considering core skills should think of these skills being applied in the use of primary and secondary knowledge and techniques
- the core skills of the quantity surveyor comprise analysis, appraisal/evaluation, communication, documentation, management, quantification, and synthesis
- the primary expert knowledge areas of the quantity surveyor are cost management and procurement management
- the secondary expert knowledge areas of the quantity surveyor are those which underpin or support the primary areas, such as law, building/construction technology, and economics
- providers of quantity surveying education should place emphasis on the development and enhancement of the core skill, together with the primary and secondary areas of expertise. Peripheral areas should be reviewed and their relevance ascertained; if no longer of fundamental significance, they should be reduced in importance or dropped from the curriculum altogether.

## CONTENTS AND CHARACTERISTICS OF QUANTITY SURVEYING DEGREE COURSES

### Contents of Quantity Surveying Courses

The contents of the quantity surveying degree courses usually embrace the common core subjects, although their treatment, detailed syllabuses and time allocations can vary significantly. Table 18.1, produced by Nottingham Trent University in 1994, gives an analysis of the courses offered by ten universities in the UK in relation to the main subject areas. It shows considerable variations in emphasis on the seven main subject areas with, for example, the time allocated to measurement varying from 0 to 25 per cent, construction economics from 5 to 16 per cent, management from 5 to 16 per cent and technology from 11 to 39 per cent.

It should however be borne in mind that it is not always possible to readily identify course subjects in a scheme in relation to the seven selected main subject areas. Furthermore, the remaining allocations of course time also vary substantially, particularly where a course has been designed to cater for a specialist market, such as contract finance and management, and possibly a double honours in quantity surveying and business management or foreign languages. Some courses may devote considerable periods

Table 18.1    Content of quantity surveying degree courses

|  | University | | | | | | | | | |
|---|---|---|---|---|---|---|---|---|---|---|
| Subject | A (%) | B (%) | C (%) | D (%) | E (%) | F (%) | G (%) | H (%) | I (%) | J (%) |
| Measurement | 10 | 13 | 10 | 13 | 12 | 0 | 2 | 18 | 19 | 25 |
| Construction Economics | 15 | 13 | 10 | 11 | 14 | 12 | 5 | 13 | 16 | 11 |
| Economics | 6 | 3 | 3 | 6 | 5 | 5 | 6 | 4 | 3 | 6 |
| Law | 9 | 10 | 6 | 3 | 5 | 5 | 0 | 5 | 2 | 3 |
| Construction Law | 3 | 7 | 6 | 11 | 2 | 11 | 1 | 8 | 3 | 11 |
| Management | 16 | 14 | 14 | 14 | 14 | 14 | 11 | 7 | 5 | 14 |
| Technology | 20 | 30 | 11 | 22 | 18 | 26 | 39 | 12 | 18 | 11 |

*Source*: K. I. Hogg (1994)

of time to personal development studies which may include statistics, computing, interpersonal skills, such as self confidence, communication, self organisation, awareness of self and group needs and of the exercise of responsibility, and possibly research methodology. Other personal development studies in the later years of courses may incorporate group based integrating projects and activity based experiential learning conducted in a supervised outdoor location away from the university campus, to develop self awareness and self confidence in a teamwork situation. Other course components often include a final year dissertation which will be examined later in the chapter, and there may be elective or optional subjects in the final year such as project management, value management and facilities management.

A RICS survey, 'Procedures and guidelines for course accreditation' prepared by Nottingham Trent University in 1993 aimed at ascertaining exactly what practitioners expected from their graduate recruits, although as described earlier in this chapter, different types of quantity surveying employers have different requirements, and hence there is a need for courses with varying characteristics and objectives. The survey indicated that many employers considered that quantity surveying education should place greater emphasis on general skills, in areas such as communication, management, research and information technology (IT), in addition to technical expertise. The survey identified two distinct types of quantity surveying expertise: technical and management skills, which are both considered vital by employers. The highest value was accorded to two skill areas, namely technical expertise in measurement, specification and preparation of bills of quantities, cost estimating and contract law; and written, oral and interpersonal skills.

The survey tended to indicate that the surveying profession and the educational courses supporting it had in the past been preoccupied with technical and practical prowess at the expense of personal transferable skills such as expertise in negotiation and business presentation. The re-

Table 18.2   Workplace activities in quantity surveyors' offices

| Activity | Never (%) PQS | Never (%) Cont. | Occasional (%) PQS | Occasional (%) Cont. | Often (%) PQS | Often (%) Cont. | Daily (%) PQS | Daily (%) Cont. |
|---|---|---|---|---|---|---|---|---|
| Administration | 10.1 | 5.2 | 54.1 | 31.0 | 25.7 | 36.2 | 10.1 | 27.6 |
| Procurement Advice | 34.9 | 44.6 | 50.9 | 37.5 | 13.2 | 17.9 | 1.0 | — |
| Contractual Advice | 23.4 | 9.1 | 52.4 | 63.4 | 21.5 | 25.5 | 2.7 | 2.0 |
| Interim Valuations | 3.6 | — | 7.2 | 8.6 | 82.0 | 82.8 | 7.2 | 8.6 |
| Measurement | — | — | 7.2 | 15.5 | 57.7 | 60.3 | 35.1 | 24.2 |
| Project Man'ment | 60.0 | 69.8 | 29.5 | 20.8 | 6.7 | 5.7 | 3.8 | 3.7 |
| Budget Estimating | 5.5 | 32.8 | 41.3 | 51.7 | 46.8 | 15.5 | 6.4 | — |
| Contract Documents | 16.7 | 10.3 | 45.4 | 34.5 | 35.2 | 48.3 | 2.7 | 6.9 |
| On-Site Activity | 8.5 | — | 39.6 | 13.6 | 47.2 | 42.4 | 4.7 | 44.0 |
| *Use of IT* | | | | | | | | |
| Measurement | 21.7 | 62.3 | 26.1 | 14.8 | 33.0 | 14.8 | 19.2 | 8.1 |
| Spreadsheets | 13.0 | 9.7 | 23.5 | 35.5 | 33.9 | 35.5 | 29.6 | 19.3 |
| Database | 47.8 | 59.3 | 34.2 | 28.8 | 12.6 | 10.2 | 5.4 | 1.7 |
| Word Processor | 21.2 | 39.7 | 29.2 | 36.2 | 18.6 | 15.5 | 31.0 | 8.6 |

*Source*: K. I. Hogg (1994)

port recognises the importance of practical surveying ability but suggests the need for a different more rounded approach, with courses covering general skills more comprehensively. The RICS considered that the new approach would help to improve the profession's image and encourage more high calibre graduates from a wider range of backgrounds, to become surveyors.

Hogg (1994) using a range of activities carried out by quantity surveyors, including those traditionally recognised as core areas and those of a more specialised nature, prepared the analysis in table 18.2, showing the frequency of involvement by graduate surveyors. The statistics have been subdivided into the two main employment categories of private practice and contracting for ease of comparison.

The statistical analysis suggests the following:

(1) the most frequently occurring activities relate to those requiring technical knowledge and skills;
(2) the advisory role of the graduate quantity surveyor in terms of procurement and contractual matters occurs less frequently but is not insignificant for the majority of respondents;
(3) the profile of activities of the private quantity surveyor and the contractor's surveyor is similar but with the different emphases which are consistent with the generally held perceptions of each role;

(4) information technology is widespread, although used more extensively by the private practice surveyor.

The principal influences on and needs of quantity surveying courses in the mid 1990s were as follows:

- clients' needs are widening with the old demarcation lines between surveying disciplines disappearing, resulting in the need for a different course approach
- procurement, cost and value management and control often extend from building into services and other forms of engineering
- need for versatility and ability to communicate
- knowledge required of principal legal concepts, business organisation and management and accounting methods
- understanding required of professional ethics and conduct and their application
- European and international implications of quantity surveying activities
- acquiring analytical skills in various forms and development of information technology (IT)
- design appreciation.
- general understanding of environmental problems and their solution.

Jaggar (1992) rightly believed that the challenge of the 1990s was to look for more innovative and effective ways of ensuring that the skill base is not eroded, nor its context lost, within the increasingly complex and varied role that the construction industry demands from the quantity surveyor.

### Characteristics of Quantity Surveying Courses

The principal aims of quantity surveying degree courses are to enable students to achieve academic excellence and the development of personal and interpersonal skills, while at the same time satisfying the educational and vocational needs of employers, the professional bodies and society. The course should ideally seek to provide a challenging and dynamic educational experience in which students are encouraged to be self reliant, questioning, creative and enterprising with ability in critical analysis, and where staff are fully committed to academic excellence, the care and well being of students and responsive to the changing demands of the profession. The underlying philosophy must be one of 'education for long term capability', rather than 'short term preparation for employment'.

In the early 1990s there were a number of major developments in the formulation and operation of surveying courses, with significant implications for students. The main areas of change were as follows:

- Common areas of study in the early years of courses and particularly a common first year in the various surveying disciplines to give the student choice of direction (described by Nottingham Trent University as 'pathway') after enrolment on the course without interfering with the progress of his/her studies.

- Multi-mode courses whereby the student has the choice of pattern of course which he/she can follow, possibly ranging from full time through to sandwich and part time and even distance learning. In these situations, the student can also transfer from one mode to another at the equivalent stage if there is a change in personal circumstances.
- Modularisation of courses whereby they are formulated in a series of modules or specific freestanding units of study which give the student some measure of choice over elective or optional modules. In principle, the core or compulsory modules convey the essential knowledge, breaking longer subjects down into smaller and readily assimilated units. Outside the core units are areas of optional choice which can expose professional oriented students to wider areas outside their core discipline, enabling them to acquire a variety of complementary skills and experience. As this approach largely eliminates the integrating nature of non modular courses, it is essential to include integrating mechanisms such as multi disciplinary projects.
- Semesterisation whereby many courses contain two semesters per year with examinations at the end of each semester. In this way students are able to concentrate their studies for examinations over shorter periods and to resit referred subjects in the first semester along with the examinations at the end of the second semester. This approach has the disadvantage of fragmentation of subjects limiting the scope for their integration, over emphasis of examinations, problems with the two semester system with examinations in the middle of the spring term, and even worse problems with the three semester approach with three sets of examinations per year and still greater fragmentation of subjects. Semesterisation is also incompatible with part time and distance learning course patterns and can lead to over concentration in some core subjects.
- Credit accumulation and transfer systems (CATS) whereby programmed units are afforded credit values which offer greater flexibility by enabling students to follow different routes through their educational programmes, to allow for changes in their aspirations or changes in their careers. They also allow units to be combined to provide general, specialist, academic and vocational programmes and to provide a variety of approaches to learning. This system also aims to broaden access to courses by offering wider educational opportunities to students without the requisite formal entry qualifications, and to provide a means of entry to higher education for returning adults (Ashworth, 1994). On quantity surveying courses, compulsory core units are normally allocated higher credit values than the wider ranging optional subjects.
- Mature entrants without the necessary formal entry qualifications are

usually permitted greater access to surveying courses taking into account their wider experience, often described as accreditation of prior learning and experience (APEL). Another alternative is the fast track approach whereby non cognate degree holders and other suitable candidates can enter fast track courses which are generally designed on a one year full time or two years part time basis. Holders of relevant BTEC higher diploma and certificate holders may be offered partial exemption from quantity surveying degree courses. In the years ahead consideration will need to be given to holders of the higher levels of relevant National Vocational Qualifications (NVQs).

• Franchising of programmes of study, or parts of programmes, by universities to colleges of higher education and further education (HE and FE) gathered momentum in the 1990s and was extended to overseas establishments. The main reason for this growth has been the perceived need to provide access to university degree courses for those students who could not normally attend the university, because the course was full, inaccessibility of its location or for financial reasons. In these situations it is desirable for the franchising university to formally monitor the HE and FE college procedures and to set the examinations and assignments and moderate them, to ensure that satisfactory standards are maintained (Ashworth, 1994).

## Course Patterns

The main types of courses operating in quantity surveying and related fields are full time, sandwich, part time and distance learning. Quantity surveying degree courses are mainly of the thick sandwich and part time varieties. Each of these course patterns will now be examined.

*Full time* courses have the merit of an unbroken programme of study normally spread over three academic years. Hence it is argued that the student's momentum is not lost and he/she does not suffer from the disruption to studies caused by the year away from university which occurs on the thick sandwich course. The student is however restricted to gaining work experience in the relatively short periods in summer vacations.

*Thick Sandwich* courses have proved very popular in quantity surveying education as they incorporate well integrated phases of educational (or academic) and office (or profession based) experience. Sandwich courses can take various forms but the most popular and, I believe, the most valuable is the thick sandwich which comprises two years in the university, one year in the profession and a final year (the fourth year) in the university. The year of professional training is almost invariably more than one year as it usually stretches from June in one year to September in the next (Seeley, 1972).

The successful operation of a thick sandwich course is dependent on a high level of cooperation from the profession, as this concept envisages professional training as a tripartite enterprise with surveyors, educational establishments and students operating together in a close and harmonious partnership for the benefit of the profession in general and the youngest recruits in particular. It is, however, appreciated that in times of severe recession, the scope for employer participation is unfortunately restricted.

By the end of the second year, students on thick sandwich courses have reached a reasonably advanced stage in their studies with a good understanding of the core quantity surveying subjects, are computer literate, have a good grounding in communicative skills and have been encouraged to use their initiative, think for themselves, make critical analyses and test hypotheses, and should thus be able to make a meaningful and useful contribution to the organisation which they join. The employer, for his part, is expected to provide the young surveyor with a suitable and varied programme of work and to give him the opportunity to undertake, under supervision, some really worthwhile and satisfying tasks.

The thick sandwich approach provides the student with a new understanding of some of his earlier studies and enables him to see the relevance and application of some of the professional techniques in a live situation. He is able to obtain experience of professional practice at first hand and to more readily appreciate and assess the restraints imposed by practical situations and the variety of factors that influence and impinge upon professional judgements. Moreover, he can measure his own ability, aptitudes and attitudes in a professional situation, learn to become part of an effective team, adjust himself to a practical environment, and become aware of the application of the factors of time and cost in practice. The student will also become familiar with the interrelationship of personnel and with the operation of organisational structure. He will also develop an understanding of the general economic and administrative factors involved, and of professional ethics (which will also be studied on the course) (Seeley, 1972). (He/she has been omitted to avoid constant repetition.)

Consequently, when the student returns to the university, he is more mature, discerning and knowledgeable than before. The combination of theoretical and practical experience ensures that the student retains a balanced outlook and approach. In addition, it has been found that, as a result of the professional training, the student gains greater meaning from his studies, secures greater motivation to learn, matures more quickly and obtains a greater understanding of people. He develops skills in human relations and is oriented to the industrial, professional and commercial world. He is more adequately fitted to undertake the more professionally based subjects on the final year of the course, as he is better able to appreciate their practical implications. He can also prepare a much more valuable and useful final year dissertation, designed to develop personal initiative and independent approach to a topic, where much of the data may have been collected in the year of professional training. It is hoped that the dissertations will be of considerable value to the profession and they are always available for reference from university libraries (Seeley, 1972).

Practising surveyors can play an even more effective role in the operation of integrated training schemes by sponsoring students on courses. They can interview school leavers or students on the first years of degree courses. The employer thus becomes associated with the student throughout the course and the student, for his part, obtains a sense of belonging to the employing organisation, with undoubted benefits to both parties.

*Part Time* courses are normally spread over five academic years and have the great advantage that the student is obtaining practical experience in the profession concurrently with carrying out his studies at an educational establishment. It entails a heavier workload for the student and will take a longer period to obtain a degree. There are situations, particularly financial ones, where this arrangement offers the most practicable way forward, provided there is a university offering the course within reasonable travelling distance of the student's home. Fortunately, a good network of part time quantity surveying degree courses has been established throughout the United Kingdom. In times of severe recession, there is unfortunately a reduction in the number of workplaces available.

*Distance Learning* courses are evolving, particularly for post graduate courses, to overcome the problems that can be associated with attendance for long periods at higher educational establishments and, in the case of part time courses, students living and/or working in remote parts of the country. This type of course requires a very exhaustive and skilfully prepared input by course designers to be successful. High quality course material supported by videos, cassettes and other visual aids, need to be made available in conjunction with detailed practical projects and self assessment work. Some face to face sessions with tutors is required to back up the distance learning material and resolve students' difficulties and problem areas. All course material requires periodic updating and extension, to keep it up to date and abreast of new developments. Designers of and tutors on these courses should be careful not to underestimate the complexities of these arrangements and the enormous amount of work involved in their formulation and operation.

## LEARNING METHODS

Ashworth (1994) has aptly described how there has been a discernable shift towards problem solving and away from a reliance upon memory recall of available knowledge, with an emphasis on understanding, development of transferable skills and competent application. It is fairly well recognised that a lecture is an inefficient method of student learning and that seminars, tutorials and project based learning are much more efficient, coupled with the student's own personal development. There has also been a change from teaching to learning with the objective of developing the student's capabilities, both in terms of the chosen profession and the wider role in society. The change towards increased student participation has created the need for greater student self study skills and better access to resource bases, libraries and computer facilities. The promotion of a learning culture has resulted in greater emphasis being placed on study methods, organising study time effectively, working to schedules, using learning resources effectively and self monitoring study behaviour, thereby optimising the self esteem of the individual and achieving improved time management of the learning process.

A principal aim should be to make studies as interesting and relevant to professional needs as practicable, and to endeavour to assist the student to see the relevance of his studies, including supporting subjects, to practical situations and his/her ultimate employment. With this in mind it is helpful to draw upon examples related to live situations, particularly with project work. The use of specialist outside lecturers is an asset as they assist in widening the students' horizons and in understanding the extension of the quantity surveyor's role into such activities as project management.

Another primary aim is to organise the various learning methods into a coherent whole, using each method for the purpose(s) for which it is best suited. Thus *lectures* are concentrated activities addressing all the students in a particular group, amply supported by visual aids and handouts. *Seminars, studios, workshops and laboratory work* back up the lectures and normally involve 12 to 18 students in large group discussions, supervised practical tasks or laboratory work. Seminars often involve the critical examination of students' submitted work which can have a significant feedback. Alternatively, students may be asked to prepare a paper for discussion, either individually or as a group. *Tutorials* usually take place in groups of about 8 students and one of their main aims is to encourage student participation. They make an excellent contribution to effective and original thought and argument. Another valuable mechanism is *problem solving exercises* in the various subjects to enhance the students' ability in critical analysis and evaluation.

*Practical projects and case studies* form an important part of learning programmes aimed at developing the students' operative skills. They assist in widening and deepening the students' general understanding of particular activities, possible situations which can arise in practice and problems to be overcome. *Dissertations* generally form an important component in the final year of degree courses, as they provide the opportunity for individual research and the writing up of a dissertation on a topic of the student's choice related to the course of study. The dissertation topic requires careful selection and the subsequent investigation, collation of material and its analysis must be conducted with care and vigour, with a positive personal contribution made to the final product.

Students' interest can be increased by visits to construction work in progress, as well as manufacturers' works and offices of organisations connected with the construction industry. Much greater benefit can be obtained from site visits if they embrace more than just a look around the site. For example, when the Queens Medical Centre, a large teaching hospital was being erected in Nottingham, final year quantity surveying degree students visited the site, supported by the following talks and discussions to obtain maximum benefit from the inspection:

(1) briefing by the architect on the design and the problems encountered;
(2) talk by the quantity surveyor for the contract on the procedures used;
(3) seminars held at the university after the visit to discuss various aspects of the project.

## GRADUATE RECRUITMENT

### Employer Approach

Lovell (1990) considered that a common mistake among quantity surveying practices was to leave their recruitment efforts until too late and some approached the task in a very haphazard way. For example, a recruitment letter displayed on a student notice board a month before graduation is unlikely to have much impact. On the other hand some practices, notably the larger firms, have realised the value of quantity surveying graduates and run concerted campaigns to attract the best graduates. Other firms think that students should apply to them and not the otherway round.

Lovell suggested that the most successful recruiters were those who made formal presentations to classes of students. The points to emphasise are training, APC success rates, uses of computers, specialist services, overseas offices where appropriate, responsibility and remuneration, age of the youngest associate/partner, and how much recruits can hope to earn in the long term. He felt that quantity surveying graduates were likely to be more convinced by talking to someone who has recently joined the firm than by a public relations person making a slick presentation. Several firms have realised this and take some recent employees who are former students of the same university.

Hogg (1994) received responses from 153 former graduates in quantity surveying of Nottingham Trent University concerning their interviews for their first posts after graduation, and these gave the following statistics:

(1) With 66 per cent of interviews being less than one hour duration and a further 29.4 per cent of 1 to 2 hours, lengthy and sophisticated

interviews are very unusual. I would expect interviews of about three quarters to one hour duration to be the norm.

(2) Relatively few employers test the candidate's area of competence or are interested in degree course content or honours degree classification, which I find surprising.

(3) The majority of employers (68 per cent) were interested in the experience gained during the professional training period.

Research carried out on behalf of the 'Graduate to Industry Journal' into employer attitude to graduate recruitment suggested that many companies are unconcerned about which university applicants have attended and that they consider the class of degree to be important as evidence of certain things, such as intelligence, self organisation and hard work, but there are other considerations which may balance the lack of a good degree.

## Job Applications

The CIOB (1995) carried out a survey of 250 senior personnel managers in construction related organisations and a summary of the more useful findings is now given. The CV was aptly described as one of life's great levellers as after being posted there is no way of knowing whether it is a winner or destined for the waste bin. Where an advertisement requests completion of an application form the insertion of 'refer to CV' detracts from the applicant's chances of success.

The most common weaknesses found in CVs are poor spelling, bad grammar and shoddy layout. Incorrect addresses and dates are less common but still find their way into many applications. A CV is an integral part of a selling process and, as such, deserves thorough preparation and a relevant, professional presentation, bearing in mind that only a small percentage of candidates can normally be interviewed, so that it is unwise to wait until the interview to enlarge on strengths, experience and assets. Too much information is regarded as being almost as bad as too little. It is sound policy to focus less on what you want to say, and more on what you think the potential employer wants to know.

The contents of the covering letter and a brief profile are widely seen as key selling points in a CV. Employers want to know why the applicant is applying for a job and what qualities and skills make them suitable for interview.

The following were identified as important points to be observed in preparing CVs:

- keep the overall length down to two pages
- always include a covering letter
- tailor the CV to the job, and use the covering letter to emphasise the specific reasons for applying for the post

- the CV should be typed, not a photocopy, look professional and be error free
- do not send a photo, use fancy bindings or gimmicks
- include scholastic and academic achievements and work experience
- list all information in reverse chronological order (most recent first)
- send CV to a named individual (ensure name, address and gender are right)
- when previously employed, include current job title, salary, responsibilities and reason for leaving
- detail strengths, achievements, skills and types of projects handled
- omit irrelevancies, such as ages of children, and details of spouse's or parents' careers
- use humour sparingly and with caution.

## PROFESSIONAL TRAINING

### General Background

Nisbet (1989) has described in considerable detail the changing patterns of professional training for quantity surveyors commencing with office training in the form of articled pupilage and includes a copy of the RICS articles of pupilage, whereby the pupil agreed to serve the principal, and the principal, in consideration of a sum of money paid by the pupil's father or guardian, agreed to teach and instruct the pupil to the best of his ability in the said profession. The pupilage system disappeared in the wake of the Education Act of 1944 and the Higher Education Report (Robbins Report) of 1963, the latter advocating that 'courses of higher education should be available for all those qualified by ability and attainment to pursue them and who wish to do so'. These measures subsequently led to a massive expansion of educational opportunities which affected many professions and particularly quantity surveying as it had relied substantially upon the recruitment of intelligent school leavers. In the early 1960s many colleges were offering full time courses up to but not including the final and qualifying examination, which could only be taken after two years practical 'training' in an approved office.

The absence of training in full time courses had a detrimental effect on the competence of newly qualified quantity surveyors. Hence in 1972 the RICS issued rules and guidance for a new test of professional competence which required candidates to receive 'approved professional training', and this was superseded in 1977 by a more realistic requirement of 'approved professional experience'. The test of professional competence was first introduced in 1974 with the aim of ensuring that each entrant had a suitable level of experience and professional judgment to ensure a competent level of practice in the profession. Adequate experience was to be satisfied by a period of three years working in an approved office – usually an office of a quantity surveyor or one which was supervised by a qualified quantity surveyor – and by compliance with detailed guidance of the subjects in which experience had to be obtained. Professional judgement of quantity surveying students was to be satisfied by a test of two working days duration, comprising a practice problem which put candidates under pressure in a supposedly office situation. With rising candidate numbers tests were held twice a year and placed great burdens on both candidates and assessors, and the arrangements were subject to considerable and often justifiable criticism because of the various weaknesses in the system. To partially overcome the problem the RICS in 1979 decided that all new members should be obliged to devote a minimum of 20 hours each year to formal continuing education, and this aspect will be examined later in the chapter.

## Weaknesses of the QS Test of Professional Competence

Latham (1995) has described how fundamental changes in the manner and content of degree education, with the greater number of entrants to the profession coming with university awards, meant that most aspiring chartered surveyors have an academic background of wide variety, flavour and knowledge base, which may well need to be underpinned during the professional training period. The original approach of recording professional experience in a diary was unstructured and random in its attainments. Furthermore, it was not until the final assessment that it was subjected to a scrutiny which seriously attempted to assess the adequacy of the candidate's progression and to rectify any deficiencies which may be identified.

Experience of the test of professional competence (TPC) over many years indicated considerable variation in the apparent level of employer involvement in the training of candidates. There was also clear evidence that the strength or weakness of candidates' experience over the required training period was reflected in their performance when completing the practice problem, described earlier. Quality of experience, therefore, had a direct influence on the TPC final pass rate, which was invariably thought to be too low, and hence the introduction of structured training in the replacement proposals in the 1990s. Some employers considered that the high failure rate reflected adversely on the credibility of their offices and the staff employed.

## Quantity Surveying Professional Training Initiative

In 1992 the Quantity Surveyors Divisional Council approved some major changes to the TPC to take effect from 1994, and to be renamed the Assessment of Professional Competence (APC). A primary objective was to convert the TPC test with its two day practice problem into an assessment of professional ability and skill. In the APC the candidate is required to attend a local assessment centre for one day to write a report on a professional topic and present it to a panel of three assessors and to be interviewed on it and on his/her experience. Hence every candidate is interviewed and this allows a better assessment of candidates' ability to present their approach to a professional topic and to react to questions. Assessors are given training in interviewing and assessment. Structured training can be introduced in firms as an alternative to the original TPC system, which required candidates to gain experience in specified areas of practice, and to maintain a diary to record that they had done so. The revised quantity surveying scheme was based on the need to achieve competence in an activity, as opposed to merely experiencing it. Candidates must achieve a minimum specified level of competence across a range of work before being eligible for the final assessment, which prevents candidates presenting themselves for the final assessment before they are adequately prepared.

Other features of the quantity surveying APC were as follows:

- diary and log book to be more effectively monitored
- experience period to remain at 33 months but reduced to 24 months where the employer has an approved structured training scheme in hand
- interim submission halfway through the training period
- prequalification structured learning as an integrated part of the training process.

Under the APC scheme the employer is expected to provide the candidate with the required training and experience, to give encouragement and to make the necessary time and facilities available. A supervisor will be appointed by the employer to give guidance on training and oversee day to day work. A chartered surveyor will be appointed as training manager and is responsible to the employer for ensuring that the candidate receives appropriately supervised training and experience and develops an acceptable level of confidence and professional independence. The candidate cannot enter the final assessment until the training manager has certified that all training requirements have been fulfilled.

### RICS Review of Assessment of Professional Competence (1994)

In 1994 the RICS Education and Membership Committee submitted its proposals embracing all divisions of the Institution and after consultation and debate was approved by the General Council in 1995. The proposals centred around the following four main conditions:

- training advisory service for employers
- planned or structured training

- continuous, progressive assessment
- quality management.

The quality management component included the introduction of regional training advisers, training agreements, formal training for all members and staff involved in the administration and assessment process, independent monitoring of assessments, and regular review of progress. The training framework shall be built upon a minimum training period of two years, day to day supervision accompanied by regular, documented progress assessment, training structured and recorded in an approved format, and supervision by a suitably qualified person. In essence there is a period of structured training concluding with a professional assessment.

With regard to structured training, the RICS produced a structured training agreement, to be signed by the employer and the candidate, with the option that employers could submit their own schemes for approval. A clearly defined training programme comprises hands-on professional experience and structured learning to cover necessary areas of required knowledge which do not conveniently fit into practical experience. There is provision for approval of trainees' employers and their training schemes and all training must be structured and will be mandatory once the latest scheme is operative, and relate to the attainment of competencies in specific activities and the progress made therein. The employer carries out continuous assessment and the Institution assesses after twelve months and at the final stage.

The concluding professional assessment comprises:

- overall assessment of candidates' training, experience and the application of their knowledge
- assessment of candidates' written skills, consisting of the submission of experience, training and responsibilities since the interim submission, and a critical analysis of a professional project or commission undertaken by the candidate
- assessment of the candidate's knowledge based upon an interview by three chartered surveyors
- assessment of the candidate's oral skills, consisting of a presentation by the candidate, based upon the critical analysis.

Regional training advisers will be appointed to advise employers on the development of training schemes, supervision of trainees, and trainee assessment; as monitors of registered training schemes; as the first point of enquiry in the areas of complaint about a training environment; and as facilitators towards the development of secondment schemes for employers whose provision of professional experience is less broad than the requirements of the APC.

## TECHNICAL SUPPORT STAFF

### Society of Surveying Technicians

The Society of Surveying Technicians (SST) was founded in 1970 to meet the need for a nationally recognised organisation for people supporting the work of chartered surveyors. The society's aims were to promote technical education and the advancement of members in the business community (SST, 1994). It was organised into divisions, of which the largest was quantity surveying, regions and branches. Society membership comprised five grades ranging from student to fellow. The qualifications for entry were founded in the BTEC series of National and Higher National certificates and diplomas. Experience gained in approved surveying offices was recognised and a joint test of competence (JTC) was the final hurdle prior to gaining membership (Burr, 1994). There was, however, a strong feeling that the RICS had not done enough to recognise and encourage the work and status of the 5000 members of the society.

In 1993 a declaration of intent was signed and partnership with the RICS realised in 1994, when the society's administration was merged with that of the RICS, and RICS services, including the journal and branch activities, were made available to SST members. It was anticipated that full integration would be achieved in 1996. The partnership brought SST members within the guidance of the RICS, making SST members subject to similar monitoring and CPD arrangements as chartered surveyors, and ensuring that professional standards are formally recognised and maintained (Burr, 1994; SST, 1994).

The proposal to merge the two bodies was in 1995 subject to the following three decisions:

(1)  the Privy Council must agree that the Institution develops a technical grade of membership
(2)  RICS members must agree to the formation of a technical grade, together with a designation and associated career structure
(3)  SST members will need to agree both the negotiated merger offer and to wind up the legal identity of the society.

The selection of an acceptable designation and designatory letters occupied a long time and many suggestions were made to the RICS. Finally, my own proposals which had been publicised in the RICS Journal and in the General Council, received the support of the SST, RICS Education and Membership Committee and General Council. The designatory letters were F or A Tech Surv (RICS) with Tech being the abbreviation for technical and Surv for surveyor. The full description being fellow (or associate) technical surveyor of the Royal Institution of Chartered Surveyors, and the designation of technical surveyor replacing technician to more adequately reflect

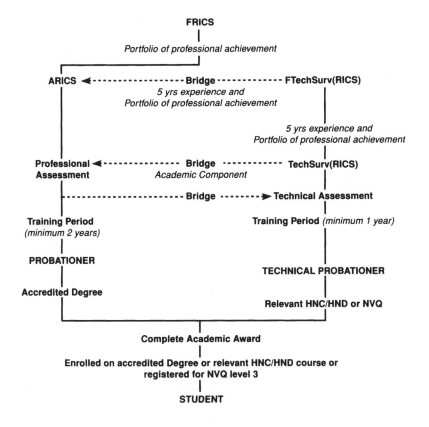

**Figure 18.2**  *RICS integrated career structure*

*Source*: RICS Education and Membership Committee (1994).

the valuable and highly skilled role that many SST members perform and their current status. In the final submission by the RICS to the Privy Council, two sets of designatory letters were put forward, namely FTechSurv (RICS) and TechSurv (RICS), to avoid confusion under an integrated career structure.

Under an integrated career structure, illustrated in figure 18.2, all recruits would join the RICS as students and enrol on either a RICS accredited course or a relevant HND/HNC or NVQ level 3 course. After achieving their academic qualification, registered students would elect for the professional or technical stream as appropriate, followed by a period of practical training concluding with an assessment (professional or technical), leading to chartered or technical status, with progression to fellow on experience and merit.

Two alternative bridges provide the means for technical surveyors to achieve chartered status:

(1) qualified technical candidates complete the academic component of the RICS qualification with the final component of the professional assessment;
(2) after a combined total of ten years experience, higher technical level candidates produce a portfolio of professional experience.

A reverse bridge could be used by PA candidates who demonstrate a high technical rating but fail, because of some other shortcoming, to achieve chartered status, by undertaking the TA for admittance to the technical class.

The RICS believed that the development of a support class within the RICS is a major step forward in the presentation of the Institution as the natural home for surveying related professionals at all levels. The identification of subdegree qualifications by the RICS/SST will raise interest in those qualifications and thereby assist recruitment on to the relevant courses.

## National Vocational Qualifications

A National Vocational Qualification (NVQ), or SNVQ in Scotland, is based upon the holder's ability, as judged by an approved assessor, to carry out a function with competence, and this could include a professional activity. The assessor observes the employee's performance in the workplace, backed up by an underlying knowledge and understanding of the task. The development of these qualifications has been progressing since the mid 1980s with substantial support from government funds.

The Local Government Management Board (1994) described how NVQs are based on national standards, have considerable flexibility, assure quality of performance, are focused on the workplace and are controlled and developed by a single national body. The Board described how the assessment is objective, rigorous and individual, using line managers in an organisation. Advantages claimed include clear links to the objectives of the employer, precise analysis of training needs, enables training to be focused on individual needs, provides precise evaluation of training effectiveness, cost effective training and development and balanced responsibility for qualification.

The RICS (1995) recognises that NVQs will impact upon the professional qualification and will be likely to contribute to academic awards and to some of the experience requirements of the APC, and to have a major role in qualification for the SST. Hence the RICS General Council decided that the Institution should be proactive in the development of NVQs, and help to ensure that they evolve in an acceptable manner, by close involvement with the Construction Industry Council (CIC), through the Construction Industry Standing Conference (CISC), the forum for NVQs at professional, managerial and technical levels in planning, construction, property and related engineering services. The RICS may consider becoming an awarding body in certain cases as, for instance, with the NVQ in construction project management.

In the construction industry NVQs for professional, technical and managerial roles start at level 3, approximating to National Certificates and Diplomas; level 4 equates to HND/HNC, while level 5 will be at professional membership and CPD levels, and it will be possible to gain further units throughout one's working life. The standards that have been developed for the construction industry place a great deal of emphasis on the underpinning knowledge required in all professional disciplines and it is the intention that the basic theory will continue to be taught and tested.

CISC delegates have drafted evidence specifications to support the standards and so lead to fully designated NVQs/SVQs, while CISC members have made their selection of units from the standards map to produce prototype NVQs/SVQs to meet their particular needs; a process described as 'templating'. One of the many strengths of NVQ/SVQ methodology is that the qualifications are a reflection of competence in functions and not merely tasks. Hence the commonality between client's surveyors and contractor's surveyors can be exploited, and individuals can transfer their skills as they change jobs both within their own companies and also across construction disciplines.

The implementation of NVQs/SVQs will be determined by what the market wants. Larcombe (1993) describes how this can be influenced by 'carrots' (added value and increased competitiveness for employers, and better career progression and marketability for individuals) and by 'sticks' (public purchasing policy and tax penalties).

## CONTINUING PROFESSIONAL DEVELOPMENT

### General Background

'Quantity Surveying 2000' upheld the decision of the RICS to make continuing professional development (CPD) mandatory for all members as from 1991 as changes in the market place and in surveying expertise and techniques make the need for CPD ever greater. Furthermore, a quantity surveyor must now develop adaptability and flexibility rather than focus on one particular skill. He will also tend to plan his own career rather than have it planned. Knowledge itself is advancing very quickly and a quantity surveyor's original degree can soon become dated, although the underlying principles remain valid. Hence clients and society can reasonably expect any profession to take active steps to ensure that all its members keep up to date and abreast of latest developments. Swanston (1994) believed that surveying professionals need a coherent, seamless system of education and training which extends throughout their professional careers, and CPD must be a valuable facilitator.

## RICS Bye-law Requirements

Bye-law 9(3) requires every Professional Associate and Fellow of the Institution to undergo CPD in each year and to provide the Institution with evidence as required. CPD is defined as the systematic maintenance, improvement and broadening of knowledge and skill and the development of personal qualities necessary for the execution of professional and technical duties throughout the member's working life. They are required to keep suitable records of qualifying activities and keep the records for three years thereafter.

By 1996 the RICS was monitoring a thousand cases a year and a small number of persistent non-conforming members faced possible expulsion.

CPD may comprise the following:

- attendance at courses and technical meetings organised by the Institution and its branches; universities or other academic institutions; employers of chartered surveyors; other professional bodies; and other course providers.
- discussion meetings on technical topics
- private studies of a structured nature on specified themes
- distance learning opportunities, or other supervised packages, being a programme of reading or recorded lectures
- research or post qualification studies
- authorship of published technical work or the time spent in preparation and delivery of lectures in connection with a qualifying CPD event or other similar professional technical meeting.

To qualify as CPD the previous activities must be related to some part of the theory and practice of surveying, and/or other technical topics related to a member's current or potential occupation, and/or personal or business skills designed to increase a member's management or business efficiency.

Professional Associates and Fellows must complete 60 hours CPD in every consecutive period of three years, of which time recorded under private studies cannot be more than two thirds of the total requirement. Members who have completely retired from practice are exempt from these requirements.

**RICS 1992 Review Document**

The RICS (1992g) made the following forward looking recommendations for future action in connection with CPD and these have received widespread support from the profession:

- increased emphasis on the output from CPD (e.g. increased performance) in contrast to the current preoccupation on input (i.e. number of hours)
- continued development of structured modular training courses which may lead to post graduate qualifications, possibly based on the Credit Accumulation and Transfer Scheme (CATS)
- production of guidelines to recognise 'structured development' as an integral part of CPD, and this concept will be examined later in the chapter
- need to assist practitioners in the diagnosis of their CPD needs by the use of performance appraisal or competence models which indicate the appropriate transferable skills at a range of career stages
- need to encourage the use of 'personal development plans' (PDP) to plan CPD activities
- need to encourage practitioners to view CPD in its broadest sense and, in particular, to emphasise the growing need for 'continuing management development' (CMD) to enhance surveyors' skill base in business and commercial management
- need to enhance guidance to practitioners on the assessment of CPD needs and the evaluation of the CPD investment
- need to recognise good practice in the wider training field including national initiatives such as 'Investors in people' and to communicate this practice within the wider membership.

**Training/CPD Views**

A study by Ashbridge Management School assessed the attitude of British business to training and identified two distinct approaches: the fragmented and the focused. The *fragmented* approach does not link CPD to business goals. It sees CPD as a cost, as short term, as being about directive training and knowledge acquisition; hence CPD is seen as unimportant, something to be ignored in the office or viewed as a reward for good performance. The *focused* approach sees CPD as an investment, concerned with skills development as well as knowledge acquisition, something to be evaluated to initiate change and flexible in the way it is done (Kennie, 1993).

**Structured Development**

Structured development to be acceptable for CPD purposes needs to demonstrate an improved performance in the form of a structured learning contract. This contract seeks answers to the following four questions:

(1) Where am I now in relation to this activity (in terms of knowledge, skills, abilities, etc.)?
(2) Where do I need to be? (What do I need to improve upon?)
(3) How will I get there? (What is my learning strategy?)
(4) How will I know when I get there? (How will I provide evidence to demonstrate how I have improved) (RICS, 1992g).

It is considered that personal development based on the approach outlined would qualify for up to two-thirds the total CPD allowance of 60 hours.

Another form of structured development could, for example, encompass the following activities:

• chairing a working party/committee
• managing the introduction of a change, such as new systems, business plans or a new computer
• merging with or acquiring a practice
• secondment to another organisation or spending time in another section of the same organisation
• performing a new type of work which requires new knowledge, skills and experience.

Thomas (1993) has described how a surveyor can apply a CPD topic to his or her current or future career, provided it is studied in a structured manner with suitably combined courses, discussion groups and private study. Examples of such topics could include the following:

• establishing a group CPD programme for your firm
• investigating professional practice methods abroad
• learning a foreign language for commercial purposes
• acquiring new skills such as accountancy, law or management functions.

**CPD and the Development of Professional Competence**

Where the initial diagnosis of the perceived CPD need proves to be inaccurate and is not focused on a real need, further work is likely to be needed in the following three areas:

(1) the development of review processes (e.g. performance appraisal/career planning reviews) as part of CPD and the development of compe-

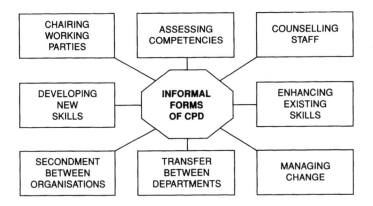

**Figure 18.3**  *Informal mode of learning*

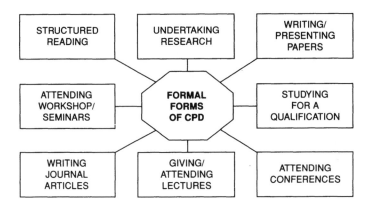

**Figure 18.4**  *Formal mode of learning*

tence models for typical career stages (ICE, 1992b);

(2) the widening of 'structured CPD' to include 'structured development' opportunities, as previously described, in addition to 'structured training', as illustrated in figures 18.3 and 18.4, showing the formal and informal modes of learning;

(3) clearer assessment of the objectives of CPD and the business/personal need to which the CPD is related and the manner in which it will be evaluated.

Hence the main changes of emphasis proposed for CPD activities are as follows:

| FROM | TO |
|------|-----|
| 1. Input (i.e. number of hours) | Output (i.e. improved performance) |
| 2. Recording | Planning |
| 3. Unstructured | Structured |
| 4. Training | Development |
| 5. Professional updating | Professional and management development (RICS, 1992g) |

### CPD Funding

In 1994/95 several RICS branches were very concerned to learn that in future CPD events arranged by branches were required to be self-funding. Many of these events comprised the normal technical meetings which had been provided previously out of branch funds, and which most members considered to be an essential service which should be financed out of their professional subscriptions without any further charge being levied. The RICS Branches Support Panel weighs up each branch budget on its merits but will not support the direct funding of CPD. Most of branch grant is now spent on administrative support within branches and branch newsletters effectively marketing CPD events, leaving members to fulfil their professional obligations by meeting the (thereby reduced) costs of meetings. Branches can organise their CPD programmes to have some meetings which attract large audiences and thus subsidise smaller events, provided the overall programme is self-financing.

### TRANSFER TO FELLOWSHIP

An applicant must have been a Professional Associate for at least five years and be able to demonstrate that he has attained a sufficient level of achievement in his professional career to be worthy of transfer by the submission of a portfolio of professional experience, which shall include the following four elements:

(1) record of employment history giving evidence of aspiration to and achievement of increasing professional responsibility culminating in a position of seniority deserving recognition by transfer to fellowship;
(2) evidence of managerial as well as professional responsibility except in certain specific cases;
(3) evidence of development of professional or managerial skills as, for example, through the attainment of a postgraduate award or by a programme of sustained structured CPD or as a result of exceptional

contribution to the knowledge base of the profession;

(4) evidence of satisfactory compliance with the byelaws and regulations relating to CPD, accounting for clients' money and compulsory professional indemnity insurance.

## RESEARCH

### General Background

The report 'Quantity Surveying 2000' described how the profession did not have a strong tradition of pure or university based research and called for an examination of fundamental issues, such as the nature of quantity surveying skills, core skills needed and the role of training. Research undertaken at Nottingham Trent University in the 1990s helped to clarify some of these issues. Furthermore, the report omits to mention the valuable research carried out by groups of quantity surveyors in years past, particularly the Wilderness Group, the Aqua Group and the RICS Building Cost Information Service. The RICS Education Policy Report (1994) described how divisional APCs encompass research as an area of experience and that limited funds were available through the Research Committee for Institutionally funded activity.

The Latham Report (1994) reiterated the findings of a Construction industry Council (CIC) working party (1994) that the 'UK spending on construction research and its dissemination is substantially below that judged necessary by a succession of authoritative studies'. An EC/W S Atkins report found that within the EU, 'The total level of construction research is low, and needs to be raised nearer to the average of other industries. The EU spends much less than Japan in construction R&D'.

### Nature of Research

Dixon (1994) has aptly described how the term 'research' is often used in an inaccurate and misleading way as for example referring to gathering information or collecting data. To constitute research systematic analysis of the data to answer specific questions is required. The Oxford Dictionary defines it as 'a systematic investigation and study in order to establish facts and reach new conclusions'. Another more detailed but useful definition is 'Research may be characterised as a systematic and critical search or enquiry with the specific purpose of extending the boundaries of existing knowledge; it embodies the elements of originality and imagination. While research entails the collection of data these two processes are not to be equated; for research to exist, the data must be collected for some purpose and it is that purpose which defines the research activity. The ultimate purpose of a research project may be to construct a theory or an explanation; the activity can be construed as research only if it is performed within the context of such a framework. Research is effective only if it is adequately disseminated'.

For academics the need for research is evident. It complements and reinforces teaching activities, promotes a spirit of enquiry and a skill in problem solving, generates effective approaches to data collection and analysis, and assists in the advancement of knowledge. It is anticipated that most of the research undertaken in surveying departments of universities will be of the applied variety and will have relevance to the needs of the surveying profession. In the 1990s, research had become a top priority in the new universities (the former polytechnics), as their credibility and funding were being judged in part by their research output.

Brandon (1990) summarised the main thrust of quantity surveying research in the 1980s as follows:

- a move to the front end of the development process
- an acceptance of uncertainty and risk
- taking a longer term view
- harnessing of computer based information technology
- reappraisal of manual techniques
- identification of market opportunities.

## RICS Research Strategy

The RICS (1992h) published an important document on research – 'A Research Strategy for the RICS, 1992–96' which contained valuable guidelines. For example, it asserted that research is concerned with efforts to increase the sum of knowledge pertinent to the activities of the profession, and to be successful the research process must be analytical, objective and critical. The principal objective was to support the work of members in a practical way, enhance the status of the profession and ensure that the profession makes a positive contribution to society.

The RICS believed that the results of research were needed to cope with challenges, largely created by major structural changes impinging on surveying. These created opportunities to be exploited (e.g. technology), threats to be overcome and turned to advantage (e.g. abolition of fee scales) and hybrids, such as the open European market in 1992. Where significantly successful, research in these areas can also create competitive advantage, as can also developments in the specialist technical base of the profession. It could be that the future of the profession depends very much upon keeping up with developments and ahead of competing professions. Ideally, the profession should push forward, regenerating its own knowledge base through sustained research, thus creating opportunities out of the radical changes occurring.

The Institution identified three principal areas in which action could be taken, namely stimulating, facilitating and communicating research using the following strategies:

- support a balanced portfolio of research and development within identified priority areas
- proactively seek to initiate and support research in areas typified by major structural changes, such as impact of information technology, management of increasingly systematic client relations (eg quality management, green issues, facilities management), and implications of changes in market structures and processes
- receptive and supportive towards research proposals in specific technical areas of particular potential
- overcome fragmentation by developing a strong research community in surveying and promoting strong links between practitioners and researchers
- communicating research material to the membership
- provide a coordinating influence for all RICS research funds.

In 1995 the Research Committee reported to the General Council on the progress made with the implementation of the research strategy. For example, stimulation of research was mainly through the targeted application of RICS research funds, including the Education Trust. Facilitating research comprised developing the research community through research conferences, promulgating best practice in research methodologies, promoting the role of the RICS in research to external audiences, and establishing links with allied research disciplines. Communicating research was carried out using Chartered Surveyor Monthly, RICS Research Paper Series, RICS Research Reports, Education News, RICS Research Register, Research in the RICS, and 'Update'. A number of the research projects funded by the research and policy budgets and Education Trust in 1991–95 included a number with direct relevance to quantity surveying. There is a need to concentrate more resources on group projects some of which could be inter disciplinary and for greater cooperation with other research bodies, such as CIC and CIRIA.

A symposium organised by the RICS (1995d) attempted to identify the key areas where research should be focused in construction economics and these were as follows:

- construction capacity and economic cycles
- construction industry organisation
- inter-sectoral links
- impact of long term changes in society
- international markets.

It will be appreciated that this list is not exhaustive.

**Funding of Research**

Researchers often require financial assistance – whether it be for equipment, travel, assistance with questionnaire surveys or other essential activity. Universities may have research funds available but they are likely to be very restricted and priority is generally afforded to the needs of research assistants. Hence it is frequently necessary to have recourse to outside bodies, such as the research councils and other organisations, such as DoE, with a particular interest in the matter being investigated. As mentioned previously, the RICS Education Trust has made payments to many individual researchers pursuing topics of relevance to the profession and subject to adequate safeguards.

Late in 1994, it was announced that the European Commission (EC) was awarding £10 billion covering as much as 50 per cent of research costs spread over a four year period, and this could include a significant amount of construction work. In the UK only 5 per cent of all research work takes advantage of EC funding and hence the CIC and CIEC launched the Construction Industry European Research Club to help clarify the process and assist potential applicants. A major deterrent for most potential UK researchers is that each research proposal seeking funding must involve collaboration between more than one country (typically five or six) and several different types of organisation.

Every year in the UK, more than £200m is spent on construction related research, of which the government contributes some £75m. With this in mind, the Construction Research and Innovation Strategy Panel (CRISP) was set up in 1995, to set targets, help and influence those who commission research, monitor progress and update strategy.

**Research Initiatives**

In the 1990s there were many research projects underway covering a range of vital quantity surveying related activities, with a significant input from the universities, often in conjunction with the RICS. One particularly significant piece of research involved collaborative action between the University of Glamorgan, Liverpool John Moores University and Nottingham Trent University, and the establishment of the International Procurement Research Group (IPRG). The aim of the Group was to investigate the procurement methods adopted in the construction industries of developed, developing and underdeveloped nations. The research, planned for completion in 1996, should assist in the identification and implementation of appropriate construction procurement strategies in developing and underdeveloped countries (Sharif, 1993).

Other valuable research projects initiated and/or supported by the RICS QS Division included Alvey projects, integrated databases, risk analysis/ assessment, computer based tool for estimating the implications of construction demands, clients' value system, core skills and knowledge base of the quantity surveyor, and improving value for money in construction.

## Conferences, Seminars and Workshops

Many conferences, seminars and workshops are organised each year both in the UK and in numerous countries overseas. Researchers could find some of these to be of considerable value where they impinge upon their research projects, as well as academics and practising surveyors.

## Postgraduate Courses

In the 1990s there was a significant growth in the establishment of taught masters' and postgraduate diploma courses. These courses usually occupy one year full time or two years part time and some are available by distance learning. The modular format of many courses also provides greater flexibility to part time students and enables them to opt for study periods which are more easily assimilated into their employment activities. Courses which incorporate business and financial management are generally welcomed by quantity surveyors.

Another type of post graduate study relates to research degrees, usually in the form of a MPhil or PhD. These encompass intensive research by the research student or fellow into a specific topic approved by the university awarding the higher degree. Assessment is by thesis and oral examination in both cases. The doctorate requires study and research of a much higher order than is required for the master's degree, must show originality on the part of the candidate and help to extend the frontiers of knowledge. An alternative route to a PhD is available to persons primarily engaged in research by substituting substantial published work for the thesis.

# 19 Future Developments and Trends

## GENERAL BACKGROUND

Clark (1987) in his RICS presidential address emphasised that the scope for the chartered surveyor to extend his skills, to broaden his services to his clients, and to enlarge his contribution to society has never been greater. Extracts from the promotional literature of three quantity surveying practices are now taken to illustrate the diversity of approaches that are being adopted when offering their services to potential clients, efficiently and quickly and in full understanding of client needs.

- As the construction industry has become more competitive in price, and contractually aggressive and innovative in its approach, it is necessary for quantity surveyors to provide a group of services that can quickly and efficiently react to ever changing needs, and to provide a cost effective and commercially viable solution for clients.
- Every successful construction project relies on the essential component of effective budgetary control. This is the central discipline of the professional quantity surveyor, and one which demands accuracy, speed, a thorough understanding of design and construction techniques and an appreciation of the client's requirements. This core discipline has been developed by combining the human contributions of skill, experience and professionalism with the accuracy and time saving potential of advanced technology, involving total commitment from inception through to completion and commissioning.
- Quality of quantity surveying service entails meeting clients' needs, efficiently and effectively, consistently and reliably, through internal quality management systems enhanced by full quality assurance documentation prepared in accordance with the requirements of BS 5750.

As long ago as 1983, a RICS report stressed the need for quantity surveyors to accept responsibility for cost, time and contractual matters, with the omnibus title of construction management. While the RICS report *Quantity Surveying 2000* (1991) believed that the profession should use increasingly automated services to provide an expanding range of high quality, value added services.

Subsequently the report of the QS Focus Study Group (1992) accepted the concept of the Lay Report (1991) that the role of chartered surveyors is to add value to existing and new property assets; notable examples of this evolution being the changes in emphasis in services from measurement to management, and from cost to value. The report also recognised that the chartered quantity surveyor has a core of skills which render him unique within the Institution, which must be preserved, aided by a versatile Quantity Surveyors Division. He is also at the forefront of all the construction professions in the use of computers in, for example, cost planning, risk analysis, project management and expert systems.

Challenges posed by the changing market will require an innovative and resourceful response from the profession in the form of new and more sophisticated services. It was argued that clients want readily purchasable design, procurement and management of construction. They question the relevance of traditional professional boundaries and challenge the worth of many functions, including quantity surveying, and so there is still plenty of innovative and promotional work required.

## THE RICS QS DIVISION 1995 REPORT

The RICS Quantity Surveying Think Tank: Chartered Quantity Surveyors in a Changing World – An Agenda for Change (1995) was produced to identify change, its consequences, implications and, most importantly, the actions which the Institution should take in order that chartered quantity surveyors are better able to respond to the demands of the market. It recognised that fundamental changes are taking place in the way in which buildings are designed, procured, executed and maintained. In 1995, construction tender prices were believed to be at a discount of up to 30 per cent to those of 1990, with fee levels similarly reduced. Greater significance was now attached to regeneration and/or renewal of former industrial land and its decontamination, and there was a noticeable shift of emphasis from out of town greenfield developments to renewal of town centres. Industry is making a massive investment in IT and patterns of employment are changing in the manner described later in the chapter.

Some of the consequences of the identified change factors were as follows:

- Developing countries, which accommodated three quarters of the world's population and whose governments aspired to emulate the wealth of developed countries, had enormous construction needs. To maximise opportunities, detailed local knowledge and the identification of sources of finance were necessary, and creation of links or partnerships with other global players is desirable.
- Awareness of environmental issues is needed, together with a knowledge of the environmental implications of the specification and use of materials, and an understanding of regeneration, renewal and decontamination of land issues.
- There is a pressing need to control cost within established budgets; for

strategic construction project financial advice to international and national investors; an understanding of the private finance initiative (PFI) and the identification of opportunities for its use; and a need for strategic alliances, relationships and partnerships to increase capability and breadth of service.

- Need for quantity surveyors to invest in IT to support their skills and services; and to better understand, with the aid of IT, the interactive relationship between time and cost to better manage the process.
- In the important area of meeting client requirements, the following needs were identified: familiarity with clients' business and culture; full involvement with establishment and/or implementation of contracting and project strategies; full understanding of alternative contract procurement strategies; definition of project risks, their scope, allocation and management; knowledgeable about value for money in the context of certainty of outcome relative to the interlinking elements of time, quality and cost; flexibility in responding to clients' requirements (eg one-stop shopping); and responding to clients' needs for responsibility and accountability.

The following core skills were re-examined and are still highly relevant:

- measurement and valuation of construction works
- assessment of the interactive relationship of project time and cost
- advice upon procurement options, their implementation and administration
- budgetary control, including the regular forecasting of expenditure against budget.

The Think Tank members also identified the following additional services which are increasingly required by clients:

- information management on construction projects, which is examined later in the chapter
- rigorous project strategy
- risk assessment and management, which is also examined later in the chapter
- costs in use (LCC) which was examined in chapter 8.

Other areas explored included technological support, education, communications, innovation and research, and presentation skills, to round off a very comprehensive investigation into the current position and likely future needs of quantity surveyors. Finally, the report identified the many areas where guidance was needed to fill the gaps that had earlier been identified and the forms that this should take.

Subsequently, a work plan was prepared and periodically updated for the attention of the QS Divisional Council. This plan identified the target audiences, proposed products, primary responsibilities for implementation and the timescale envisaged, and the January, 1996 issue contained 45 separate initiatives. The primary responsibility for taking the necessary action often lay with sections of the Institution outside the QS Division, such as market and skills (practice) panels, Research Committee, Education Committee and RICS Business Services, with dissemination by articles in Chartered Surveyor Monthly (CSM), research reports and publications by RICS Books. Hence much will depend on extensive cooperation, coordination and monitoring to achieve a successful outcome.

## BUILDING PROCUREMENT

### Predominant Use of the Traditional/Lump Sum Contracting Method

Surveys of construction contracts in use (DLE/RICS, 1994) indicated that the traditional/lump sum contracting system predominated in the UK in the mid 1990s. Saito (1994) explained this phenomenon by deducing that in the UK construction clients were generally more interested in quality and cost than in speed of project completion. Turner (1990) has identified the many advantages of using this procedure, which would also have had their impact.

Clients requiring a faster project completion were often advised to use design and build or management procurement methods although, as Morledge (1994) has described, the facility for some increase in speed in the traditional approach can be provided through provisional and prime cost sums, approximate quantities and two stage tendering, but with a less certain outcome.

### Problems of Choice of Procurement Method

The review of the construction industry by Latham (1994) illustrated the extent of the adversarial processes operating in construction and the need for change. While Masterman (1992) considered that 'enormous pressure is usually needed to change any procedures, such as the conventional (traditional) procurement method, which become institutionalised within the industry'. NEDO reports in 1983 and 1989 were critical of the practice in the 1980s, declaring it unnecessarily long and difficult when related to industrial and commercial building.

Morledge (1995) contended that a mismatch between selected procurement methods and client expectations and/or characteristics frequently occurs in the UK, and this can be created by 'institutionalised' attitudes and a lack of strategic overview. Strategic overviews of this kind need to consider client characteristics (needs and expectations), project typology (location, complexity, and attitude to risk), bespoke selection and risk distribution (taking into account time/cost trade off and methods of implementation).

The RICS has rightly acknowledged the need for the development and implementation of appropriate procurement strategies and published 'The Procurement Guide' in 1996 to assist in refining the process, identifying risk and rationalising procurement selection in relation to client needs and project type.

The Business Round Table (1995) identified the problems facing a client when faced with deciding a procurement method. It rightly emphasises that some procurement routes are better suited to certain types of buildings and no single procurement route is best in all circumstances. A great deal must depend on the relative importance of risk, flexibility, cost certainty and speed of provision when making the assessment. The extent to which a client wishes to pay others to accept liability for risks such as time and cost overruns will influence his decision making. The checklist in chapter 3 (figure 3.1) will aid the client in making his choice.

**Procurement of Goods and Services**

Quaife (1992) has shown how many of the quantity surveyor's core skills can be used in the procurement of construction related goods and services. These skills include drafting/adapting conditions of contract, analysis of information, establishing the adequacy of raw data and preparing performance specifications. He emphasises that the quantity surveyor must have a detailed knowledge of the specific goods and services and that many provisions of the building contract will be appropriate. For example, relevant clauses in a contract for cleaning services in schools would include: insurances, staff welfare, foreman in charge, interim payments, and determination.

Following establishment of the client's brief, the procurement route can be chosen. When sufficient design or development work has been undertaken, a feasibility study can be completed. This may result in further development of the design and quantification of the work or the preparation of a performance specification and subsequently contractor selection and tendering. The contract may embrace the maintenance of the facility, which should ideally be reviewed periodically, to permit the consultant and client to make changes where necessary. The complete process is shown diagrammatically in figure 19.1.

## IMPROVING VALUE FOR MONEY IN CONSTRUCTION

The RICS (1995e) produced an important, comprehensive report which effectively illustrates best practice in ensuring the best value for the client, which must always be a primary objective of the quantity surveyor. The report aims to promote a better understanding of client needs and the

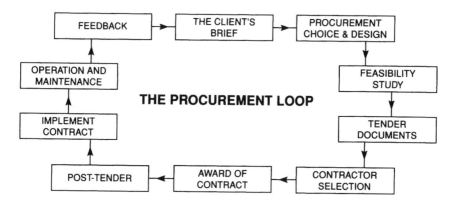

**Figure 19.1**   *The procurement loop*

*Source*: Bucknall Austin.

way in which they can be satisfied. It was found that clients generally wanted higher quality buildings at lower prices and which were produced more quickly, coupled with a better service from the construction industry. The report showed that to achieve value for money inevitably means searching for it, not by chance discovery.

A major part of the study behind the report is an investigation of those factors which could contribute to the cost reduction target of 30 per cent advocated by Latham (1994). In terms of areas where real cost savings can be found, the study identified over 20 potential areas under three major groupings

(1)  client's business case
(2)  development of the design
(3)  management of the project

Each of these groupings and their potential for savings is now explored and they are illustrated in table 19.1.

(1)  The business case involves a rigorous analysis of the client's business needs and the range of options available, and where choices and decisions offer the greatest potential for cost savings, possibly as much as 10–20 per cent.
(2)  The development of the design offers further opportunities through the link between cost and quality needs with its tactical response giving possible savings of 5–15 per cent.
(3)  Managing the project entails putting plans (and designs) into action and could produce savings in the order of 1–10 per cent, looking particularly at the quality of the end product.

Table 19.1   How to make cost savings

**Strategic level: client's business case (10–20% forecast saving)**
- rigorous briefing
- implementing value management at the outset of a project, and
- selecting a procurement method based on thorough appraisal of options.

**Tactical level: development of the design (5–15% forecast saving)**
- awareness of applicable standards, designing to them and eliminating over-specification
- appropriate design life(s) for different elements of the building with matched levels of quality
- undertaking value engineering as normal practice
- increasing off-site manufacture/ prefabrication
- introducing more standardised components/design, and
- increasing attention on, and coordination of, environmental services.

**Operational Level: management of the project (1–10% forecast saving)**
- clear strategy for managing the project under an appointed

manager or leader (including management of design)
- appointing someone at the outset of a project to advise on buildability
- introducing collaborative contracts – partnering
- off-site design coordination with manufacturers and fabricators involved in the process
- supply chain management with emphasis on materials' checking and elimination of waste
- build right, first time – a focus on total quality
- documentation to reach a minimum level of completeness before construction starts
- contract arrangements which ensure (prompt) payments and which allow for off-site work
- raising profitability for everyone
- increasing business and management training
- raising expertise of specialist and trades contractors, especially in their management and generally improving management down-the-line
- keeping successful teams together, and
- creating good site conditions for people.

*Source*: RICS (1995e)

It is essential that the client's needs be detailed and acted upon. The report also advocates the greater use of value management, as described in chapter 9, to ensure that the need for a building is thoroughly analysed and justified before the client is committed. Encouragement is given to finding better ways of delivering what the client wants at lower cost, and hence providing added value.

Another valuable approach is that of 'checking agencies', which are firms or individuals who examine documentation for errors or omissions, as well as for compliance with the brief and applicable regulations, before tenders are invited. The use of checking agencies offers the following advantages:

- to concentrate the minds of the project team members on getting the job right first time
- to compel team members to achieve greater cooperation with each other
- to engender better team working
- to encourage contractors to submit lower bids because of the reduced risks
- to help obtain building insurance on more favourable terms
- to reduce the risk to the client of extra costs or delays resulting from inadequate design details.

It was considered, with justification, that if projects are to be completed to the client's satisfaction, then the following factors are critical:

- Briefing must be comprehensive and realistic and undertaken at the highest strategic level within the client and project organisations.
- Value management should be undertaken to ensure that the project addresses real needs and adds value, by providing the same or better performance at lower cost.
- Procurement options should be carefully evaluated encompassing the right reasons for choosing a particular method and for consultant selection; risks should be carefully evaluated, managed and passed to the party best able to absorb them.
- Full cooperation between the project team members and with the client; partnering dealt with later in the chapter, can help in this regard.
- Decision making should be open and decisive and communicated to all concerned with the project.
- Checking of designs and supporting documents should be undertaken to focus everyone's mind on getting the job right first time.
- Health and safety considerations must become part of the decision making processes and of each project member's core skills.
- Problems should be avoided not solved.
- Clients must be able to understand what they are buying, and the relationship between performance and price, if their expectations are to be matched by results.

## REVIEW OF GOVERNMENT PROCUREMENT PROCEDURES

A Treasury white paper (1995) recommended that procurement of government construction projects should focus on quality and long term value for money instead of lowest prices in the short term. The report revealed that on 803 government construction schemes in 1993–94, more than one quarter finished over budget and nearly one fifth were late, while on civil engineering schemes nearly two thirds finished late.

Government departments were advised to base decisions on whole life costs (LCC) rather than on the initial tenders, which would make it easier to operate quality assessment at the tender stage. The selection and evaluation of tenderers should take into account the experience, financial viability and track record of contractors and consultants. It was also considered that the appointment of well trained project sponsors was vital to the success of construction projects.

A National Audit Office (NAO) report in 1995, based on two large building projects for the Benefits Agency and DoH headquarters in Leeds and the British Library in London, together estimated to cost £600m, revealed errors in the way they were procured. The British Library due for completion in late 1996, was in the mid 1990s about £380m over budget. The report concluded that government construction procurement should be improved in the following fundamental ways:

- provide enough time for design development when going for a fast track strategy, to ensure that the final design meets the project's specification
- ensure that the size and structure of occupying organisations are known
- set priorities in terms of time, cost and quality for building projects
- give the project team clearly defined roles and responsibilities
- promote open discussion between parties to avoid confrontation
- draw up specifications that include performance targets
- use value engineering at the concept design stage.

In 1995 the NAO also started a third major investigation into a government building project, this time the 35 000 m$^2$ Inland Revenue headquarters building in Nottingham, designed by Michael Hopkins & Partners. There is no doubt that if the NAO recommendations are implemented on future government building projects, then they will benefit from more efficient and economical buildings.

The NAO report was followed in late 1995 by the Levene report issued by the Cabinet Efficiency Office, based on an investigation of 20 major government construction projects, where costs had increased by £500m (24%) and many were over time. It reported that the government, which spends £6bn per year on building projects, was far from blameless and that the construction industry had a long way to go before it matched the best practice in other sectors. The findings, which were fully accepted by the prime minister, called for all government departments, agencies and trusts to undertake a thorough review of their construction procurement policies by April 1996.

Departments are required to use the report's recommendations as a 'template' to change their relationship with industry by working with its best and most cooperative practitioners, but making no compromises with the incompetent or adversarial. This sobering and far reaching report aims to rectify the lack of accountability, communications failures and overoptimism on budgets that have plagued government construction projects in the past. It set out the basis of a new deal between government and industry, whereby government should learn to be a better client and the construction industry should be less adversarial and more customer focused. The report's recommendations include the following:

- shorter tender lists – three firms for design and build schemes, six for other projects, with the power to reimburse some tender costs
- establish by December 1995 the Government's 'investment decision maker', 'project owner' and 'project sponsor' on each contract over £1m and give them clear, short lines of communication and responsibility
- set up a Construction Procurement Council inside government to coordinate policy
- ensure all contracts detail how variations shall be dealt with
- settle disagreements soon after construction and agree final accounts within weeks rather than months of project completion, except in situations of 'genuine dispute'
- departments are to put in place risk management initiatives to help staff understand the risk involved in capital works projects.

The report also commented that the main reasons for cost increases were over-optimistic estimating, inadequate briefing, incomplete design or design errors, ambiguous risk allocation and inadequate management control. Contingency funds were often misunderstood and poorly used, being seen as pools in which to dip, rather than sums calculated to cover the potential financial impact of retained risks as estimated in risk analysis.

There is much of interest and value to quantity surveying students in each of these government sponsored reports, which deserve careful study.

## MULTI-DISCIPLINARY PRACTICES AND MERGERS

### Multi-disciplinary Practices

Multi-disciplinary practices (MDPs) provide clients with a range of different professional services from one source. They can take different forms including partnerships of individuals each offering different skills, professional service groups with each discipline operating as a separate subsidiary company, or a federation of independent professional firms. Such practices can also include professionals outside construction related activities, such as accountants and lawyers, which are conditional upon harmonisation of codes of conduct and professional indemnity insurance. As long ago as 1983 an RICS QS Division report considered the further move towards multi-disciplinary teams offering a design and management service as a continuing challenge to the quantity surveyor's expertise.

An example of a very large and successful multi-disciplinary practice is

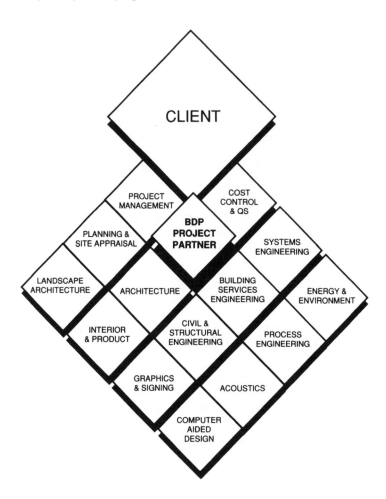

**Figure 19.2**   *Services offered by Building Design Partnership*

Building Design Partnership (BDP) whose staff embraces all the construction related professions as illustrated in figure 19.2 and listed in table 19.2. They offer a full service of professional design, engineering, cost control, planning and management skills, backed by extensive CAD technology, from a single source. The quantity surveyors employed by BDP are virtually assured of a constant flow of work from the commissions obtained by the Partnership.

## Mergers of Quantity Surveying Practices

The period 1994–96 saw a number of mergers of large quantity surveying practices with a variety of other organisations which largely came as a surprise to most of the profession. It results in the breaking down of the long established professional barriers and opens up a new era. Some have questioned how quantity surveyors can remain truly independent financial advisers under these arrangements. The main characteristics of the mergers will be examined, although it will be appreciated that others could follow.

The first changes to quantity surveying practices came with the move from partnerships to limited companies as described in chapter 15. It was claimed by the firms involved that the changes were made to reflect modern business practices, with partnerships being considered more cumbersome. It was also considered that the change in status would make it easier for the firm to expand and acquire other businesses.

In 1994 Beard Dove, a quantity surveying and project management practice, was acquired by the Capita Group making the new organisation the third largest multi-disciplinary consultant in the UK. The Capita Group was formed in 1987 as a management buyout of the information technology division of the Chartered Institute of Public Finance and Accountancy (CIPFA). It has the three divisions of outsourced services, advisory services and property services and has worked with over 1000 clients, and wishes to go further into construction. It previously acquired an architect, building services consultant, and a facilities manager.

Also in 1994 Bucknall Austin, management and cost consultant, bought property consultant Frost Bevan to offer a wider range of services. In another surprising development, the quantity surveying practice of Currie and Brown formed a management and consulting company with office equipment giant Rank Xerox, to offer a range of services that has rarely been offered before by a quantity surveyor, including an expansion of its facilities management operation. This was followed by property consultant Chesterton acquiring the quantity surveying practice of Cyril Sweett, the unusual step of a general practice surveyor buying a quantity surveying practice. Cyril Sweett believed that there was a need for a 'super surveying' practice covering all disciplines. Chesterton's new division of facilities management and construction services, which encompassed the 190 strong quantity surveying practice, aims to take a major share of Britain's estimated £12bn facilities management market. This indicates how quantity surveying skills may be moving in the future.

In 1996 the engineering consultancy firm of W. S. Atkins acquired the quantity surveying practice of Faithful and Gould, to create the largest facilities management consultancy in the UK with a combined staff of around 5000, including engineers, architects, quantity surveyors and property managers. Faithful and Gould, founded in 1947, was the largest firm of quantity surveyors in the UK with a staff of 850, which in 1995 became a limited company and had a 5 year plan to double its size through acquisitions with increased support from banks, and to increase its profitability through economies of scale. Forty per cent of the firm's work was in facilities management with the remainder in quantity surveying and

Table 19.2  Largest quantity surveying and multi-disciplinary practices

| Organisation | Total chartered staff (UK) 1995 | Total staff worldwide 1995 | Analysis of chartered staff (UK) | | | | | Non-chartered staff | Admin and finance staff | Offices | |
| --- | --- | --- | --- | --- | --- | --- | --- | --- | --- | --- | --- |
| | | | Engineers | Architects | Surveyors | Project Managers* | Other | | | UK | Outside UK |
| W. S. Atkins | 1927 | 4840 | 1506 | 149 | 272 | 85 | | 513 | 645 | 32 | 19 |
| Building Design Partnership | 377 | 780 | 109 | 222 | 31 | 6 | 9 | 201 | 149 | 5 | 3 |
| Capita Property Services | 299 | 508 | 121 | 26 | 122 | 30 | | 96 | 74 | 15 | 2 |
| E. C. Harris | 282 | 719 | | | 233 | 29 | 20 | 97 | 113 | 17 | 22 |
| Faithful & Gould | 268 | 861 | 1 | 1 | 266 | | | 456 | 60 | 10 | 8 |
| Gleeds | 243 | 604 | 21. | | 219 | 3 | | 201 | 98 | 15 | 9 |
| Davis Langdon & Everest | 242 | 1515 | | | 242 | 35 | 6 | 178 | 119 | 19 | 31 |
| Gardiner & Theobald | 213 | 365 | 12 | | 161 | 40 | | 41 | 71 | 11 | 16 |
| Bucknall Group | 208 | 450 | 25 | | 183 | 40 | 25 | 96 | 91 | 18 | 7 |
| Currie & Brown | 173 | 536 | | | 173 | 34 | | 183 | 82 | 17 | 17 |
| Turner & Townsend | 161 | 412 | 1 | 1 | 142 | 17 | | 140 | 64 | 16 | 10 |
| Franklin & Andrews | 135 | 558 | | | 125 | 10 | | 22 | 51 | 12 | 40 |
| James R. Knowles | 124 | 280 | 19 | 2 | 100 | 5 | | 130 | 109 | 22 | 6 |
| The MDA Group | 115 | 672 | 6 | 2 | 107 | 23 | | 61 | 57 | 14 | 39 |
| WT Partnership | 112 | 822 | 3 | 1 | 91 | 17 | | 65 | 26 | 12 | 45 |
| Northcroft | 96 | 330 | 1 | | 87 | 8 | | 32 | 38 | 8 | 22 |
| Ridge & Partners | 72 | 132 | 5 | 2 | 54 | 9 | | 21 | 26 | 3 | 0 |
| Widnell | 56 | 515 | 2 | | 49 | 5 | | 37 | 30 | 13 | 52 |
| Walfords | 53 | 126 | | | 51 | 2 | | 31 | 28 | 5 | 4 |
| Crosher & James | 46 | 104 | | | 46 | | | 31 | 21 | 13 | 2 |
| Henry Riley & Son | 40 | 87 | | | 36 | 4 | | 32 | 15 | 7 | 0 |
| Edmond Shipway | 37 | 84 | | | 37 | 19 | | 30 | 17 | 7 | 0 |

* May be included in other categories.

Source: Building, 29 September 1995.

project management. Atkins dominated the public sector having purchased large sections of the government's privatised Property Services Agency (PSA), whereas Faithful and Gould's clients were mainly in the private sector, thus complementing each other. Both companies considered their merger to be a positive response to their respective clients' needs, and that providing extra services and specialist skills would bring added value to their businesses. For their part, Faithful and Gould saw the opportunity to offer services such as management consultancy and risk analysis, both being activities at the forefront of RICS QS Division thinking in 1996.

These mergers could be the forerunner of deeper and wider changes with a profound effect on the professions by year 2000 and beyond. Leading professionals in years past were primarily experts in their respective disciplines with management skills being a secondary attribute, but this is now changing with success frequently depending on the ability to manage. Increasingly, clients are seeking single point responsibility (one stop shopping) for overall project delivery from their professional advisers. To meet this need, quantity surveyors and other professionals may join forces to broaden the scope of their services.

Table 19.2 gives details of the larger quantity surveying firms in 1995, showing staffing figures, range of disciplines employed and number of offices. Building Design Partnership and W. S. Atkins have been included for purposes of comparison.

## QUANTITY SURVEYING PACKAGES

Quantity surveyors have the expertise to diversify and provide a package of financial services which could be of great value to building owners and financial institutions. In addition, they should be able to provide a building package of consultants and specialists to obtain the best combination of design, cost and construction that satisfactorily meets the client's needs. This approach could have significant advantages over contractors' design and build systems where no loyalty is owed to the client and could form a suitable approach throughout Europe.

Blakeley (1986) has rightly emphasised that quantity surveyors should ideally provide overall packages to suit existing portfolios as well as new building works. Such packages could include life cycle costing (LCC), advice on maintenance expenditure programming, capital allowances, specialist funding techniques and insurance services. One example could be the management of capital expenditure programmes for large organisations and a complete review of their monitoring systems to improve accuracy, speed and management of expenditure.

Very large sums of money are spent on the refurbishment or development of sophisticated city buildings, in particular those required by the financial services industry. These complex buildings with their extensive dealing floors and intricate services installations encapsulate substantial capital tax allowable plant and associated work. Hence very large tax savings can be generated in these developments and the quantity surveyor, with his specialist knowledge of building costs, statute and precedent case law, is well fitted to maximise these allowances to the benefit of the client. He can also advise non-taxpaying institutions who are unable to take advantage of these allowances.

The quantity surveyor can also advise funding institutions on the feasibility of development proposals, overall budget limits and realisable economies, and ensure that the developer's requirements for stage payments are being met at the correct level and at the appropriate time. He is also well placed to advise on replacement costs and, in addition, through liaison or association with insurance brokers, can provide an attractive package of consultancy and valuation (Blakeley, 1986).

Other quantity surveying packages could usefully encompass maintenance management/life cycle costing, value management/engineering, and project management services, as described in chapters 8, 9 and 12.

## PARTNERING IN CONSTRUCTION

### Nature of the Concept

The United States Construction Institute defined partnering as 'a long term commitment between two or more organisations for the purpose of achieving business objectives by maximising the effectiveness of each participant's resources'.

While the Reading Construction Forum (1995) defined it as 'a management approach used by two or more organisations to achieve specific business objectives by maximising the effectiveness of each participant's resources. The approach is based on mutual objectives, an agreed method of problem resolution and an active search for continuous measurable improvements'. The Forum also stated that partnering can be based on a single project (project partnering) but greater benefits are available when it is based on a long term commitment (strategic partnering).

This new concept was introduced largely to eliminate the adversarial conditions which often prevail on other forms of contractual arrangement and to achieve an improved service for the client. Partnering requires changing traditional relationships to a shared culture without regard to organisational boundaries. The relationship is based on trust, dedication and common goals, and an understanding of each other's individual expectations and values, thereby reducing the necessity for protracted and expensive dispute resolution procedures.

Thus suggestions are made to keep costs within budget, minimise time overruns, and the best way to approach the project is openly discussed.

In this way disputes can be avoided and, in return for the satisfactory completion of the project, the client ensures that profit margins are maintained at a level acceptable to all parties to the agreement. Quantity surveyors can play an important role in partnering agreements by helping to negotiate appropriate procurement methods and contract prices, and by finding ways of ensuring improved programme assurance. They may also help in advising clients on partner selection (Davis, 1995).

**Benefits of Partnering**

There are many benefits to be derived from the use of partnering agreements and Meara (1995) has identified the following:

* maximise the effectiveness of each participant's resources and expertise
* reach mutually agreed and measurable targets for improving the services and works
* through cooperation, exchange of ideas and adoption of milestone benchmark techniques, all participants strive for the common objectives of achieving the best value for money, highest quality relative to cost, and satisfaction of the client.

Further important benefits have also been recognised by Mills (1995) and these are now listed:

* motivate innovation
* improve cooperation between design and implementation teams to produce a project better suited to client needs
* increase willingness to solve design and site problems, cutting delays and inefficiencies
* encourage sharing of identified savings in time and cost
* reduce potential claims
* encourage good service and improve subcontract quality and timeliness
* speed decision making
* establish a relationship between parties which may lead to future work
* helpful impact on health and safety issues.

The Reading Construction Forum (1995) considered that significant cost savings were achievable and estimated savings of 2–10 per cent on project partnering schemes and as high as 30 per cent over time with strategic partnering, as against a likely cost of less than one per cent in undertaking partnering. It may be that the upper cost savings will prove overoptimistic. In addition to reducing costs it was believed that partnering can also improve service quality, deliver better designs, make construction safer, meet earlier completion deadlines and provide everyone with larger profits.

## Selection of Consultants and Contractors

Davis (1995) has described how the requirements of individual organisations will differ widely when looking for a partner under a partnering agreement, but the following basic factors deserve consideration:

*Financial issues*: assessment of the financial stability and outlook of the potential partner.

*Company culture*: management and personnel structures must be compatible.

*Human resources*: both parties must be able to provide personnel in the numbers and of the calibre required for the project in hand.

*IT*: adequate and compatible IT systems are in place.

*Other factors*: these may include health and safety record, track record, and current client base.

It is likely that a client will invite consultants and contractors to submit tenders for the partnering activities on the basis that the selected firms will be appointed under separate contracts for the individual projects that make up the programme. The tenderers are required to submit details of how they would resource the partnering activities, the personnel they would deploy and hourly rates. In addition, they may be required to provide, for the purposes of negotiating the individual projects, details of allowances for preliminary items, overheads and profit and their procedures for inviting subcontractors' tenders (Meara, 1995).

In 1995 the British Airports Authority (BAA) concluded the first stage of a 12-month process to select consultants for a partnering system. Seven quantity surveyors, from the initial selection list of 13, were invited to enter into a five year framework agreement with BAA, guaranteeing them part of the work from the £450m per year capital spending programme. Under the agreement, the quantity surveyor's first option is to compete for BAA contracts under a simplified tender procedure. Pre-agreed 'benchmarks' on fees were expected to reduce, but not eliminate, competitive fee bidding on individual projects. An open book system meant that the shortlisted seven had to justify their fees to BAA, which expected strong commitment from its future partners. The quantity surveyors responded to an advertisement in the EU *Official Journal*. The seven were chosen after completing three questionnaires and undergoing a series of interviews and office visits.

**Partnering Agreements**

Partnering agreements will contain undertakings requiring all parties to:

- cooperate to their mutual advantage
- work to achieve best quality and value
- endeavour to achieve continuous improvement
- agree to a non-adversarial form of contract for the individual projects (Meara, 1995).

While there is no universal partnering agreement, and every project needs an agreement tailored to its individual requirements, the Latham Report (1994) indicated the type of clauses that may be appropriate and included the following.

- a specific duty for all parties to deal fairly with each other and with their subcontractors, specialists and suppliers in an atmosphere of mutual trust
- firm duties of teamwork, with shared financial motivation to pursue those objectives
- an interrelated package of documents that clearly defines the roles and duties of all involved
- a choice of allocation of risks to be decided appropriate to each project, but then allocated to the party best able to manage, estimate and carry it
- the taking of all possible steps to avoid conflict on site, and providing for speedy dispute resolution if any conflict arises
- incentives for exceptional performance (Mills, 1995).

**Working Arrangements**

The Reading Construction Forum (1995) emphasised that to make partnering work, those involved needed to agree a set of mutual objectives, which are usually established in the first workshop. Other workshops often follow to innovate and test new ways of working together and to generate action plans.

In partnering, instead of putting a project out to tender, the client chooses a contractor and the two agree price, programme and contract terms between them. By fostering a spirit of cooperation the likelihood of problems occurring is much reduced. Important aims are to negotiate a reasonable price, ensure greater programme assurance and a smooth running project free of claims.

The contractor is asked by the client to come up with ideas to improve the cost, quality, programme and certainty of final account. All aspects of the project are discussed between the parties and together they will come up with solutions to any problems that arise. An essential difference of partnering as opposed to a standard negotiated contract is the open book approach, whereby there is no prescribed set of rules. For example, when agreeing a price, all matters affecting the price are examined. Ways of improving the contractor's cashflow and decreasing the price of the project are thoroughly examined. There is also a great difference in attitude accompanied by a feeling of goodwill, as all are asked to contribute (Davis, 1994b).

### Partnering Case Studies

In the UK in 1995, private sector clients had led the way in promoting the partnering approach for construction. Leaders in this field included BAA, Marks & Spencer, Rover (vehicle manufacturer) and Norwich Union. By virtue of the Treasury White Paper (1995), referred to earlier in the chapter, public sector clients are likely to follow this lead.

Extensive case studies are contained in the publications by Reading Construction Forum (1995) and Hellard (1995), to which the reader is referred for more detailed information.

## MANAGEMENT CONSULTANCY

The RICS publication *QS 2000* (1991) identified management skills as applied to construction projects as an area for the development of quantity surveying services. The emphasis of the quantity surveyor's advice could in the future change from that of supplying the information for others to act upon to the control of the information and giving direct advice to the client.

It seems evident that quantity surveyors need to involve themselves closely with the business community, understand their core objectives, identify their needs and understand how their built assets relate to their overall business strategies. Quantity surveyors will be better able to promote their services if they have a good understanding of the role that property and construction plays within the business organisation. This could lead to quantity surveyors becoming major construction consultants giving advice, guidance and management services within the client's built resource and property asset strategy.

The RICS (1995f) found that the market for management consultancy advice increased from £168m in 1985 to over £1bn in 1994. Furthermore, a diverse range of management advice, such as human resource management and IT management, was being provided by management consultants, accountants, bankers and other professional disciplines. Construction and property management advice or property resource management (PRM) consultancy, is a newly emerging market, which the surveying profession must address if it wishes to maintain its position as leading advisers to the construction and property industry.

The RICS (1995f) believed that clients want PRM consultancy at both strategic and business level and at operational and property level. Increasing opportunity exists for surveyors to operate at the second level, as more property occupiers and end users of property are seeking ways of improving their management practice in relation to their property resources. To meet this challenge surveyors need to acquire new management skills and a broader appreciation of the current and future needs of clients in the context of their core businesses. It is necessary to remove many clients' perceptions of surveyors as providing a narrow technical service, with limited business management skills and little experience of the consultancy process.

The RICS (1995f) considered that the skills required to deliver effective management consultancy advice at the strategic and business level embrace breadth of knowledge and experience in business and management, as well as the ability to communicate effectively with client organisations and team members from diverse backgrounds. Whereas the skills required for PRM consultancy at the operational and property level were more likely to require problem solving and process management skills, together with an understanding of the construction and property markets and a knowledge of the appropriate technical skills.

The RICS investigations found that undergraduate degree programmes from recognised accredited institutions provided adequate business management education but lacked attention to problem solving in a business and management context. Postgraduate education programmes, particularly the MBA courses, together with mid-career training assisted in extending surveyors' business and management skills, but there was a lack of evidence of the support given and importance attached to these acquisitions by surveying organisations.

In 1994, Davis found that some private quantity surveyors were turning away from traditional construction work, which had been depleted by the recession, and turning towards the more lucrative business of business management consultancy to boost income and broaden the client base. For example, one practice pursuing consultancy business was involved in tender evaluation work for clients commissioning consultants for major projects on a Europe wide basis. While another firm was offering such services as revising and rationalising in-house purchasing contracts for local authorities, giving advice on information technology, and even advising urban development corporations on tourism and leisure strategies. Yet another practice adopted a different approach to management consultancy by combining it with information technology (IT) to offer 're-engineering' advice to clients.

It was, however, considered vital by one quantity surveyor to develop areas in which the practice had expertise. It was also believed that success in these activities had more to do with individual attitudes and approaches to problem solving than with the skills of specific disciplines.

Readers wishing to explore this topic more fully are referred to the construction management books by Langford *et al.* (1995) and Harris and McCaffer (1995).

## RISK ANALYSIS AND MANAGEMENT

### General Background

*Risk* can be defined as the possibility of a forecast of an event not being accurately fulfilled by the actual occurrence, and often having cost or revenue implications. *Uncertainty,* on the other hand, relates to the possibility of the occurrence of an event which cannot be budgeted for and which could in extreme cases defeat the project's aims or for which there is no reliable basis for a forecast (Hutchinson, 1993). In practice, an item can on occasions change from one category to the other and they can often be interrelated. For convenience, the construction industry uses the term risk to encompass both risk and uncertainty.

The construction industry is subject to greater risk and uncertainty than probably any other industry. Furthermore, risk can manifest itself in many ways, varying over time and across activities. Essentially, it stems from uncertainty which, in turn, is caused by a lack of information. For example, the quantity surveyor often has to determine the budget price for the foundations to a new building without knowing the ground conditions or the loading of the building. Hence there is an element of risk in forecasting this price, but past cost data and experience will help the quantity surveyor to assess that it is achievable within the budget price with some degree of certainty (Flanagan and Stevens, 1990).

Typical risks on construction projects include:

- failure to complete within the stipulated design and construction times
- unforeseen adverse weather conditions
- exceptionally severe weather
- *force majeure*
- a claim from the contractor for loss and expense caused by the late supply of design details or other relevant information
- failure to complete the project within the client's budget.

Construction projects and contracts are commercial ventures. Both the client promoting a project and the contractor employed by him are investing money and taking financial risks in order to achieve some desired benefit or return. The risks are likely to be extensive, interactive, sometimes cumulative and to extend over the entire life cycle of the project. Identification, evaluation and management of these risks are performed through risk analysis and risk management. *Risk analysis* is the methodical examination of all estimates and uncertainties associated with a project, while *risk management* is a positive process and team effort which aids understanding of the nature and likely outcome of the investment (Thompson and Norris, 1993).

## Operation of Risk Analysis

Yates (1986) held the view that risk analysis is not so much a technique as a way of looking at a problem. For example, it involves identifying the key factors that might affect an estimate and then assessing the probability and extent of the effect. Thus in construction cost estimates the information assembled is presented in probabilistic rather than deterministic terms, such as by showing both the most likely value and a range of other values with the chances of each separate value being achieved.

The way in which risk analysis is applied to construction cost estimates will vary but the following characteristics are universal:

- *Risk factors*: the key factors affecting the estimate are listed and could include the following: adverse weather, ground conditions, industrial action, effects of inflation, variations in interest rates and availability of resources
- *Limits of risk analysis*: these are established but should exclude factors that have say a less than one per cent chance of occurring
- *Forum for risk analysis*: a formal structure is developed and a common approach is to establish a workshop comprising key members of the team, and when the limits of the forum have been decided, risk analysis can be undertaken (Yates, 1986).

Risk analysis involves developing an attitude of mind to think in terms of the chances of events occurring. Statistics and calculations are used to assist in the assessment of chance but, alternatively, judgements can be made based on the experience of the team. When the assessment has been made for each component of an estimate, it is necessary to combine the results in a way that reflects the overall probability, but disregarding results whose probability falls outside the agreed limits, such as 1 in 100. The task is eased by using a simulation approach, as described later, aided by a computer and using the Monte Carlo method. The results are expressed as a percentage chance of the figures being exceeded, with a graph showing the probability attached to the estimate. The client can then gauge the risk associated with the estimate and a zone of acceptability produced.

Thus risk analysis is concerned with assessing the effect of uncertainties and these occur in every construction project. Their significance will, however, vary as between projects. For example, in a public sector scheme the uncertainties in time and cost factors are likely to have their main impact on occupational and functional aspects, whereas in private sector schemes they will have a critical influence on the profitability of the scheme. Construction projects vary also in their distribution of risks and this influences the client's demand for risk analysis. Thus conventional fixed price contracts place a higher degree of risk for costs on the contractor than on the client, in contrast to management contracts. Life cycle cost estimates are more variable than capital cost estimates (Yates, 1986).

### Risk Management Process

The main purpose of the tasks performed by the quantity surveyor must be to enable business, whether on behalf of the client or the contractor, to take the right risks. A simplified risk management process is shown in figure 19.3, illustrating the three main stages and the factors to be considered. Pasquire (1995) considered that a structured approach to risk management was important if the intention was to reduce loss arising from unforeseen circumstances. She also believed that a fully implemented risk management strategy enabled optimisation, allowing organisations to face risk deliberately, acquiring maximum benefit for limited and controlled exposure.

Perry and Hayes (1986) advocated a systematic approach to risk management by breaking the task down into the following three stages, as illustrated in figure 19.3:

(1) risk sources are identified
(2) their effects are assessed or analysed
(3) responses and policies are developed which lead to risk reduction and control.

However, it must be recognised that risk management does not remove all risk from the project, rather it aims to ensure that risks are managed most efficiently. Certain risks will remain to be carried by the client, sometimes described as residual risk, and these are allowed for in the client's estimate of time and cost.

*Risk identification*: The major decisions are made early in the project life, at appraisal and sanction stages, thus it is important to identify all potential risks and uncertainties in these early stages.

*Risk analysis*: The purpose of risk analysis is to quantify the effects on the project of the identified major risks. The impacts of each risk are then accumulated to provide a minimum/maximum range of estimated values. Greater sophistication, realism and confidence can be achieved by using more complex techniques incorporating probabilities and interdependence of risks. Finally, judgements are needed on the impact of each risk and, in some cases, of the probability of occurrence of each risk and of the various possible outcomes (Perry and Hayes, 1986).

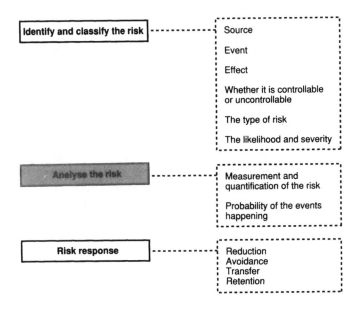

**Figure 19.3** *The risk management process*

*Source*: Flanagan and Stevens, 1990.

*Responses to risk*: The responses to risk can be considered in terms of avoidance or reduction, and transfer or retention. In extreme cases, risks may have such serious consequences as to justify a reappraisal of the project or possibly its replacement by an alternative scheme. However, it is more likely that risk identification and analysis will indicate the need for redesign, more detailed design, further site investigation, different packaging of the work content, alternative contract strategies, or different methods of construction. In the case of transfer of risk, the consequences of any risks that occur are shared with or carried by a party other than the client. Furthermore, risks which are retained by either party may be controllable or uncontrollable (Perry and Hayes, 1986).

## Simulation Techniques

Flanagan and Stevens (1990) used a Monte Carlo simulation to develop a risk analysis technique which studies the relationship between an estimated price and the probability of the tender price deviating from this amount. Since the price prediction is the sum of many parts, an objective evaluation of its precision can only be achieved by using statistical techniques. Monte Carlo analysis is a form of stochastic simulation. Stochastic or probabilistic means that the technique is concerned with controlling factors that cannot be estimated with certainty. It is called Monte Carlo because it uses random numbers to select outcomes.

Monte Carlo simulation generates hypothetical mean unit price rates for each elemental category in the cost plan of the proposed building. The hypothetical rates are taken from probability distributions with the same statistical characteristics (probability density functions) as those in the original sample data from which the mean unit price rates were estimated. A total price forecast for the proposed building is then built up from the hypothetical rates, and by repeating the exercise a large number of times, the most likely total price will be obtained. A series of simulations gives price predictions for the project which are plotted as a cumulative frequency curve and as a histogram (Flanagan and Stevens, 1990).

A sophisticated risk modelling technique has been developed by Berny and Townsend (1993), who developed a methodology using a software package called VISIER for analysing risks, taking account of time and dependence with the risk event modelled as one ·of a number of types of probability distribution.

## PRACTICAL APPLICATIONS OF RISK MANAGEMENT

### Broad Approach

E. C. Harris Risk Management aims to quantify latent as well as actual risks; assess the sensitivity and vulnerability of major assets; and devise specific strategies to reduce and contain risks and their associated costs.

E. C. Harris believe that alongside the development of a computerised database must be the development of expertise and techniques to interpret the information. For example, risk analysis identifies factors which affect, or could affect, an estimate or budget and quantifies the probability and extent of their effect. Sensitivity analysis is a further refinement of this process and monitors the relative effect of changes in any use of these key factors. Hence both are damage limitation exercises and can be carried out at various levels on construction schemes.

At the core of risk and sensitivity analysis is the identification of the key variables in the estimate or budget, and an estimate of their probable movement. When used as a tool for estimating, risk analysis identifies the main items in the building and then assesses the possible range of costs of each of the key variables. Not all items will carry the same degree of risk, but if each element is considered in turn and its risk assessed in financial terms, the original estimate can be compared with the total maximum and minimum likely costs. The main advantage of risk and sensitivity analysis is in providing guidance on those variables that can have the greatest effect on performance, thus allowing the client and design team to concentrate their efforts on ensuring that they are the most favourable for the project.

E. C. Harris use these techniques to project different ways of optimising profit levels and/or reducing risk. This is valuable where a scheme is particularly sensitive to variables. Eliminating some might make the difference between success and failure as, for example, preletting, forward funding or forward selling could remove variables from one side of the equation. While phased completion, fixed price building tenders and value engineering might remove variables on the other side.

## Project Management

The principal aims of risk management in project management are:

(1) to ensure that only projects which are genuinely worthwhile are sanctioned
(2) to avoid excessive overruns.

Risk management can be one of the most creative tasks of project management. This is achieved through generating realism and thereby increasing commitment to sound control, and through the encouragement of problem solving, which can open the way to innovative solutions to project implementation.

A major advantage of the early identification of risk and uncertainty is that it focuses project management attention on policies and strategies for the control and allocation of risk, such as the choice of an appropriate contract strategy. In particular, it highlights those areas where further design, development work, investigation or clarification are most needed (Perry and Hayes, 1986).

## Construction Tendering

Raftery (1985) has described how when risk analysis is applied to tendering, it focuses thought on the possible pattern of costs in a systematic and rigorous way. Thus estimators can make full use of available information and of their skill and experience. The correlation technique enables estimators to isolate the various sources of uncertainty into reasonably independent categories, as an aid to clear thinking. The form of output helps management by communicating the estimator's confidence and considered judgement on the estimate in a concise way.

## General Conclusions

- The RICS Quantity Surveyors Division in 1995 highlighted the increasing need for quantity surveyors to acquire skills in risk analysis and management.
- Often risk is either ignored or dealt with in an arbitrary way – simply adding a 10 per cent contingency sum on to the estimated cost of a project is unsatisfactory.

- The greatest uncertainty is in the earliest stages of a project, which is when decisions of greatest impact are being made. Risk must be assessed and allowed for at this stage.
- Risks change during the execution of most projects and hence risk management should also be a continuous process throughout the life of a project.
- The quantitative assessment of risk requires analysis of the likely extent and interaction of variable factors. The need for judgement should not be used as a reason for failing to adequately consider project or contract risk.
- Delay in completion can be the greatest cause of extra cost and loss of financial return and other benefits from a project. The first estimate of costs and benefits should be based on a realistic programme for a project, so that the potential effects of delays can be predicted realistically.
- The overriding conclusion is that all parties involved in construction projects and contracts would benefit greatly from a reduction in uncertainty prior to financial commitment. In this context, money spent early buys more than money spent late. Willingness to invest in anticipating risk is a test of a client's wish for a successful project (Thompson and Norris, 1993).

## NATIONAL LOTTERY, GOVERNMENT AND EC FUNDING INITIATIVES

### National Lottery

The National Lottery offers the prospect of a stream of commissions for those construction consultants, including quantity surveyors, who become involved with projects which have a realistic chance of securing funding. In 1996 the lottery was expected to increase the size of the UK construction industry by 3 per cent, with spending of about £1.9bn on capital projects over eight years. Potential projects ranged from cycleways and scout halls to a new art gallery beside the Thames. The larger sports applications relate to new indoor swimming pool centres and multi-sports projects.

However there is also a downside in that firms commissioned by unsuccessful National Lottery applicants could find themselves wasting resources on non-fee paying speculative work. To avoid this situation, consultants are advised to market their skills actively, examine carefully the lottery distributors' selection criteria and make sure appointments are legally watertight.

Another difficulty is that there is no database of lottery applicants, which makes it harder to find out about schemes in a given area, and it is suggested that firms contact the regional offices of distributors and/or target local sports and art groups. Consultants should ideally acquire an expert knowledge of the selection criteria, instead of relying on clients' second hand instructions. It is worth noting that at least one quantity surveyor guides promoters of small sports and leisure schemes worth up to £400 000 through the complicated project assessment procedure.

Furthermore, consultants bear an extra burden of risk and responsibility on lottery projects because they often have complicated partnership structures and inexperienced clients. They need to be commercially astute and able to separate out viable projects from weak ones and insist on fees in advance for the latter. Unfortunately, the announcement of an award does not necessarily end the uncertainty as the EU tendering requirements apply to consultants and contractors on National Lottery work, where it exceeds prescribed price baselines. Hence quantity surveyors need to approach this work with care and ensure that a competitive selection process precedes their initial appointment. This need not incorporate full fee bidding, but must include some costing element and be fair to a third party (Froomberg, 1996).

Another major problem encountered in 1996 was the difficulty experienced by some promoters in raising 25 to 50 per cent of the capital cost from non-lottery sources, as there was increasing concern that donors do not have sufficient funds to contribute to all the schemes that have lottery funding, which could result in delays and downsizing. During the period 1996–99 it seemed likely that £750m per year could be generated for construction projects, but this would need to be matched by £400m to £600m per year in donations. For example in 1995 the Royal Opera House in London's Covent Garden was awarded a £55m grant from the lottery towards the redevelopment cost of £213m, but six months later the promoters were still £65m short of matching money from donations.

### UK Government Grants

Government grants are available for certain types of development/redevelopment and the quantity surveyor must be fully cognisant of them and how they operate. City Challenge schemes were introduced in 1991 whereby urban programme authorities competed for funds to undertake a wide range of much needed schemes in partnership with private enterprises and community involvement, over a five year period. However, it was a mixed success as it was subject to onerous conditions and procedures, and no new schemes were accepted after March 1995.

*English Partnerships* was established in 1995 and adopted a proactive and strategic approach by participating actively in joint ventures and offering considerable flexibility with loans, guarantees and grants (deficit funding) from the EP Investment Fund. The Agency is concentrating on coalfield closures, City Challenge, inner city areas, other assisted areas and rural areas of severe economic need. The Agency (EP) can however respond to urgent needs outside its priority areas and to structural shifts in local economies. The Agency will judge bids against standard criteria, namely location, strategic context, scale, practicality, value for money and risk sharing. English Partnerships also highlighted three other issues that it will be considering when assessing bids: environmental issues, community involvement and design quality. The Agency will focus on physical regeneration initiatives which show a significant return in terms of jobs created, inward investment attracted and the provision of new housing and employment space. In considering environmental projects, it will concentrate on those which are urgently required for health and safety reasons and form part of a wider strategy of development and reclamation (English Partnerships, 1994).

In 1994 various regeneration programmes were combined in a single £1.4bn budget (increasing in later years) administered by government regional offices and known as the *single regeneration budget* (SRB), and this included EP, Urban Development Corporations (UDCs) and Housing Action Trusts (HATS). The objectives of the SRB were similar to City Challenge, ranging from employment and economic development to improvement of the environment and infrastructure, and maximising public sector services by coordination with other programmes.

**European Assistance**

The European Union (EU) has three structural funds that will, as money for regeneration in Europe gets tighter, assume greater importance. These funds are the European Social Fund, the European Regional Development Fund (ERDF), and part of the European Agriculture Guidance and Guarantee Fund, of which the ERDF is most relevant to the construction industry. The purpose of the funds is to promote development of the EU's poorer regions, and employment/training of disadvantaged groups throughout the Community.

In the five years up to the end of 1993, the EC allocated £4500m to projects designed to assist the most depressed or backward regions. Large areas of the UK were eligible for financial help from these funds, yet our reticence in taking up funding opportunities surprised our European partners. EC grants are given on the basis that it is matched by domestic spending, but the UK government and local authorities were generally unwilling/unable to do this because of the shortage of public money. Furthermore, the private sector was mainly unaware of the potential EC funding largely it is suggested because of lack of publicity.

The four main grant programmes which together make up the structural funds were allocated more than £2250m for the five year period 1994–99. These funds are region specific and almost totally concerned with economic and physical regeneration through development and are allocated to the public sector. A UK government 'competitiveness' white paper published in 1994 announced a Regional Challenge exercise, along City Challenge lines, inviting public/private consortia to bid for £150m to £200m of EC money, representing a relatively small proportion of the available EC funding.

## PRIVATE FINANCE PROJECTS

### General Background

In 1992 the government launched a private finance initiative aimed at involving the private sector more fully in financing and managing projects and services which have traditionally been the responsibility of government. The term 'private finance' is possibly rather misleading as it amounts to more than just the provision of private finance. The government wished to use private sector management and expertise to secure improved public services and better value for the taxpayer. The concept was to be applied not only to infrastructure projects such as railways or roads, but also to the provision of specific services such as patient care or prisons.

There are, broadly, three types of private finance projects:

(1) *Financially freestanding projects* where the private sector contractor undertakes the project on the basis that costs will be recovered by charging the users as with toll roads or bridges.
(2) *Joint ventures* which consist of projects to which both the public and private sectors contribute but where the private sector has overall control. Private sector partners will normally be chosen through competition and the government's contribution should be clearly defined and limited, with the allocation of risk also clearly defined and agreed in advance.
(3) *Services sold to the public sector* usually where a significant part of the cost is capital expenditure, such as the provision of prison places by the private sector designing, constructing, managing and financing new prisons. The public sector purchaser will need to be assured that better value for money will be obtained than by any alternative (Treasury, 1994a).

In the period between 1992 and 1995 progress on implementing the private finance initiative (PFI) was far less than had been expected largely resulting from the lengthy testing procedures adopted in government departments and the complex and costly bidding process. Hence following the 1995 autumn budget, the Treasury (1995) introduced an 'action plan' to simplify procedures and reduce the costs of bidding for projects. The government aimed to approve £14bn of private finance contracts by 1998–99. The Treasury pledged to:

- restrict tender lists to a maximum of four bidders and preferably three
- exclude low value projects where a government department's business case showed the scheme to be unsuitable for private funding
- set targets for government departments by prioritising projects.

The Treasury also published a list of 1500 possible PFI schemes subdivided into two categories according to their relative suitability. The A list contained 150 projects collectively worth £9bn, which the Government wished to implement without delay and included the Norfolk and Norwich hospital, which was already at the bid stage.

The Treasury 'action plan' identified a range of benefits generated by the PFI, mainly through securing better value for money, including:

- consideration and better control of risk from the outset
- better incentives to perform
- close integration of service need with design and construction.

The RICS (1995g) has shown the relevance of the PFI, how it operates, the benefits to be gained from its use, and how to avoid the various pitfalls. It describes how there are two essentials in the PFI, namely value for money, and risk transfer. To ensure value for money the public sector needs to carry out the following tests:

- value the costs and benefits to confirm that the value exceeds the cost
- use techniques for testing value for money as set out in the Treasury Green Book (1991)
- appraise the options
- discount to present value (PV)
- choose the appropriate discount rate; usually 6 per cent, rising to 8 per cent for industrial and economic development proposals where central government is involved
- evaluate intangible costs and benefits such as environmental and social factors, normally evaluated by scoring the outputs
- select the best option which shows the highest net present value (NPV)
- present the results.

This will be followed by an assessment of risk transfer from the public to the private sector. Thus the private sector should expect to bear all of the risk in design, construction (cost and quality), maintenance, operation of the service contract and facilities management, and finance. Other risks of a procedural or systematic type remain with the public sector and trading risks should be shared (RICS, 1995g).

The Treasury (1994b) whilst confirming its preference for competition in the selection of promoters, accepted that a case could be made for alternatives to stimulate innovation, but subject to EC rules. In like manner use of the PFI enables the service required to be specified (the output) rather than the means of supplying it (the input). The PFI route envisages capital being amortised over 15 to 18 years, as compared with the more usual period in excess of 50 years for publicly funded schemes (RICS, 1995g).

The RICS (1995h) investigated the use of the PFI in other countries and compared their approaches to those adopted in the UK. For example, in France, the value of private sector management skills in the delivery of public services has been recognised for a long time and influenced a large section of the economy. In Germany, the government buys a large part of the social housing service in the private sector, rather than the dwellings themselves. While in the United States it proved difficult to select examples of public/private partnerships which had relevance to the UK.

## PFI in Practice

Some of the more important ongoing PFI projects in the UK in early 1996 are now considered to illustrate the range, nature and size of the schemes.

*Roads*   Much higher than anticipated bids were received for the construction and maintenance for 30 years on a design-build-finance-operate basis for the A1/M1 link road bypassing Leeds. This scheme together with three others indicated that it can be more expensive to build roads privately than with public money. Bridges have included the £300m Second Severn Crossing.

*Railways*   PFI railway projects included the urgently needed £2.8bn high speed Channel Tunnel rail link, £1bn worth of improvements to Britain's main west coast rail line (London to Glasgow), £500m lease of new tube trains for the Northern Line, £130m Docklands Light Railway and the Lewisham extension, £300m Heathrow Express from Paddington, and new light rail or tram systems for Croydon and Guildford.

*Hospitals*   Three consortia were shortlisted for the largest PFI project in the health sector, namely to design, build, finance and operate the new 950 bed Edinburgh Royal Infirmary for a concession period of around 15 years, with construction alone expected to cost £150m. There were six other hospital projects worth more than £40m at various bidding stages and a further 12 schemes, worth more than £20m each, advertised in the European Union's *Official Journal*. The NHS private finance database operated by Newchurch showed more than 500 schemes registered with a capital value of £1.5bn, although NHS figures showed £235m in private finance between 1995–96 and 1998–99. Problems with hospital projects ranged from the 49 week initial PFI testing period, to complaints by contractors undertaking costly feasibility studies for schemes which were unlikely

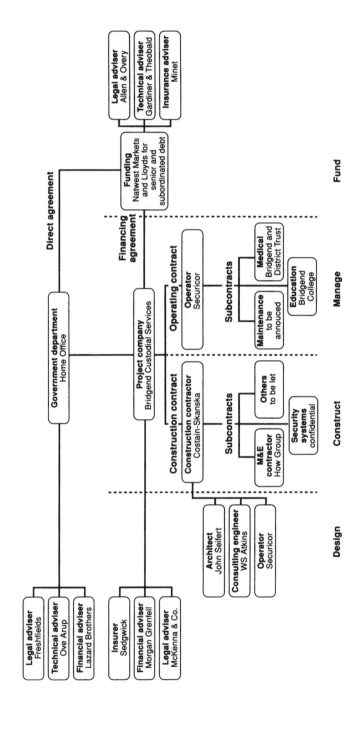

**Figure 19.4** *Anatomy of Bridgend: design–construct–manage–finance contract*

*Source:* Building, 20 October 1995.

to go ahead, and delays by funding banks because they believed that they were being asked to take on too much risk.

*Prisons*    Three new prisons at a cost of £50m each were to be privately designed, constructed, managed and financed on a 25 year operational phase following the two year construction period. It is argued that it is easier to deal with one organisation, the Home Office, and identify the revenue stream, than would be the case with with a health project. The first two projects were at Fazakerly on Merseyside and Bridgend in South Wales. Figure 19.4 shows the complex organisation required for the Bridgend contract.

*Public Offices*    Public office PFI projects included the £150m Post Office new IT system for benefit payments and the £200m redevelopment of the Treasury's Whitehall headquarters.

## Conclusions

In 1995 the obligation to consider private finance applied only to central government but local authorities were being strongly encouraged to do so. There is significant scope for the use of the PFI in such areas as re-placing and renewing schools and outsourcing local authority aspects of care for the elderly. Doubtless market needs will change over time as will also PFI solutions. Russell (1995) rightly identified as good candidates for PFI the hundreds of government office buildings built in the 1960s, 1970s and 1980s often constructed to the lowest initial cost, and now faced with enormous bills on upkeep under obligations flowing from ownership or full repairing and insuring (FRI) leases.

As identified earlier the NHS is likely to place a heavy load on the PFI and will conceivably face difficulty in attracting suitable partners. The oversupply of business opportunities is likely to be compounded by a critical shortage of skilled operators and facilities managers who are pre-pared or able to accept the risk. It is clear that these developments will create work for many surveyors including quantity surveyors.

Many large contractors were already facing difficulties in coping with the increasing supply of PFI projects by 1995. The sheer volume of work is likely to stretch to the limit the bidding and operational resources of the limited number of large operators, and this could result in more mergers of large contracting organisations, as each major contracting organisation can probably only risk stakes in one or two PFI projects at a time. The cost of fees involved in preparing bids has been estimated at £1m or more for a £100m project. Furthermore, most of these fees are incurred in the pre-tender period before any contracts are signed and before contrac-tors have a chance to recoup their investment. Hence the BEC were in 1995 attempting to persuade the Treasury to reimburse contractors for their tendering costs for PFI projects.

A meeting of the environment secretary with the construction industry in November 1995 highlighted the following matters where action was needed:

- need for departments to be more selective of PFI projects
- need for successful projects to be brought to fruition
- concern about tender costs
- training for the PFI
- risk transfer, with particular reference to prisons
- delegation of decision making to departments
- limits on the industry's resources to finance PFI projects.

In an excellent summing up in 1995 Simon Pott, President of the RICS, commented that 'There is now widespread recognition that the only way governments can achieve the renewal of the country's infrastructure, and accommodate the huge range of public services for which they are responsible, is in partnership with the private sector. It is not only motorways and other transport links which are required but new hospitals, schools, prisons and other facilities.

'Although the number of PFI schemes increased markedly in 1995, progress was slower than had been hoped for. One of the reasons for this had been widespread ignorance about what the PFI is all about. Many companies, particularly small and medium sized ones, remain unaware of how the PFI works.

'Companies at all levels need to understand that the PFI represents an enormous opportunity for the private sector and that failure to come to terms with the PFI is going to become a major handicap. Equally those in the public sector – ranging from major government departments to governors of schools – need to have a sound understanding of the concept.'

## INFORMATION TECHNOLOGY IN CONSTRUCTION

### Effect of IT on Office Working

Young (1990) and the RICS Quantity Surveying Division (1995) have identified the rapid advances in information technology (IT) and the need for the quantity surveyor to be in the forefront of these developments. This section builds on the introduction to the subject in chapter 15.

There is likely to be an increase in the numbers working part time or at home and economising on office space by sharing workplaces, using telephone, fax and on-line links. As Weatherhead (1997) has described, laptop computers, modems and advances in telecommunications make it much easier for many people to work where it is most convenient, affecting not only quantity surveyors but also influencing clients' office requirements. Furthermore, hot desking requiring an infrastructure of computer databases and telecom networks, may be provided in open plan offices with some quiet rooms for undisturbed work. The travelling worker can use a laptop computer and mobile phone to link up to the office (workplace) computer and its local area network (LAN) to work on files and download or send data, E-mail and faxes over much of Europe.

Workflow installations, often driven by the need to save space by reducing storage and office accommodation, have led to qualitative improvements in terms of higher productivity, better retrieval, better flow of information, improved efficiency, greater security and improved service to clients. Sophisticated systems can determine what documents will be needed and prefetch them (Weatherhead, 1997).

The development of CAD and virtual reality has meant that proposals for new buildings can be presented to clients so that they can 'walk' through their proposed building and gain a much better impression of it before making a final commitment. Other important developments are geographical information systems (GIS) and automated storage and retrieval systems (ASRS) with immense advantages in use.

## Construct IT

In 1995, the DoE published the above report, recognising that information technology is a highly powerful tool which has a major role to play in the coordination of the construction process. It provides an IT strategy for the UK construction industry and is intended to give focus to a common IT infrastructure to support fully integrated systems and secure the rapid, accurate and auditable flow of information related to projects. Initially, the strategy will have four main components – an integrated project database, a monitored best practice programme, a common knowledge base, and the creation of a Centre of Excellence. The strategy was developed in collaboration with BT, representatives of the construction industry and experts from other industries. A new group was established to ensure that all the activities carried out under the strategy remain firmly focused on the industry's priorities.

The Government is committed to the success of the strategy and was supporting more than 20 sponsored projects, each focusing on a different strategy recommendation. Examples included demonstrating the use of integrated project databases on real construction projects; a prototype system to demonstrate the knowledge database, linking several existing information systems; construction design visualisation through reality/virtual environment; intelligent decision support system for managing assets; quantifying the benefits of IT; and Building IT 2005, a multi-media presentation of experts' views on IT in the construction industry up to the year 2005. The strategy is a good example of the Government and the construction industry working closely together – this time in partnership with British Telecom (BT) – to press forward with improvements envisaged in the Latham report (1994).

The strategy presented an overall vision for the improved use of technology in construction, encompassing the following three main components:

- Integrated project communications framework based on a project database, supporting closer teamwork and focused on satisfying all client requirements.
- Integrated industry-wide information, including standard component listings, building performance benchmarks and best practices, available to improve and inform construction projects.
- Specific improvements to the construction process, including better retrieval and appraisal of client requirements, integration of design and analysis activities and better supply chain integration.

It is envisaged that the development of a project database supporting an integrated suite of software applications will improve the project brief process, design and analysis, project estimating process, site based communications, informal project communications through consideration of E-mail and video conferencing, and facilities management.

Other proposed important roles of IT contained in the report, of particular interest to quantity surveyors, included the following:

- modelling of client requirements to include performance benchmarks and cost trade offs
- life cycle cost modelling including in use cost and performance data
- visualisation modelling techniques to allow the client to visualise different design options
- project database provides facility to assess impact of design changes, allowing 'what-if' analysis by consultants, contractor and client, and inform design team members of changes which affect parts of the design for which they are responsible
- parallel costing attached to design packages allowing cost implications of design decisions to be assessed
- IT based estimating models to support human knowledge
- store standard contract information on project database or provide access to this through industry knowledge base
- site communications infrastructure could support E-mail, video links and electronic purchasing to create the necessary links between site management, design team, head offices, suppliers, etc.
- site communications could use wireless networks to control transmission of information around the site
- site information could be captured electronically using simple interfaces, such as bar codes and smart cards
- use of CAD supported directly by project database will allow optimisation of refurbishment and maintenance schedules

- building performance and monitoring of asset performance and depreciation can be modelled using software
- performance information and life time costs can be captured for use on future projects
- performance information can be used to produce predictive maintenance schedules.

### BRE: Performance and Cost Managed Building

An important advance in the provision of better managed buildings was made with the development by BRE (1995) of a new approach to building procurement and operation, through the use of improved IT techniques. This approach – *Performance and Cost Managed Building* (PCMB) – makes it possible to optimise the performance and costs in use of a building throughout its life. A bank of data about the building and linked software tools are used in making all the key decisions in design, management, maintenance, repair and refurbishment. The system should save time and money in procuring and operating the building and, in the longer term, reduce the need for maintenance and repair.

Two practical versions of the system have been prepared: an interactive software version and a paper version. The system comprises the following four components:

- *Building Data Spine* (BDS): a structured database for each specific building providing a comprehensive and continuous record of its performance and costs in use. Data is fed into it throughout the design and construction stages by all involved and the occupier records actual performance and passes this on to future owners. A BDS can also be constructed for existing buildings (see figure 19.5).
- *Performance and Costs in Use Database*: information on different building types, elements, systems, materials and products, their life expectancies, maintenance needs and costs, and modes of failure. It provides data for compiling a BDS.
- *Data System*: checklists of design and facilities management issues, decisions, activities and sources of information to support decision making.
- *Software Environment Support System*: the essential link between the aforementioned tools, which enables calculations and comparisons to be made.

### EXPERT SYSTEMS

Expert systems, which are more correctly termed intelligent knowledge based systems, are sophisticated computer programs which manipulate a

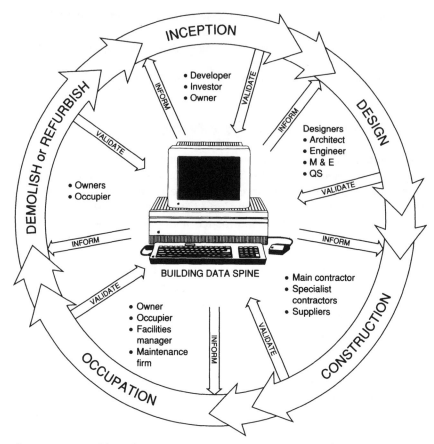

**Figure 19.5**   *Building data spine*

*Source*: BRE (1995)

stored body of knowledge in order to mimic the reasoning experience and judgement of human experts. An expert system may also be able to deal with the end-user inputting incomplete information, by making reasonable assumptions on the basis of other supplied information. In addition, an expert system can often cope with uncertainty in the knowledge it manipulates to find solutions to the end-users' problems (Brandon *et al.,* 1990).

However, expert systems do have limitations, as, for example, they can only cope with well-defined areas of knowledge, which are often of a technical nature. They cannot handle everyday knowledge, are incapable of intuitive thinking and cannot give inspired solutions to problems. They must be focused on a relatively small set of knowledge, are not often proficient at quickly processing numerical information, and cannot handle sensory or pictorial information.

Expert systems differ from conventional computer programs in the following ways:

(1) they involve the representation and manipulation of knowledge rather than data;
(2) heuristic as opposed to algorithmic procedures are used to solve problems;
(3) problems are solved by using inferential methods rather than the repetitive processes of conventional programs.

A survey of quantity surveying organisations carried out by the University of Salford for the RICS identified the areas which commonly generated major problems, requiring specialist knowledge for their solution and which it was felt could be resolved by expert systems. The following areas related to the general field of building economics: feasibility studies; investment appraisal; funding of projects; cashflow forecasting and monitoring; life cycle forecasting and control; cost estimating; and risk analysis and management.

## Imaginor Systems

Imaginor Systems was formed in 1989 as a partnership between a subsidiary limited company of the Royal Institution of Chartered Surveyors (RICS) and Salford University Business Services Ltd (SUBSL). It was established to develop and market expert knowledge based software (KBS) in the property and construction industry. Its first product called ELSIE-Commercial, evolved from a UK Government Alvey initiative awarded to the two parties in 1986. In 1992 the installed base was in excess of 350 expert packages comprising a mix of ELSIE-Commercial and its later sister product called ELSIE-Industrial. The software operates on any IBM PC and can achieve a high standard of feasibility estimating using KBS at the beginning and early stages of a building project, and packages are mainly purchased by UK quantity surveyors, although the market has extended to architectural practices, design and build contractors, property developers and corporate companies.

The main advantages claimed for both systems are the production of estimates without the existence of a design; save feasibility costs; give very fast results; are easy and intuitive; provide advice from an expert quantity surveying knowledge bank; provide high quality detailed printed reports; are very adaptable using a 'what-if?' feature; and they complement most building software. Imaginor Systems offer a major step forward in the refinement of cost planning techniques, within certain limitations.

Readers requiring more information on the application of expert systems to construction projects are referred to Brandon *et al.* (1990).

## COST–BENEFIT ANALYSIS

### Nature of Technique

Cost–benefit analysis (CBA) is a developing technique which could become more refined and widely used in the future, with important implications for the quantity surveyor. Cost–benefit analysis aims at setting out the factors which need to be taken into account in making economic choices. Most of the choices to which it has been applied involve investment projects and decisions – whether or not a particular project is worthwhile financially, which is the best of several alternative projects, or even when to undertake a particular project. The aim is generally to maximise the present values of all benefits less that of all costs, subject to specified restraints. Cost–benefit analysis has been defined as 'a technique of use in either investment appraisal or the review of the operation of a service for analysing and measuring the costs and benefits to the community of adopting specified courses of action and for examining the incidence of these costs and benefits between different sections of the community'.

It has the basic objective of identifying and measuring the costs and benefits which stem from either the investment of monies or the operation of a service, but in particular it is concerned with examining not only those costs and benefits which have a direct impact on the providing authority but also those which are of an external nature and accrue to other persons. Furthermore, the costs and benefits to be measured are those which accrue throughout the life of the project.

The principal criteria to be determined are:

(1) Which costs and benefits are to be included?
(2) How are they to be valued?
(3) At what interest rate are they to be discounted?
(4) What are the relevant constraints?

A suitable methodology for use in a cost–benefit study is as follows:

(1) define the problems to be studied;
(2) identify the alternative courses of action;
(3) identify the costs and benefits, both to the providing authority and to external parties;
(4) evaluate the costs and benefits; and
(5) draw conclusions as to the alternative to be adopted.

### Externalities

These are the costs and benefits which accrue to bodies other than the one sponsoring a project. The promoters of public investment projects should take into account the external effects of their actions in so far as they alter the physical production possibilities of other producers or the satisfactions that consumers can obtain from given resources; they should not take *side-effects* into account if the sole effect is through the prices of products or factors.

An example of an external effect to be taken into account would be the construction of a reservoir by the upstream authority of a river basin which results in more dredging by a downstream authority. An example of a side-effect is where the improvement of a road leads to a greater profitability of the garages and restaurants on that road and the employment of more labour by them and higher rent payments. Any net difference in profitability and any net rise in rents and land values is simply a reflection of the benefits of more journeys being undertaken, and it would be double counting if these were included too.

### Applications of Cost–Benefit Analysis

Cost–benefit analysis techniques have been applied to a wide range of projects.

In the UK one of the first cost–benefit studies was that conducted on the economic assessment of the London – Birmingham motorway (M1). This attempted to provide the economic justification for the expenditure of large sums of public money on motorway construction by showing the benefits which would flow from their development. It was restricted inasmuch as it was concerned with the benefits directly attributable to the construction of motorways and did not take account of the effect these new roads would have on neighbouring communities.

A further important study was undertaken on the construction of the Victoria Underground line in London. This study took place after the decision to construct the line had been taken and attempted to show the benefits which would accrue to different sectors of the population when it became operational. It is a particularly useful study for the way it illustrates the difficulty of placing measurements on certain intangible items, such as time savings during leisure hours, and also for the way it emphasises the importance of exercising extreme care in deciding the cut-off points in a practical situation.

The action of public authorities often has a ripple effect – the costs and benefits spread out from the centre and become more diffuse and difficult to measure as they become more remote from the direct action taken by the authority. A good example is that of a local authority providing a housing estate which affects the tenants, shopkeepers and others who provide them with services. It could also affect servicing arrangements in neighbouring communities and a decision has to be taken as to the extent to which attempts are made to measure these indirect effects. Determining cut-off points is often one of the most difficult problems because if it is too tightly circumscribed major effects will be omitted from the study.

Lichfield evolved a methodology known as the *planning balance sheet* by which he applied cost benefit analysis to a wide range of town and regional planning problems, although this has been little used in practice.

## Application to Development Decisions

Cost–benefit studies suitably refined could make a useful contribution in arriving at decisions on a wide range of development problems, of which the following are typical examples:

(1) Should a new town programme be carried out as a small number of large developments or a larger number of more modest size?
(2) Should a new town be built on a 'greenfield' site or be grafted on to an existing town of substantial size?
(3) What funds should be allocated to redeveloping a central area?
(4) How should the choice be made between improving old houses and clearing the site and building anew?
(5) What economic criteria are there for choosing which areas of old housing to improve when improvement is the preferred course rather than replacement?

The general principles of choice between replacement and improvement of old housing is one of the more obvious applications but even here there is the major practical problem of evaluating the difference in the standard of accommodation provided by new houses or flats and old houses improved. Scoring the accommodation according to the presence or absence of specific features would be one way, albeit not very precise. Another approach would be to use free-market rents (fair rents) as a basis for evaluating the difference in standards, but there are rarely sufficient to do this. Furthermore, as incomes rise occupiers are likely to become more exacting in their requirements and attach increased importance to features which do not exist in the improved older dwelling.

Similarly town development is a complex process with far-reaching physical, social and economic consequences. Its impact varies as between public and private developers and the community at large, and so a method is needed for determining the consequences of decisions in this field from the various points of view. Thus the need arises for evaluating alternative solutions to the same problem and alternative ways of investing the same resources, with due regard to appearance, amenity and costs. The term amenity as used in this context includes the factors of safety and convenience. It is much easier to reach rational decisions when the consequences of alternative solutions have been quantified and due consideration must be given to all social and economic benefits. Both costs and benefits are spread over long periods of time involving difficult future predictions and discounting, some costs and benefits are indirect and are not easily costed and some intangible benefits become a matter of opinion.

A cost–benefit study of the provision of a tunnel or bridge to cross the Thames at Thamesmead indicated a close balance between the quantified factors (traffic and housing benefits being roughly equal to the higher cost of a tunnel). The unquantified environmental and delay factors favoured a tunnel, so that the final decision would have to be based on subjective matters, always provided that the extra money for the tunnel was available.

## Application to Building Proposals

There is little evidence of cost–benefit analysis being applied to building projects to any great extent and difficulties can arise if subjective evaluations of benefits are included as, for example, when comparing different floor finishes. One useful example was a study carried out by the Department of the Environment and published in 1971, to compare the costs and benefits of planned open offices with air conditioning, and traditional cellular offices without air conditioning. In the study it was assumed that the open offices would be planned to certain essential standards of space, lighting, acoustic treatment, layout and furnishing for functional reasons. Differences were calculated on a cost per capita basis, and discounted to a net present value (NPV) in the year of building at eight per cent, and also at three per cent and sixteen per cent. It was assumed that the life of the buildings would be sixty years and 1966 prices were used throughout. An allowance was made for a relative increase in the cost of labour-intensive industries in calculating maintenance and cleaning costs. The following costs and benefits were quantified.

*Costs* Capital costs of building, mechanical and electrical depreciation and running costs, and initial provision, maintenance and renewal of furniture.

*Staff benefits* Increased productivity due to teamwork and better communications (one per cent staff salaries and overheads); increased productivity from air conditioning (1.5 per cent staff salaries, etc.), less sick leave because of better working conditions (two per cent of sick leave); saving on messenger services (two messengers per 100 staff); increased productivity from better lighting (one per cent staff salaries etc.); and better recruiting and staff satisfaction (0.1 per cent of non-career grade staff salaries).

*Building benefits* Greater flexibility – avoid remodelling internally after forty years (at cost of ten per cent of original cost of building), more economical use of space (valued at $0.1m^2$/head rising to $0.5m^2$/head during forty years before remodelling), no removal of partitions (valued at one per cent of partitions moved each year); low maintenance materials –

reduction in replacement costs and decoration and in staff disturbance; and reduction in cleaning costs (owing to easier and fewer surfaces and building kept cleaner through air conditioning).

The study concluded that the higher costs of open plan arrangements with air conditioning were more than offset by the additional benefits at interest rates of eight per cent and three per cent; at sixteen per cent the cellular offices were slightly better. At eight per cent interest air conditioned planned open offices were calculated to give a capitalised saving of about £260 per head compared with traditional non-air conditioned cellular offices and £200 per head compared with air conditioned cellular offices, at 1966 prices. Unfortunately no updated costs have been published and doubts have been expressed about some of the assumptions made in the study.

### Conclusions

Wide divergences of view have been expressed about the role and usefulness of cost–benefit analysis. Certainly problems arise from uncertainties about the consequences of various courses of action and the difficulties of measuring the costs and benefits. Diverse types of benefit, avoidance of double counting, dealing with externalities and choice of discount rates pose a formidable range of problems.

An important advantage of cost–benefit study is that it compels those responsible to quantify costs and benefits as far as possible, rather than resting content with vague qualitative judgements or personal hunches. Furthermore, quantification and evaluation of benefits, however rough, does give some indication of the charges which consumers are willing to pay. Its limitations are clearly shown in the Roskill Commission's report on the third London airport, where after costing all the intangibles relating to the four sites and arriving at total figures in excess of £4000 million, the cheapest site at Cublington was only five per cent less than the most expensive at Foulness. This is hardly a large enough margin to be conclusive in making a choice.

Even where 'shadow prices' cannot be computed for all the main costs and benefits, cost–benefit analysis may still prove useful to the decision maker. In some cases, the valued and unvalued factors may both point in the same direction. If not, and if there is only one unvalued factor, such as noise, it is possible to calculate the value which the decision maker would have to put on this to justify a particular decision. Where there are several unvalued costs or benefits, the problem becomes more complex, but it may still be possible to show that such projects are worse than others on all counts; for instance, one transport strategy might be more expensive, save less travelling time, and have worse distributional effects on the environment than another. In some cases it might be more meaningful to assign limiting values (maxima and minima) to factors that cannot be valued more precisely.

The technique of cost–benefit analysis is still in an early stage of development and its application to building economics requires many more realistic case studies in order to develop and refine the process, so that it can become a positive aid in the decision-making process. The best results are obtained from cost–benefit analysis when it is used for choosing between a limited range of alternatives with few intangibles, because of the difficulties of definition and evaluation (Balchin *et al.*, 1995) as was highlighted when attempting to apply it to the resiting of Covent Garden Market.

## POST LATHAM DEVELOPMENTS

### Construction Industry Board

It was most encouraging to see the large measure of support that the Latham report (1994) received from all sections of the construction industry and the action taken by the new *Construction Industry Board* (CIB) and other bodies following up recommendations in the 1994 report. The Government agreed to provide one half of the CIB's running costs. The Board members represent government, contractors, subcontractors, professionals and clients, with materials producers being added in 1996.

The remit of the CIB was to assist the Latham working parties to complete their work and to encourage good practice throughout the industry. It has also been very much concerned with improving the performance of the industry by reducing conflict and construction costs. It issued a report in 1996 which criticised architects and engineers for increasing construction costs by failing to understand the components and materials they specify, failing to keep up with new technologies and ignoring a building's life cycle costs. It believed that better procurement methods and the calculation of building costs for the whole life cycle were the keys to cutting costs by 30 per cent by the year 2000. Important as these items are, there are many other ways of reducing construction costs although they will vary from one project to another, and a number were identified in the RICS report (1995e).

A number of CIB subgroups also produced useful reports in 1995. For example the total management subgroup proposed the idea of holding a construction quality awards scheme to encourage organisations to adopt the principles of TQM, a management system designed to ensure that a project team carries out its work properly first time. The benchmarking subgroup identified an urgent need for a separate body to review this key area, which involves seeking out the best way of doing a piece of work and testing systems against it. While the concept subgroup identified the main areas of avoidable cost as changes, overspecification, waste and duplication, unnecessary complexity and conflict. To rectify these problems the subgroup's report advocated the adoption of risk management techniques, clear management structures, and the establishment of an industry standard for information technology. The report also stressed the importance of comparing UK standards with international practice.

### Contributions from Other Bodies

The RICS 1995 report *Improving value for money in construction: guidance for chartered surveyors and their clients* and the Reading Construction Forum 1995 report *Trusting the team* gave plenty of sound advice on how to improve efficiency and cut construction costs. Another RICS report in 1995 *Securing the services of a chartered surveyor* will assist those involved in the fee tendering process, while the Treasury 1995 White Paper *Setting new standards – a strategy for government procurement* stressed the importance of considering lifetime costs.

In 1995 the *Whole industry research strategy for construction* was launched aimed at focusing efforts of the construction industry, its clients and government into research that is relevant and meaningful to all those involved in the construction process. It was believed that a coordinated approach to research and innovation could help achieve the aims of reduced costs and improved quality, which are the key aims of the Latham report (1994) *Constructing the team*.

### The Construction Bill 1996

The 1996 Construction Bill, which was expected to become law in July 1996, introduced adjudication, encouraged milestone payments, banned 'set-off' and 'pay when paid' and allowed the main contractor to stop work if the client ceased paying. Three other Latham recommendations were however omitted, namely trust funds because of industry disagreements, while latent defects insurance and reform of the liability law were passed to the Law Commission. After the bill becomes law, standard forms of contract will have to be amended to comply with the legislation and 'bespoke' contracts will have to comply with a DoE scheme for construction contracts.

The effect of each of the four main provisions will now be examined:

- Under the adjudication provision all contracts over a certain level will require the appointment of a named expert who can be called in by either party to settle disputes on the spot. If a party disagrees with the adjudicator's findings, it must wait until practical completion before going to arbitration or law.
- The ability of a contractor to set off money from a subcontractor working

on one contract to settle a dispute on another will be severely restricted.
- Contractors and subcontractors will find it easier to sue for late payment or retention monies under a clause that will make the 'establishment of a debt' easier to prove.
- It will no longer be possible for a contractor to include a clause in a subcontract stating that the subcontractor will only be paid when the main contractor receives payment from the client.

## GENERAL CONCLUSIONS

### Changing Role of the Quantity Surveyor

Quantity surveyors have an important part to play in increasing the efficiency and competitiveness of construction work and in reducing confrontation, by securing a comprehensive brief from the client, ensuring that the most suitable procurement route is followed, providing effective cost control procedures and ensuring value for money at all times, coupled with meeting satisfactorily the client's requirements in regard to time, quality, function and allocation of risk, and reducing to a minimum the scope for errors and omissions and giving added value to the project.

Quantity surveyors are likely to acquire and operate a range of diverse skills outside the normal traditional quantity surveying activities as they proceed into the 21st century, and the needs and expectations of clients change. The new and extended skills may also require an extension of the quantity surveyor's knowledge base, as portrayed in the changing content of relevant degree courses. They must also be versatile, adaptable and innovative, and prepared to meet challenges, such as by joining or associating with other disciplines to offer a more comprehensive package of services and thus provide clients with a single point responsibility service. The extension of quantity surveying services outside the UK is likely to gather momentum, particularly in eastern Europe and the Far East, and into fast expanding and important service areas such as project management, facilities management and value management, as described in earlier chapters, and updating IT systems as appropriate.

As funding is likely to become ever more restricted, public clients are faced with a seemingly easy option of devaluing their new buildings, such as by reducing architectural quality and ceasing to consider likely future maintenance costs. Unfortunately, they do this at their peril, and the quantity surveyor as an integral member of the design team, should bring pressure to resist policy decisions of this type.

Although quantity surveyors are primarily concerned with urban developments, the needs, opportunities and challenges in the countryside should not be overlooked. The countryside should be viewed as a constantly changing, working environment, capable of generating the jobs needed to sustain the rural community, adequately served by affordable housing, readily accessible schools, employment opportunities and social services. Rural workshops and other facilities need to be carefully and skilfully provided. Another essential feature is the proper integration of land use planning and transport, coupled with adequate provision for visitors to the country-side. The RICS *et al.* (1995) have rightly emphasised the importance of conserving the rural environment as, for example, by environmental accounting of rural programmes and projects, to enable the wider environmental benefits and costs to be estimated and built into the overall costs of carrying out the programmes and projects.

## Comparison of UK and Continental Practices and Performance

We should also be prepared to examine and learn from construction related operations in other countries where practices differ from those in the UK. For example, the Business Round Table (1994) in *Controlling the upwards spiral* compared the constructional methods and costs of similar buildings on the mainland of Europe with those pertaining to the UK. The study covered production, industrial and office buildings, retail units and shopping malls, and on average the UK costs were some 60 per cent higher than those in the other EU member states. The principal areas for action to reduce UK costs were identified as engineering services, site supervision and contractual relationships.

For instance, air conditioning was shown to be more expensive in the UK, mainly because of overspecification, a preference for tailor made solutions rather than standard items, and space restrictions resulting from lack of coordination between structural and services design.

The number of managerial and supervisory staff on Continental sites was found to be about half those on similar UK sites and they were more effective, had higher status and were better paid. This arrangement also reduced the number of misunderstandings and mistakes on site and produced much less paperwork.

On the Continent, the *Napoleonic code civil*, with its decree of strict liability results in far fewer problems in the operation of contracts. European contractors have far less protection and are made more responsible for their own actions as opposed to the more uncertain and litigious arrangements operating in the UK and to which Latham (1994) drew attention.

The differences in superstructure costs were largely explained by UK clients requiring a more precise specification and form of installation, while on the Continent there was a more liberal approach with a wider use of system building. Higher services costs in the UK mainly stemmed from oversophistication. Continental site working benefited from fewer lines of communication and two or three times as many drawings.

Hence the author considers that this report can have important implications for the UK construction industry and also for the quantity surveyor, although the difficulties of making precise cost comparisons have to be recognised. The report highlighted ways in which constructional costs can be reduced by changing contractual arrangements, rationalising the design of buildings and services installations, improving the efficiency, responsibility and status of site personnel and increasing site documentation.

# Appendix A  Preliminary Enquiry for Invitation to Tender

Applicable where the JCT Standard Form of Building Contract, Intermediate Form of Building Contract or Agreement for Minor Building Works is to be used

Dear Sirs,

Heading

I am/We are authorised to prepare a preliminary list of tenderers for construction of the works described below.

Please indicate whether or not you wish to be invited to submit a tender for these works. Your acceptance will imply your agreement to submit a wholly bona fide tender in accordance with the principles laid down in the NJCC 'Code of Procedure for Single Stage Selective Tendering' 1994, and not to divulge your tender price to any person or body before the time for submission of tenders. Once the contract has been let, I/we undertake to supply all tenderers with a list of the tender prices.

You are requested to reply by .... Your inability to accept will in no way prejudice your opportunities for tendering for further work under my/our direction; neither will your inclusion in the preliminary list at this stage guarantee that you will subsequently receive a formal invitation to tender for these works.

Yours faithfully...

a  Project...
b  Type and function of the building, e.g. commercial, industrial, housing, etc., with any other details...
c  General description of the project...
d  Employer...
e  Employer's professional team...
f  Location of site... (Site plan enclosed)
g  Approximate cost range of project £... to £...
h  Number of tenderers it is proposed to invite...
i  Nominated sub-contractors for major items... [1]
j  Form of Contract:
   List here Form of Contract, e.g. JCT 80 / IFC 84 / MW 80, together with
   (i) Appendix items to the appropriate form completed (not applicable to MW 80)
   (ii) The JCT Amendments proposed to the standard form.
   Note: Attention is drawn to the advice given in clause 4.2.3]
k  Examination and correction of priced bill(s) [Section 6 of Code] [2]
   Alternative 1 / Alternative 2 [3] will apply
l  The contract is to be executed as a deed / a simple contract
m  Anticipated date for possession...
n  Period for completion of works...
o  Approximate date for despatch of all tender documents...
p  Tender period... weeks
q  Tender to remain open for... weeks [4]
r  Details of guarantee requirements...
s  Particular conditions applying to the contract are...

References
[1] Not appropriate for IFC 84 or MW 80.
[2] Only appropriate for JCT 80 With Quantities or IFC 84 if the priced document is to be contract bills.
[3] Delete as appropriate before issuing.
[4] This period should be as short as possible.

*Code: Single Stage* (NJCC, 1994a)

# Appendix B Formal Invitation to Tender

Dear Sirs,

Heading

Following your acceptance of the invitation to tender for the above, I/we now have pleasure in enclosing the following:

a   two copies of the bill(s) of quantities/specifications/schedules; [1]
b   (i) two copies of the location drawings, component drawings, dimension drawings and information schedules (for JCT 80 With Quantities or IFC 84 where Contract Bills are provided); [1]
     (ii) two copies of all drawings (for JCT 80 Without Quantities, IFC 84 with Specification and Drawings or MW 80); [1]
c   two copies of the form of tender;
d   addressed envelopes for the return of the tender (and priced bill(s)(†) and instructions relating thereto.

Will you please also note:

1   drawings and details may be inspected at ...
2   the site may be inspected by arrangement with the employer/architect [1]
3   tendering procedure will be in accordance with the principles of the NJCC 'Code of Procedure for Single Stage Selective Tendering' 1994
4   examination and adjustment of priced bill(s) (Section 6 of the Code), Alternative 1 / Alternative 2 [1] will apply.

The completed form of tender is to be sealed in the endorsed envelope provided and delivered or sent by post to reach ... not later than ... hours on ... the ... day of ... 19 ...

*   *The completed form of tender and the priced bill(s) of quantities sealed in separate endorsed envelopes provided are to be lodged not later than ... hours on ... the ... day of ... 19 .... The envelope containing the tender should be endorsed with the job title: that containing the bill(s) of quantities should be endorsed with the job title and tenderer's name.*

Will you please acknowledge receipt of this letter and enclosures and confirm that you are able to submit a tender in accordance with these instructions.

Yours faithfully,
Architect/Quantity Surveyor [1]

References    [1] Delete as appropriate, before issuing.
              *    *Applicable in Scotland only in which case the preceding sentence would not be used.*
              †    Applicable in Scotland only.

*Code: Single Stage* (NJCC, 1994a)

# Appendix C    Form of Tender

This form of tender is suitable for use only when a formal contract is
entered into

Tender for . . . (description of Works)

To . . . (Employer)

Sir(s),

I/We having read the conditions of contract and bill(s) of quantities/specifications/
schedules [1] delivered to me/us and having examined the drawings referred to therein
do hereby offer to execute and complete in accordance with the conditions of contract the
whole of the works described for the sum of . . . £ . . . and within . . . weeks [2] from the
date of possession.

* *I/We undertake in the event of your acceptance to execute with you a formal
contract embodying all the conditions and terms contained in this offer.*

I/We agree that should obvious errors in pricing or errors in arithmetic be discovered
before acceptance of this offer in the priced bill(s) of quantities submitted by me/us these
errors will be dealt with in accordance with Alternative 1 / Alternative 2 [1] contained in
Section 6 of the NJCC 'Code of Procedure for Single Stage Selective Tendering' 1994.

This tender remains open for consideration for . . . days [3] from the date fixed for the
submission or lodgement of tenders [4].

Dated this.................................................................day of........................................ 19..............

Name.............................................................................................................................................

Address.........................................................................................................................................

Signature...................................................... Witness*.......................................................

Witness*.......................................................

* *The completed form of tender and the priced bill(s) of quantities sealed in separate
endorsed envelopes are to be lodged not later than . . . hours on . . . the . . . day of . . .
19 . . . . The envelope containing the tender should be endorsed with the job title: that
containing the bill(s) of quantities should be endorsed with the job title and the
tenderer's name.*

**References**

[1]  Delete as appropriate, before issuing.
[2]  To be completed before Form of Tender is sent out.
[3]  Any period quoted should not normally exceed 28 days and only in exceptional
     circumstances  should it extend beyond 56 days.
[4]  Refer to Clause 4.5 of this Code.
*    Applicable in Scotland only.

*Code: Single Stage* (NJCC, 1994a)

# Appendix D

---

Project: Office Block, Haslebury New Town

Project No: O/H/21          Date: 7 May 1997

---

1. Minutes of last meeting
2. Matters arising and action taken
3. Report on weather since last meeting and time lost
4. Labour force on site at date of meeting (by trades)
5. Queries on architect's instructions – serial number of latest instruction and verbal instructions requiring confirmation
6. Dayworks general report and date of last sheet passed to quantity surveyor
7. Main contractor's progress report
8. Main contractor's report and queries on nominated subcontractors:
   - (i) Mechanical services
   - (ii) Electrical services
   - (iii) Structural work
   - (iv) Other services
9. Main contractor's report and queries on nominated suppliers
10. Comments from those in attendance
11. Any other business
12. Date of next meeting

---

*Distribution*: Employer
Architect
Quantity Surveyor
Consultants
Contractor
Clerk of Works
Subcontractors and suppliers

# Appendix E

## Valuation

Chartered Quantity Surveyor
A.F. Guy FRICS
14 The Square
Hopton New Town

Works  Newacre Shopping Centre
        Refurbishment

Valuation No: 4
Date of Issue:  5 June 1997
QS Reference:  264 SR

To Architect/Contract Administrator
J.P. Smithbury RIBA
25 The Close
Southbury

I/We have made, under the terms of the Contract, an Interim Valuation

as at 30 May 1997              * and I/we report as follows:—

Gross Valuation
(excluding any work or material notified to me/us by the Architect/The Contract Administrator in
writing, as not being in accordance with the Contract.

£  3,532,860

Less total amount of Retention, as attached Statement.

£  99,968

£  3,232,892

Less total amount stated as due in Interim Certificates previously issued by the
Architect/The Contract Administrator up to and including Interim Certificate No....3............

£  2,325,113

Balance (in words)     Nine Hundred and Seven Thousand,
                       One Hundred and Seventy Nine Pounds

£  907,179

Signature: [signature]

Chartered Quantity Surveyor FRICS/ARICS
                              (delete as applicable)

Notes:
(i)    All the above amounts are exclusive of V.A.T.
(ii)   The balance stated is subject to any statutory deductions which the Employer may be obliged to make under the provisions of the
       Finance (No. 2) Act 1975 where the Employer is classed as a "Contractor" for the purposes of the Act.
(iii)  It is assumed that the Architect/The Contract Administrator will—
       (a)  satisfy himself that there is no further work or material which is not in accordance with the Contract.
       (b)  notify Nominated Sub-Contractors of payments directed for them and of Retention held therein by the Employer.
       (c)  satisfy himself that the previous payments directed for Nominated Sub-Contractors have been discharged.
* (iv) The Architect's/The Contract Administrator's Interim Certificate should be issued within seven days of the date indicated thus
(v)    Action by the Contractor should be taken on the basis of figures in, or attached to, the Architect's/The Contract
       Administrator's Interim Certificate.

Employer
Good Service Development Company
83 High Street
Great Withington

Contractor
G.E.W. (Contractors) Ltd
34C The Cedars Industrial Estate
Great Shoreham

Contract sum    £15 645 270

568

# Appendix F

## Statement of Retention and of Nominated Sub-Contractors' Values

Chartered Quantity Surveyor
A.F.Guy FRICS
14 The Square
Hopton New Town

Works Newacre Shopping Centre Refurbishment

This Statement relates to:
Valuation No : 4
Date of issue : 5 June 1997
QS Reference: 264 SR

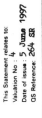

| | Gross Valuation | Amount subject to: | | | Amount of Retention | Net Valuation | Amount Previously Certified | Balance |
|---|---|---|---|---|---|---|---|---|
| | | Full Retention of 3 % | Half Retention of – % | No Retention | | | | |
| | £ | £ | £ | £ | £ | £ | £ | £ |
| Main Contractor | 2 939 470 | 2 939 470 | – | – | 88 184 | 2 851 286 | 2 325 113 | 526 173 |
| Nominated Sub-Contractors:- | | | | | | | | |
| Prestige Engineering | 195 640 | 195 640 | – | – | 5 869 | 189 771 | – | 189 771 |
| G.P. Precast Concrete | 135 870 | 135 870 | – | – | 4 076 | 131 794 | – | 131 794 |
| B & C Electrical | 61 280 | 61 280 | – | – | 1 839 | 59 441 | – | 59 441 |
| TOTAL | 3 332 260 | 3 332 260 | – | – | 99 968 | 3 232 292 | – | 907 179 |

No account has been taken of any discounts for cash to which the Contractor may be entitled if discharging the balance within 17 days of the issue of the Architect's/Contract Administrator's Interim Certificate. The sums stated are exclusive of V.A.T.

© 1980 RICS

569

# Appendix G

Issued by: J.P. Smithbury RIBA
address: 25 The Close Southbury

Works: Refurbishment
situated at: Newacre Shopping Centre

**Statement of Retention**
and of Nominated Sub-Contractors' Values

Job reference: NSC: 147
Relating to Certificate no: 4
Issue date: 9 June 1997

| | Gross valuation | Amount subject to: | | | Amount of retention | Net valuation | Previously certified | Balance due |
| | | Full retention of 3 % | Half retention of - % | Nil retention | | | | |
|---|---|---|---|---|---|---|---|---|
| | £ | £ | £ | £ | £ | £ | £ | £ |
| Main Contractor | 2 939 470 | 2 939 470 | – | – | 88 184 | 2 851 286 | 2 325 113 | 526 173 |
| Nominated Sub-Contractors: | | | | | | | | |
| Prestige Engineering | 195 640 | 195 640 | – | – | 5 869 | 189 771 | – | 189 771 |
| G.P. Precast Concrete | 135 870 | 135 870 | – | – | 4 076 | 131 794 | – | 131 794 |
| B & C Electrical | 61 280 | 61 280 | – | – | 1 839 | 59 441 | – | 59 441 |
| **Total** | 3 332 260 | 3 332 260 | – | – | 99 968 | 3 232 292 | 2 325 113 | 907 179 |

All amounts are exclusive of VAT

F802 for JCT 80

No account has been taken of any discounts for cash to which the Contractor may be entitled if discharging the balance within 17 days of the issue of the Interim Certificate.

We are grateful to the Royal Institution of Chartered Surveyors for allowing us to adapt their copyright form of the same title.
© RIBA Publications 1992

# Appendix H

Issued by: J.P. Smithbury RIBA
address: 25 The Close
Southbury

**Interim Certificate and Direction**

Employer: Good Service Development Company
address: 83 High Street
Great Withington

Contractor: G.E.W. (Contractors) Ltd
address: 34C The Cedars Industrial Estate
Great Shoreham

Works: Refurbishment
situated at: Newacre Shopping Centre

Contract dated: 23 January 1997

Serial no: **C**

Job reference: NSC: 147

Certificate no: 4

Issue date: 9 June 1997

Valuation date: 5 June 1997

Contract sum: £15 645 270

**Original to Employer**

This Interim Certificate is issued under the terms of the above-mentioned Contract.

Gross valuation inclusive of the value of works by Nominated Sub-Contractors ..................................................... £ 3 332 260

*Less* Retention which may be retained by the Employer as detailed on the Statement of Retention ....................................... £ 99 968

Sub-total £ 3 232 292

*Less* total amount stated as due in Interim Certificates previously issued up to and including Interim Certificate no: 3 .......... £ 2 325 113

Net amount for payment ......................................... £ 907 179

I/We hereby certify that the amount for payment by the Employer to the Contractor on this Certificate is (in words)

Nine Hundred and Seven Thousand, One Hundred and

Seventy Nine Pounds

I/We hereby direct the Contractor that this amount includes interim or final payments to Nominated Sub-Contractors as listed in the attached *Statement of Retention and of Nominated Sub-Contractors' Values*, which are to be discharged to those named in accordance with the Sub-Contract.

*All amounts are exclusive of VAT*

To be signed by or for the issuer named above

Signed _J P Smithbury_

[1] Relevant only if clause 1A of the VAT Agreement applies. Delete if not applicable.

[1] The Contractor has given notice that the rate of VAT chargeable on the supply of goods and services to which the Contract relates is _____ %

[1] _____ % of the amount certified above is .......................... £

[1] Total of net amount and VAT amount (for information) ............... £

*This is not a Tax Invoice*

F801 for JCT 80

©RIBA Publications Ltd 1990

571

# Appendix J

Issued by: J.P. Smithbury RIBA
address: 25 The Close
Southbury

Employer: Good Service Development Company
address: 83 High Street
Great Withington

Job reference: NSC: 147

Notification no: 1

Main Contractor: G.E.W. (Contractors) Ltd
address: 34C The Cedars Industrial Estate
Great Shoreham

Issue date: 9 June 1997

Works: Refurbishment
situated at: Newacre Shopping Centre

Contract dated: 30 January 1997

Original to Nominated
Sub-Contractor

Nominated
Sub-Contractor: Prestige Engineering
address: Cowsmeadow
Bilston

Under the terms of the above-mentioned Main Contract,

I/we hereby inform you that I/we have directed the Contractor that

Interim Certificate no. 4     dated 9 June 1997

*Delete as
appropriate

includes *an interim/~~a final~~ payment of ....... .... .................... £ 189 771
which is to be discharged to you.

To be signed by or for
the issuer named
above

Signed *J P Smithbury*

------------------------------------------------------------

Nominated Sub-Contractor's

Main Contractor: G.E.W. (Contractors) Ltd
address: 34C The Cedars Industrial Estate
Great Shoreham

**Acknowledgement
of Discharge**
of payment due

Works: Refurbishment
situated at: Newacre Shopping Centre

Job reference: NSC: 147

In accordance with the terms of the relevant Sub-Contract,
we confirm that we have received from you discharge of the amount of ..... £ 189 771

included in Interim Certificate no. 4     dated 9 June 1997

as stated in Notification no. 1     dated 9 June 1997

Please complete
acknowledgement slip
and send to
Contractor

Signed *S M Maltby*     Date 12 June 1997

For Prestige Engineering

# Appendix K

Issued by: J.P. Smithbury RIBA
address: 25 The Close
Southbury

**Final
Certificate**

Employer: Good Service Development Company
address: 83 High Street
Great Withington

Serial no: **A**

Contractor: G.E.W. (Contractors) Ltd
address: 34C The Cedars Industrial Estate
Great Shoreham

Job reference: NSC: 147

Issue date: 16 October 1998

Contract sum: £15 645 270

Works: Refurbishment
situated at: Newacre Shopping Centre

Contract dated: 23 January 1997

| Original to Employer |
| --- |

This Final Certificate is issued under the terms of the above-mentioned Contract.

The Contract Sum adjusted as necessary is .............................. £ 15 783 510

The total amount previously certified for payment to the Contractor is .... £ 15 626 210

The difference between the above-stated amounts is .................... £     157 300

I/We hereby certify the sum of (in words)
One Hundred and Fifty Seven Thousand, Three Hundred

Pounds

as a balance due:

*Delete as
appropriate

*to the Contractor from the Employer.

*~~to the Employer from the Contractor.~~

*All amounts are exclusive of VAT*

To be signed by or for
the issuer named
above

Signed _J P Smithbury_

The terms of the Contract provide that, subject to any amounts properly deductible by the Employer, the said balance shall be a debt payable from the one to the other as from the

[1] Delete as
appropriate. See cover
notes for provision in
particular contract.

[1] 14th / ~~21st / 28th~~ day after the date of this Certificate.

[2] Relevant only if
clause 1A of JCT 80
VAT Agreement,
clause A1-1 of IFC 84
Supplemental
Conditions or clause
B1-1 of MW 80
Supplementary
Memorandum applies.
Delete if not
applicable.

[2] The Contractor has given notice that the rate of VAT chargeable on the supply of goods and services to which the Contract relates is _____ %

[2] _____ % of the amount certified above is ........................ £

[2] Total of balance due and VAT amount (for information) .............. £

*This is not a Tax Invoice*

F852 for JCT 80 / IFC 84 / MW 80

© RIBA Publications Ltd 1990

# Appendix L  Initial Cost Plan and Record of Cost Checks of Social Club

Gross floor area: 470 m²

| Element | | Initial cost plan | | Cost check 1 | | Cost check 2 | |
|---|---|---|---|---|---|---|---|
| | | Total cost of element £ | Cost of element/m² of gross floor area £ | Date | Cost of element/m² of gross floor area £ | Date | Cost of element/m² of gross floor area £ |
| 1 | Substructure | 24 703 | 52.56 | 4.7.94 | 52.32 | | |
| 2A | Frame | 2 312 | 4.92 | 4.7.94 | 5.04 | | |
| 2B | Upper floors | 1 429 | 3.04 | 4.7.94 | 2.92 | | |
| 2C | Roof | 42 563 | 90.56 | 7.7.94 | 84.80 | 4.8.94 | 54.24 |
| 2D | Stairs | 780 | 1.66 | 7.7.94 | 1.66 | | |
| 2E | External walls | 28 219 | 60.04 | 20.7.94 | 59.92 | 4.8.94 | 57.41 |
| 2F | Windows and external doors | 9 616 | 20.46 | 20.7.94 | 23.00 | | |
| 2G | Internal walls and partitions | 10 819 | 23.02 | 13.7.94 | 22.76 | 4.8.94 | 8.59 |
| 2H | Internal doors | 7 266 | 15.46 | 13.7.94 | 14.24 | | |
| 3A | Wall finishes | 10 660 | 22.68 | 18.7.94 | 22.44 | | |
| 3B | Floor finishes | 17 202 | 36.60 | 18.7.94 | 36.48 | 4.8.94 | 33.51 |
| 3C | Ceiling finishes | 8 187 | 17.42 | 18.7.94 | 19.32 | | |

574

| | | | | |
|---|---|---|---|---|
| 4 | Fittings | 6 495 | 13.82 | 20.7.94 | 3.70 |
| 5A | Sanitary appliances | 4 418 | 9.40 | 25.7.94 | 9.40 |
| 5C | Disposal installations | 517 | 1.10 | 25.7.94 | 1.10 |
| 5D | Water installations | 2 538 | 5.40 | 25.7.94 | 5.52 |
| 5E | Heat source ⎫ | | | | |
| 5F | Space heating | 35 429 | 75.38 | 25.7.94 | 75.36 |
| 5G | Ventilating system ⎭ | | | | |
| 5H | Electrical installations | 25 841 | 54.98 | 25.7.94 | 54.72 |
| 5I | Gas installations | 263 | 0.56 | 26.7.94 | 0.56 |
| 5K | Protective installations | 2 143 | 4.56 | 26.7.94 | 0.96 |
| 5M | Special installations | 3 252 | 6.92 | 27.7.94 | 3.38 |
| 5N | Builder's work in connection with services | 2 322 | 4.94 | 27.7.94 | 4.94 |
| 5O | Builder's profit in connection with services | 4 183 | 8.90 | 27.7.94 | 8.80 |
| 6A | Siteworks | 21 319 | 45.36 | 28.7.94 | 8.80 |
| 6B | Drainage | 11 524 | 24.52 | 28.7.94 | 34.80 |
| 6C | External works | 5 198 | 11.06 | 28.7.94 | 24.40 |
| | Preliminaries | 9 804 | 20.86 | 28.7.94 | 11.00 |
| | Contingencies | 3 196 | 6.80 | 29.7.94 | 20.80 |
| | Price and design risk | 5 997 | 12.76 | 29.7.94 | 6.80 |
| | | | | | 12.76 |
| | Totals | £308 195 | £655.74 | | £623.90 |

Source: Seeley (1996)

# Appendix M    BCIS Detailed Elemental Cost Analysis of Office Building

CI/SfB - 320.
Offices - 3

**New build**

## Detailed Elemental Analysis

BCIS *Online* analysis number: 14452

BCIS code:  A - 3(4) - 2,510

| | |
|---|---|
| **Job title:** | Financial Services Centre, Glasgow Caledonian University |
| **Location:** | Glasgow, Strathclyde |
| **Client:** | University Court of Glasgow Caledonian University |
| **Dates:** | Receipt:  3-May-1994    Base:  22-Apr-1994    Acceptance:    Possession: |

**Project details:** 3-storey, high specification office building with part basement 14 x 10 x 3m.  Floor plate areas decrease from ground upwards, creating a stepped elevation and roof terraces.  External works include brick paving, water features to internal courtyard, boundary walls, landscaping, services, drainage and some external lighting.

**Site conditions:** City centre infill site with poor ground conditions.

**Market conditions:** Very competitive.
Project tender price index was 138 on a base of 1985 BCIS Index Base
Indices used to adjust costs to base price level:  TPI for 2Q94 120; location factor 1.10

| | |
|---|---|
| **Tender documents:** | Bill of Quantities |   **Contract:**    JCT Private 1980 contractors designed portion
| **Procurement:** | Selected competition |   **Cost fluctuations:**    Firm
| **Number of tenders:** | Issued: 6    Received: 6 |   **Contract period:**    Stipulated: 12    Offered:    Agreed: 12

Altered - no details

| Contract breakdown | | | Competitive tender list | | |
|---|---|---|---|---|---|
| | Contract £ | Analysis £ | | Tender £ | % above lowest |
| **Measured work** | 2,223,460 | 2,223,460 | 1 | 2,646,532 | - |
| **Provisional sums** | 174,000 | 174,000 | 2 | 2,685,520 | 1.5 |
| **PC sums** | 119,000 | 119,000 | 3 | 2,718,043 | 2.7 |
| **Preliminaries** | 104,351 | 104,351 | 4 | 2,733,118 | 3.3 |
| **Contingencies** | 73,725 | 73,725 | 5 | 2,853,086 | 7.8 |
| **Contract sum** | 2,694,536 | 2,694,536 | 6 | 2,955,301 | 11.7 |

576

**Accommodation and design features:** 3-storey office block with part basement. Concrete bored piles, beams, and upper floors, PCC suspended ground floor. Steel frame. Stone faced cavity walls; glass cladding; double glazed curtain walling. Pitched and flat steel roofs with aluminium and PVC coverings. Double glazed steel windows; revolving door. Metal stud partitions; glass screens. Flush doors. Paper and tiles to walls; carpet and tiles to floors; suspended ceilings. Fittings. Gas HW central heating. Sanitary, ventilation, electrics. Lift, alarms, emergency lights, BMS, data.

| Storeys as a % of gross floor area | | Average Storey Heights | |
|---|---|---|---|
| basmt | 4% | Below ground floor | 3.30 |
| 3 | 96% | At ground floor | 3.30 |
| | | Above ground floor | 3.30 |

**Areas**

| | | |
|---|---|---|
| Basement floors | 90 m2 | |
| Ground floor | 850 m2 | |
| Upper floor | 1,570 m2 | |
| Gross floor area | 2,510 m2 | |
| Usable area | 1,865 m2 | |
| Circulation area | 455 m2 | |
| Ancillary area | 125 m2 | |
| Internal Divisions | 65 m2 | |
| Gross floor area | 2,510 m2 | |
| Area not enclosed | - m2 | |
| External wall area | 2,786 m2 | |
| Wall to floor ratio | 1.11 | |
| Internal cube | 8283 m3 | |

| Element | Percentage | Total cost of element £ | Functional unit £ per m2 | £/m2 incl Preliminaries | |
|---|---|---|---|---|---|
| | | | | Tender prices | Rate 1995 constant prices |
| Substructure | 7% | 201,075 | 80.11 | 83.43 | 82.17 |
| Superstructure | 50% | 1,350,414 | 538.01 | 560.32 | 551.83 |
| Internal finishes | 9% | 240,892 | 95.97 | 99.95 | 98.44 |
| Fittings | 2% | 47,428 | 18.90 | 19.68 | 19.38 |
| Services | 20% | 553,615 | 220.56 | 229.71 | 226.23 |
| Building sub-total | 88% | 2,393,424 | 953.56 | 993.10 | 978.05 |
| External works | 5% | 123,036 | 49.02 | 51.05 | 50.28 |
| Preliminaries | 4% | 104,351 | 41.57 | - | - |
| Contingencies | 3% | 73,725 | 29.37 | -29.37 | 28.93 |
| Total | | 2,694,536 | 1,073.52 | 1,073.52 | 1,057.25 |

Submitted by: Wilkinson & Lowe

Appendix M  (*continued*)

**CI/SfB - 320.**
**Offices - 3**

■■■■ BCIS ■■■■■■■■■■■■■■■■■■■■■■■■■■■■■■■■■■■■■■

| Element | | Preliminaries shown separately | | | |
|---|---|---|---|---|---|
| | | Total cost | Cost per m2 | Element unit quantity | Element unit rate |
| 1 | Substructure | 201,075 | 80.11 | 850 m2 | 236.56 |
| 2A | Frame | 155,202 | 61.83 | 2,370 m2 | 65.49 |
| 2B | Upper floors | 24,584 | 9.79 | 1,760 m2 | 13.97 |
| 2C | Roof | 150,313 | 59.89 | 891 m2 | 168.70 |
| 2D | Stairs | 19,498 | 7.77 | 5 No | 3,899.60 |
| 2E | External walls | 528,567 | 210.58 | 2,000 m2 | 264.28 |
| 2F | Windows and external doors | 273,105 | 108.81 | 786 m2 | 347.46 |
| 2G | Internal walls and partitions | 156,345 | 62.29 | 2,839 m2 | 55.07 |
| 2H | Internal doors | 42,800 | 17.05 | 110 No | 389.09 |
| 2 | Superstructure | 1,350,414 | 538.01 | | |
| 3A | Wall finishes | 42,345 | 16.87 | 6,100 m2 | 6.94 |
| 3B | Floor finishes | 126,168 | 50.27 | 2,510 m2 | 50.27 |
| 3C | Ceiling finishes | 72,379 | 28.84 | 2,510 m2 | 28.84 |
| 3 | Internal finishes | 240,892 | 95.97 | | |
| 4 | Fittings | 47,428 | 18.90 | | |
| 5A | Sanitary appliances | 19,692 | 7.85 | 41 No | 480.29 |
| 5B | Services equipment | - | | | |
| 5C | Disposal installations | 10,724 | 4.27 | | |
| 5D | Water installations | 15,706 | 6.26 | 41 No | 383.07 |
| 5E | Heat source | 21,559 | 8.59 | | |
| 5F | Space heating and air treatment | 136,866 | 54.53 | | |
| 5G | Ventilating systems | 14,386 | 5.73 | | |
| 5H | Electrical installations | 228,949 | 91.21 | | |
| 5I | Gas installations | - | | | |
| 5J | Lift and conveyor installations | 19,500 | 7.77 | 1 No | 19,500.00 |
| 5K | Protective installations | 2,045 | 0.81 | | |
| 5L | Communications installations | 64,492 | 25.69 | | |
| 5M | Special installations | - | | | |
| 5N | Builder's work in connection | 19,696 | 7.85 | | |
| 5O | Builder's profit and attendance | - | | | |
| 5 | Services | 553,615 | 220.56 | | |
| | Building sub-total | 2,393,424 | 953.56 | | |
| 6A | Site works | 91,237 | 36.35 | | |
| 6B | Drainage | 10,303 | 4.10 | | |
| 6C | External services | 21,496 | 8.56 | | |
| 6D | Minor building works | - | | | |
| 6 | External works | 123,036 | 49.02 | | |
| 7 | Preliminaries | 104,351 | 41.57 | | |
| | Total (less Contingencies) | 2,620,811 | 1,044.15 | | |
| 8 | Contingencies | 73,725 | 29.37 | | |
| | Contract sum | 2,694,536 | 1,073.52 | | |

# Appendix N

LIST OF PROBABLE ACTIVITIES
TO BE PERFORMED IN A QUANTITY
SURVEYING PRACTICE ADMINISTRATION

*FINANCIAL*

Staff salaries
Income tax returns and
payments
National insurance
Staff pension scheme
Staff luncheon vouchers
VAT returns and payments
Petty cash
Banking
Books of account
Project costing
Fee preparation
Cashflow control
Expenditure control

*SECRETARIAL*

Acknowledgements/confirmations/
covering letters
Organising appointments
Distribution of incoming mail
Photocopying letters and documents
Dealing with faxes, E-mail and telephone
messages
Filing
Posting of outgoing mail
Initiating standard procedures, such
as valuations and subcontract
nominations
Attending office meetings–keeping
minutes and action thereon
Enquiries for office equipment and
other facilities
Repairs and renewals
Recording and storage of completed
projects

*STOREKEEPING*

Stationery
Writing materials
Storage facilities
Furniture
Equipment
Computing facilities
Site signboards

*OTHER*

Holiday and sickness records
Tea/coffee service
Delivering letters and errands
Cleaning of premises including
purchasing materials

# References

Abdul B. O. (1993). Trafford Park Development Corporation – reclamation of part of a former British Steel site. *Proc. Instn Civ Engrs, Mun. Engr, June, 99–105.*

Akenhead R. (1994). A tortuous path of liability. *Building, 7 October, 29.*

Allsop M. (1979). Management in the professions. *Estates Gazette, 15 December.*

Aqua Group (1990a). *Tenders and Contracts for Building.* BSP.

Aqua Group (1990b). *Contract Administration for the Building Team.* BSP.

Aqua Group (1992). *Precontract Practice for the Building Team.* BSP.

Ashworth A. (1986). Cost models – their history, development and appraisal. *CIOB Technical Information Service nr 64.*

Ashworth A. (1994). Education and training of quantity surveyors. *CIOB Construction Paper 37.*

Association of Consultant Architects (1982). *Form of Building Agreement.*

Association of Metropolitan Authorities (1983). *Defects in Housing. Part 1: Non-traditional dwellings for the 1940s and 1950s.*

Association of South African Quantity Surveyors (1995). *Annual Report 1994/95.*

Australian Institute of Quantity Surveyors (1994). *1994 Report to Members.*

Balchin P. N., Bull G. H., and Keive J. L. (1995). *Urban Land Economics.* Macmillan.

Ballantyne J. K. (1986). *The Resident Engineer.* Telford.

Banwell Committee (1964). *The Placing and Management of Contracts for Building and Civil Engineering Works.* HMSO.

Barrett F. R. (1992). *Cost Value Reconciliation.* CIOB.

Barton P. K. *et al.* (1978) *Coordination of Mechanical and Engineering Services.* CIOB Occasional Paper 16.

BCIS (1990). *Schedule of Basic Plant Charges for use in connection with Dayworks under a Building Contract.*

BCIS (1993). *Quarterly Review of Building Prices.*

BCIS (1994). *Guide to Daywork Rates.*

Beard Dove (1990). *Value Engineering pamphlet.*

Bennett J. (1986). *Cost Management and the Chartered Quantity Surveyor.* RICS.

Bennett J. and Grice T. (1990). Procurement systems for building. *Quantity Surveying Techniques: New Directions.* BSP.

Bennett J. and Jayes S. (1995). *Trusting the Team: the best practice guide to partnering in construction.* Reading Construction Forum.

Benson J., Evans B., Colomb P. and Jones G. (1980). *The Housing Rehabilitation Handbook.* Architectural Press.

Berny J. and Townsend P. R. F. (1993). Macrosimulation of project risks – a practical way forward. *International Journal of Project Management 11(4), 201–8.*

Bevan O. A. (1991), *Marketing and Property People*, Macmillan.

Blakeley P. (1986). QSs: go forth and diversify, *Chartered Surveyor Weekly, 16 October, 277.*

BMI (1993). *BMI Special Report 222: Information Guide for Facilities Management.*

BMI (1995). *BMI Special Report 239: Condition Assessment Surveys.*

Bolton A. (1995). Culture clash. *New Builder, 3 March, 20.*

Bowen P. *et al.* (1987), Cost modelling: a process modelling approach. *Building Cost Modelling and Computers.* Spon.

Bowman P. (1986). Where to find, *Promoting the Professions: which way do we go?* Eds Eldridge and Carvell. Surveyors Publications.

Boyce T. (1993). *Successful Contract Negotiation,* Hawkesmere.

Brandon P. S. (1990), The challenge and the response, *Quantity Surveying Techniques: New directions,* BSP.

Brandon P. S. and Moore R. G. (1983). *Microcomputers in Building Appraisal,* BSP.

Brandon P. S., Stafford B. and Atkin B. (1990). *The potential for expert systems within quantity surveying.* RICS.

BRE (1978). *A Survey of Quality and Value in Building.*

BRE (1990). *BREEAM. An Environmental Assessment for New Office Designs, 1/90.*

BRE (1993). *Information Paper IP2/93: Industrial building refurbishment: opportunities for energy efficiency.*

BRE (1995). *The performance and cost in use of buildings: a new approach.* British Property Federation (1983). *Manual of the BPF System.*

Brown R. P. (1992). Value management: a new service for the client. *RICS QS Bulletin, April, iii.*

BS 3811: 1984 *Glossary of Maintenance Management Terms in Technology.*

BS 5750: 1987/BS EN ISO 9000: 1994. *Quality Systems.*

BS 7750: 1992. *Specification for Environmental Management Systems.*

Bullock A. (1994) . Divine meditation. *Chartered Builder, June, 17, 19.*

Burke H. T. (1976). *Claims and the Standard Form of Building Contract.* CIOB.

Burr D. (1994). This is the SST. *Chartered Surveyor Monthly, September, 45.*

Business Round Table (1994). *Controlling the Upwards Spiral: construction performance and cost in the UK and mainland Europe.*

Business Round Table (1995). *Thinking about Building.*

Cabinet Efficiency Unit (1995). *Construction Procurement by Government.* HMSO.

Cadman D. (1990). The environment and the property market. *Town and Country Planning, October, 267–70.*

Cambridge Econometrics Ltd (1981). *Policies for Recovery and Evaluation of Alternatives.*

Cantaouzine S. and Brandt S. (1980). *Saving Old Buildings.* Architectural Press.

Carter T. G. (1990). *Lecture on project management to staffs of East Midlands housing associations in Nottingham.*

Carter T. G. (1991/92). *Value Management selected papers.* Davis Langdon Management.

Carter T. G. (1992a). Value Engineering: a comparison between the 40 hour workshop and a one/two day study. *RICS QS Bulletin, April, v.*

Carter T. G. (1992b). *Project Management selected papers.*

CCPI (1987). *Coordinated Project Information for Building Works; a guide with examples.*

Central Unit on Procurement (1994). *The Use of Bonds and Guarantees in Government Contracts.*

Centre for Strategic Studies in Construction, University of Reading (1988). *Building Britain 2001.*

Centre for Strategic Studies in Construction, University of Reading (1991). *Construction Management Forum Report.*

Chalkley R. (1994). *Professional Conduct: A handbook for chartered surveyors.* RICS Books.

Chambers H. T. (1983). Furnish and file in style. *Public Service and Local Government, February.*

Chappell D. (1993). *Understanding JCT Standard Building Contracts.* Spon.

Churchill Winston. (Undated) *Foreword to The Chartered Surveyor: His Training and his Work.* Chartered Surveyors' Institution.

CIOB (1980a). *Building for Industry and Commerce: Client's Guide.*

CIOB (1980b). *Programmes in Construction: A Guide to Good Practice.*

CIOB (1981). Weather and construction. *Site Management Information Service nr 87.*

CIOB (1983). *Code of Estimating Practice.*

CIOB (1987). The control of materials and waste. *Technical Information Service nr 87.*

CIOB (1989). *Quality Assurance in the Building Process.*

CIOB (1990). *Maintenance Management.*

CIOB (1994). *Handbook of Facilities Management.*

CIOB (1995). CVs: what's hot and what's not. *Construction Manager, October, 23, 25.*

CIOB and IES (Institute for Employment Studies) (1995). *Balancing the building team: gender issues in the building professions.*

CIRIA (1983). *Buildability: An Assessment.*

Civic Trust (1988). *Urban Wasteland Now.*

Civic Trust for Wales/Civic Trust Regeneration Unit (1992). *Llanidloes: A Town Study.*

Clark M. (1987). RICS Presidential Address. *Chartered Quantity Surveyor, December, 34–7.*

Cole-Morgan J. (1986). The eye of the beholder. *Promoting the Professions: which way do we go?* Eds. Eldridge and Carvell. Surveyors Publications.

Constrado (1973). *Project Study 3: University Hospital and Medical School, Nottingham.*

Construction Industry Council. (1992). *The procurement of professional services – guidelines for the application of competitive tendering.*

Construction Industry Council, Research and Development Committee (1994). *Private Funding for Construction Innovation and Research: options for a national initiative.*

Cook G. K. and Hinks A. J. (1992). *Appraising Building Defects.* Longman.

Cooke B. (1992). *Contract Planning and Contractual Procedures.* Macmillan.

Cooke B. and Walker G. (1994). *European Construction Procedures and Techniques.* Macmillan.

Cottrell G. P. (1979). *The Builder's Quantity Surveyor.* CIOB.

Couch C. (1990). *Urban Renewal: Theory and Practice.* Macmillan.

Cox and Hamilton. (1994). *Clerk of Works Manual.* NJCC.

Creswell H. B. (1930). *The Honeywood File* and *The Honeywood Settlement.*

Dann C. (1983). The responsibility of a profession. *Chartered Quantity Surveyor, December.*

Darlow C. and Morley S. (1982). New views on development appraisal. *Estates Gazette, 26 June.*

Davey N. (1983). Partnership capital: how much is needed? *Chartered Surveyor Weekly, 8 September.*

Davidson N. (1994). QS offshore – the need for change. *Chartered Surveyor Monthly, March.*

Davies I. H. (1979). Professional liabilities. *Chartered Quantity Surveyor, January.*

Davies T., Hay H. and Sneden J. (1980). Processing civil engineering claims. *Chartered Quantity Surveyor, November.*

Davis L. (1994a). Insurance: a risky business. *Building Economist, March, 8–10.*

Davis L. (1994b). Union benefits. *Building Economist, July, 7.*

Davis L. (1994c). Softening the blow. *Building Economist, June, 6–7.*

Davis L. (1995). Two's company. *Chartered Surveyor Monthly, May, 30–1.*

Davis Langdon Everest/RICS (1994) *RICS Contracts in Use Survey.*

Day D. W. J. (1994). *Project Management and Control.* Macmillan.

Dixon T. (1988). *Computerised Information Systems for Surveyors.* RICS.

Dixon T. (1994). The benefits of property research. *Chartered Surveyor Monthly. November/December, 28.*

DoE (1971). Whitehall Development Group. *Planned Open Offices: CBA.*

DoE (1978). *Value for Money in Local Authority Housebuilding Programmes.*

DoE (1987). *Reusing Redundant Buildings: Good Practice in Regeneration.* HMSO.

DoE (1989). *Defects in Buildings.*

DoE (1991a). *General Conditions of Contract for Building and Civil Engineering, Mechanical and Electrical Works and Services; Standard Forms of Contract.*

DoE (1991b). *Environmental Action Guide for Building Purchasing Managers.* HMSO.

DoE (1995a). *Users Guide: Price adjustment formulae for construction contracts: 1990 series of indices.*

DoE (1995b). *Construct IT: Bridging the gap.* HMSO.

DoE, Audit Inspectorate (1983). *Control of Capital Projects.* HMSO.

Doyle N. (1993). Straight talking. *New Builder, 22 January, 12–13.*

Dugdale A. M. and Stanton K. M. (1982). *Professional Negligence.* Butterworths.

Economic Development Committee for Building (1967). *Action on the Banwell Report.*

Economist Intelligence Unit Ltd (1981). *Capital Spending and the UK Recovery.*

Eggleston B. (1993). *ICE Conditions of Contract: Sixth edition.* BSP.

Eldridge N. (1987). Marketing and presentation. *Project Manual 1: The Management and Marketing of Professional Practices.* CASLE.

Eldridge N. and Carvell P. Eds. (1986). *Promoting the Professions: which way do we go?* Surveyors Publications.

English Partnerships (1994). *Investment Guide.*

Equal Opportunities Commission (1986). *Men's jobs, women's jobs.* HMSO.

Equal Opportunities Commission (1987a). *Equality between sexes in industry; how far have we got.* HMSO.

Equal Opportunities Commission (1987b). *Women and men in Britain: a research profile.* HMSO.

Fearson W. (1990). *Collateral Warranties: A practical guide for the construction industry.* BSP.

Feenan R. and Dixon T. J. (1992). *Information and Technology Applications in Commercial Property.* Macmillan.

Fellows R. F. (1995). *1980 JCT Standard Form of Building Contract: A commentary for students and practitioners.* Macmillan.

Fiber A. (1986). Sign of the times. *Chartered Quantity Surveyor, November, 20–1.*

Field A. J. (1992). *BRE Information Paper IP/92: Energy Audits and Surveys, May.*

Fish R. (1981). Tendering and contract procedures. *Chartered Quantity Surveyor, July.*

Flanagan R. (1990). *Risk Management and Construction,* BSP.

Flanagan R. and Norman G. (1983). *Life Cycle Costing for Construction,* RICS.

Flanagan R. and Stevens S. (1990). Risk analysis. *Quantity Surveying Techniques: New Directions.* Ed. P. S. Brandon. BSP.

Flint R. F. (1979). Heavy engineering at home and overseas. *Chartered Quantity Surveyor, June.*

Fraser W. D. (1993). *Principles of Property Investment and Pricing.* Macmillan.

Froomberg J. (1996). Winning the lottery. *Building, 12 January, 29.*

Fryer B. (1983). Managing to lead. *Building Technology and Management, May.*

Fryer B. (1990). *The Practice of Construction Management.* BSP.

Garnett J. (1979). Motivating people in professional organisations. *Proc. Ann Conf. RICS, Jersey, October.*

Gerrity H. B. (1980). Variation orders, site works orders and daywork records. *The Practice of Site Management.* CIOB.

Gillespie B. (1994). Procurement route. *Building, 29 July, 46.*

Glancey J. (1989). *New British Architecture.* Thames and Hudson.

Gleeds (1994a). Eurofile: France. *Building Economist, June, 10–11.*

Gleeds (1994b). Eurofile: Germany. *Building Economist, September, 10–11.*

Gray C., Hughes W. and Bennett J. (1994). *The Successful Management of Design: A handbook of building design management.* Centre for Strategic Studies in Construction, University of Reading.

Greater London Council. (1978). *An Introduction to Housing Layout.* Architectural Press.

Greater London Council. (1981). *GLC Preferred Dwelling Plans.* Architectural Press.

Greed C. (1991). *Surveying Sisters: women in a traditional male profession.* Routledge.

Green S. and Moss G. (1993). Value for money from SMART management. *Chartered Builder, October, 5–7.*

Greenhalgh B. (1987a). Marketing and the quantity surveyor. *Chartered Quantity Surveyor, July, 24.*

Greenhalgh B. (1987b). Promoting the quantity surveyor. *Chartered Quantity Surveyor, October, 16.*

Greenstreet B. (1994). *Legal and Contractual Procedures for Architects.* Butterworth Architecture.

Grierson J. (1995). Competitive tendering: strange bedfellows. *Estates Times Review, 9 June, 29–30.*

Griffith A. (1990). *Quality Assurance in Building.* Macmillan.

Griffith A. (1994). *Environmental Management in Construction.* Macmillan.

Griffith A. and Sidwell A. C. (1995). *Constructability in Building and Engineering Projects.* Macmillan.

Gunning J. G. (1983). Site management – mainly a process of control. *Building Technology and Management, December.*

Haley G. N. (1994). An introduction to joint venture tendering. *CIOB Construction Paper nr 40.*

Harmer C. and Camp P. (1982). *Practical Partnership.* Oyez Longman.

Harris E. C. (undated). *Risk Management.*

Harris E. C. (undated). *Outlook.*

Harris F. and McCaffer R. (1995). *Modern Construction Management.* Blackwell Science.

Harrison R. S. (1981). Estimating and tendering – some aspects of theory and practice. *CIOB Estimating Information Service nr 41.*

Hartigay S. and Dixon T. (1991). *Software Selection for Surveyors.* Macmillan.

Harvey R. C. and Ashworth A. (1993). *The Construction Industry of Great Britain.* Butterworth Heinemann.

Hawkings G. (undated). Making better use of technology. E. C. *Harris: Outlook.*

Heap D. (1991). *An Outline of Planning Law.* Sweet & Maxwell.

Hellard R. B. (1995). *Project Partnering: Principle and Practice.* Telford.

Hemsley A. (1995). The perils of auditing. *Building, 14 July, 39.*

Herbert J. (1991). Construction management: what's in it for us. *Chartered Quantity Surveyor, October, 20.*

Hogg K. I. (1994). *The diversity of quantity surveying degree courses and trends in recruitment.* Nottingham Trent University.

Hollis M. (1991) *Surveying Buildings.* Surveyors Publications.

Hong Kong Institute of Surveyors. (1995). *Directory and Annual Report 1995.*

HRH The Prince of Wales. (1989). *A Vision of Britain: A personal view of architecture.* Doubleday.

Hughes G. A. (1983) *Building and Civil Engineering Claims in Perspective.* Construction Press.

Hughes W. P. (1991). An analysis of construction management contracts. *CIOB Technical Information Service nr 135.*

Hutchinson K. (1993). *Building Project Appraisal.* Macmillan.

ICE (1986). *Civil Engineering Procedure.*

ICE (1991). *The New Engineering Contract.*

ICE (1992a). *Design and Construct Conditions of Contract.*

ICE (1992b). *Management Development in the Construction Industry – Guidelines for the Professional Engineer.* Telford.

ICE (1993). *Conditions of Contract for Minor Works.*

ICE (1994). *Contaminated Land: investigation, assessment and remediation.* Telford.

ICE, Conditions of Contract Standing Committee (1983). *Guidance on the preparation, submission and consideration of tenders for civil engineering contracts for use in the UK.*

ICE, Infrastructure Policy Group (1988). *Urban Regeneration.* Telford.

ICE and FCEC (1991). *Civil Engineering Standard Method of Measurement: CESMM3.* Telford.

ICE, ACE and FCEC (1991). *ICE Conditions of Contract, Sixth Edition.*

Isaac D. (1994). *Property Finance.* Macmillan.

Jaggar D. (1992). Opinion: quantity surveying education – mediocrity or excellence? *RICS QS Bulletin, January, i.*

James J. N. C. (1980). The challenge of the 1980s: independence and responsibility. *The Chartered Surveyor, December.*

James M. F. (1994). *Construction Law.* Macmillan.

Janssens D. E. L. (1991). *Design-Build Explained.* Macmillan.

JCT (1980a). *Standard Form of Building Contract: Private with Quantities, with amendments.*

JCT (1980b). *Agreement for Minor Building Works.*

JCT (1981). *Standard Form of Building Contract with Contractor's Design.*

JCT (1984). *Intermediate Form of Building Contract.*

JCT (1987). *Management Contract.*

JCT (1989). *Measured Term Contract.*

JCT (1992). *Standard Form of Prime Cost Contract.*

Jelley A. and Povall W. (1979). A multi-headed client at Sullum Voe. *Chartered Quantity Surveyor, June.*

Jess D. C. (1982). *A Guide to the Insurance of Professional Negligence Risks.* Butterworths.

Johnson S. (1993). *Greener Buildings: Environmental impact of property.* Macmillan.

Johnston N. G. (1978). Directors in construction companies – another role for the quantity surveyor. *Proc. RICS CQS Eleventh Triennial Conf.*

Jonas C. (1992). Of those to whom much is given, much is expected. *Chartered Surveyor Monthly, November, 4–8.*

Jones D. (1983). Changing requirements keep architecture flexible. *Estates Times Review, February.*

Just R. and Williams D. (1997). *Urban Regeneration: A practical guide.* Macmillan.

Keating D. (1967). *Insolvency and the building contract. Chartered Surveyor, September, 16.*

Kelly J. and Male S. (1991). *The Practice of Value Management: Enhancing Value or Cutting Cost?* Heriot-Watt University/RICS QS Division.

Kelly J. and Male S. (1992). Functional analysis method. *RICS QS Bulletin, April, iv.*

Kelly J. and Male S. (1993). *Value Management in Design and Construction.* Spon.

Kennie T. (1993). Continuing professional development: who needs it? *The Surveying Technician, November, 7.*

Kennie T. (1995). Professional education and change. *Chartered Surveyor Monthly, May, 25.*

Knowles R. (1990). Managing the risk. *Chartered Quantity Surveyor, September, 16.*

Kwayke A. A. (1993). Alternative dispute resolution in construction. *CIOB Construction Paper 21.*

Langford D. A. and Rowland V. R. (1995). *Managing Overseas Construction Contracting.* Telford.

Langford D., Hancock M. R., Fellows R. and Gale A. W. (1995). *Human Resources Management in Construction.* Longman.

Larcombe R. A. (1993). *The Surveying Technician, September, 19.*

Latham Report. (1994). *Constructing the Team.* HMSO.

Latham C. (1995). Education report. *Chartered Surveyor Monthly, March, 33.*

Latchmore A. *et al.* (1992). Collateral Warranties 2. Warranties: in practice. *CIOB Construction Paper 10.*

Lavers A. P. (1983). Towards strict liability? *Portsmouth Polytechnic, now University.*

Lavers A. P. (1996). *Professional Negligence in the Construction Industry.* Longman.

Lay R. (1991). *Market Requirements of the Profession.* RICS.

Lee H. S. and Yuen G. C. S. (1993). *Building Maintenance Technology.* Macmillan.

Lee N. W. and Hayles B. J. (1989). Contractual administration. *Mun. Engr, 6 June, 137–44.*

Lever J. (1981). 12 Great George Street: why choose Waterhouse? *Chartered Quantity Surveyor, September.*

Local Government Management Board (1994). *The ABC of NVQs.*

Lovell J. (1990a). Tools of the trade. *Chartered Quantity Surveyor, January, 14–15.*

Lovell J. (1990b). Management. *Chartered Quantity Surveyor, September, 27.*

Luff R. (1982). A question of balance. *Chartered Surveyor Weekly, 11 November.*

Macneil J. (1994). The ban plays on. *Building, 8 July, 31–3.*

Marshall R. (1983). Managing seabed operations. *Chartered Quantity Surveyor, August.*

Masterman J. W. E. (1992). *An Introduction to Building Procurement Systems.* Spon.

Mayo-Chandler B. (1980). *Estimating for the Engineering Services.* Electrical Contractors' Association.

Meara C. (1995). Choose your partners . *Building, 3 November, 35.*

Meike J. L. and Hillebrandt P. M. (1989). *The French Construction Industry: A guide for UK professionals.* CIRIA.

Meopham B. (1983). Cost control for the civil engineering contractor. *Chartered Quantity Surveyor, June.*

Metra Martech (1991). *European Environmental Legislation and its Application.*

Meyer W. T. (1983). *Energy Economics and Building Design.* McGraw-Hill.

Miller S. (1992). Going for the green option. *Chartered Quantity Surveyor, March, 9–11.*

Mills E. (1995). Teams with common goals. *Building, 18 August, 25.*

Millwood A. M. (1983). Changing methods of placing contracts. *Chartered Quantity Surveyor, December.*

Morledge R. (1987). The effective choice of building procurement method. *Chartered Quantity Surveyor, July, 26.*

Morledge R. (1995). Building procurement: UK and international models (attitudes and expectations). *RICS Joint Branch CPD Conference, Loughborough University, 5 April.*

Morledge R. and Sharif A. (1994). A functional approach to the modelling of international procurement systems. *Chartered Surveyor Monthly, June, 30–1.*

Morledge R. and Hogg K. (1995). Institutional and undergraduate perceptions of ethics in the practice of quantity surveying. *Nottingham Trent University Paper.*

Muir J. (1983). Effective management through delegation. *Chartered Surveyor Weekly, 4 August.*

National Federation of Housing Associations (1989). *Maintenance Planning.*

NEDO (1975). *The Public Client and the Construction Industries.*

NEDO (1983). *Faster Building for Industry.*

NEDO (1985). *Thinking about Building.*

NEDO (1989). *Faster Building for Commerce.*

Neil J. (1982). *Construction Cost Estimating for Project Control.* Prentice-Hall.

Newens J. (1994). *Building, 15 July, 10.*

Nisbet J. (1981). After the contract is over. *Chartered Quantity Surveyor, May.*

Nisbet J. (1989). *Called to Account, Quantity Surveying 1936–86.* Stoke Publications.

Nisbet J. (1990). *Design and Build Contracts: Client's Guide to the JCT Standard Form of Building Contract with Contractor's Design.*

NJCC (1986). *Guidance Note 2: Performance Bonds.*

NJCC (1992). *Guidance Note 6: Collateral Warranties.*

NJCC (1994a). *Code of Procedure for Single Stage Selective Tendering.*

NJCC (1994b). *Code of Procedure for Two Stage Selective Tendering.*

Norton B. (1992a). Value added. *Chartered Quantity Surveyor, June, 21–3.*

Norton B. (1992b). A value engineering case study. *Chartered Quantity Surveyor, September, 7.*

Norton B. (1993) Living the American dream. *Chartered Quantity Surveyor, May,* *10–11.*

Norton B. and McElligott W. (1995). *Value Management in Construction: A practical guide.* Macmillan.

Nottinghamshire County Council (1993). *Listed Buildings at Risk in Nottinghamshire.*

Noy E. A. (1995). *Building Surveys and Reports.* Blackwell Science.

Oddy I. V. (1994). Professional indemnity insurance. *RICS Professional Practice Group Workshop, Oxford, 2–3 September.*

Oddy I. V. (1995). *Professional Competency.* St Quintin.

Oxley R. (1978). Incentives in the construction industry – effects on earnings and costs. *CIOB Information Service nr 74.*

Park J. A. (1992). Facilities for growth. *Chartered Quantity Surveyor, February,* *17–19.*

Park J. A. (1994). *Facilities Management: An Explanation.* Macmillan.

Parker D. (1993). Back to reality. *New Builder, 16 April, 22–3.*

Pasquire C. (1995). *Risk in Construction.* Loughborough University of Technology, Department of Civil Engineering.

Pearce P. (1992). *Construction Marketing: A professional approach.* Telford.

Perry J. G. and Hayes R. W. (1986). Risk management for project managers. *Building Technology and Management, August/September, 8–11.*

Pettipher M. (1994). Modular magic. *New Builder, 11 November.*

Pigg D. R. (1993). An introduction to competitive tendering for professional engineering services. *Proc. Instn Civ. Engrs Mun. Engr., 98, June, 69–78.*

Pilcher R. (1992). *Principles of Construction Management.* McGraw-Hill.

Potts S. (1995). PFI – RICS Guide. *RICS Press Release, 22 September.*

Powell-Smith V. (1990). *Problems in Construction Claims.* BSP.

Powell-Smith V. and Sims J. (1988). *The JCT Management Contract: A Practical Guide.* Kluwer.

Powell-Smith V. and Sims J. (1990). *Contract Documentation for Contractors.* BSP.

Price J. (1994). *Subcontracting under the JCT Standard Forms of Building Contract.* Macmillan.

Property Helpline (1994). The guide to better decisions in facilities management. *The Facilities Handbook Vol 1.* CML Data Ltd.

PSA (1988). *The Conservation Handbook.*

PSA (1990). *Schedule of Rates for Building Works.*

PSA (1994). *General Conditions of Contract for Building and Civil Engineering Works (PSA/1).*

Quaife G. (1992). Opportunities for the quantity surveyor in procurement. *RICS QS Bulletin, May, iv.*

Raftery J. (1985). *Risk Analysis in Construction Tendering: a guide for building contractors.* Technical Research Centre of Finland.

Raftery J. (1994). *Risk Analysis in Project Management.* Spon.

Raftery J. and Newton S. (1995). *Building Cost Modelling.* BSP.

Reading Construction Forum (1995). *Trusting the Team: The best practice guide to partnering in construction.*

Reynolds D. and Sheppard S. G. F. (1989). *The Iberia Construction Industry: A guide for UK professionals.* CIRIA.

Reynolds and King (1992). *The Expert Witness and His Evidence.* BSP.

RIBA (1973). *Handbook of Architectural Practice and Management.*

RICS (1960). *Review of Educational Policy.* (The Wells Report).

RICS (1967). *Report of the Educational Policy Committee.* (The Eve Report).

RICS (1970). *Surveyors and their Future.*

RICS (1973). *Quantity Surveying in the Public Sector.*

RICS (1974). *A Study of Quantity Surveying Practice.*

RICS (1975). *Definition of Prime Cost of Daywork carried out under a Building Contract.*

RICS (1978). *Review of Education Policy.* (Brett-Jones Committee).

RICS (1979). *Principles of Measurement (International) for Works of Construction.*

RICS (1980a). *Definition of Daywork carried out under a Heating, Ventilating, Air Conditioning, Refrigeration, Pipework and/or Domestic Engineering Contract.*

RICS (1980b). Report of RICS Review Committee. *Surveying in the Eighties.*

RICS (1981). *Definition of Prime Cost of Daywork carried out under an Electrical Contract.*

RICS (1982). *Refurbishment and Alteration Work: Quantity Surveying Documentation.*

RICS (1983). *Guidance Note on Structural Surveys of Commercial and Industrial Property.*

RICS (1989a). *Quality Assurance: Introductory Guide.*

RICS (1989b). *Future Education and Training Policies.*

RICS (1990). *Quality Assurance: Guidelines for the Interpretation of BS 5750 for Use by Quantity Surveying Practices and Certification Bodies.*

RICS (1991). *Structural Surveys for Residential Property: A Guidance Note.*

RICS (1992a). *Living Cities.*

RICS (1992b). *Construction Management in Engineering Construction Projects: Guidance Notes.*

RICS (1992c). *Competing for Quality Competition in the Provision of Local Services.*

RICS (1992d). *Client Guide to the Appointment of a Quantity Surveyor.*

RICS (1992e). *Form of Enquiry and Fee Quotation for the Appointment of a Quantity Surveyor.*

RICS (1992f). *Form of Agreement, Terms and Conditions for the Appointment of a Quantity Surveyor.*

RICS (1992g). *CPD: Review of policy and future development.* discussion document.

RICS (1992h). *A Research Strategy for the RICS, 1992–96.*

RICS (1993a). *Environmental Management and the Chartered Surveyor.*

RICS (1993b). *Energy Efficiency in Buildings: Energy Appraisal of Existing Buildings: A Handbook for Surveyors.*

RICS (1993c). *Procedures and Guidelines for Course Accreditation: Supplementary Advisory Notes: Postgraduate Programmes.* (Nottingham Trent University research).

RICS (1994a). *The RICS Salary Survey 1994.*

RICS (1994b). *Setting up in Practice: Guidance Notes.*

RICS (1994c). *Professional Practice Group Workshop, Oxford: Conference report and speakers' papers, 2–3 September.*

RICS (1994d). *Review of the Constitution: Methods of Investigating Inadequate Professional Services.* (green paper).

RICS (1995a). *The Problems of Practical Completion.*

RICS (1995b). *Methods of Investigating Inadequate Professional Services.* (report).

RICS (1995c). NVQs and SVQs. *Chartered Surveyor Monthly, January,* 49.

RICS (1995d). *An Agenda for Construction Economics Research.*

RICS (1995e). *Improving Value for Money in Construction: Guidance for Chartered Surveyors and their Clients.*

RICS (1995f). *The Chartered Surveyor as Management Consultant: An Emerging Market.* (research by University of Reading and DTZ Debenham Thorpe).

RICS (1995g). *The Private Finance Initiative: The Essential Guide.*

RICS (1995h). *Considering Private Finance for Public Sector Works.*

RICS (1996). *The Procurement Guide.*

RICS, Building Surveyors Division (1991). *Dilapidations: Guidance Note.*

RICS, Education and Membership Committee (1994a). *Education Policy: A Strategy for Action: Trends in Recruitment, Education. Training and Development.*

RICS, Education and Membership Committee (1994b). *The Assessment of Professional Competence: A Review and Proposals for Reform.*

RICS, Facilities Management Skills Panel (1993). Facilities management: what you should know. *Chartered Surveyor Monthly. July/August, 4.*

RICS Insurance Services (1989). *Caveat Surveyor II.* Surveyors Publications.

RICS, Junior Organisation (1980). *Partnership and the Chartered Surveyor.*

RICS, Junior Organisation (1991). *RICS fit for the 21st century.*

RICS, Project Management Skills Panel (1995). *Selection and Appointment of the Project Manager: Guidance Notes.*

RICS, Quantity Surveyors Committee (1971). *The Future Role of the Quantity Surveyor.*

RICS, Quantity Surveyors Division (1979). *UK and US Construction Industries: A Comparison of Design and Contract Procedures.*

RICS, Quantity Surveyors Division (1982). *Precontract Cost Control and Cost Planning.*

RICS, Quantity Surveyors Division (1983). *The Future Role of the Chartered Quantity Surveyor.*

RICS, Quantity Surveyors Division (1984). *A Study of Quantity Surveying Practice and Client Demand.*

RICS, Quantity Surveyors Division (1991). *QS 2000: The Future Role of the Chartered Quantity Surveyor.*

RICS, Quantity Surveyors Division (1992a). *The Core Skills and Knowledge Base of the Quantity Surveyor.*

RICS, Quantity Surveyors Division (1992b). *Report of the Focus Study Group.*

RICS, Quantity Surveyors Division (1995). *Chartered Quantity Surveyors in a Changing World: An Agenda for Change.*

RICS and BEC (1988a). *Standard Method of Measurement of Building Works (SMM7).*

RICS and BEC (1988b). *SMM7 Code of Procedure for Measurement of Building Works (CPI).*

RICS/English Heritage (1993). *Investment Performance of Listed Buildings.*

RICS/NFBTE (1979). *SMM6: Standard Method of Measurement of Building Works.*

RICS and PMDA (1992) *Project Management Agreement and Conditions of Engagement and Guidance Note.*

RICS, RTPI, CPOS and DPOS (1995). *Tomorrow's Countryside: A Rural Strategy.*

Robertson J. D. M. (1983). *Maintenance Audit.*

Robinson T. H. (1977). *Establishing the Validity of Contractual Claims.* CIOB.

Robson P. (1991). *Structural Appraisal of Traditional Buildings.* Gower.

RTPI (1993). *The Character of Conservation Areas.*

Russell A. (1995). Understanding the PFI. *Chartered Surveyor Monthly, November/December, 18–19.*

Saito T. (1994). A comparative study of procurement systems in the UK and Japan. *CIB Working Commission W92, HK.*

Seeley I. H. (1972). Integrated professional training. *Chartered Surveyor, June, 621–2.*

Seeley I. H. (1974). *Planned Expansion of Country towns.* Godwin/Longman.

Seeley I. H. (1976). Education and training of the quantity surveyor. *Chartered Surveyor, Building and Quantity Surveying Quarterly, 4.2, Winter, 19–24.*

Seeley I. H. (1978). Conference on avoiding a dispute at Trent Polytechnic. *The Quantity Surveyor, September.*

Seeley I. H. (1985). *Building Surveys, Reports and Dilapidations.* Macmillan.

Seeley I. H. (1987). *Building Maintenance.* Macmillan.

Seeley I. H. (1989). *Advanced Building Measurement.* Macmillan.

Seeley I. H. (1992). *Public Works Engineering.* Macmillan.

Seeley I. H. (1993a). *Civil Engineering Quantities.* Macmillan.

Seeley I. H. (1993b). *Civil Engineering Contract Administration and Control.* Macmillan.

Seeley I. H. (1995). *Building Technology.* Macmillan.

Seeley I. H. (1996). *Building Economics.* Macmillan.

Sharif A. (1993). Better built in the Third World. *Chartered Quantity Surveyor. September, 12.*

Shash M. (1993). Factors considered in tendering decisions by top UK contractors. *Construction Management and Economics, March, 111–18.*

Simmonds D. T. (1979). Evaluating contractors' claims: presentation of claims by contractors. *Chartered Quantity Surveyor, February.*

Skitmore R. M. and Marsden D. E. (1988). Which procurement system? – Towards a universal procurement selection technique. *Construction Management and Economics, 6.1, 71–89.*

Skoyles E. R. (1982). Waste and the estimator. *CIOB Technical Information Service nr 15.*

Smith A. J. (1995). *Estimating, Tendering and Bidding for Construction: Theory and Practice.* Macmillan.

Smith G. (1980). Cost planning the design process. *Chartered Quantity Surveyor, August.*

Smith G. (1981). Tendering procedures scrutinised: Essex costs the alternatives. *Chartered Quantity Surveyor, June.*

Smith M. (1993). VE: is it withering on the vine? *Chartered Quantity Surveyor, February, 18–19.*

Sneden J. (1979). First steps to a closer bond in Great George Street? *Chartered Quantity Surveyor, January.*

Society of Chief Quantity Surveyors in Local Government (1981). *The Insolvency of Building Contractors.*

Society of Chief Quantity Surveyors in Local Government (1993). *Assessment of Liquidated and Ascertained Damages on Building Contracts.*

SST. (1994). *Annual Review.*

Stephenson D. (1993). *Arbitration Practice in Construction Contracts.* Spon.

Stoy Hayward (1991). *IT Rulebook: using information technology in mid-sized businesses.*

Swanston R. (1994). The nation and its buildings – closing the output gap. *Address by RICS President, 12 July.*

Tapping L. (1995). Teleworking – implications for the future. *Chartered Surveyor Monthly, April, 23.*

Tate B. (1992). The 1991 procedure. *Chartered Quantity Surveyor, May, 20.*

Taylor N. P. (1991). *Development Site Evaluation.* Macmillan.

TBV Consult (1994). *PSA/1 General Conditions of Contract for Building and Civil Engineering Works.* HMSO.

Telling and Duxbury (1993). *Planning Law and Procedure.* Butterworths.

Thomas R. (1993). *Construction Contract Claims.* Macmillan.

Thompson F. M. L. (1968). *Chartered Surveyors: The Growth of a Profession.* Routledge and Kegan Paul.

Thompson J. C. (1994). Japanese housebuilding: Orient express. *Building, 2 December.*

Thompson P. and Norris C. (1993). The perception, analysis and management of financial risk in engineering projects. *Proc. Instn Civ. Engrs, Civil Engineering, 93, February, 42–7.*

Trade Union Congress (1981). *The Reconstruction of Britain.*

Treasury (1991). *Green Book: Economic Appraisal in Central Government.* HMSO.

Treasury (1994a), The UK recession 1990–92. *Economic Briefing nr 6, February.*

Treasury (1994b). *Competition and the Private Finance Initiative.*

Treasury (1995). *White paper: Setting New Standards: a Strategy for Government Procurement.*

Treasury, Private Finance Unit (1995). *Private Opportunity, Public Benefit: Progressing the Private Finance Initiative.*

Trickey G. (1979). Evaluating contractors' claims: the professional quantity surveyor's approach. *Chartered Quantity Surveyor, February.*

Trickey G. (1983). *The Presentation and Settlement of Contractors' Claims.* Spon.

Trotter P. (1983). Are architects what they seem? *Building Technology and Management, January.*

Truman G. (1989). CPI up to tender stage. *Chartered Quantity Surveyor, December, 17–19.*

Turner A. (1997). *Building Procurement.* Macmillan.

Turner D. F. (1983). *Quantity Surveying Practice and Administration.* Longman.

Turner D. F. (1994). *Building Contracts: A practical guide.* Longman.

Wallace I. N. D. (1995). *Hudson's Building and Engineering Contracts.* Sweet & Maxwell.

Warren M. (1993). *Economics for the Built Environment.* Butterworth Heinemann.

Waterhouse R. (1992). Project management. *College of Estate Management memo.*

Waterhouse R. (1994). Implementing project management: is the client the problem? *Chartered Surveyor Monthly, May, 26–7.*

Watkins P. (1981). Looking at ourselves – now. *Chartered Surveyor, December.*

Waters A. B. (1980). Arbitration in building disputes. *The Practice of Site Management.* CIOB.

Watts T. (1992). Facilities management typifies shift to demand led market. *RICS keynote address, 21 February.*

Weatherhead M. (1997). *Real Estate and Corporate Strategy.* Macmillan.

Wheatley G. (1989). When in France. *Chartered Quantity Surveyor, October, 9.*

Wheeler C. (1983). Who? – coming out of the shell. *Chartered Quantity Surveyor, October.*

Whitfield J. (1994). *Conflicts in Construction: avoiding, managing, resolving.* Macmillan.

Williams P. J. (1979). The growth and style of private practice in the eighties. *Proc. RICS Ann. Conf., Jersey, October.*

Wilson A. J. (1984). Introductory report on cost modelling. *CIB Ottawa, vol 1*, *61–2*.

Wilson R. M. (1979). The profession's gift to the future. *Chartered Surveyor, December.*

Wood R. D. (1978). *Builders' claims under the JCT Form of Contract.* CIOB.

Wordsworth P. (1992). The greening of building maintenance. *The Building Surveyor, June.*

Yamamoto K. (1994) Japanese procurement. *Design and Build, 8 July, 9.*

Yates A. (1986) . Assessing uncertainty. *Chartered Quantity Surveyor, November, 27.*

Yates A. and Gilbert B. (1989). *The Appraisal of Capital Investment in Property.* Surveyors Publications.

Yates B. (1987). Marketing the practice. *Project Manual 1: The Management and Marketing of Professional Practices.* CASLE.

Young B. A. (1990). *Information Technology: Its impact on job content, skills and knowledge for construction management.* RICS.

# Index